Molecular Neuroendocrinology

Molecular Neuroendocrinology

From Genome to Physiology

EDITED BY

David Murphy
University of Bristol, Bristol, UK

Harold Gainer
National Institutes of Health, Bethesda, Maryland, USA

WILEY Blackwell

Contents

List of Contributors

Greti Aguilera
Section on Endocrine Physiology, National Institute of
Child Health and Human Development
National Institutes of Health
Bethesda, Maryland, USA

Ferdinand Althammer
Schaller Research Group on Neuropeptides
German Cancer Research Center
Heidelberg, Germany

Jose Antunes-Rodrigues
Department of Physiology
Faculty of Medicine of Ribeirão Preto
University of São Paulo
Ribeirão Preto, SP, Brazil

George J. Augustine
Center for Functional Connectomics
Korea Institute of Science and Technology
Seoul, Korea;
Lee Kong Chian School of Medicine
Nanyang Technological University
Singapore;
Institute of Molecular and Cell Biology
Singapore

Shifra Ben-Dor
Department of Biological Services
Weizmann Institute of Science
Rehovot, Israel

Steeve Boulant
Schaller Research Group of Viral Infection Dynamics
German Cancer Research Center and University of
Heidelberg
Heidelberg, Germany

J. Peter H. Burbach
Brain Center Rudolf Magnus
Department of Translational Neuroscience
University Medical Center Utrecht
Utrecht, the Netherlands

David A. Carter
School of Biosciences
Cardiff University
Cardiff, UK

Samuel J. Clokie
West Midlands Regional Genetics Laboratories
Birmingham Women's NHS Foundation Trust
Birmingham, UK

Matthew Concannon
Cellular and Molecular Physiology
Institute of Translational Medicine
University of Liverpool
Liverpool, UK

Steven L. Coon
Molecular Genomics Laboratory
Eunice Kennedy Shriver National Institute of
Child Health and Human Development
National Institutes of Health
Bethesda, Maryland, USA

Judy M. Coulson
Cellular and Molecular Physiology
Institute of Translational Medicine
University of Liverpool
Liverpool, UK

Andrii Domanskyi
Molecular Biology of the Cell I
German Cancer Research Center
Heidelberg, Germany

Marina Eliava
Schaller Research Group on Neuropeptides
German Cancer Research Center
Heidelberg, Germany

Lloyd D. Fricker
Department of Molecular Pharmacology
Albert Einstein College of Medicine
Bronx, New York, USA

Cong Fu
State Key Lab of Theoretical and Computational Chemistry
Jilin University
Changchun, Jilin, China

Harold Gainer
Molecular Neuroscience Section, Laboratory of Neurochemistry
National Institutes of Health
Bethesda, Maryland, USA

Martha U. Gillette
Department of Cell and Developmental Biology and the
Beckman Institute
University of Illinois at Urbana-Champaign
Urbana, Illinois, USA

Yoav Gothilf
Department of Neurobiology, George S. Wise Faculty of
Life Sciences and Sagol School of Neuroscience
Tel-Aviv University
Tel-Aviv, Israel

Valery Grinevich
Schaller Research Group on Neuropeptides
German Cancer Research Center
Heidelberg, Germany

Christian W. Gruber
Center for Physiology and Pharmacology
Medical University of Vienna
Vienna, Austria

Johann Guillemot
Laboratory of Biochemical Neuroendocrinology
Institut de Recherches Cliniques de Montréal
Montréal, Canada

Stephen Hartley
Genome Technology Branch, National Human Genome
Research Institute
National Institutes of Health
Bethesda, Maryland, USA

Georgina G. J. Hazell
School of Clinical Sciences
University of Bristol
Bristol, UK

Charles K. Hwang
Department of Ophthalmology
Department of Pharmacology
Emory University School of Medicine
Atlanta, Georgia, USA

Samuel J. Irving
Department of Molecular and Integrative Physiology
University of Illinois at Urbana-Champaign
Urbana, Illinois, USA

P. Michael Iuvone
Department of Ophthalmology
Department of Pharmacology
Emory University School of Medicine
Atlanta, Georgia, USA

Jinsook Kim
Lee Kong Chian School of Medicine
Nanyang Technological University
Singapore;
Institute of Molecular and Cell Biology
Singapore

Lanikea B. King
Center for Translational Social Neuroscience
Department of Psychiatry and Behavioral Sciences
Yerkes National Primate Research Center
Emory University
Atlanta, Georgia, USA

David C. Klein
Office of the Scientific Director, Eunice Kennedy Shriver
National Institute of Child Health and Human Development
National Institutes of Health
Bethesda, Maryland, USA

H. Sophie Knobloch-Bollmann
Schaller Research Group on Neuropeptides
German Cancer Research Center
Heidelberg, Germany

Gil Levkowitz
Department of Molecular Cell Biology
Weizmann Institute of Science
Rehovot, Israel

Yunfei Li
Department of Orthopaedics
General Hospital of Pingmeishenma Group
Pingdingshan, Henan, China

Zita Liutkevičiūtė
Center for Physiology and Pharmacology
Medical University of Vienna
Vienna, Austria

Stephen J. Lolait
School of Clinical Sciences
University of Bristol
Bristol, UK

Alejandro Lomniczi
Division of Neuroscience
Oregon National Primate Research Center/Oregon Health
& Science University
Beaverton, Oregon, USA

Edward A. Mead
Department of Animal Sciences, Department of Genetics
Rutgers University
New Brunswick, New Jersey, USA

Jennifer W. Mitchell
Department of Cell and Developmental Biology
and the Beckman Institute
University of Illinois at Urbana-Champaign
Urbana, Illinois, USA

James C. Mullikin
Genome Technology Branch, National Human Genome
Research Institute
National Institutes of Health
Bethesda, Maryland, USA

Peter Munson
Mathematical and Statistical Computing Laboratory,
Center for Information Technology
National Institutes of Health
Bethesda, Maryland, USA

Chris Murgatroyd
School of Healthcare Science
Manchester Metropolitan University
Manchester, UK

David Murphy
School of Clinical Sciences
University of Bristol
Bristol, UK;
Department of Physiology, Faculty of Medicine
University of Malaya
Kuala Lumpur, Malaysia

Ryuichi Nakajima
Center for Functional Connectomics
Korea Institute of Science and Technology
Seoul, Korea

Anne-Marie O'Carroll
School of Clinical Sciences
University of Bristol
Bristol, UK

Sergio R. Ojeda
Division of Neuroscience
Oregon National Primate Research Center/Oregon Health
& Science University
Beaverton, Oregon, USA

Jim Pickel
National Institute of Mental Health
National Institutes of Health
Bethesda, Maryland, USA

Andrzej Z. Pietrzykowski
Department of Animal Sciences, Department of Genetics
Rutgers University
New Brunswick, New Jersey, USA

Elena V. Romanova
Department of Chemistry and the Beckman Institute
University of Illinois at Urbana-Champaign
Urbana, Illinois, USA

James A. Roper
School of Biochemistry
University of Bristol
Bristol, UK

Lena C. Roth
Schaller Research Group on Neuropeptides
German Cancer Research Center
Heidelberg, Germany

Nabil G. Seidah
Laboratory of Biochemical Neuroendocrinology
Institut de Recherches Cliniques de Montréal
Montréal, Canada

Leming Shi
Center for Pharmacogenomics, School of Pharmacy
Fudan University
Shanghai, China

Megan Stanifer
Schaller Research Group of Viral Infection Dynamics
German Cancer Research Center and University of
Heidelberg
Heidelberg, Germany

Jonathan V. Sweedler
Department of Chemistry and the Beckman Institute
University of Illinois at Urbana-Champaign
Urbana, Illinois, USA

Austin P. Thekkumthala
Department of Animal Sciences,
Department of Genetics
Rutgers University
New Brunswick, New Jersey, USA

Fiona J. Thomson
Institute of Cancer Sciences
University of Glasgow
Glasgow, UK

Sachiko Tsuda
Lee Kong Chian School of Medicine
Nanyang Technological University
Singapore;
Institute of Molecular and Cell Biology
Singapore

Ilya A. Vinnikov
Molecular Biology of the Cell I
German Cancer Research Center
Heidelberg, Germany

Yongping Wang
Department of Animal Sciences, Department of Genetics
Rutgers University
New Brunswick, New Jersey, USA

Einav Wircer
Department of Molecular Cell Biology
Weizmann Institute of Science
Rehovot, Israel

Ning Yang
Department of Chemistry and the Beckman Institute
University of Illinois at Urbana-Champaign
Urbana, Illinois, USA

Giles S.H. Yeo
MRC Metabolic Diseases Unit
University of Cambridge Metabolic Research Laboratories
Wellcome Trust-MRC Institute of Metabolic Science
Addenbrooke's Hospital
Cambridge, UK

Larry J. Young
Center for Translational Social Neuroscience
Department of Psychiatry and Behavioral Sciences
Yerkes National Primate Research Center
Emory University
Atlanta, Georgia, USA

Series Preface

This Series is a joint venture between the International Neuroendocrine Federation and Wiley-Blackwell. The broad aim of the Series is to provide established researchers, trainees and students with authoritative up-to-date accounts of the present state of knowledge, and prospects for the future across a range of topics in the burgeoning field of neuroendocrinology. The Series is aimed at a wide audience as neuroendocrinology integrates neuroscience and endocrinology. We define neuroendocrinology as a study of the control of endocrine function by the brain and the actions of hormones on the brain. It encompasses study of normal and abnormal function, and the developmental origins of disease. It includes study of the neural networks in the brain that regulate and form neuroendocrine systems. It also includes study of behaviors and mental states that are influenced or regulated by hormones. It necessarily includes understanding and study of peripheral physiological systems that are regulated by neuroendocrine mechanisms. Clearly, neuroendocrinology embraces many current issues of concern to human health and well-being, but research on these issues necessitates reductionist animal models. Contemporary research in neuroendocrinology involves use of a wide range of techniques and technologies, from subcellular to systems and whole organism level. A particular aim of the Series is to provide expert advice and discussion about experimental or study protocols in research in neuroendocrinology, and to further advance the field by giving information and advice about novel techniques, technologies and interdisciplinary approaches.

To achieve our aims each book is on a particular theme in neuroendocrinology, and for each book we have recruited an editor, or pair of editors, expert in the field, and they have engaged an international team of experts to contribute Chapters in their individual areas of expertise. Their mission was to give an up-date of knowledge and recent discoveries, to discuss new approaches, 'gold-standard' protocols, translational possibilities and future prospects. Authors were asked to write for a wide audience to minimize references, and to consider use of video clips and explanatory text boxes; each Chapter is peer-reviewed, and has a Glossary, and each book has a detailed index. We have been guided by an Advisory Editorial Board. The Masterclass Series is open-ended: books in preparation include *Neuroendocrinology of Stress; Computational Neuroendocrinology; Molecular Neuroendocrinology* and *Neuroendocrinology of Appetite*. The first book in the Series, *Neurophysiology of Neuroendocrine Neurons*, was published in December 2014.

Feedback and suggestions are welcome.

John A. Russell, University of Edinburgh,
and William E. Armstrong, University of Tennessee

Advisory Editorial Board:
Ferenc A. Antoni, Egis Pharmaceuticals PLC, Budapest
Tracy Bale, University of Pennsylvania
Rainer Landgraf, Max Planck Institute of Psychiatry, Munich
Gareth Leng, University of Edinburgh
Stafford Lightman, University of Bristol

International Neuroendocrine Federation: www.isneuro.org

About the Companion Website

This book is accompanied by a companion website:

www.wiley.com/go/murphy/neuroendocrinology

The website includes:
- End-of-chapter references and glossary
- Powerpoint figures and PDF tables from the book for downloading
- Supplementary videos and data sets

Introduction

David Murphy[1,2] and Harold Gainer[3]

[1] School of Clinical Sciences, University of Bristol, Bristol, UK
[2] Department of Physiology, Faculty of Medicine, University of Malaya, Kuala Lumpur, Malaysia
[3] Molecular Neuroscience Section, Laboratory of Neurochemistry, National Institutes of Health, Bethesda, Maryland, USA

Recent deep exploration of the mammalian genome and its expression has produced a huge amount of new knowledge that will form the foundation of the next generation of studies in neuroendocrinology. These data have also revealed a number of surprises, not least of which has been the revelation that the number of protein-coding genes needed to make a mammal is surprisingly small – maybe as few as 22,333. However, this small number belies an unanticipated complexity that is evident at a number of levels. Firstly, it has been found that *cis*-regulatory sequences evolve much faster than the genes they control, and the cognate transcription factors that recognize them have expanded, suggesting that the divergence between species is driven by mutations in these elements. Secondly, we now know that much of the genome is transcribed not only into protein-coding mRNAs but also into non-coding transcripts, most of which have no known function. However, we now appreciate the role of small (22nt) microRNAs (miRNAs) in the coordinated regulation of the translation of classes of target mRNAs. Thirdly, we now are aware of the extent of alternative splicing in the generation of multiple, subtly different mature mRNAs from the precursor transcript encoded by a single gene. These different mRNAs are translated into proteins that contain different functional domains, thereby contributing to further physiological complexity. Fourth, the increasing recognition that regulation of chromatin structure can determine gene expression has led to the emerging field of epigenetics. Finally, the important role of posttranslational events in the generation of functional protein diversity is now widely recognized.

In this book, distinguished authors from across the globe critically examine the above and other issues in the context of the expression, regulation, and physiological functions of genes in neuroendocrine systems. The 19 chapters in this book are written so as to be accessible and of value to graduate students as well as experienced researchers in the field, and include annotated references. There is also a glossary at the end of the book. The chapters are divided into four parts.

First, in Chapters 1–6 (Part A), the genome and the regulation of its expression are considered, and the molecular mechanisms that contribute to protein, and hence biological, diversity are explored. These issues are discussed within the context of neuroendocrine systems, for example, by focusing on the genes encoding neuropeptides and their receptors.

Second, in Chapters 7–10 (Part B), the mechanisms that enhance peptide and protein diversity beyond what is encoded in the genome by posttranslational modification are

Molecular Neuroendocrinology: From Genome to Physiology, First Edition. Edited by David Murphy and Harold Gainer.
© 2016 John Wiley & Sons, Ltd. Published 2016 by John Wiley & Sons, Ltd.
Companion website: www.wiley.com/go/murphy/neuroendocrinology

described and discussed. In this section, the chapter on neuropeptide receptors should be of particular interest to neuroendocrinologists.

Third, in Chapters 11–14 (Part C), the molecular tools that today's neuroendocrinologists can use to study the regulation and function of neuroendocrine genes within the context of the intact organism are described and discussed. These chapters compare and contrast germline and somatic *in vivo* gene transfer technologies, with a particular emphasis on novel methods such as optogenetics, which through manipulation of gene expression is able to provide extraordinary insights into neuronal function and circuitry. The contributions of non-mammalian model systems to the understanding of neuroendocrine mechanisms are also considered.

Ultimately, neuroendocrinology is about the things that really matter to us as human beings. It is the study of how the brain makes hormones that drive urges such as sex, love, eating, and drinking. Thus, in our final section (Part D, Chapters 15–19), we describe case studies that exemplify the state-of-the-art application of genomic technologies in physiological and behavioral experiments that seek to better understand these complex biological processes.

Genome and Genome Expression

CHAPTER 1

Evolutionary Aspects of Physiological Function and Molecular Diversity of the Oxytocin/Vasopressin Signaling System

Zita Liutkevičiūtė and Christian W. Gruber

Center for Physiology and Pharmacology, Medical University of Vienna, Vienna, Austria

1.1 Evolution of peptidergic signaling

One of the major eukaryotic signal transduction machineries is composed of G protein-coupled receptors (GPCRs) and their associated signaling molecules (see Chapter 10). GPCRs share a seven α-helical transmembrane architecture with an extracellular N-terminus and an intracellular C-terminus, and are able to sense a diverse set of ligand molecules, including proteins, peptides, amino acids, nucleosides, nucleotides, ions, and photons. GPCRs are involved in many processes such as cell growth, migration, density sensing or neurotransmission (Gruber *et al.*, 2010).

Proteins comprising a seven-transmembrane topology have been identified as far back in the evolutionary timeline as prokaryotes; these are, for instance, light-sensitive proteo-, bacterio- and halorhodopsin proteins that are involved in non-photosynthetic energy harvesting in Archaea and bacteria. Structurally similar sensory rhodopsin proteins can also be found in eukaryotes, but to determine the phylogenetic relationship between prokaryotic and eukaryotic GPCRs is rather difficult, since (i) the prokaryotic and eukaryotic proteins have evolved independently for approximately 1.2 billion years, which resulted in low sequence conservation, and (ii) the occurrence of lateral gene transfer between prokaryotes and eukaryotes has been reported for certain microbial rhodopsins, which further complicates the analysis of phylogenic relations (Strotmann *et al.*, 2011).

Recent studies on the evolution of GPCR signaling systems in eukaryotes – covering not only the receptors and their cognate G proteins, but also upstream and downstream regulators of the system – concluded that the last eukaryotic common ancestor must have already expressed a complex repertoire of GPCRs. Furthermore, it has been suggested that different parts of the GPCR signaling system evolved independently, and that some of them have been lost or became simplified without disrupting overall signaling functionality. For instance, most organisms contain most of the known GPCR signaling components, but certain species have retained only a subset of those, whereas others are completely reduced. These findings suggest that the GPCR signaling system is modular and that during evolution, drastic rearrangements can occur without complete loss of functionality. Analyses of protein domain architectures additionally suggest that domain shuffling is a major mechanism of signaling system evolution (de Mendoza *et al.*, 2014).

Molecular Neuroendocrinology: From Genome to Physiology, First Edition. Edited by David Murphy and Harold Gainer.
© 2016 John Wiley & Sons, Ltd. Published 2016 by John Wiley & Sons, Ltd.
Companion website: www.wiley.com/go/murphy/neuroendocrinology

Gene families and protein domain architectures of cytoplasmic transduction elements (for example, G proteins, arrestins, regulators of G protein signaling, guanine nucleotide exchange factors) are largely conserved between unicellular holozoans and metazoans. In contrast, receptors underwent a dramatic expansion in metazoans compared to their closest unicellular relatives. For instance, the human and mouse genomes code for more than 800 and 1300 GPCRs, respectively, which equals more than 1% of the total predicted genes, while yeast has as little as 10 GPCR genes, less than 0.2% of the total predicted genes (de Mendoza *et al.*, 2014; Fredriksson and Schioth, 2005). This could be due to adaptation of GPCR signaling systems for new functions, such as cell–cell communication, developmental control, and complex environmental sensing, from light to odor and taste (de Mendoza *et al.*, 2014). However, most GPCRs do not play a primary vital role in these organisms. Only 8% of GPCR genes in mice responded to gene disruption by embryonic or perinatal lethality; about 41% exhibited an obvious phenotype and more than 50% of knockout mice of individual GPCRs display no obvious phenotypical change (Schoneberg *et al.*, 2004). However, in humans, mutations in genes encoding GPCRs and G proteins result in pathological conditions, for instance severe vision impairment and blindness, and many other retinal, endocrine, metabolic or developmental disorders (Schoneberg *et al.*, 2004).

1.1.1 Evolution and diversity of peptide G protein-coupled receptors and their endogenous ligands

Of particular interest for this chapter are peptidergic systems, which are generally defined as a functional complex consisting of a cell that synthesizes and releases a peptide mediator, a cell that responds to that peptide by a certain physiological change, and the process of transferring the peptide from the site of synthesis to the site of action. In particular, we use the term **peptidergic signaling** for pathways that are mediated by peptides, their endogenous receptors and associated signaling molecules, which commonly belong to the family of G protein-coupled receptors. Many signaling peptides are released by the central nervous system and these neuropeptides are closely associated with the emergence of the first nervous system. Neuropeptides and the nervous system probably evolved in the common ancestor of cnidarians since sponges (the evolutionary older animal group) do not exhibit any physiological or anatomical signs of a nervous system. Neuropeptides are expressed in brains and are involved in the complex regulation of homeostatic processes and neuronal activity in metazoans. They may act as neurotransmitters, if released within synapses, or as neurohormones to activate receptors distal from the site of release. **Neuropeptides** are short (<50 amino acids) secreted polypeptides derived from larger precursor proteins which share defining features at the level of their primary sequence, which is useful for evolutionary studies, because short peptides sometimes lack sequence similarities (Mirabeau and Joly, 2013).

Recently, the evolutionary history of **bilaterian** neuropeptides and receptors was reconstructed to clarify the relationships between **protostomian** and **deuterostomian** peptidergic systems (Mirabeau and Joly, 2013). The results clearly indicated that the majority of peptidergic systems were present in the last common ancestor of bilaterians (the **urbilaterian**) (Table 1.1). This further supports the theory that the urbilaterian was an animal with a sophisticated physiology and a complex nervous system, capable of integrating sensory information. Another conclusion of this study was the existence of co-evolution between the majority of receptors and their ligands, although previously it has been suggested that, during evolution, novel ligands may outcompete existing ones for a given receptor (Mirabeau and Joly, 2013).

Table 1.1 Inferred evolutionary relationships between the different ancestral bilaterian peptidergic systems.

Family	Peptide name (Deuterostome)	V	T	B	A	L	D	I	N	Peptide name (Protostome)
1.	Vasopressin									Vasopressin
2.	Tachykinin									Tachykinin
3.	GnRH									AKH/corazonin
4.	Cholecystokinin									Sulfakinin
5.	Neuromedin U									Capability/pyrokinin
6.	Neuropeptide Y									Neuropeptide F
7.	Corticoliberin									Diuretic hormone 44
8.	Calcitonin									Diuretic hormone 31
9.	Orexin									Allatotropin
10.	Neuropeptide S									CCAP
11.	NPFF									SIFamide
12.	Endothelin/GRP									CCHamide
13.	Galanin									Allatostatin A
14.	Thyroliberin									Uncharacterized
15.	Kiss1									Uncharacterized
16.	QRFP									Uncharacterized
17.	PTH/blucagon/PACAP									Uncharacterized
18.	Uncharacterized									Leucokinin
19.	Uncharacterized									Ecdysis-triggering hormone
20.	Uncharacterized									RYamide/luqin
21.	GPR139/142									Allatostatin B/proctolin
22.	Uncharacterized									PDF/cerebellin
23.	GPR19									Uncharacterized
24.	GPR83									Uncharacterized
25.	GPR150									Uncharacterized
26.	Uncharacterized									Uncharacterized
27.	Uncharacterized									Uncharacterized
28.	Uncharacterized									Uncharacterized
29.	Uncharacterized									Uncharacterized

Mirabeau and Joly suggested that 29 peptidergic systems (here shown as different families from 1 to 29) were present in the last common ancestor of bilaterians (the urbilaterian). A dark gray square denotes the presence of both peptides and receptors from a given peptidergic system, a light gray square denotes the presence of receptor or peptide and the white square shows that both peptides and receptors are absent for a given peptidergic system in the phylogenetic group. Subphylum Vertebrata (**V**) is composed of *Homo sapiens* and *Takifugu rubripes*, phylum Tunicata (**T**) of *Ciona intestinalis* and *Ciona savignyi*, superphylum Ambulacraria (**A**) of *Strongylocentrotus purpuratus* and *Saccoglossus kowalevskii*, Lophotrochozoa (**L**) of *Capitella teleta* and *Lottia gigantea*, class Insecta (**I**) of *Drosophila melanogaster*, *Tribolium castaneum*, and *Acyrthosiphon pisum*, phylum Nematoda (**N**) of *Caenorhabditis elegans* and *Pristionchus pacificus*, Branchiostoma (**B**) of *Branchiostoma floridae*, and Daphnia (**D**) of *Daphnia pulex* (Mirabeau and Joly, 2013).

1.1.2   Origin of the oxytocin (OXT)/arginine vasopressin (AVP) signaling system

One of the best known peptidergic systems is OXT/AVP signaling. Homologs of OXT/AVP receptors and ligands have been identified in diverse organisms such as hydra, worms, insects, and vertebrates. Across evolutionary lineages, the OXT/AVP neuropeptide signaling system shows conserved functions in water homeostasis, reproductive behavior, learning, and memory (see section 1.4 later in this chapter) (Gruber, 2014).

Phylogenetic grouping indicated that invertebrate crustacean cardioactive peptide (CCAP) receptor, neuropeptide S (NPS) receptor, and AVP-like receptors form a monophyletic family which is phylogenetically the closest to the gonadotropin-releasing hormone (GnRH) receptor superfamily (Figure 1.1). Although CCAP, NPS, and AVP-like peptides are not similar in sequences, analyses of precursors encoding those peptides showed the presence of neurophysin domains in some of the genes, which consolidates the common origin of the peptides. Furthermore, AVP and neuropeptide S are found in neighboring tandem position in the amphioxus (*Branchiostoma floridae*) genome, indicating that they are the product of ancient duplication. Accordingly, in an ancestor of bilaterians, the duplication of a single gene (AVP/NPS/CCAP-like) must have occurred, which gave rise to AVP-like and NPS/CCAP-like genes (Mirabeau and Joly, 2013; Pitti and Manoj, 2012).

The origin of the OXT-like system can be dated back to a common ancestor of bilateral animals – more than 600 million years ago. Invertebrates, with few exceptions, have only one OXT/AVP peptide homolog, whereas vertebrates have two. Although sharing high sequence similarities, historically the peptides were given different names in different species (for instance, lysipressin, phenypressin, vasotocin, mesotocin, isotocin, annepressin, conopressin, and inotocin – all of them belong to the family of OXT/AVP peptides) (Table 1.2). The ancestral vasotocin gene was duplicated before vertebrate divergence approximately 450 million years ago, forming OXT-like peptides. Within these lineages, the genes of the peptides are found in close proximity on the same chromosome. In all taxa, the short nine-amino acid peptide first is translated as longer precursor proteins, which contain a signal peptide, followed by the neuropeptide domain, a dibasic amino acid cleavage site, and a neurophysin domain; additionally, in some genes a copeptin sequence completes the precursor. Precursor genes are also characterized by similar intron sites and lengths (Acher *et al.*, 1995; Mirabeau and Joly, 2013).

Besides the peptide precursor proteins, there are four receptors in the vertebrate OXT/AVP receptor family: one OXT receptor (OTR) and three AVP receptors ($V_{1a}R$, $V_{1b}R$, and V_2R), whilst invertebrates commonly have only one receptor: VPR (invertebrate AVP-like receptor). OXT and AVP V_1 receptors are more similar to each other compared to the AVP V_2 receptor, which indicates that V_2R arose before the V_1R/OTR split. At the base of vertebrate lineage, two rounds of whole genome duplication (*2R*) have been described. It has been proposed that local duplication of an ancient vertebrate AVP receptor before *2R* gave rise to the V_2R and V_1R/OTR genes. After *2R*, these genes were quadrupled, although there have been several losses in different vertebrate classes. Interestingly, the spotted gar (*Lepisosteus oculatus*) has three V_2 receptors (V_2A, V_2B, and V_2C), indicating that three subtypes of the V_2 receptor arose early in vertebrate evolution, but since then they have been lost in some vertebrates. Eleost-specific tetraploidizations (*3R*) gave teleost fish additional gene family members of OXT/AVP receptors (Lagman *et al.*, 2013).

In mammals, OXT and AVP differ from each other at only two amino acid positions. Contrary to many other signaling systems, selectivity of the different receptors is not achieved via the ligands (because they are very similar and can cross-react), but mainly via interplay of factors including receptor up- or downregulation, release of specific ligand-degrading enzymes, local ligand production, and receptor clustering (Goodson, 2008; Gruber and Muttenthaler, 2012). However, ligand-receptor selectivity also has important implications for drug design and development of selective compounds for pharmacological applications.

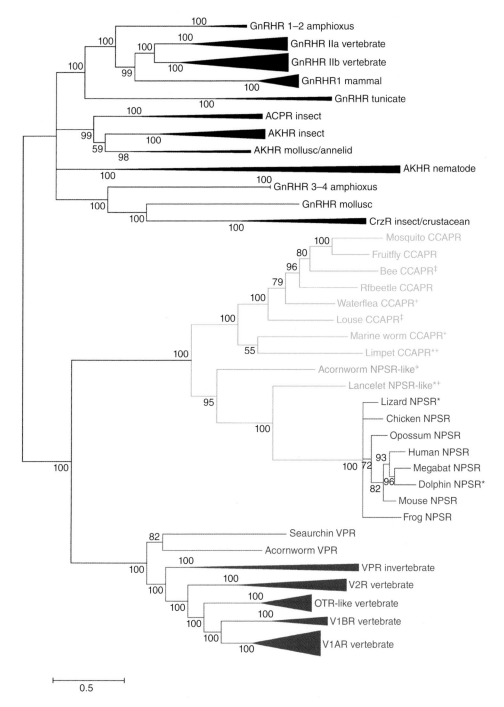

Figure 1.1 Phylogenetic relationship of the neuropeptide S, cardioacceleratory peptide, gonadotropin-release hormone, and OXT-/AVP-like receptors from vertebrates and invertebrates. Phylogenetic relationship is plotted as a Bayesian tree of neuropeptide S (NPSR; *red*), cardioacceleratory peptide (CCAPR; *green*), gonadotropin-release hormone (GnRHR; *black*), and OXT-/AVP-like (VPR, V_2R, $V_{1A}R$, $V_{1B}R$, OTR; *blue*) receptors from vertebrates and invertebrates. **Bayesian posterior probabilities** are marked near branches as a percentage, and are used as confidence values of tree branches. Nodes were compressed to represent the animal lineages. The scale bar represents the number of estimated changes per site for a unit of branch length. The sequence names marked with asterisk (*) and plus (+) symbols represent manually corrected sequences at the N-terminus and C-terminus, respectively. Sequence names marked with the double-cross (‡) symbol in this figure represent fragmented sequences. Adapted and modified from Pitti and Manoj (2012).

Table 1.2 OXT/AVP-like peptide sequences across the animal kingdom.

Animal groups		Peptide	Sequence
Vertebrates	Humans, mammals	Oxytocin	CYIQNCPLG*
		[Arg8]-Vasopressin	CYFQNCPRG*
	New World monkeys	[Pro8]-Oxytocin	CYIQNCPPG*
	Pigs, marsupials	[Lys8]-Vasopressin (Lysipressin)	CYFQNCPKG*
	Marsupials	[Phe2]-Vasopressin (Phenypressin)	CFFQNCPRG*
	Birds, reptiles, amphibians, fish	Vasotocin	CYIQNCPRG*
	Birds, reptiles, amphibians	Mesotocin	CYIQNCPIG*
	Fish	Isotocin	CYISNCPLG*
Annelids	Earthworms	Annetocin/annepressin	CFVRNCPTG*
	Leeches	Lys-conopressin G	CFIRNCPKG*
Molluscs	Cephalopods	Cephalotocin	CYFRNCPIG*
		Octopressin	CFWTSCPIG*
	Gastropods	Lys-conopressin G	CFIRNCPKG*
Nematodes	*Caenorhabditis elegans*	Nematocin	CFLNSCPYRRY*
Arthropods	Insects	Inotocin	CLITNCPRG*
		Inotocin	CLIVNCPRG*
	Arachnids	Arachnotocin[†]	CFITNCPPG*
		Arachnotocin[†]	CFITNCPIG*
	Myriapods	Myriatocin[†]	CYITNCPPG*
	Crustacea/brachiopods	Vasotocin-like	CFITNCPPG*

*indicates C-terminal amidation.
[†]Sequences were discovered by genome mining according to Gruber and Muttenthaler (2012).
Table modified and adapted from Gruber (2014).

It is therefore intriguing to analyze the diversity of the OXT/AVP system by discovery of novel peptide sequences and to study evolutionary aspects of the molecular interaction of different ligand-receptor pairs.

1.2 The discovery of neuropeptide signaling components in the era of genomics

The increasing number of genome projects is generating an unprecedented amount of sequence information, which is, and will be, available in public databases for the identification of genes, such as the National Center for Biotechnology Information (www.ncbi.nlm.nih.gov/). 'Genome mining' is a term that has been used to describe the exploitation of genomic information for the discovery of new processes, targets, and products. This technique is particularly valuable in the genomic era, since the number of available genomes is steadily increasing as whole-genome sequencing is becoming affordable and achievable. Furthermore, there are an even greater number of transcriptome projects, which offer the possibility to identify and annotate not only genes but also messenger RNA (mRNA) transcripts. According to the central dogma of molecular biology, when synthesizing a protein, the genetic information of DNA is transcribed into mRNA which is then translated into a protein. Therefore, 'transcriptome mining' provides information about the transcribed and translated genetic pool of a cell, tissue, organ or, indeed, the whole organism at any one time. In addition, transcriptome sequencing is quicker and cheaper and the genetic information is easier to annotate.

1.2.1 Brief overview of finished and ongoing genome projects

The vast amount of publicly available genomic or RNA sequence information offers the opportunity to search for novel peptidergic signaling components, due to the steadily increasing number of ongoing genome sequencing projects as well as advances in bioinformatics tools. For example, one can mine these sequences for identification of neuropeptide precursor and receptor genes or transcripts. The number of total genome sequencing projects currently (August 2015) stands at 69,949, of which 7210 are listed as completed (www.genomesonline.org). Most of these sequences will be deposited in public databases mainly as non-annotated data. Hence it is the responsibility of the scientific community to make sense of this information. The first step is to identify and annotate genes or transcripts, and use them for comparative *in silico* studies or for functional genomics with the aim of elucidating their biological and physiological purpose.

1.2.2 Data mining for the non-expert computational biologist

Genetic sequence data mining, in particular large-scale genome analysis or systems biology approaches, can be difficult and laborious, which requires expertise, powerful computing hardware and specialist analysis software that are usually only accessible to the expert computational biologist. However, many tools to mine genetic sequence data sets are readily available via online databases and are accessible for the non-expert biologist. The following section provides a simple description of this workflow, which is summarized in Figure 1.2. We would like to place particular focus on the identification of novel invertebrate neuropeptide precursor and receptor sequences (see section 2.3).

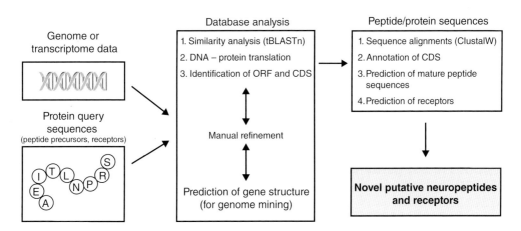

Figure 1.2 Flowchart of genome mining for the discovery of nonapeptides and their receptors. Whole genome shotgun or transcriptome data, and amino acid sequences of precursor proteins from nonapeptides or receptors of interest (for example, OXT-like neuropeptides and their receptors) were used for database analysis. This included similarity analysis of target DNA sequence and query protein amino acid sequence using tBLASTn, DNA to protein translation of discovered hit sequences and identification of open reading frames and coding sequence. The obtained automated results were refined and confirmed manually and used for gene structure prediction using the GeneWise algorithm. Database analysis yielded precursor protein and peptide sequences that were further annotated and analyzed by sequence alignments and similarity comparison to identify signal sequences, propeptides, and mature peptide chains. Using this genome-mining methodology, it is possible to predict the amino acid sequences of bioactive peptides in ants. Adapted and modified from Gruber and Muttenthaler (2012).

1.2.2.1   tBLASTn similarity search

Prior to the **BLAST** search, a number of suitable query sequences (peptide or protein sequences) need to be selected. It is advisable to choose query sequences that are related to the gene/transcript of interest of the target species, i.e. phylogenetic neighboring species, as well as the most homologous protein sequences. Sometimes including more distant and less related query sequences, i.e. sequences of another animal class, can provide additional information or can be used as a control. These query sequences will be used for initial tBLASTn analysis via GenBank (http://blast.ncbi.nlm.nih.gov/Blast.cgi?PROGRAM=tblastn&BLAST_PROGRAMS=tblastn&PAGE_TYPE=BlastSearch&SHOW_DEFAULTS=on&LINK_LOC=blasthome) using the search set of interest, for example, whole genome shotgun contigs or transcriptome shotgun assemblies. The most similar hits (provided as genomic sequences) will be extracted and translated into their respective amino acid sequence (six-frame translation) using appropriate web tools (http://web.expasy.org/translate/). Open reading frames (ORFs) of the respective genes and coding sequences of transcripts can be identified via manual sequence assignments. For larger data sets, there are automated computing tools available, e.g. ORFPredictor, Dragon TIS, and MetWAMer. The annotated and identified amino acid sequences should be confirmed by another tBLASTn analysis against the search set and species of interest. These peptide/protein sequences will then be used for further identification and annotation.

1.2.2.2   Identification and annotation of novel genes

In case of genomic information, annotation of tBLASTn hits can be performed using the GeneWise2 algorithm (www.ebi.ac.uk/Tools/psa/genewise/) or gene structure prediction tools (for example, FGENESH available via Softberry Inc. http://linux1.softberry.com/berry.phtml?topic=fgenesh&group=programs&subgroup=gfind). In particular, for non-model organisms, these prediction tools may not be accurate enough and manual refinement will be necessary. Often gene structure analysis (intron/exon determination) fails completely, in which case a combined genome and transcriptome mining approach (if available for the same species) may be useful. If successful, the automated gene structure analysis will yield full or partial protein sequences, their open reading frames, and the corresponding DNA sequences. The predicted protein-coding sequences as well as the intron/exon structure from genomic data can be used for further annotation and similarity alignments, as well as functional genomics or physiological *in vitro* and *in vivo* studies (Gruber and Muttenthaler, 2012).

1.2.3   Case study: the discovery of novel inotocin peptides in ant genomes

Following in the footsteps of many successful genome-sequencing projects in animals, plants, and microbes, the genomes of seven ant species have recently been reported. These include the invasive Argentine ant *Linepithema humile*, the red harvester ant *Pogonomyrmex barbatus*, the fire ant *Solenopsis invicta*, the carpenter ant *Camponotus floridanus*, a basal ant *Harpegnathos saltator*, the leaf-cutter ant *Atta cephalotes*, and the farming ant *Acromyrmex echinatior*.

   The aim of this study was to analyze three representative ant genomes for the discovery of peptide-encoding genes and their sequences using the above described workflow. We were able to identify numerous putative peptide sequences corresponding to partial or full-length precursors of ant neuropeptides. This included the first annotation of mature sequences of inotocin nonapeptides in social insects, which are OXT/AVP-related neuropeptides (Figure 1.3).

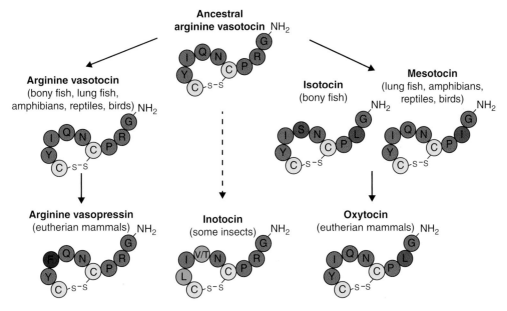

Figure 1.3 Discovery and origin of insect inotocin nonapeptides. (A) Arginine vasotocin is the presumed ancestral peptide of OXT and AVP. Mammalian OXT evolved via intermediate forms of isotocin (bony fish) and mesotocin (lung fish, amphibians, reptiles, and birds) (simplified illustration for clarity). It is yet to be determined whether invertebrate OXT/AVP-related peptides in insects or snails (e.g. conopressins, not shown) have also evolved from ancestral vasotocin (indicated as dashed line). The peptide sequences are shown in one-letter amino acid code. The highly conserved cysteine residues and disulfide bonds are colored in yellow. Residues in the ancestral arginine vasotocin and those that are identical to vasotocin are in dark gray. Residues that have changed during AVP evolution are colored in red, residues that have changed during OXT evolution are colored in purple and residues that are unique to insect inotocins are colored in green. Adapted and modified from Gruber and Muttenthaler (2012).

This study provided proof of concept for the use of a simple **genome mining** workflow to analyze endogenous neuropeptides from insects (Gruber and Muttenthaler, 2012). The newly identified ant inotocin peptide sequences displayed high similarity to vasotocin and OXT/AVP-like sequences from other species. In all analyzed ant genomes, the short mature nonapeptides are generated as a longer precursor protein which share molecular features with precursors of other insect inotocin proteins, and even human OXT and AVP precursors (see section 1.1). Besides the sequence of the mature nonapeptides, all precursors contain conserved protein domains for cellular secretion, enzymatic processing and physiological transport, and are characterized by similar position and lengths of introns (Gruber and Muttenthaler, 2012). Besides the mature nonapeptides, we also reported putative receptor sequences in ant genomes that share high similarity to other insects, like the beetle *Tribolium castaneum* (Gruber and Muttenthaler, 2012; Stafflinger *et al.*, 2008), indicating that possibly not only the genetic structure but also the function of these receptors and their nonapeptide ligands may be conserved across species. Recently, this work has been extended by characterizing OXT- and AVP-like precursors and receptor sequences from the genomes of several arthropod species, such as the red spider mite *Tetranychus urticae*, the predatory mite *Metaseiulus occidentalis*, and the centipede *Strigamia maritima* (Gruber, 2014).

These results offer the possibility to interpret the phylogenetic relationship and evolution of insect peptide hormone systems but, most importantly, the predicted mature peptide sequences could provide novel drug leads or tools to study similar and conserved receptor systems in humans.

1.3   Evolutionary aspects of OXT/AVP diversity

1.3.1   Diversity of OXT-like peptides

As introduced in section 1.1, the origin of the OXT and AVP signaling system is considered to date back at least 600 million years. OXT-like peptides are today present in many different species, including invertebrate animals, such as mollusks, annelids, nematodes and insects, non-mammalian vertebrates, such as amphibians, reptiles and birds, fish, mammals, and humans (Figure 1.4; see Table 1.2).

Invertebrates are considered to possess only one single OXT-/AVP-like nonapeptide form. However, the published information about OXT/AVP signaling in invertebrates as compared to vertebrates or mammals is vast and requires more detailed analysis. For example, it is not clear whether nematocin (= OXT homolog of *C. elegans*) is expressed as a single peptide or whether there exists a processing variant (Beets *et al.*, 2012). The earliest vertebrates probably possessed only a single nonapeptide, namely arginine vasotocin, although the vasotocin gene duplicated at about the same time that jaws evolved, so all jawed vertebrates now exhibit two nonapeptide forms in the brain – an OXT-like form and either vasotocin or AVP. The most common OXT-like forms are isotocin, which is found in bony fish, and mesotocin, which is found in birds, lungfish, reptiles, amphibians, and some marsupials. Interestingly, cartilaginous fish have evolved at least six OXT-like homologs (Donaldson and Young, 2008; Hoyle, 1999).

All members of the OXT/AVP/vasotocin peptide family share high sequence similarity, namely a N-terminal six-residue ring, formed by a disulfide bond between the two cysteines at positions 1 and 6, and a flexible C-terminal three-residue tail (exceptions are nematocin and tunicate vasotocin-like peptides). When comparing the intercysteine and tail sequences of those peptides from different organisms, it is obvious that certain positions are highly variable whereas others are highly conserved. For instance, positions 2 and 3 (hydrophobic or aromatic residues), positions 4 and 5 (polar or charged residues), as well as position 7 (proline) and position 9 (glycine) are conserved, whereas position 8 is highly variable (Figure 1.5; see Table 1.2). These amino acid variations are presumably responsible for species-selective recognition, binding, and activation of the different receptors. Subtle differences in the amino acid sequence of the peptide ligands may have significant effects on binding (K_d) and efficacy (EC_{50}) of the ligand to its receptor. Using an *in silico* approach, we recently attempted to correlate these interspecies differences of endogenous ligand with receptor sequence variations with the aim of providing a molecular understanding of recognition, binding, and activation of OXT and AVP receptors by their native ligands. This in turn could assist the design and development of novel selective ligands (Koehbach *et al.*, 2013c).

1.3.2   Molecular diversity of OXT-like receptor residues involved in ligand binding and activation

We previously compared 69 known OXT-like receptor sequences from a wide range of species. Figure 1.5 illustrates the comparison of the ligand sequence variation with the receptor sequence evolution. Consistent with the high receptor sequence similarity across

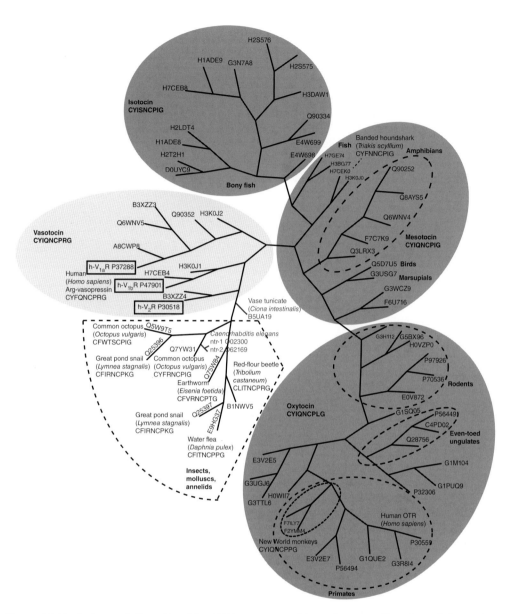

Figure 1.4 Evolution and diversity of the OXT receptor and its endogenous ligands. A phylogenetic tree consisting of 69 OXT, OXT-like, vasotocin, and human AVP receptors and their respective ligands is shown and has been prepared by sequence alignment using ClustalW. Receptor clusters for OXT, isotocin, mesotocin, and vasotocin/AVP receptors display high sequence similarity and are highlighted in dark/light gray. Receptors of species within the same class or order are highlighted by dashed lines. The UniProtKB entry numbers of the receptors are shown at the end of each branch. Originally published in Koehbach *et al.* (2013c).

closely related species (e.g. bony fish, amphibians or primates), it was found that also the respective ligands within these clusters were highly conserved. This is in agreement with the evolution of OXT- and AVP-related nonapeptides (Acher *et al.*, 1995; Goodson, 2008). Residues believed to be responsible for OXT/AVP ligand-receptor binding were analyzed and their degree of conservation was compared.

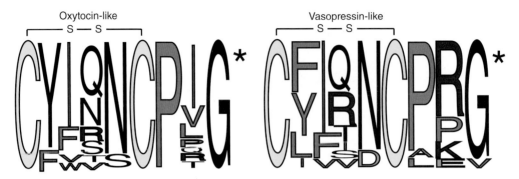

Figure 1.5 Molecular diversity of the OXT/AVP nonapeptide family. Sequence diversity of OXT/ AVP-like nonapeptides is illustrated as sequence plots. Relative frequency sequence logo plots with conserved cysteine residues are highlighted in yellow, hydrophobic, aromatic and non-polar amino acids are labeled in gray, negatively charged residues are labeled in blue and positively charged residues are labeled in red. Modified and adapted from Gruber *et al.* (2012).

Previously, the residues arginine R34 (*numbering of all residues according to human OXT receptor*), phenylalanine F103, tyrosine Y209, and phenylalanine F284 were considered to be important for ligand binding and selectivity at the human OXT receptor. Additionally, aspartic acid D85 and lysine K270 are involved in receptor signaling (Zingg and Laporte, 2003). Interestingly, all OXT-like receptors share the latter two residues, which were reported to be important for receptor activation.

The N-terminal arginine residue (R34) is highly conserved in all OXT-like receptors, indicating its importance for ligand binding. Phenylalanine (F103) in the first extracellular loop has previously been demonstrated to be important for ligand selectivity in the OXT receptor and is considered to interact with the residue at position 8 of the peptide ligand (leucine in human OXT). However, a comparative study of 69 receptor-ligand pairs indicated that a direct interaction of this residue in the receptor and the residue at position 8 of the ligand may not generally be important for ligand selectivity (Koehbach *et al.*, 2013c). The two functionally important receptor residues located in the transmembrane region (Y209 and F284) have been considered as other important residues for ligand-receptor binding (Zingg and Laporte, 2003). Importantly, the residue in position 209 is always an aromatic residue and hence is highly conserved amongst all OXT-like receptors. Accordingly, all native peptide ligands contain an aromatic residue (Tyr or Phe) in position 2 and a hydrophobic or aromatic residue in position 3 (Ile, Phe, Val or Trp) of the peptide ligand (Gruber *et al.*, 2012), which confirms that this ligand-receptor interaction may indeed be conserved throughout evolution.

Overall, the analysis of ligand-receptor evolution with respect to the certain sequence variation of OXT ligand-receptor pairs is in agreement with previous biochemical studies (Zingg and Laporte, 2003), but it is likely that these molecular contacts may not be conserved in all native ligand-receptor pairs. Therefore, we developed an *in silico* approach to describe molecular interaction of OXT ligands and their receptors from an evolutionary perspective (Koehbach *et al.*, 2013c).

1.3.3 OXT ligand-receptor interaction: an *in silico* approach

Using publicly available GPCR crystal structures, sequence similarity alignments, ligand-receptor contact frequency plots, and receptor homology modeling, we were able to define a common binding site, deep within the **vestibule** of the OXT receptors. The receptor residues with the highest frequency of direct contact with ligands included Q119 and M123 (in helix III), I204 and V208 (in helix V), W288, F291, F292 and Q295 (in helix VI), and residue M315 (in helix VII; *numbering according to the human OXT receptor*). Taken together with information from the literature, i.e. the four residues described to be important for human OXT ligand-receptor interaction (R34, F103, Y209, and F284), we hypothesize that it is likely that OXT binds to a common binding site that is located in between those four residues as proposed in the OXT receptor model.

Based on the three largest clusters of receptor-ligand evolution (see Figure 1.4) and with a focus on the residues forming the common binding site in GPCR-ligand structures, we also compared the sequences of OXT, isotocin, and mesotocin receptors and their respective ligands. Residues of the transmembrane helices that were oriented towards the ligand-binding site were found to be more conserved compared to the membrane-exposed residues. Interestingly, the degree of conservation was unevenly distributed: helices II, III, and VI were the most conserved, while residues in helix VII displayed the largest variability.

Biochemical studies with the human OXT receptor and its native ligand have identified four residues important for ligand binding and recognition (Zingg and Laporte, 2003). We compared 69 OXT, OXT-like, and vasotocin receptor sequences to gain further insight into peptide ligand binding. We utilized an *in silico* approach to map the common ligand interaction sites of recently published GPCR structures to a model of the human OXT receptor and compared the interacting residues within different receptor sequences. Our analysis indicates the existence of a binding site for OXT peptides within the transmembrane core region. Previous evolutionary studies of OXT receptors pointed out a number of conserved residues in helices II, III, IV, VI, and VII that may be important for ligand binding, but their analysis included only a few different receptor sequences. It is evident from our comparison that transmembrane helix VII displays the greatest sequence variability, which might be the source for interspecies selectivity. This helix also experiences the biggest structural movements when the receptor is activated and certainly is important for G protein signaling (Koehbach *et al.*, 2013c). While this information brings together empirical data, molecular sequence comparison, and homology modeling, there is still a lack of structure activity and mutagenesis studies to propose a working OXT receptor model that could explain selectivity differences observed in binding studies. In summary, only a ligand-bound crystal structure of OXT and its receptor will be able to shed light on the mechanism of interaction and provide the means for future design of novel and selective ligands.

To conclude, the quality and accuracy of such phylogenetic and evolutionary analysis are determined by the number and distribution of available genetic information about OXT and AVP signaling systems in different animal classes. To date, there are only a few examples of invertebrate peptide and receptor genes that have been characterized. Hence there is a need to study these genes and their function in the invertebrate animal class, which will be described in the following section.

1.4   Physiology of OXT and AVP signaling: from worm to man

Following the description of recent efforts aimed at the discovery of OXT- and AVP-like peptides in several kingdoms of life, as well as the molecular analysis of OXT/AVP diversity, we will summarize available information about the physiology and behavioral role of the OXT and AVP signaling systems, focusing on invertebrate animals, which are by far the biggest group of animals living on Earth.

The function of OXT- and AVP-like peptides has been studied across several animal groups. The effects of both peptides – OXT/OXT homologs and AVP/AVP homologs – appear

Table 1.3 Overview of OXT and AVP nonapeptide physiology.

Animal groups		Functional role of OXT/AVP-like peptide signaling
Mammalian vertebrates	Humans, mammals, marsupials	*Peripheral:* water homeostasis, blood pressure regulation, and reproduction (uterus and mammary gland contraction; ejaculation) *Central:* regulation of learning, memory, stress and anxiety, and complex social behavior (maternal care, bonding)
Vertebrates (non-mammalian)	Birds, reptiles, amphibians, fish	Reproductive physiology, osmoregulation, social communication, affiliation behaviors, aggression, and multiple aspects of stress responses
Invertebrates	Annelids	Reproductive behavior, osmoregulation, metabolism
	Mollusks	Reproductive behavior, memory and metabolism (osmoregulation)
	Nematodes	Reproductive behavior and associative learning
	Insects	Regulation of water homeostasis, other unknown functions(?)

to be distinctive in different species. For example, AVP-like peptides have a major role in regulating reproductive behavior in fish and amphibians, while in birds and mammals OXT-like peptides are more important for this. This situation becomes even more complicated in invertebrate animals, where often only one peptide precursor and only one or two OXT-like receptors exist as compared to humans and mammals which express two peptide ligands and up to four receptors. In the following, we will provide a brief overview of the physiology of OXT and AVP signaling in invertebrates, non-mammalian vertebrates, and mammals, including man (see Table 1.3 for a summary).

1.4.1 Function of OXT and AVP homologs in invertebrates: annelids, arthropods, nematodes, and mollusks

The function of OXT-like peptides has been studied across several invertebrate model organisms, such as annelids, arthropods, nematodes, and mollusks (see Table 1.3). OXT- and AVP-like signaling seems to be involved in neurotransmission, metabolism, and osmoregulation in mollusks and annelids, and in osmoregulation and possibly neurotransmission in insects. In nematodes, this neuropeptide system facilitates gustatory associative learning and coordinates reproductive behavior. Moreover, OXT and AVP show their function during reproduction of leeches, earthworms, and snails (Gruber, 2014). The most comprehensive and detailed studies regarding OXT/AVP physiology in invertebrates came from studies of the model worm *Caenorhabditis elegans* and the red flour beetle *Tribolium castaneum*. Whereas there is only little information on the function of those neuropeptides in insects (Aikins *et al.*, 2008; Stafflinger *et al.*, 2008), there is compelling evidence that this signaling system has a crucial function during reproduction and behavior in nematodes.

The recent discovery and functional characterization of two so-called nematocin receptors and their endogenous peptide ligand showed that this signaling system is involved in the worm's gustatory associative learning (chemotaxis towards NaCl and gustatory plasticity; Beets *et al.*, 2012) and reproductive behavior (regulation of reproductive efficiency, mating behavior, mate searching, brood size, and locomotion behavior; Garrison *et al.*, 2012). However, most studies in invertebrates used molecular biology, immunohistochemistry or *in vitro* pharmacology to describe the physiology of the OXT/AVP neuropeptide system. Functional genomics (for example, gene knock-down) and detailed biochemical analysis have generally not been carried out yet (except for *C. elegans* and *T. castaneum*), but are required to obtain *in vivo* understanding of the physiological and biological role of this neuropeptide system in invertebrates.

In summary, the OXT and AVP neuropeptide signaling system shows conserved functions in invertebrate physiology (water homeostasis) and reproductive behavior, but more detailed studies are required for comparative analysis of OXT/AVP physiology across invertebrates. Especially from an evolutionary perspective, it is interesting to note that the OXT/AVP peptide hormone system is present in some **holometabolous** insects (beetles and ants), whilst these genes are absent in related insect species, such as the honeybee (*Apis mellifera*) and fruit fly (*Drosophila melanogaster*) (Gruber, 2014; Gruber and Muttenthaler, 2012).

1.4.2   Function in non-mammalian vertebrates: fish, reptiles, amphibians, and birds

In non-mammalian vertebrates, OXT, AVP, and their homologs modulate reproductive behavior. OXT-like peptides are involved in the induction of vocalization, courtship behavior, female sexual receptivity, alternative mating, and many more types of behaviors. Based on comparative studies of non-mammalian vertebrates, it is evident that this nonapeptide system has effects on reproductive physiology, osmoregulation, social communication, affiliation behaviors, aggression, and multiple aspects of stress responses (Donaldson and Young, 2008; Gimpl and Fahrenholz, 2001; Goodson *et al.*, 2012; Gruber *et al.*, 2010; Knobloch and Grinevich, 2014). Generally, the OXT/AVP signaling system exhibits extensive conservation of function, but there is at least some variation. For instance, whereas these OXT- and AVP-like peptides are anxiogenic in rodents, they are strongly anxiolytic in male zebra finches (Goodson *et al.*, 2012). In addition, the independent evolution of multiple behavioral characters is associated with evolutionary convergence in the anatomy of nonapeptide systems and their behavioral effects. This is observed in (i) the convergent roles of mesotocin (an OXT homolog of birds, amphibians, and reptiles) and OXT in the extended maternal care of mammals and most birds, (ii) the independently derived effects of mesotocin and OXT on pair bonding in female prairie voles and zebra finches, and (iii) the convergent patterns of nonapeptide receptor distributions in certain finch species that have independently evolved similar patterns of grouping behavior (Goodson *et al.*, 2012).

1.4.3   Function in mammals and man

Since the 1920s it has been known that an 'extract of the posterior lobe and pituitary gland' can be separated into two fractions, rich in oxytocic and pressor activity, respectively. Subsequent comparative studies between numerous species conducted in the first half of the 20th century revealed that in both mammalian and non-mammalian species, OXT stimulates the activity of smooth muscle in reproductive tracts, sperm movement and ejaculation, as well as uterus contraction during parturition and milk ejection from the mammary glands for breastfeeding. Besides these peripheral effects, it is now evident that OXT is a key player for orchestrating complex reproductive, prosocial, and in-group supporting behavior in the central nervous system, such as trust, maternal care and bonding, as well as stress and anxiety (Knobloch and Grinevich, 2014). However, AVP regulates peripheral fluid balance and blood pressure (see Chapter 15). For instance, the lack of AVP production or mutations of the AVP 2 receptor located in the kidney that prevent its correct folding and secretion to the membrane can cause diabetes insipidus. Centrally AVP is involved in memory and learning, stress-related disorders, and aggressive behavior. Both peptides act on GPCRs: OXT signals via one OTR and AVP via three AVP receptors (vasopressor $V_{1a}R$, pituitary $V_{1b}R$, renal V_2R) and these receptors are expressed in various tissues and organs (Frank and Landgraf, 2008; Gruber *et al.*, 2010; Meyer-Lindenberg *et al.*, 2011).

Due to this physiological importance, ligands of OXT and AVP receptors have potential therapeutic applications for novel treatment approaches in mental disorders characterized by social dysfunction, such as autism, social anxiety disorder, borderline personality disorder, and schizophrenia (Meyer-Lindenberg *et al.*, 2011), childbirth-related conditions, such as premature labor and postpartum hemorrhage (Gruber and O'Brien, 2011), osmoregulatory dysfunction, such as diabetes insipidus, as well as cardiovascular disorders, such as congestive heart failure or vasodilatory shock states (Gruber *et al.*, 2010; Manning *et al.*, 2008).

OXT and AVP are closely related, highly conserved, multifunctional neurohypophysial peptides. The high extracellular receptor homology and ubiquitous receptor distribution constitute a major hurdle for the development of selective ligands and therapeutics (Gruber *et al.*, 2012). For example, OXT is still the ligand of choice in the clinic, although it is well established that OXT also signals via the AVP receptors, which can lead to unwanted side effects. A selective OXT/AVP ligand hence has enormous potential for therapeutic development and it will be intriguing to explore natural OXT-like peptides (and their modifications) for selectivity and potency on human receptors in the future (Gruber *et al.*, 2012).

1.4.4   Excursus: Peptides from nature to mimic OXT and AVP peptides

1.4.4.1   Cyclotides from plants
We recently identified the first OXT-like peptide from plants (Koehbach *et al.*, 2013b). The peptide kalata B7 belongs to the family of plant **cyclotides**. They were originally discovered in the Rubiaceae species *Oldenlandia affinis* based on its use in traditional African medicine to accelerate labor. Since then, cyclotides have been identified in numerous plant species of the coffee, violet, cucurbit, pea, potato, and grass families (Koehbach *et al.*, 2013a). Chemically, cyclotides are peptides with an average length of approximately 30 amino acids that contain three conserved disulfide bonds linked together in a knotted arrangement. In addition, they are characterized by a head-to-tail cyclized backbone. This unique structural topology is known as a cyclic cystine-knot motif, which makes them extremely stable against biochemical degradation.

Using bioactivity-guided fractionation of a herbal peptide extract, the cyclotide kalata B7 was identified as a strong uterotonic agent on human uterine smooth muscle cells. Radioligand displacement and second messenger-based reporter assays confirmed that kalata B7 is a partial agonist of human OXT and AVP V$_{1a}$ receptors. In addition, cyclotide sequences have been used as templates for the design of selective peptide ligands by generating nonapeptides with nanomolar affinities to the human OXT receptor (Koehbach *et al.*, 2013b).

We designed several OXT-like nonapeptides and the peptide [G5,T7,S9]-OXT turned out to be a potent and selective agonist of the human OXT receptor. This is of great interest from a drug design point of view, since OXT, AVP, and many analogs synthesized to date have lacked receptor selectivity (Gruber *et al.*, 2012; Manning *et al.*, 2012). Our findings thus highlight the potential of exploiting cyclotides as templates for the design of peptide GPCR ligands. Given the diversity of plant cyclotides as well as the overall number of GPCRs (the human genome encodes for at least 800 different GPCRs; Gruber *et al.*, 2010), many of which are promising drug targets, our results provide the proof of principle for the development of cyclotide-based peptide ligands using a combination of two disciplines, ethnopharmacology and chemical biology (Koehbach and Gruber, 2013).

1.4.4.2   Conopressins from cone snail venom
Rich sources of venom peptides can be found in spiders, scorpions, snakes, and marine animals, in particular cone snails. Evolutionary pressures have afforded a preoptimized, structurally sophisticated collection of disulfide-rich peptide toxins that have been produced

in a combinatorial fashion and fine-tuned over millions of years. Hence it does not come as a surprise to find that within the venom of these animals, structural scaffolds are found that give rise to a very large number of agonists and antagonists, which act on functionally diverse targets such as ion channels, transporters, and GPCRs (Gruber *et al.*, 2010).

The two AVP-like conopressins from the venom peptides of the predatory marine cone snail are a good example. The original discovery of two of these AVP analogs was later characterized by the observation of grooming and scratching behavior upon intracerebral injection into mice. Although the sequences of conopressins are similar to human AVP, they have an additional positive charge in position 4, which is only found in two other endogenous AVP analogs, cephalotocin and annetocin. Conopressin-S was isolated from *Conus striatus*, whereas conopressin-G was first isolated from *Conus geographus* venom, but later also found to be present in the venom of *Conus imperialis* as well as in tissue extracts of the nonvenomous snails *Lymnea stagnalis* and *Aplysia californica*, and the leech *Erpobdella octoculata*. It is not clear what evolutionary advantage is conferred by the presence of these peptides in the venom of the cone snail. Nevertheless, the discovery and characterization of conopressin-T in comparison with the human neuropeptides AVP and OXT led to the identification of an interesting agonist/antagonist switch, which is currently being investigated towards novel antagonist design for the human receptors (Dutertre *et al.*, 2008).

1.5 Perspectives

OXT- and AVP-like peptides mediate a range of physiological functions that are important for osmoregulation, reproduction, complex social behaviors, memory, and learning. The origin of the OXT/AVP signaling system is considered to date back more than 600 million years. Today these nonapeptides are present in vertebrates, including mammals, birds, reptiles, amphibians, and fish, as well as several invertebrate species, including mollusks, annelids, nematodes, and arthropods. Members of this peptide family share high sequence similarity, and it is possible that they are functionally related across the entire animal kingdom. In humans, OXT and AVP are structurally very similar; they differ only by two amino acids. Both nonapeptides contain an N-terminal cyclic six-residue ring structure, stabilized by an intramolecular disulfide bond, and a flexible C-terminal three-residue tail. They mediate their distinct functions by signaling through four GPCRs, which share ~80% sequence homology.

The structural similarity of OXT and AVP together with the high sequence conservation of their receptors results in significant cross-reactivity. This **selectivity dilemma** constitutes a major burden for the development of receptor-specific agonists and antagonists. In addition, it is known that OXT and AVP receptors signal via multiple G protein-coupling modes, and they can form functional homo- and heterooligomers, which further complicates the quest for selective and biased ligands. Although over the last decades, over 1000 OXT and AVP peptide ligands have been synthesized and characterized for therapeutic applications, there is still a great demand for selective and biased ligands that activate or block a specific receptor or a specific cellular pathway. As demonstrated with two examples, i.e. cyclotides and conopressins, peptide sequences identified from natural sources can provide evolutionary advantage over random chemical synthetic approaches and may yield novel lead compounds for therapeutic applications.

An even more promising and very timely approach will be the discovery of endogenous OXT and AVP-like nonapeptides in invertebrate species and in particular in insects, which represent more than half of all known living organisms. We are convinced that the discovery and functional characterization of OXT- and AVP-like neuropeptides from natural sources, even plants, will not only have applications in drug development to target human OXT and AVP receptors, but may also provide novel information for comparative studies to identify common features of OXT and AVP physiology throughout the animal kingdom that may yield translational insights into evolutionary aspects of human behavior.

Acknowledgments

Work on bioactive peptides and OXT physiology in our laboratory has been funded by the Austrian Science Fund – FWF (P22889-B11 and P24743-B21) and the Vienna Science and Technology Fund (WWTF) through project LS13-017.

References

Acher, R., Chauvet, J. and Chauvet, M.T. (1995) Man and the chimaera. Selective versus neutral oxytocin evolution. *Advances in Experimental Medicine and Biology*, **395**, 615–627.

Aikins, M.J., Schooley, D.A., Begum, K., *et al.* (2008) Vasopressin-like peptide and its receptor function in an indirect diuretic signalling pathway in the red flour beetle. *Insect Biochemistry and Molecular Biology*, **38**, 740–748.

Beets, I., Janssen, T., Meelkop, E., *et al.* (2012) Vasopressin/oxytocin-related signalling regulates gustatory associative learning in *C. elegans*. *Science*, **338**, 543–545. [*Characterization of OXT-like signaling in invertebrates (see also Garrison* et al.*)*.]

de Mendoza, A., Sebe-Pedros, A. and Ruiz-Trillo, I. (2014) The evolution of the GPCR signalling system in eukaryotes: modularity, conservation, and the transition to metazoan multicellularity. *Genome Biology and Evolution*, **6**, 606–619.

Donaldson, Z.R. and Young, L.J. (2008) Oxytocin, vasopressin, and the neurogenetics of sociality. *Science*, **322**, 900–904. [*Comprehensive review article about the function and genetics of the OXT/AVP system*.]

Dutertre, S., Croker, D., Daly, N.L., *et al.* (2008) Conopressin-T from *Conus tulipa* reveals an antagonist switch in vasopressin-like peptides. *Journal of Biological Chemistry*, **283**, 7100–7108.

Frank, E. and Landgraf, R. (2008) The vasopressin system-from antidiuresis to psychopathology. *European Journal of Pharmacology*, **583**, 226–242.

Fredriksson, R. and Schioth, H.B. (2005) The repertoire of G-protein-coupled receptors in fully sequenced genomes. *Molecular Pharmacology*, **67**, 1414–1425.

Garrison, J.L., Macosko. E.Z., Bernstein, S., *et al.* (2012) Oxytocin/vasopressin-related peptides have an ancient role in reproductive behaviour. *Science*, **338**, 540–543.

Gimpl, G. and Fahrenholz, F. (2001) The oxytocin receptor system: structure, function, and regulation. *Physiological Reviews*, **81**, 629–683.

Goodson, J.L. (2008) Nonapeptides and the evolutionary patterning of sociality. *Progress in Brain Research*, **170**, 3–15.

Goodson, J.L., Kelly, A.M. and Kingsbury, M.A. (2012) Evolving nonapeptide mechanisms of gregariousness and social diversity in birds. *Hormones and Behavior*, **61**, 239–250.

Gruber, C.W. (2014) Physiology of invertebrate oxytocin and vasopressin neuropeptides. *Experimental Physiology*, **99**, 55–61.

Gruber, C.W. and Muttenthaler, M. (2012) Discovery of defense- and neuropeptides in social ants by genome-mining. *PLoS One*, **7**, e32559. [*Methodology for the discovery and in silico identification of novel neuropeptides in invertebrates*.]

Gruber, C.W. and O'Brien, M. (2011) Uterotonic plants and their bioactive constituents. *Planta Medica*, **77**, 207–220.

Gruber, C.W., Muttenthaler, M. and Freissmuth, M. (2010) Ligand-based peptide design and combinatorial peptide libraries to target G protein-coupled receptors. *Current Pharmaceutical Design*, **16**, 3071–3088.

Gruber, C.W., Koehbach, J. and Muttenthaler, M .(2012) Exploring bioactive peptides from natural sources for oxytocin and vasopressin drug discovery. *Future Medicinal Chemistry*, **4**, 1791–1798.

Hoyle, C.H. (1999) Neuropeptide families and their receptors: evolutionary perspectives. *Brain Research*, **848**, 1–25. **[Review about the evolution and diversity of neuropeptide signaling systems.]**

Knobloch, H.S. and Grinevich, V. (2014) Evolution of oxytocin pathways in the brain of vertebrates. *Frontiers in Behavioral Neuroscience*, **8**, 31.

Koehbach, J. and Gruber, C.W. (2013) From ethnopharmacology to drug design. *Communicative & Integrative Biology*, **6**, e27583.

Koehbach, J., Attah, A.F., Berger, A., *et al.* (2013a) Cyclotide discovery in Gentianales revisited Identification and characterization of cyclic cystine-knot peptides and their phylogenetic distribution in Rubiaceae plants. *Biopolymers*, **100**, 438–452.

Koehbach, J., O'Brien, M., Muttenthaler, M., *et al.* (2013b) Oxytocic plant cyclotides as templates for peptide G protein-coupled receptor ligand design. *Proceedings of the National Academy of Sciences of the United States of America*, **110**, 21183–21188. **[Characterization of the first OXT-like peptide from plants.]**

Koehbach, J., Stockner, T., Bergmayr, C., Muttenthaler, M. and Gruber, C.W. (2013c) Insights into the molecular evolution of oxytocin receptor ligand binding. *Biochemical Society Transactions*, **41**, 197–204.

Lagman, D., Ocampo Daza, D., Widmark, J., *et al.* (2013) The vertebrate ancestral repertoire of visual opsins, transducin alpha subunits and oxytocin/vasopressin receptors was established by duplication of their shared genomic region in the two rounds of early vertebrate genome duplications. *BMC Evolutionary Biology*, **13**, 238.

Manning, M., Stoev, S., Chini, B., Durroux, T., Mouillac, B. and Guillon, G. (2008) Peptide and non-peptide agonists and antagonists for the vasopressin and oxytocin V_{1a}, V_{1b}, V_2 and OT receptors: research tools and potential therapeutic agents. *Progress in Brain Research*, **170**, 473–512.

Manning, M., Misicka, A., Olma, A., *et al.* (2012) Oxytocin and vasopressin agonists and antagonists as research tools and potential therapeutics. *Journal of Neuroendocrinology*, **24**, 609–628. **[Comprehensive review on the design and synthesis of selective ligands for the OXT and AVP receptor family.]**

Meyer-Lindenberg, A., Domes, G. Kirsch, P. and Heinrichs, M. (2011) Oxytocin and vasopressin in the human brain: social neuropeptides for translational medicine. *Nature Reviews. Neuroscience*, **12**, 524–538.

Mirabeau, O. and Joly, J.S. (2013) Molecular evolution of peptidergic signalling systems in bilaterians. *Proceedings of the National Academy of Sciences of the United States of America*, **110**, E2028–2037. **[Provides a detailed analysis and comparison of protostomian and deuterostomian peptidergic GPCR systems.]**

Pitti, T. and Manoj, N. (2012) Molecular evolution of the neuropeptide S receptor. *PLoS One*, **7**, e34046.

Schoneberg, T., Schulz, A., Biebermann, H., Hermsdorf, T., Rompler, H. and Sangkuhl, K. (2004) Mutant G-protein-coupled receptors as a cause of human diseases. *Pharmacology & Therapeutics*, **104**, 173–206.

Stafflinger, E., Hansen, K.K., Hauser, F., *et al.* (2008) Cloning and identification of an oxytocin/vasopressin-like receptor and its ligand from insects. *Proceedings of the National Academy of Sciences of the United States of America*, **105**, 3262–3267.

Strotmann, R., Schrock, K., Boselt, I., Staubert, C., Russ, A. and Schoneberg, T. (2011) Evolution of GPCR: change and continuity. *Molecular and Cellular Endocrinology*, **331**, 170–178.

Zingg, H.H. and Laporte, S.A. (2003) The oxytocin receptor. *Trends in Endocrinology and Metabolism*, **14**, 222–227. **[A highly cited and comprehensive review that describes the function and molecular pharmacology of the OXT/AVP receptor system.]**

CHAPTER 2

The Neuroendocrine Genome: Neuropeptides and Related Signaling Peptides

J. Peter H. Burbach

Brain Center Rudolf Magnus, Department of Translational Neuroscience, University Medical Center Utrecht, Utrecht, the Netherlands

The **neuroendocrine system** links the nervous system to the endocrine system, and thus exerts important integrative functions in the control of perhaps the entire physiology of an organism. In this way, internal and external information about the physiological state and the environment of the organism can be processed into adequate responses to maintain homeostasis, to adapt or to initiate and propagate required physiological reactions. In almost all physiological systems, even the immune system, a direct or indirect involvement of the neuroendocrine system can be recognized.

The basic principle of the neuroendocrine system is the communication between neurons and endocrine cells. Either neurons instruct endocrine cells to change their hormonal output to the blood, or the peripheral endocrine state is sensed and processed by the nervous system to provide neural responses. This communication is mediated by chemical signals that are released by one cell and received by another. All vertebrate and invertebrate species engage a multitude of signal molecules to meet the complexity of their nervous systems. The chemical nature of signal molecules ranges from simple gaseous molecules, like nitric oxide, to aminergic and fatty molecules, amino acids, and proteins. Each of these classes of chemical messengers has its own biochemical pathway and cell biological system for synthesis, release, action, and degradation. A prerequisite of their signaling function is the presence of a receptor on the receiving cell that transduces the signal into a response.

The 'neuroendocrine genome' would encompass all genes that are required to develop, maintain, and operate neuroendocrine systems. These include not only the genes that code for **neuropeptides** but also the transcription and growth factors that help to specify the cells that produce them, the constituents of the secretory routes and vesicles that transport and store neuropeptides, the receptors and signal transduction components that confer responses to neuropeptides signals. This chapter focuses on the gene products that are primarily responsible for chemical signaling, namely the neuropeptides, defined as small proteinaceous substances produced, stored, and released through the regulated secretory route by neurons and acting on neural substrates.

Molecular Neuroendocrinology: From Genome to Physiology, First Edition. Edited by David Murphy and Harold Gainer.
© 2016 John Wiley & Sons, Ltd. Published 2016 by John Wiley & Sons, Ltd.
Companion website: www.wiley.com/go/murphy/neuroendocrinology

2.1 The discovery of neuropeptides

Four scientific milestones mark the history of neuropeptides. These are, first, the discovery of peptide hormones as mediators of the endocrine system. Second, the description of the **neurosecretion** of peptidergic substances. Third, the recognition that nerve cells themselves are responsive to peptides. Fourth, the unraveling of the mammalian genome.

Neuropeptides were encountered long before they were designated as such, even before chemical signaling existed as a physiological concept. In ancient times, it was thought that organs of animals, including man, could affect the body and the mind when eaten. In the 19th century when experimental life sciences developed, extracts of organs were systematically investigated for their effects on physiological systems and the fundamentals of endocrinology were determined. Substances like secretin, insulin, vasopressin (AVP), and oxytocin (OXT) were encountered as biological activities in crude extracts, detected by bioassays in whole animals or organ systems. In 1905, such substances were named 'hormones,' a term introduced by the English physiologist Ernest Starling (1866–1927). Starling derived the term from the Greek verb *ormao* (to arouse or to excite), during a Croonian Lecture at the Royal College of Physicians in London on the chemical control of the functions of the body in 1904. He defined hormones in terms of chemical messengers produced recurrently to answer the physiological needs of the organism, and carried from the organ where they are produced to the organ which they affect by means of the bloodstream. Amongst hormonally active substances were molecules that we now call neuropeptides, like AVP, OXT, and substance P.

Only in the 1950s and beyond were the chemical identities of hormones elucidated; almost all of them appeared to be short chains of amino acids (peptides). Du Vigneaud (1901–1978) received the Nobel Prize in Chemistry for the structural elucidation and chemical synthesis of OXT in 1955, and von Euler (1905–1983) received the Nobel Prize in Physiology and Medicine for the identification of noradrenaline and substance P in 1970. In the 1960s, guided by bioassays, large-scale purifications of releasing factors from hypothalamic extracts were undertaken. This resulted in the elucidation of the peptide structures of releasing factors in the late 1960s, and a shared Nobel Prize in Physiology and Medicine for Guillemin (1924) and Schally (1926) in 1977. From the late 1950s onward, Victor Mutt (1923–1998) identified many new peptide hormones from gut organs, and later also from the brain based on chemical properties (Tatemoto and Mutt, 1980). Guided by biological activity, specifically opioid activity, John Hughes and Hans Kosterlitz (1903–1996) identified the enkephalins in 1975 (Hughes *et al.*, 1975). It was a seminal moment, bringing neuropeptides from endocrinology into the neurosciences.

While one historical line leading to neuropeptides is the conception of peptide hormones and their chemical identification, the second is that of neurosecretion. Neurosecretion of peptidergic substances is a fundamental aspect of neuropeptides (Hökfelt *et al.*, 1980). Preceded by the studies of Carl Speidel in 1917, demonstrating intrinsic secretion in nervous tissue, Ernst Scharrer (1905–1965) proposed the concept of neurosecretion in 1934 (Scharrer, 1987). Early on, it was recognized by physiologists and histologists that nerve cells in the brain stored substances that could be secreted, not into the bloodstream but locally in the nervous system. For example, by chemical staining of disulfide bridges, peptides could be detected in the hypothalamus and ventral brain in neuronal fibers to the brain separate from the posterior pituitary gland, the canonical site of release for peripheral hormonal actions. These peptides were AVP, OXT, and their neurophysins, later demonstrated by immunocytochemistry.

A third line in the history of neuropeptides is the discovery of biological actions of peptides on the nervous system, and particularly of cognitive actions in the brain. Bioassays were instrumental in the discovery of neuropeptides. Bioassays for activity on the enteric nervous system already existed but for the central nervous system, this was new. David de Wied (1925–2004) explored the central activity of peptide hormones on rat behavior from the 1950s onward and discovered that ACTH, MSH, and AVP acted on the brain and affected learning and memory processes (de Wied, 1971). In the 1970s he used the term 'neuropeptides' to designate neuroactive peptide hormones and fragments thereof. Since then, the term 'neuropeptides' has been adopted and has marked an exciting field in the neurosciences. At that time, receptors were still undiscovered mediators of the biological effects, but with the development of radioligand binding assays and molecular technologies, receptors for many neuropeptides were identified in the 1980s and 1990s (see Chapter 10).

Coming right up to date, a fourth line of investigation has emerged, namely the mining of genome sequences for novel neuropeptides. Two bioinformatic strategies are being used. The first is relatively straightforward. It compares sequences from a novel genome to that of a known one, and scores the homologies to known neuropeptides. It needs tools with strong algorithms for sequence comparison. In this way, orthologs of neuropeptides in the new genome can be annotated. This strategy is not aimed at identifying new peptides with unknown function. However, the second approach is designed to mine known genomes for previously unidentified peptides. This approach requires new algorithms that select sequences based on the structural prerequisites and common architectural features of neuropeptides and their precursors (see section 2.4), and exclude other features and domains that are never encountered in peptide **precursor proteins**, such as transmembrane domains. A hidden Markow model tool for screening for peptide hormone-like characteristics in all proteins predicted from the human genome (the human proteome) had been successfully applied (Mirabeau *et al.*, 2007). This study reported the identification of two peptide precursors that had only been annotated as an open reading frame in the human genome. These were spexin (C12orf39 gene) and augurin (C2orf40 gene). Another machine learning protocol (NeuroPID) was recently developed and used to identify neuropeptides in metazoan genomes (Ofer and Linial, 2014). Such approaches provide perspectives towards a bioinformatics-driven discovery of novel regulatory peptides in neuroendocrine systems.

2.2   Characteristics of neuropeptides

Neuropeptides are small proteinaceous substances produced, stored, and released through the regulated secretory route by neurons which act on neural substrates. This definition imposes several prerequisites: (1) gene expression and **biosynthesis** by neurons, (2) storage and regulated release by neurons, and (3) receptor-mediated activation of modulation of neuronal activity.

2.2.1   Gene expression and biosynthesis by neurons

Neuropeptides are used by neurons to signal to other cells, and for that purpose genes encoding neuropeptides are expressed by neurons, and neuropeptide biosynthesis takes place in these neurons. For most 'classic' neuropeptides, studies using *in situ* hybridization and immunocytochemistry have unambiguously shown that the transcript and peptide products are produced by neurons. In a few cases, biosynthesis of peptides could actually be demonstrated by incorporation of radioactive amino acids in pulse-chase experiments. Such experiments in the hypothalamo-neurohypophysial system (HNS) demonstrated the synthesis of a precursor protein, its axonal transport to the posterior pituitary gland and the accumulation of smaller peptides, in this case for the AVP and OXT neurons (Brownstein *et al.*, 1980). Neurons display an extraordinary strict cell specificity in expressing neuropeptides genes. In several cases, the expression is limited to one or a few related neuronal types. Examples of neuropeptide genes with highly restricted expression are the OXT, AVP, proopiomelanocortin (POMC), corticotropin-releasing hormone (CRH), agouti-related protein (AGRP), calcitonin-related polypeptide (CALCB), neuromedin B (NMB), and urotensin 2 domain-containing (UTS2D) genes illustrated in Table 2.1.

Table 2.1 Neuropeptide gene families and receptors. Neuropeptide genes are listed according to family relationships. For each gene, chromosomal localization, brain expression, encoded precursor structure, biologically active peptide products, and their established receptors are presented. Hyperlinks provide additional information to the locus in the human chromosome through http://genome.ucsc.edu/, to neuroanatomy of mouse brain expression through the Allen Brain Atlas, or GenePaint, structural comparison to related precursors and those of other species through the BLINK tool of NCBI, and to receptor properties through the IUPHAR database.

Gene (gene symbol)	Chromosomal localization	Gene expression	Precursor	Active peptide(s)	Receptor(s)
Opioid gene family					
Pro-enkephalin gene (PENK)	8q12.1		Prepro-enkephalin	Leu-enkephalin, Met-enkephalin, amidorphin, adrenorphin, peptide B, peptide E, peptide F, BAM22P	δ receptor (OPRD1) κ receptor (OPRK1) μ receptor (OPRM1) MAS-related GPR-X1 (*MRGPRX1*)
Pro-opiomelanocortin gene (POMC)	2p23.3		POMC	α-melanocyte-stimulating hormone (α-MSH), γ-melanocyte-stimulating hormone (γ-MSH), β-melanocyte-stimulating hormone (β-MSH), adrenocorticotropic hormone (ACTH), β-endorphin, α-endorphin, γ-endorphin, β-lipoprotein (β-LPH), γ-lipoprotein (γ-LPH), corticotropin-like intermediate peptide (CLIP)	MC$_1$ receptor (MC1R) MC$_5$ receptor (MC5R) MC$_3$ receptor (MC3R) MC$_4$ receptor (MC4R) MC$_2$ receptor (MC2R) μ receptor (OPRM1)

Gene	Location		Precursor	Peptides	Receptors
Pro-dynorphin gene (PDYN)	20p13		Prepro-dynorphin	Dynorphin A, dynorphin B, α-neo-endorphin, β-neo-endorphin, dynorphin-32, leu-morphin	δ receptor (OPRD1) κ receptor (OPRK1) μ receptor (OPRM1)
Orphanin gene, pre-pronociceptin gene (PNOC)	8p21.1		Prepro-nociceptin, prepro-orphanin	Nociceptin (orphanin FQ), neuropeptide1, neuropeptide 2	NOP receptor (OPRL1)
Vasopressin/ oxytocin gene family					
Vasopressin gene (AVP)	20p13		Prepro-vasopressin-neurophysin II	Vasopressin (VP), neurophysin II (NP II), C-terminal glycopeptide CPP	V_{1A} receptor (AVPR1A) V_{1B} receptor (AVPR1B) V_2 receptor (AVPR2) OT receptor (OXTR)
Oxytocin gene (OXT)	20p13		Prepro-oxytocin-neurophysin I	Oxytocin (OT), neurophysin I (NP 1)	OT receptor (OXTR)
CCK/gastrin gene family					
Gastrin gene (GAST)	17q21.2		Prepro-gastrin	Gastrin-34, gastrin-17, gastrin-4	CCK_1 receptor (CCKAR) CCK_2 receptor (CCKBR)
Cholecystokinin gene (CCK)	3p22.1		Prepro-CCK	CCK-8, CCK-33, CCK-58	CCK_1 receptor (CCKAR) CCK_2 receptor (CCKBR)

(continued)

Table 2.1 (Continued)

Gene (gene symbol)	Chromosomal localization	Gene expression	Precursor	Active peptide(s)	Receptor(s)
Somastostatin gene family					
Somastostatin gene (SST)	3q27.3		Prepro-somatostatin	SS-12 (SRIF-12), SS-14 (SRIF-14), SS-28 (SRIF-28), antrin, neuronostatin	sst$_1$ receptor (SSTR1) sst$_2$ receptor (SSTR2) sst$_3$ receptor (SSTR3) sst$_4$ receptor (SSTR4) sst$_5$ receptor (SSTR5) GPR107
Cortistatin gene (CST)	20p11.21		Prepro-cortistatin	Cortistatin-29, cortistatin-17	sst$_1$ receptor (SSTR1) sst$_2$ receptor (SSTR2) sst$_3$ receptor (SSTR3) sst$_4$ receptor (SSTR4) sst$_5$ receptor (SSTR5) MAS-related GPR-X2 (*MRGPRX2*)
F- and Y-amide gene family					
Gonadotrophin inhibitory hormone gene, RF-amide related peptide gene (RFRP)	7p15.3		Prepro-NPRF	QRF-amide (neuropeptide RF-amide, GnIH, p518, RF-related peptide-2), RF-related peptide-1, RF-related peptide-3, neuropeptide VF 26RF-amide	QRFP receptor (QRFPR) NPFF1 receptor, GPR103 (NPFFR1) NPFF2 receptor (NPFFR2)
Neuropeptide FF gene (NPFF)	12q13.13		Prepro-NPFF	Neuropeptide FF, neuropeptide AF, neuropeptide SF	NPFF1 receptor, GPR103 (NPFFR1) NPFF2 receptor (NPFFR2)
Neuropeptide Y gene (NPY)	7p15.3		Prepro-NPY	NPY, C-flanking peptide CPON	Y$_1$ receptor (NPY1R) Y$_2$ receptor (NPY2R) Y$_4$ receptor (NPY4R) Y$_5$ receptor (NPY5R)

Gene	Location		Precursor	Peptide(s)	Receptor(s)
Pancreatic polypeptide gene (PPY)	17q21.31		Prepro-PPY	PPY	Y_1 receptor (NPY1R) Y_2 receptor (NPY2R) Y_4 receptor (NPY4R) Y_5 receptor (NPY5R)
Peptide YY gene (PYY)	17q21.31		Prepro-PYY	PYY, PYY-(3–36)	Y_1 receptor (NPY1R) Y_2 receptor (NPY2R) Y_4 receptor (NPY4R) Y_5 receptor (NPY5R)
Prolactin-releasing peptide (PRLH)	2q37.3	not available	Prepro-PrRP	PrRP-31, PrRP-20	PRRP receptor (PRLHR1)
Calcitonin gene family					
Calcitonin I gene (CALCA)	11p15.2		Prepro-CALC	Calcitonin, katacalcin	CT receptor (CALCR)
			Prepro-CGRP-α	Calcitonin gene-related peptide I (α-CGRP)	AMY_1 receptor (CALCR) CGRP receptor (CALCRL)
Calcitonin II gene (CALCB)	11p15.2		Prepro-CGRP-β	Calcitonin gene-related peptide II (β-CGRP)	AMY_1 receptor (CALCR) CGRP receptor (CALCRL)
Islet amyloid polypeptide gene (IAPP), amylin gene	12p12.1		Prepro-IAPP	IAPP (amylin, amyloid polypeptide)	AMY_1 receptor (CALCR) AMY_2 receptor (CALCR) AMY_3 receptor (CALCR)
Adrenomedullin gene (ADM)	11p15.4		Prepro-adrenomedullin	Adrenomedullin, AM, PAMP	AM_1 receptor (CALCRL)

(continued)

Table 2.1 (Continued)

Gene (gene symbol)	Chromosomal localization	Gene expression	Precursor	Active peptide(s)	Receptor(s)
Adrenomedullin-2 gene (ADM2)	22q13.33		Prepro-adrenomedullin-2	Adrenomedullin-2, intermedin-long (IMDL), intermedin-short (IMDS)	AM$_1$ receptor (CALCRL) AM$_2$ receptor (CALCR)
Natriuretic factor gene family					
Atrial natriuretic factor gene (NPPA)	1p36.22		Prepro-ANP	Atrial natriuretic factor (natriuretic peptide A, ANF, ANP, natriodilatine, cardiodilatine-related peptide)	NPR-A (NPR1) NPR-C (NPR3)
Brain natriuretic factor gene (NPPB)	1p36.22		Prepro-BNP	Brain natriuretic factor (natriuretic peptide B, BNF, BNP)	NPR-A (NPR1)
Natriuretic peptide precursor C gene (NPPC)	2q37.1		Prepro-CNP	C-type natriuretic peptide (CNP-23), CNP-29, CNP-53	NPR-B (NPR2)
Osteocrin (OSTN)	3q28		Osteocrin precursor	Osteocrin (musclin)	NPR-C (NPR3)
Bombesin-like peptide gene family					
Gastrin-releasing peptide gene (GRP)	18q21.32		Prepro-GRP-1	GRP-27, GRP-14, GRP-10 (neuromedin C)	BB$_2$ receptor (GRPR) BB$_1$ receptor (NMBR)

Gene	Location		Prepro-peptide	Peptide(s)	Receptor(s)
			Prepro-GRP-2	GRP-27, GRP-14, GRP-10 (neuromedin C)	BB$_2$ receptor (GRPR), BB$_1$ receptor (NMBR)
			Prepro-GRP-3	GRP-27, GRP-14, GRP-10 (neuromedin C)	BB$_1$ receptor (NMBR)
Neuromedin B gene (NMB)	15q25.2-q25.3		Prepro-neuromedin B1	Neuromedin B (ranatensin-like peptide, RLP)	BB$_1$ receptor (NMBR)
			Prepro-neuromedin B2	Neuromedin B (ranatensin-like peptide, RLP)	BB$_1$ receptor (NMBR)
Endothelin gene family					
Endothelin 1 gene (EDN1)	6p24.1		Prepro-endothelin 1 (PPET1)	Endothelin 1 (ET-1)	ET$_A$ receptor (EDNRA), ET$_B$ receptor (EDNRB)
Endothelin 2 gene (EDN2)	1p34.2		Prepro-endothelin 2 (PPET2)2	Endothelin 2 (ET-2)	ET$_A$ receptor (EDNRA), ET$_B$ receptor (EDNRB)
Endothelin 3 gene (EDN3)	20q13.32		Prepro-endothelin 3 (PPET3)	Endothelin 3 (ET-3)	ET$_B$ receptor (EDNRB)
Glucagon/secretin gene family					
Glucagon gene (GCG)	2q24.1		Prepro-glucagon	Glicentin; glicentin-related polypeptide (GRPP); oxyntomodulin (OXY) (OXM); glucagon; glucagon-like peptide 1 (GLP-1); glucagon-like peptide 1(7-37) (GLP-1(7-37)); glucagon-like peptide 1(7-36) (GLP-1(7-36)); glucagon-like peptide 2 (GLP-2)]	GLP-1 receptor (GLPR1), Glucagon receptor (GCGR), GLP-2 receptor (GLPR2)

(continued)

Table 2.1 (Continued)

Gene (gene symbol)	Chromosomal localization	Gene expression	Precursor	Active peptide(s)	Receptor(s)
Secretin gene (SCT)	11p15.5		Prepro-secretin	Secretin	Secretin receptor (SCTR)
Vasoactive intestinal peptide gene (VIP)	6q25.2		Prepro-VIP-1	VIP, PHM-27/PHI-27, PHV-42	VPAC$_1$ receptor (VIPR1) VPAC$_2$ receptor (VIPR2)
			Prepro-VIP-2	VIP, PHM-27/PHI-27, PHV-42	VPAC$_1$ receptor (VIPR1) VPAC$_2$ receptor (VIPR2)
Pituitary adenylcyclase-activated peptide gene (ADCYAP1)	18p11.32		Prepro-PACAP	PACAP-38, PACAP-27, PRP-48	PAC$_1$ receptor (ADCYAP1R1) VPAC$_1$ receptor (VIPR1) VPAC$_2$ receptor (VIPR2)
Growth hormone-releasing hormone gene (GHRH)	20q11.23		Prepro-GHRH	GHRH (somatoliberin, GRF, somatocrinin, somatorelin, sermorelin)	GHRH receptor (GHRHR)
Gastric inhibitory peptide gene (GIP)	17q21.32		Prepro-GIP	GIP (gastric inhibitory peptide, glucose-dependent insulinotropic polypeptide)	GIP receptor (GIPR)

CRH-related gene family

Corticotropin-releasing hormone gene (CRH)	8q13.1		Prepro-CRH	CRH	CRF$_1$ receptor (CRHR1)
Urocortin gene (UCN)	2p23.3		Prepro-UNC I	UNC I	CRF$_1$ receptor (CRHR1)
Urocortin II gene (UCN2)	3p21.31	not available	Prepro-UNC II	UNC II, stresscopin-related peptide	CRF$_2$ receptor (CRHR2)
Urocortin III gene (UCN3)	10p15.1	 not available	Prepro-UNC III	UNC III, stresscopin	CRF$_2$ receptor (CRHR2)
Urotensin-II (UTS2)	1p36.23		Prepro-urotensin-2, isoform a	Urotensin-2	UT receptor (UTS2R)
			Prepro-urotensin-2, isoform b	Urotensin-2	UT receptor (UTS2R)
Urotensin-II domain containing (UTS2D)	3q28	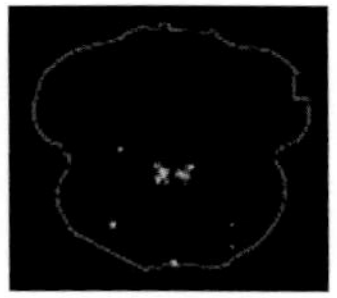	Prepro-urotensin-2B	Urotensin-2-related peptide, urotensin-2B	UT receptor (UTS2R)

Kinin and tensin gene family

Preprotachykinin A gene (TAC1)	7q21.3	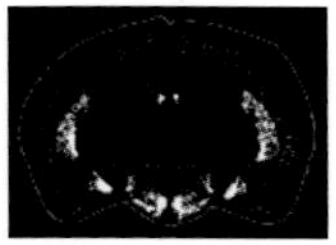	α-PPTA	Substance P, neurokinin A (NKA, substance K, neuromedin L), neuropeptide K, neuropeptide γ	NK$_1$ receptor (TACR1) NK$_2$ receptor (TACR2) NK$_3$ receptor (TACR3)

(continued)

Table 2.1 (Continued)

Gene (gene symbol)	Chromosomal localization	Gene expression	Precursor	Active peptide(s)	Receptor(s)
			β-PPTA	Substance P, neuropeptide K, neurokinin A	NK_1 receptor (TACR1) NK_2 receptor (TACR2) NK_3 receptor (TACR3)
			γ-PPTA	Substance P, neurokinin A, neuropeptide γ	NK_1 receptor (TACR1) NK_2 receptor (TACR2) NK_3 receptor (TACR3)
			δ-PPTA	Substance P, neuropeptide K, neurokinin A	NK_1 receptor (TACR1) NK_2 receptor (TACR2) NK_3 receptor (TACR3)
Preprotachykinin B gene (TAC3)	12q13.3		PPTB, isoform 1	Neuromedin K, neurokinin B	NK_1 receptor (TACR1) NK_2 receptor (TACR2) NK_3 receptor (TACR3)
			PPTB, isoform 2	Neuromedin K, neurokinin B	
Neuromedins					
Neuromedin S gene (NMS)	2q11.2	not available	Prepro-neuromedin S	Neuromedin S	NMU1 receptor (NMuR1) NMU2 receptor (NMUR2)
Neuromedin U gene (NMU)	4q12		Prepro-neuromedin U, multiple isoforms	Neuromedin U	NMU1 receptor (NMuR1) NMU2 receptor (NMUR2)
Tensins and kinins					
Kininogen-1 gene (KNG1)	3q27.3		Kininogen-1 precursor, isoform 1	Bradykinin, kallidin, LMW-K-kinin, HMW-K-kinin	B_1 receptor (BDKRB1) B_2 receptor (BDKRB2)
			Kininogen-1 precursor, isoform 2	Bradykinin, kallidin, LMW-K-kinin, HMW-K-kinin	B_1 receptor (BDKRB1) B_2 receptor (BDKRB2) cck

Gene	Location		Precursor	Peptide products	Receptor
Angiotensin gene (AGT)	1q42.2		Angiotensinogen preprotein	Angiotensin-I, angiotensin-II, angiotensin-IIII, angiotensin-(1–7)	AT_1 receptor, AT_2 receptor
Neurotensin gene (NTS)	12q21.31	not available	Prepro-neurotensin	Neurotensin (NT), neuromedin N	NTS_1 receptor (NTSR1), NTS_2 receptor (NTSR2)
Motilin family					
Motilin gene (MLN)	6p21.31		Prepro-motilin isoform 1	Motilin, motilin-associated peptide	Motilin receptor (MLNR)
			Prepro-motilin isoform 2	Motilin, motilin-associated peptide	
Ghrelin gene (GHRL)	3p25.3		Prepro-ghrelin isoform 1	Ghrelin, obestatin (appetite-regulating hormone)	Ghrelin receptor (GHR), *GPR39* (putative)
			Prepro-ghrelin isoform 2	Ghrelin, obestatin	Ghrelin receptor (GHR), *GPR39* (putative)
			Prepro-ghrelin isoform 3, pro-obestatin	Obestatin	*GPR39* (putative)
			Prepro-ghrelin isoform 4, pro-obestatin	Obestatin	*GPR39* (putative)
			Prepro-ghrelin isoform 5	Obestatin	*GPR39* (putative)
Galanin family					
Galanin gene (GAL)	11q13.3		Prepro-galanin	Galanin, galanin message-associated peptide (GMAP)	GAL_1 receptor (GALR1), GAL_2 receptor (GALR2), GAL_3 receptor (GALR3)

(continued)

Table 2.1 (Continued)

Gene (gene symbol)	Chromosomal localization	Gene expression	Precursor	Active peptide(s)	Receptor(s)
Galanin-like peptide precursor gene (GALP)	19q13.43		Galanin-like peptide precursor	Galanin-like peptide (GALP)	GAL$_1$ receptor (GALR1) GAL$_2$ receptor (GALR2) GAL$_3$ receptor (GALR3)
GnRH family					
Gonadotropin-releasing hormone gene (GnRH1)	8p21.2		Prepro-GnRH1	GnRH (LHRH, gonadoliberin)	GnRH receptor (GNRHR) GnRH2 receptor (GNRHR2)
Gonadotropin-releasing hormone gene (GnRH2)	20p13	not in mouse	Prepro-GNRH2, isoform-a	GnRH2 (LHRH II, gonadoliberin II)	GnRH receptor (GNRHR) GnRH2 receptor (GNRHR2)
		not in mouse	Prepro-GNRH2, isoform-b	GnRH2 (LHRH II, gonadoliberin II)	GnRH receptor (GNRHR) GnRH2 receptor (GNRHR2)
		not in mouse	Prepro-GNRH2, isoform-c	GnRH2 (LHRH II, gonadoliberin II)	GnRH receptor (GNRHR) GnRH2 receptor (GNRHR2)
Parathyroid hormone (PTH) family					
Parathyroid hormone-like hormone gene (PTHLH)	12p11.22		Prepro-PTH-like hormone, isoform CRA_a	PTHrP-(1-36), PTHrP-(38-94), PTHrP-(107-139) (osteostatin)	PTH1 receptor (PTH1R) PTH2 receptor (PTH2R)

Gene	Location		Precursor	Peptide	Receptor
		not available	Prepro-PTH-like hormone, isoform CRA_b	PTHrP	PTH1 receptor (PTH1R), PTH2 receptor (PTH2R)
Parathyroid hormone 2 (PTH2)	19q13.33		Prepro-PTH2	PTH2, TIP39	PTH2 receptor (PTH2R)
Neuropeptide W gene (NPW)	16p13.3		Prepro-neuropeptide W, PPL8	Neuropeptide W-23 (peptide L8), neuropeptide W-30, neuropeptide B-23, neuropeptide B-29	NPBW1 receptor (NPBWR1), NPBW2 receptor (NPBWR2)
Neuropeptide S gene (NPS)	10q26.2	not available	Prepro-neuropeptide S	Neuropeptide S	NPS receptor (NPSR1)
Insulin/relaxins					
Relaxin-1 gene (RLN1)	9p24.1		Prepro-relaxin-1	Relaxin-1, H1-relaxin	RXFP1 receptor (RXFP1), RXFP2 receptor (RXFP2)
Relaxin-2 gene (RLN2)	9p24.1	not in mouse	Prepro-relaxin-2, isoform 1	Relaxin-2, H2 relaxin	RXFP1 receptor (RXFP1), RXFP2 receptor (RXFP2), RXFP3 receptor (RXFP3)
			Prepro-relaxin-2, isoform 2	Relaxin-2, H2-relaxin	RXFP1 receptor (RXFP1), RXFP2 receptor (RXFP2), RXFP3 receptor (RXFP3)

(continued)

Table 2.1 (Continued)

Gene (gene symbol)	Chromosomal localization	Gene expression	Precursor	Active peptide(s)	Receptor(s)
Relaxin-3 gene (RLN3)	19p13.12		Prepro-relaxin-3	Relaxin-3, H3 relaxin	RXFP1 receptor (RXFP1) RXFP2 receptor (RXFP2) RXFP3 receptor (RXFP3) RXFP4 receptor (RXFP4)
No-family neuropeptides					
Thyrotropin-releasing hormone gene (TRH)	3q22.1		Prepro-TRH	TRH (thyroliberin)	TRH_1 receptor (TRHR)
Melanin-concentrating hormone gene (PMCH)	12q23.2		Prepro-MCH	MCH, neuropeptide Glu-Ile (NEI), neuropeptide Gly-Glu (NGE)	MCH_1 receptor (MCHR1) MCH_2 receptor (MCHR2)
Hypocretin gene (HCRT)	17q21.2		Prepro-hypocretin	Hypocretin-1 (orexin A), hypocretin-2 (orexin B)	OX_1 receptor (HCRTR1) OX_2 receptor (HCRTR2)
Cocaine- and amphetamine-regulated transcript gene (CART)	5q13.2		Prepro-CART	CART-(1-39), CART-(42-89)	
Agouti-related protein homolog gene (AGRP)	16q22.1		AGRP precursor isoform 1	AGRP	MC_4 receptor (MC4R)

Gene	Location	Precursor	Peptide/Ligand	Receptor
		Prepro-AGRP isoform 2	AGRP	
Prolactin (PRL)	6p22.3	Prolactin precursor	Prolactin	Prolactin receptor (PRLR)
Apelin gene (APLN)	xq26.1	Prepro-apelin	Apelin-13, apelin-17, apelin-36 (APJ ligand, AGTRL1 ligand)	Apelin receptor (APLNR)
Metastasis suppressor KiSS (KISS1)	1q32.1	Kiss-1	Metastin (kisspeptin-54), (Golgi transport 1 homolog A, golt1a), kisspeptin-14, kisspeptin-13, kisspeptin-10	Kisspeptin receptor (KISS1R)
Diazepam-binding inhibitor (DBI)	2q14.2	DBI isoform 1	Diazepam-binding inhibitory peptide	
		DBI isoform 2	Diazepam-binding inhibitory peptide	
		DBI isoform 3	Diazepam-binding inhibitory peptide	
Prokineticin-1 (PROK1)	1p13.3	Prokineticin-1 precursor	Prokineticin-1 (PK1), endocrine gland-derived VEGF (EGVEGF)	PKR_1 (PROKR1) PKR_2 (PROKR2)
Prokineticin-2 (PROK2)	3p13	Prokineticin-2 precursor isoform 1	Prokineticin-2 (PK2)	PKR_1 (PROKR1) PKR_2 (PROKR2)
		Prokineticin-2 precursor isoform 2	Prokineticin-2 (PK2)	PKR_1 (PROKR1) PKR_2 (PROKR2)
Augurin (C2orf40)	2q12.2	Augurin precursor	Augurin (esophageal cancer-related gene 4, ECRG-4)	PKR_1 (PROKR1) PKR_2 (PROKR2)

(continued)

Table 2.1 (Continued)

Gene (gene symbol)	Chromosomal localization	Gene expression	Precursor	Active peptide(s)	Receptor(s)
Spexin (C12orf39)	12p12.1		Spexin precursor	Spexin	
Periostin (POSTN)	13q13.3		Periostin precursor, isoform 1	Periostin (osteoblast-specific factor 2 (OSF-2), fasciclin I-like)	αV-integrin (ITGAV)
			Periostin precursor isoform 2	Periostin (osteoblast-specific factor 2 (OSF-2), fasciclin I-like)	
			Periostin precursor isoform 3	Periostin (osteoblast-specific factor 2 (OSF-2), fasciclin I-like)	
			Periostin precursor isoform 4	Periostin (Osteoblast-specific factor 2 (OSF-2), fasciclin I-like)	
			Periostin precursor isoform 5	Periostin (Osteoblast-specific factor 2 (OSF-2), fasciclin I-like)	
			Periostin precursor isoform 6	Periostin (Osteoblast-specific factor 2 (OSF-2), fasciclin I-like)	
			Periostin precursor isoform 7	Periostin (Osteoblast-specific factor 2 (OSF-2), fasciclin I-like)	
C4Orf48	4p16.3		Neuropeptide-like protein C4orf48 isoform 1	Unknown	
			Neuropeptide-like protein C4orf48 isoform 2	Unknown	

The neuropeptides expressed by such a limited number of neurons can still exert important functions. For instance, in the nervous system, the POMC gene is expressed only by a subset of neurons located in the arcuate nucleus (ARC) (see Table 2.1). These neurons are essential in sensing the feeding state of the organism through a peripheral leptin signal, and controlling neuronal circuits of feeding by releasing POMC products like αMSH (see Chapter 16). POMC peptides can exert such an important physiological function since the POMC-expressing neurons are the focal first station of a wide neuronal circuit that reaches many brain regions (Gao and Horvath, 2007). Likewise, the OXT gene is restricted in expression to a subset of magnocellular neurons in the supraoptic (SON) and paraventricular (PVN) nuclei of the hypothalamus, but exerts important functions in reproduction and behavior through central circuitry and release to the peripheral circulation (Burbach *et al.*, 2006). These neuropeptide systems illustrate the principle that important integrative physiological functions of neuropeptides rely on gene expression in a highly restricted set of neurons which, in turn, can widely reach other brain centers or the peripheral blood. The set of released neuropeptides provides exclusive signals to which a defined physiological function can be assigned.

However, assigning a single defined physiological function is not possible for many other neuropeptides. The genes for these neuropeptides have a broad expression pattern that extends to multiple neuronal cell types in many brain areas. As a consequence, the neuropeptides expressed from genes with this widespread expression pattern often have more local functions in the area of expression, rather than integrative physiological functions. Examples are the cholecystokinin (CCK), enkephalin (PENK), somatostatin (SST), neuropeptide Y (NPY), vasoactive intestinal polypeptide (VIP), and tachykinin1 (TAC1) genes, illustrated in Table 2.1. These neuropeptides are often expressed in interneurons contributing to the modulation of the local circuits in a brain nucleus. For example, several different types of interneurons of the cerebral cortex can be distinguished by expression of the SST, NPY, CCK, and VIP genes. These neuropeptides participate in the control of the excitation-inhibition balance of circuitry of the entire cerebral cortex. Thus, their role is part of the function of the circuitry, which is expressed differently by different domains of the cortex. Furthermore, the same genes are also expressed in many other brain centers where their modulatory actions become apparent in other functions, such as releasing factors of the hypothalamus.

These different types of functions can often been read from the anatomy of gene expression in the brain, which is presented in Tables 2.1–2.5. It is notable that one group of neuropeptide genes in particular has an extremely broad, almost panneuronal expression pattern: the granin family (Table 2.2). This family comprises a group of seven genes which share some homology and are all expressed widely in the brain. Often their expression coincides with the expression of other neuropeptide genes. Although biological activities have been found for some granin-derived peptides, the general function of granins has been proposed to be a chaperone function for other neuropeptide precursors in the **regulated secretory pathway** (RSP) rather than being a precursor for neuropeptides themselves (see later in this chapter).

Concerning cell-specific expression of neuropeptide genes, an issue arises in cases of non-neuronal expression. Glial cells are well known for their production and release of growth factors and **chemokines** (see section 2.4.6). According to the definition, these are not neuropeptides. Moreover, glial cells were believed to lack an RSP, thus releasing intact or partly processed precursors in a non-regulated, constitutive fashion. However, recent data from astrocytes and glial cell lines show that these cells sometimes display characteristics of an RSP (Hur *et al.*, 2010). This would allow them to further process, store, and control the release of peptides. Therefore, putative neuropeptides may be recognized in peptide families expressed by glial cells.

Table 2.2 The granin gene family. Genes encoding granins and granin-like proteins and peptides are listed. These peptides share similarities with neuropeptides, but have ambiguous biological activities. A receptor has been established for proSAAS peptides only (GPR171). For each gene, chromosomal localization, brain expression, encoded precursor structure, and peptide products are presented. Hyperlinks provide additional information to the locus in the human chromosome through http://genome.ucsc.edu/, to neuroanatomy of mouse brain expression through the Allen Brain Atlas, or GenePaint, structural comparison to related precursors and those of other species through the BLINK tool of NCBI, and to receptor properties through the IUPHAR database.

Gene (symbol)	Chromosomal localization	Mouse brain expression	Gene (gene symbol)	Chromosomal localization	Receptors
Chromogranin A gene (CHGA)	14q32.12		Chromogranin A precursor	Chromogranin A, β-granin, vasostatin	
Chromogranin B gene (CHGB)	20p12.3		Chromogranin B precursor	Chromogranin B (secretogranin I), CCB peptide, GAWK peptide	
			Chromogranin B precursor variant	Chromogranin B (secretogranin I)	
Secretogranin II gene (SCG2)	2q36.1		Secretogranin II precursor, chromogranin C precursor	Secretogranin II (chromogranin C), EM66, secretoneurin	
Secretogranin III gene (SCG3)	15q21.2		Secretogranin III precursor	Secretogranin III	
Secretory granule neuroendocrine protein 1, 7B2 gene (SGNE1)	15q13.3		Secretory granule neuroendocrine precursor	Secretory granule neuroendocrine protein-1 (7B2, secretogranin 5)	
VGF nerve growth factor inducible protein (VGF)	7q22.1		VGF-precursor	VGF (NGF-inducible protein, neurosecretory protein), TLPQ-62, TLPQ-21, AQEE-30, LQEQ-19	C3a receptor (C3AR1) C1q receptor
Proprotein convertase subtilisin/kexin type 1 inhibitor (PCSK1N)	Xp11.23		Pro-SAAS	BigLEN, PEN, Little SAAS, LittleLEN	GPR171

2.2.2 Neuropeptide biosynthesis and regulated release

Regulated secretion is the basis for rapid responsive chemical communication by neuropeptide systems. Neurons expressing neuropeptides use the RSP. This pathway, in contrast to the constitutive secretory pathway, allows appropriate biosynthesis and storage of biosynthesized peptides in large dense-cored vesicles, and controlled release upon a stimulus (Brownstein *et al.*, 1980). The constitutive and RSPs are schematically compared in Figure 2.1. The RSP is a complex cell biological system consisting of different compartments each with its own specializations. It requires specific components that allow it to operate. Thus, genes encoding these components formally should be considered to be part of the 'neuroendocrine genome' but an inventory of these is beyond the focus of this review.

In all cells, proteins destined for secretion are by default secreted through the constitutive pathway. This requires retention of the newly synthesized protein in the lumen of the endoplasmic reticulum (ER), transport through the Golgi apparatus, packaging in vesicles, and plasma membrane fusion of vesicles. The RSP deviates from this generic theme at the following levels. First, newly synthesized proteins are selected and sorted for entry into the RSP. Here, interplay between the conformation of the protein and sorting receptors that recognize proteins for **regulated secretion** is required. Second, in the RSP compartments of the ER and Golgi, proteins are exposed to **processing** enzymes that are exclusive for the RSP, and thus produce peptide products that would not be formed in the constitutive route. These enzymes are the prohormone convertases (PCs), which belong to a subfamily of the subtilisine proteases. There are seven genes coding for PCs, each with a cell-specific expression pattern: PCSK1 to PCSK7. One of these, PCSK3, codes the enzyme furin that is operating already before sorting in the Golgi, and thus acts on both products of the RSP and the constitutive secretory pathway. Precursors and peptide products are subject to other enzymatic modifications such as glycosylation, phosphorylation, sulfation, C-terminal amidation, and N-terminal acetylation. Third, products are stored in specialized granules in which products are condensed: the dense-cored vesicles. During passage through the compartments of the RSP, the environment becomes more acidic. This promotes the activity of enzymes to process and modify the peptides and compact the vesicle cargo. Fourth, dense-cored vesicles are in a wait state until they are instructed to dock to and fuse with the plasma membrane. The instructive cellular signal is an elevation of the intracellular Ca^{2+} concentration, caused by depolarization of the neuronal plasma membrane. Finally, neuropeptides can be metabolized by proteolytic cleavage during vesicle storage or after secretion. Several metabolized neuropeptides have been idientifed that carry new or modified biological activities and may be responsible for biological effects (Burbach *et al.*, 1980, 2003).

Neuropeptides often co-exist with amino acid and amine neurotransmitters in nerve terminals, but are released only after intense or prolonged stimulation (Hökfelt *et al.*, 1980). This delayed response of neuropeptide secretion is due to storage in dense-cored vesicles that are not docked at the cellular release site, e.g. the synapse or synaptic button, requiring recruitment first, unlike the 'fast neurotransmitters' which are ready for release. This happens at elevated Ca^{2+} levels, which require enhanced or repeated stimulation.

2.2.3 Architecture of neuropeptide precursors

The RSP imposes specific structural characteristics upon a neuropeptide precursor in order for it to be sorted away from the default constitutive route, to be processed and to be modified posttranslationally. Thus, a generic architecture for a neuropeptide precursor can be recognized to meet these requirements. The structural prerequisites for sorting ('sorting signals') are thought to reside in the 3D structure of the precursor and serve to interact with proteins that allow sorting ('sortases'), which have remained largely uncovered (Tooze and Huttner, 1990; Tooze *et al.*, 2001). Different roles for the precursor conformation and sortases have been proposed. Hypotheses propose a specific sorting domain in the precursor, an overall conformational presentation of the precursor required for sorting, or the involvement of a chaperone (McGirr *et al.*, 2013). In only a few cases is there evidence for conformational sorting signals of neuropeptide precursors. It has been shown that the AVP precursor (pro-AVP-neurophysin) folds the AVP domain in the AVP-binding pocket of

neurophysin and that this conformation is required to be sorted into the RSP. If this conformation is disturbed, for instance by mutation of amino acid residues in AVP or the AVP-
binding pocket of the neurophysin domain, precursors do not reach the RSP and often
aggregate in the ER (de Bree *et al.*, 2003). Finally, the proteolytic processing of the precursor
into active peptides and posttranslational modifications is directed by short motifs of amino
acids. Thus, neuropeptide precursors possess salient structural elements, illustrated in
Figure 2.1 and discussed below.

2.2.3.1 The signal peptide

A **signal peptide** sequence is key to the entry of newly synthesized proteins (pre-
proneuropeptides) into secretion routes. This requires the proteins that are being synthesized on the ribosomes at the cytoplasmic side of the ER to enter the lumen of the ER.
Generally, the signal peptide is a short 20–25 amino acid extension at the N-terminal of
the precursor, the pre-proneuropeptide. It is directly translocated to the lumen of the ER
due to its hydrophobic nature and the help of a translocon, a channel. During translocation, the signal peptide is removed from the nascent pre-proneuropeptide during protein
synthesis by a signal peptidase and thus never found on the precursor or in the stored or
secreted peptide pool. Stripped of the signal sequence, the remaining protein becomes
the proneuropeptide.

2.2.3.2 The basic motifs

The proneuropeptide needs to be cleaved into defined peptides that will serve as chemical
signals upon secretion, and are destined to bind to receptors. Pairs of the basic amino
acids lysine (Lys, K) and arginine (Arg, R), or more rarely a single Arg in the appropriate
structural environment of the precursor serve as recognition sites and substrates of PCs.
Cleavage of the proneuropeptide leads to the generation of specific sets of peptides. Not
all cleavage products of a precursor contain biological activity. In some precursors, a single peptide sequence has biological activity, for example OXT, pancreatic polypeptide
(PPY), natriuretic peptide A (NPPA), and neuropeptide S (NPS). Many others contain
multiple copies of the same or related neuropeptides, for example PENK, TAC1, tachykinin 3 (TAC3), and thyrotropin-releasing hormone (TRH). Others give rise to different
peptides which, acting on different receptors, exert different biological activities, for
example, POMC. Its adrenocorticotropin hormone (ACTH)/melanocyte-stimulating hormone (MSH) peptides act on melanocortin receptors (MC1–3) and its co-synthesized
endorphins on μ-opiate receptors. Moreover, cell-specific differences in PCs can generate
different sets of peptides from the same precursor in different systems. Again, POMC
undergoes limited processing in the corticotropes of the anterior pituitary gland into
ACTH, β-lipotropin (βLPH), and a large form of γMSH. In ARC neurons the main products
are αMSH, γMSH, and β-endorphin, neuropeptides each having their own actions
(Burbach and Wiegant, 1990).

2.2.3.3 The C-terminal amide

Prohormone convertase-generated peptides can be subject to further modification by
peptidyl-aminotransferase (PAM). This enzyme uses a C-terminal glycine (Gly, G) as amide
donor for the preceding amino acid. This results in a peptide with an amidated C-terminus.
This amide is present on many neuropeptides. It makes the peptide more resistant to C-terminal degradation and is often essential for biological activity. Examples are OXT, AVP,
αMSH, NPY, and TRH. PAM is expressed by **peptidergic neurons** and is present in the RSP.
The presence of an amidated C-terminal in neuropeptides has been elegantly exploited by
Mutt to isolate novel neuropeptides and peptide hormones on the basis of a chemical assay

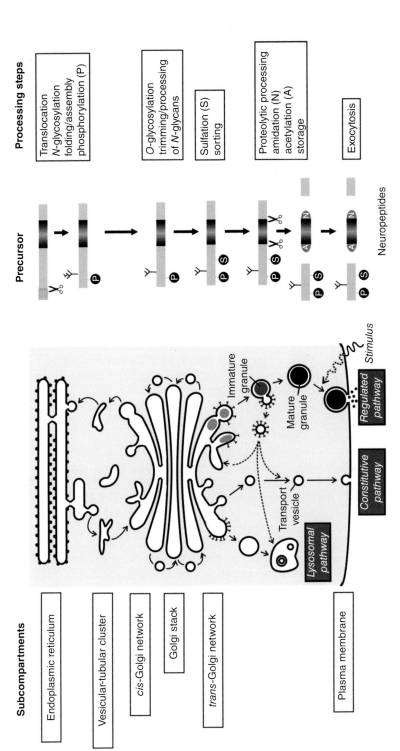

Figure 2.1 Biosynthesis and transport of neuropeptides in the regulated secretory pathway: The flow of newly biosynthesized secretory proteins through the cell is schematically depicted with emphasis on the subcompartments that are passed and the processing steps that occur. Neuropeptide precursors are sorted in the *trans*-Golgi network into vesicles of the secretory pathway. These vesicles acidify, condense the protein content and activate proteolytic enzymes, resulting in processing of the precursor protein into neuropeptides. Mature granules (dense-cored vesicles) are stored and await release upon a stimulus. Posttranslational modifications and their cellular sites are indicated. This pathway is distinguished from the default pathway which is the constitutive pathway. In the latter, proteins are packed in small clear-cored secretory vesicles that immediately fuse for secretion upon arrival at the plasma membrane. Growth factors are typically released through the constitutive pathway, while neuropeptides are stored and secreted in the regulated secretory pathway. Courtesy of Gerard Martens, Nijmegen.

to detect amidated peptides, such as neuropeptide Y, peptide YY, and peptide HI (all named after the C-terminal amidated amino acid) (Tatemoto and Mutt, 1980).

2.2.3.4 Other modified amino acids

Chemical modifications of specific amino acid residues in neuropeptides include *N*- and *O* glycosylation, phosphorylation, tyrosyl *O*-sulfation, N^α-acetylation, and N-terminal pyroglutamic acid (pGlu) formation. Except for the latter, these are enzymatic modifications that take place posttranslationally. pGlu formation is a spontaneous cyclization of a free N-terminal Gln in an acidic environment. These modifications are a consequence of passage through the RSP, and are found on stored and secreted peptides. In separate cases, a clear-cut biological function has been assigned to a modification. For example, N^α-acetylation of β-endorphin results in loss of opiate activity, the pGlu residue is required for the biological activity of TRH, tyrosyl *O*-sulfation is crucial for the cholecystokinetic activity of CCK (but has no effect on gastrin peptides), and differences in potency of glycosylated and phosphorylated forms of POMC peptides have been reported.

2.2.4 Neuropeptide receptors

Neuropeptides have biological activities in neural systems that can be observed at all levels of neural functioning. Effects of neuropeptides have been established at the genetic, biochemical, electrophysiological, cellular, behavioral, and organismal levels. Typical of the cellular action of neuropeptides are the relatively slow responses and actions, compared to 'fast neurotransmitters' like excitatory amino acids and amines. The relatively slow response to released neuropeptides is due to the type of receptor that almost all neuropeptides act on: G protein-coupled receptors (**GPCRs**) (see Chapter 10). These receptors have a common architecture characterized by seven transmembrane domains that form a pocket, an N-terminal extracellular domain that contributes to ligand binding or stabilization, and a C-terminal intracellular domain that participates in signal transduction. The latter involves an intracellular cascade of enzymatic events that result in cellular responses. The time span over which such a response occurs (seconds or longer) is considerably longer than that of neurotransmitters that modulate ion fluxes directly through action on ion channels (milliseconds). The only exception is the receptor system for natriuretic peptides ANF, BNF, and CNF. These peptides act on natriuretic peptide receptors A, B, and C (NPR-A, NPR-B, NPR-C; see Table 2.1 for links). These receptors are transmembrane proteins with a single membrane-spanning region and an intracellular guanylyl cyclase activity.

There are more neuropeptides known than there are receptors. Receptors are often responsive to multiple related neuropeptides. For example, there are four opioid receptors, but over 15 forms of opioid peptides. Many examples can be found in Table 2.1, that lists established matches between neuropeptides and receptors. Furthermore, there are several peptides with reported biological effects but with unknown receptors, and there are GPCRs with unknown ligands, the so-called orphan receptors. Databases of neuropeptide receptors are available: www.gpcr.org/7tm/ and www.guidetopharmacology.org/.

2.3 Neuropeptide genes in the genome

This section provides an overview on all neuropeptides that have been identified in the human genome. Several differences in gene content in the mouse are indicated. The focus is on established neuropeptides. However, since distinctions between neuropeptides, peptide hormones, and other peptides in the nervous system are sometimes overlapping, attention is also given to peptide hormones, **cerebellins**, **granins**, **neurotropins**, and **chemokines**.

2.3.1 Classic neuropeptides

About 60 classic neuropeptide genes can be recognized in the human genome (see Table 2.1). Their gene expression by neurons has been established and biological activities of peptides have been documented. In most cases, regulated release or vesicular storage of the neuropeptides has been demonstrated. These can be considered as 'classic neuropeptides.' Many were discovered some years ago, and have been studied extensively during the last decades. Others stem from more recent dates, such as augurin encoded by the C2orf40 gene, spexin by the C12orf39 gene, osteocrin by the OSTN gene, and peptides of the KISS1, DBI, PROK1, and PROK2 genes. Some caution is needed here with respect to their fit into all of the criteria for neuropeptides defined above. Furthermore, there is an emerging class of regulatory peptides originally found in adipose tissues (**adipose peptides**) for which gene expression in the brain has been demonstrated or assumed. These are presented as putative neuropeptides in Table 2.3.

The genes encoding these classic neuropeptides can be grouped in 20 subfamilies rather arbitrarily according to shared precursor and peptide structure, or according to related function. Clear structural similarities are found in, for example, the CRH-related **gene family** (CRH, CRH, UCN, UCN2, UCN3, UTS2, and UTS2D) and the glucagon/secretin gene family (CGC, SCT, VIP, ADCYAP1, GHRH, and GIP). In the opioid gene family (PENK, POMC, PDYN, and PNOC), grouping is based more on a functional relationship than a structural relationship, although the amino acid motif YGGF is part of all opioid peptides in this gene family. A relatively large group of miscellaneous peptide genes remains (see Table 2.1). Structurally and functionally, they seem to stand alone.

Considering the genomic distribution of neuropeptide genes, there is no general clustering of genes of the same neuropeptide gene family. This indicates that neuropeptides with related structural or functional features are not the result of local gene duplication. There are a few notable exceptions, however. The AVP and OXT genes are linked head to tail in all species investigated. The unrelated prodynorphin (PDYN) gene is near this locus. In the F- and Y-amide gene family, two clustered pairs of genes are found: the RPRF and NPY genes are close together but not linked on human chromosome 7p15.3, and the PYY and PPY genes are on 17q21.31 (see Table 2.1). The CALCA and CALCB genes are next to each other on chromosome 11p15.2, with the ADM gene nearby. Other clustered neuropeptide genes are RLN1 and RLN2 on chromosome 9p24.1, and NPPA and NPPB on chromosome 1p36.22.

2.3.2 Granins

Granins form a family of seven genes with products that do not fully fulfill all criteria of neuropeptides, particularly in terms of their biological activities (see Table 2.2). Members of this family, including chromogranins and secretogranins, display the structural features of neuropeptide precursors, and they are synthesized by neurons. An odd feature is that their expression is extremely wide, almost panneuronal for several genes (see Table 2.2). Granins have been identified as co-synthesized, co-stored or co-released proteins with neuropeptides. One view on these peptides is that they are chaperones that assist neuropeptide precursors with sorting and transport through the RSP and interfere with the activity of PCs (Braks and Martens, 1994). Biological activities of some processed forms of granins have been reported, for instance for vasostatin, catestatin, pancreastatin, secretoneurin, EL35, EM66, WE14, and TLPQ-21. However, a clear-cut biological function as signaling molecules is lacking for these peptides, and receptors have remained unidentified. For TLPQ-21, low affinity for complement factors has been found. A notable exception in this respect is the PCSK1N gene which encodes a protein with all the structural features of a neuropeptide precursor, called proSAAS, and gives rise to several processed peptides that are released from hypothalamic neurons and affect body weight and feeding. Recently, for one of these

Table 2.3 Genes coding for putative neuropeptides originally identified in adipose tissues with presumed expression in the brain. The chromosomal localization, brain expression, encoded precursor structure, and peptide products are presented. Hyperlinks provide additional information to the locus in the human chromosome through http://genome.ucsc.edu/, to neuroanatomy of mouse brain expression through the Allen Brain Atlas, or GenePaint, and to structural comparisons to related precursors and those of other species through the BLINK tool of NCBI.

Gene (gene symbol)	Chromosomal localization	Mouse brain expression	Precursor	Active peptide(s)
Leptin/ob gene (LEP)	7q32.1		Prepro-leptin	Leptin (obesin)
Adiponectin gene (ADIPOQ)	3q27.3		Adiponectin precursor	Adiponectin (Acpr30, adipocyte complement-related protein; adipocyte, C1Q and collagen domain containing)
Visfatin gene (PBEF1)	7q22.2		Visfatin precursor	Visfatin-1 (pre-B cell colony enhancing factor-1 (PBEF1), nicotinamide phosphoribosyltransferase)
Resistin gene (RETN)	19p13.2		Resistin precursor	Resistin (cysteine-rich secreted protein FIZZ3, adipose tissue-specific secretory factor, cysteine-rich secreted protein A12-α-like 2, ADSF, Xcp4)
			Resistin-δ2 precursor	Resistin-δ2
Resistin-like α gene (RETLA)	**Not in human**		Resistin-like molecule α precursor (RELMα)	Resistin-like molecule α (found in inflammatory zone 1; FIZZ1, hypoxia-induced mitogenic factor, Xcp2)
Resistin-like β gene (RETLB)	**Not in human**	not available	Resistin-like molecule β precursor (RELMβ)	Resistin-like molecule β (cysteine-rich secreted protein FIZZ2, colon and small intestine-specific cysteine-rich protein, cysteine-rich secreted protein A12-α-like 1, colon carcinoma-related gene protein, Xcp3)
Resistin-like γ gene (RETLG)	**Not in human**		Resistin-like molecule γ precursor (RELMγ)	Resistin-like molecule γ (cysteine-rich secreted protein FIZZ3, Xcp1)
Nucleobindin-2/ NEFA gene (NUCB2)	11p15.1		Nucleobindin-2	Nesfatin-1

Table 2.3 (Continued)

Gene (gene symbol)	Chromosomal localization	Mouse brain expression	Precursor	Active peptide(s)
Ubiquitin-like 5 (UBL5)	19p13.2		Beacon precursor	Beacon
Prokineticin-1 (PROK1)	1p13.3		Prokineticin-1 precursor	Prokineticin-1 (endocrine gland-derived VEGF, EG-VEGF)
Prokineticin-2 (PROK2)	3p13		Prokineticin-2 isoform a precursor	Prokineticin-2 isoform a
			Prokineticin-2 isoform b precursor	Prokineticin-2 isoform b
Serpin peptidase inhibitor, clade A, member 12 (SERPINA12)	14q32.13		Serpin A12 precursor	Vaspin, Serpin A12, visceral adipose-specific serpin
Retinoic acid receptor responder 2 (RARRES2)	7q36.1		Retinoic acid receptor responder protein 2	Chemerin
Intelectin1 (ITLN1)	1q23.3		Intelectin1 precursor	Intelectin1, omentin

peptides, BigLEN, a GPCR (GPR171) was identified. Thus, the role of granins as precursors for neuropeptides should remain under consideration.

2.3.3 Cerebellins

Cerebellins are encoded by four related genes (Table 2.4). The classification of these peptides is complicated. CBLN1 and CBLN3 are expressed by neurons of the cerebellum and a biologically active peptide is produced and released. It is not actually established if they are secreted in a regulated manner. Their mode of action is extraordinary. Unlike other peptides, cerebellin-1 and -2 form a molecular bridge between presynaptic neurexin with postsynaptic Grid, thereby modulating synaptic communication (Uemura *et al.*, 2010).

2.3.4 Neurotropins

Growth factors are proteineous signaling molecules that are produced and secreted by cells. They provide extracellular support for proliferation, differentiation, and maintenance of other or the same cells in a paracrine or autocrine fashion, respectively. Many growth factors are expressed in the brain, where they act during embryonic development, and during neuronal adaptation and maintenance states of the adult nervous system. Growth factors come in many families, often with an expanded and diverse family tree. Examples are the neurotropins, fibroblast growth factors (FGFs), Wnts, bone morphogenic factors (BMPs), epidermal growth factor (EGF), transforming growth factors α and β (TGFs), and others. Neurotropins are particularly expressed, synthesized, and released by neurons.

Table 2.4 The cerebellins gene family. The four genes encoding cerebellins are listed. For each gene, chromosomal localization, brain expression, encoded precursor structure, and peptide products are presented. Hyperlinks provide additional information to the locus in the human chromosome through http://genome.ucsc.edu/, to neuroanatomy of mouse brain expression through the Allen Brain Atlas, or GenePaint, and structural comparison to related precursors and those of other species through the BLINK tool of NCBI.

Gene (gene symbol)	Chromosomal localization	Mouse brain expression	Precursor	Active peptide(s)
Cerebellin-1 gene (CBLN1)	16q12.1		Cerebellin-1 precursor	Cerebellin-1 (Cbln1)
Cerebellin-2 gene (CBLN2)	18q22.3		Cerebellin-2-precursor	Cerebellin-2 (Cbln2)
Cerebellin-3 gene (CBLN3)	14q12		Cerebellin-3 precursor	Cerebellin-3 (Cbln3)
Cerebellin-4 gene (CBLN4),	20q13.2		Cerebellin-4 precursor	Cerebellin-4 (Cbln4, cerebellin-like glycoprotein-1)

The neurotropins encompass a small protein family including NGF, NT-3, NT-4/-5, and BDNF that regulate a wide variety of neural functions through action on four major growth factor-like tyrosine kinase receptors: TrkA, TrkB, TrkC, and p75NTR. Generally, neurotropins are not considered as neuropeptides, since growth factors are generally released by the constitutive secretory pathway. Therefore, their biological availability is primarily controlled at the level of gene expression which is relatively slow and long-lasting. This contrasts with the rapid release of stored neuropeptides upon an acute stimulus, which is required for rapid chemical signaling. Growth factors generally act through receptor protein kinases rather than GPCRs. However, some peptides and proteins historically classified as growth factors may fulfill the criteria of neuropeptides (Table 2.5). A prominent example is BDNF, one of the four neurotropins. The primary gene product is pre-pro-BDNF, a precursor having a signal peptide and cleavage sites for typical prohormone convertases. The BDNF gene is widely expressed in neurons, and BDNF is generated by processing and present in dense-cored vesicles, and is subject to stimulated release. There is a separate role for pro-BDNF, since it acts on a different receptor from mature BDNF (Yang *et al.*, 2009). Thus, BDNF fulfills all criteria to be a neuropeptide, although it is historically considered to be a growth factor. These properties may also appear for other neurotropins, NGF, NT-3, NT-4/5, and even GDNF.

2.3.5 Chemokines

Chemokines, from 'chemotactic cytokines,' were originally known as secreted factors that mediate leukocyte migration. They form a family of about 50 proteins of 8–14 kDa and are subclassified by structural similarities in cysteine configuration in the C, CC, CXC, and CXXXC subfamilies (www.copewithcytokines.de/cope.cgi?key=Chemokines). The nomenclature of individual chemokine genes follows this configuration. The C subfamily comprises only two genes: XCL1 and XCL2. The CC chemokine subfamily contains 28

Table 2.5 The neurotropin genes. The chromosomal localization, brain expression, encoded precursor structure, and peptide products are presented. Hyperlinks provide additional information to the locus in the human chromosome through http://genome.ucsc.edu/, and to structural comparisons to related precursors and those of other species through the BLINK tool of NCBI.

Gene (gene symbol)	Chromosomal localization	Precursor	Active peptide(s)
Nerve growth factor (NGF)	1p13.2	NGF precursor	NGF
Neurotropin-3 (NTF3)	12p13.31	Neurotropin-3 precursor, isoform 1	NT3 (HDNF, NGF2)
		Neurotropin-3 precursor, isoform 2	NT3 (HDNF, NGF2)
Neurotropin-4 (NTF4)	19q13.33	Neurotropin-4 precursor	NT4 (NT5)
Brain-derived nerve growth factor (BDNF)	11p14.1	BDNF precursor, isoform a	BDNF
		BDNF precursor, isoform b	BDNF
		BDNF precursor, isoform c	BDNF
		BDNF precursor, isoform d	BDNF
		BDNF precursor, isoform e	BDNF
Glial-derived neurotropic factor (GDNF)	5p13.2	GDNF precursor, isoform 1	GDNF (ATF1, ATF2, HSCR3)
		GDNF precursor, isoform 2	GDNF (ATF1, ATF2, HSCR3)
		GDNF precursor, isoform 3	GDNF (ATF1, ATF2, HSCR3)
		GDNF precursor, isoform 4	GDNF (ATF1, ATF2, HSCR3)
		GDNF precursor, isoform 5	GDNF (ATF1, ATF2, HSCR3)
Artemin (ARTN)	1p34.1	Artemin precursor, isoform CRA_a	Artemin
		Artemin precursor, isoform CRA_b	Artemin
		Artemin precursor, isoform CRA_c	Artemin
Neurturin (NTRN)	19p13.3	Neurturin precursor	Neurturin
Persephin (PSPN)	19p13.3	Persephin precursor, isoform CRA_a	Persephin
		Persephin precursor, isoform CRA_a	Persephin

genes: CCL1 to CCL28 (CCL9 and -10 are the same). The CXCL chemokine subfamily comprises a group of 17 genes: CXCL1 to CXCL17. There is only one CXXXCL gene (CX3CL1).

Chemokines are engaged in inflammatory processes. In diseases of the nervous system, chemokines have been implicated as part of the inflammatory response, for instance in multiple sclerosis (MS). Chemokine expression is often coupled with microglia activation. However, evidence has been obtained for neuronal expression of chemokines under normal physiological conditions, for chemokine receptor expression by neurons, and for electrophysiological effects of chemokines on neurons (Miller *et al.*, 2008). Neural activities of chemokines have been implicated in neuronal migration, neuroinflammation, and neuronal excitability. Chemokines for which arguments accumulate that they can function as endogenous neuropeptides of the nervous system are presented in Table 2.6. A well-illustrated example is CCL21. This chemokine was shown to be expressed by neurons and to be sorted to vesicles of the RSP (de Jong *et al.*, 2008). A notable feature concerning neurally expressed CCLs is the aggregation of a number of these genes in the same chromosomal locus. CCL2, CCL3, CCL4, and CCL5 are localized on chr17q12 in a locus that also contains CCLs of which brain expression has not been demonstrated, like CLL14, CCL15, CLL16, CLL18, and CLL23 as well as several CCL4- and CCL3-like proteins. The significance of this grouped localization for gene expression is not known.

Table 2.6 Genes coding for brain-expressed chemokines. The chromosomal localization, brain expression, encoded precursor structure, and peptide products are presented. Hyperlinks provide additional information to the locus in the human chromosome through http://genome.ucsc.edu/, to neuroanatomy of mouse brain expression through the Allen Brain Atlas, or GenePaint, and to structural comparisons to related precursors and those of other species through the BLINK tool of NCBI.

Gene (gene symbol)	Chromosomal localization	Mouse brain expression	Precursor	Active peptide(s)
C-C motif chemokine ligand 2 (CCL2)	17q12		CCL2 precursor	CCL2
C-C motif chemokine ligand 3 (CCL3)	17q12		CCL3 precursor	CCL3
C-C motif chemokine ligand 4 (CCL4)	17q12		CCL4 precursor	CCL4
C-C motif chemokine ligand 5 (CCL5)	17q12		CCL5 precursor	CCL5
C-C motif chemokine ligand 20 (CCL20)	2q36.3		CCL20 precursor isoform 1	CCL20
			CCL20 precursor isoform 2	CCL 20
C-C motif chemokine ligand 21 (CCL21)	9p13.3		CCL21 precursor	CCL21
CXC motif chemokine 10 (CXCL10)	4q21.1		CXCL10 precursor	CXCL10 (small inducible cytokine B10)
CXC motif chemokine ligand 12 (CXCL12)	10q11.21		CXCL12 precursor, SDF-1α	CXCL12 (stromal cell-derived factor 1, SDF-1a)
			SDF-1β	Stromal cell-derived factor 1, SDF-1b
			SDF-1γ	Stromal cell-derived factor 1, SDF-1c
			SDF-1δ	Stromal cell-derived factor 1, SDF-1d
CX3C motif chemokine ligand 1 (CX3CL1)	16q21		Fractalkine precursor, CX3CL1 precursor	CX3CL1 (fractalkine)

2.4 Perspectives

Neuropeptides have emerged as a distinct class of chemical signaling molecules used by neurons to participate in neurotransmission and neuromodulation. Encoded by at least 60 genes, the classic neuropeptides form the largest class of neuronal transmitters. Classic neuropeptides fulfill the criteria of synthesis and regulated release by neurons and action on brain receptors. In a broader perspective, further peptides usually classified as peptide hormones, granins, neurotropins or chemokines, together with a still growing group of putative neuropeptides, expand the repertoire of chemical signaling in the nervous system. The neuroendocrine system preeminently employs a large proportion of the neuropeptides encoded by the genome as it requires the fine-tuning provided by this peptide repertoire.

Neuropeptides are only one molecular element of the neuroendocrine system. Other elements, such as receptors, cell surface molecules, the release machinery, ion channels and molecules that shape the cell biological properties of neurons of the neuroendocrine system, are as important. Currently, knowledge about the mammalian genome, not only what proteins it codes but particularly how it encodes the instructions to specify development, function and integration of cell and organs, is expanding at a great rate. It may be expected that we can approach the neuroendocrine system in unprecedented experimental ways in the near future to gain further insight into the mechanisms and functions of this highly integrative system.

Acknowledgments

Parts of this chapter were previously published in 'What are neuropeptides?' in Merighi, A. (ed.) *Neuropeptides: Methods and Protocols, Methods in Molecular Biology*, vol. 789. Springer Science+Business, New York, 2011. Permission was obtained from www.neuropeptides.nl for database data.

References

Braks, J.A. and Martens, G.J. (1994) 7B2 is a neuroendocrine chaperone that transiently interacts with prohormone convertase PC2 in the secretory pathway. *Cell*, **78**, 263–273.

Brownstein, M.J., Russell, J.T. and Gainer, H. (1980) Synthesis, transport and release of posterior pituitary hormones. *Science*, **207**, 373–378. [***This paper demonstrated the biosynthesis of peptides through precursor proteins and how neurons cleave and transport the peptide products to nerve terminals.***]

Burbach, J.P.H. and Wiegant, V.M. (1990) Gene expression, biosynthesis and processing of proopiomelanocortin peptides and vasopressin, in Neuropeptides: Basics, and Perspectives (ed. D. de Wied), Elsevier, Amsterdam, pp. 45–106.

Burbach, J.P.H., Loeber, J.G., Verhoef, J., *et al.* (1980) Selective conversion of beta-endorphin into peptides related to gamma- and alpha-endorphin. *Nature*, **283**, 96–97.

Burbach, J.P.H., Kovács, G.L., de Wied, D., van Nispen, J.W. and Greven, H.M. (2003) A major metabolite of arginine vasopressin in the brain is a highly potent neuropeptide. *Science*, **221**, 1310–1312. [***This paper demonstrated that vasopressin is metabolized in peptides that have a many-fold higher potency in behavioral tests for central activity than vasopressin.***]

Burbach, J.P.H., Young, L.J. and Russell JA (2006) Oxytocin: synthesis, secretion, and reproductive functions, in *Knobil and Neill's Physiology of Reproduction*, 3rd edn (ed. J.D. Neill), Elsevier, Amsterdam, pp. 3055–3128.

De Bree, F.M., van der Kleij, A.A., Nijenhuis, M., Zalm, R., Murphy, D. and Burbach, J.P.H. (2003) The hormone domain of the vasopressin prohormone is required for the correct prohormone trafficking through the secretory pathway. *Journal of Neuroendocrinology*, **15**(12), 1156–1163. [***This paper showed that conformation of the precursor is essential for sorting.***]

De Jong, E.K., Vinet, J., Stanulovic, V.S., *et al.* (2008) Expression, transport and axonal sorting of neuronal CCL21 in large dense-core vesicles. *FASEB Journal*, **22**, 4136–4145.

De Wied, D. (1971) Long term effect of vasopressin on the maintenance of a conditioned avoidance response in rats. *Nature*, **232**, 58–60. [***This paper showed that peptides act on the brain and can affect behavior.***]

Gao, Q. and Horvath, T.L. (2007) Neurobiology of feeding and energy expenditure. *Annual Review of Neuroscience*, **30**, 367–398.

Hökfelt, T., Johansson, O., Ljungdahl, A., Lundberg, J.M. and Schultzberg, M. (1980) Peptidergic neurones. *Nature*, **284**, 515–521.

Hughes, J., Smith, T.W., Kosterlitz, H.W., Fothergill, L.A., Morgan, B.A. and Morris, H.R. (1975) Identification of two related pentapeptides from the brain with potent opiate agonist activity. *Nature*, **258**, 577–580. **[A milestone in neuropeptide research: the identification of enkephalins.]**

Hur, Y.S., Kim, K.D., Paek, S.H. and Yoo, S.H. (2010) Evidence for the existence of secretory granule (dense-core vesicle)-based inositol 1,4,5-trisphosphate-dependent Ca2+ signaling system in astrocytes. *PLoS One*, **5**, e11973. **[This paper showed how neuropeptides and classic neurotransmitters cooperate in neurotransmission by peptidergic neurons.]**

McGirr, R., Guizzetti, L. and Dhanvantari, S. (2013) The sorting of proglucagon to secretory granules is mediated by carboxypeptidase E and intrinsic sorting signals. *Journal of Endocrinology*, **217**, 229–240.

Miller, R.J., Rostene, W., Apartis, E., *et al.* (2008) Chemokine action in the nervous system. *Journal of Neuroscience*, **28**, 11792–11795.

Mirabeau, O., Perlas, E., Severini, C., *et al.* (2007) Identification of novel peptide hormones in the human proteome by hidden Markov model screening. *Genome Research*, **17**, 320–327.

Ofer, D. and Linial, M. (2014) NeuroPID: a predictor for identifying neuropeptide precursors from metazoan proteomes. *Bioinformatics*, **30**, 931–940.

Scharrer, B. (1987) Neurosecretion: beginnings and new directions in neuropeptide research. *Annual Review of Neuroscience*, **10**, 1–17. **[A classic paper that presented the concept of neurosecretion by peptidergic neurons.]**

Tatemoto, K. and Mutt, V. (1980) Isolation of two novel candidate hormones using a chemical method for finding naturally occurring polypeptides. *Nature*, **285**, 417–418.

Tooze, S.A. and Huttner, W.B. (1990) Cell-free protein sorting to the regulated and constitutive secretory pathways. *Cell*, **60**, 837–847.

Tooze, S.A., Martens, G.J. and Huttner, W.B. (2001) Secretory granule biogenesis: rafting to the snare. *Trends in Cell Biology*, **11**, 166–122.

Uemura, T., Lee, S.J., Yasumura, M., *et al.* (2010) Trans-synaptic interaction of GluRdelta2 and Neurexin through Cbln1 mediates synapse formation in the cerebellum. *Cell*, **141**, 1068–1079.

Yang, J., Siao, C.J., Nagappan, G., *et al.* (2009) Neuronal release of proBDNF. *Nature Neuroscience*, **12**, 113–115.

Further reading

Klavdieva, M.M. (1995) The history of neuropeptides I. *Frontiers in Neuroendocrinology*, **16**, 293–321.

Klavdieva MM (1996) The history of neuropeptides II. *Frontiers in Neuroendocrinology*, **17**, 126–153.

Klavdieva MM (1996) The history of neuropeptides III. *Frontiers in Neuroendocrinology*, **17**, 155–179.

Klavdieva MM (1996) The history of neuropeptides IV. *Frontiers in Neuroendocrinology*, **17**, 247–280.

Wade, N. (1981) *The Nobel Duel*, Doubleday, New York.

CHAPTER 3

Transcriptome Dynamics

David A. Carter,[1] Steven L. Coon,[2] Yoav Gothilf,[3] Charles K. Hwang,[4] Leming Shi,[5] P. Michael Iuvone,[4] Stephen Hartley,[6] James C. Mullikin,[6] Peter Munson,[7] Cong Fu,[8] Samuel J. Clokie,[9] and David C. Klein[10]

[1] School of Biosciences, Cardiff University, Cardiff, UK
[2] Molecular Genomics Laboratory, Eunice Kennedy Shriver National Institute of Child Health and Human Development, National Institutes of Health, Bethesda, Maryland, USA
[3] Department of Neurobiology, George S. Wise Faculty of Life Sciences and Sagol School of Neuroscience, Tel-Aviv University, Tel-Aviv, Israel
[4] Department of Ophthalmology, Department of Pharmacology, Emory University School of Medicine, Atlanta, Georgia, USA
[5] Center for Pharmacogenomics, School of Pharmacy, Fudan University, Shanghai, China
[6] Genome Technology Branch, National Human Genome Research Institute, National Institutes of Health, Bethesda, Maryland, USA
[7] Mathematical and Statistical Computing Laboratory, Center for Information Technology, National Institutes of Health, Bethesda, Maryland, USA
[8] State Key Lab of Theoretical and Computational Chemistry, Jilin University, Changchun, Jilin, China
[9] West Midlands Regional Genetics Laboratories, Birmingham Women's NHS Foundation Trust, Birmingham, UK
[10] Office of the Scientific Director, Eunice Kennedy Shriver National Institute of Child Health and Human Development, National Institutes of Health, Bethesda, Maryland, USA

3.1 Approaching transcriptome dynamics

3.1.1 What do we mean by 'transcriptome dynamics'?

The transcriptome is the complete set (see Box 3.1) of RNA **transcripts** that are expressed in a particular cell type or tissue. Three important characteristics of this pool of RNA should be emphasized.

- In addition to the (coding) mRNAs, the transcriptome also includes many non-coding, regulatory RNAs, for example, long non-coding RNAs (lncRNA) and microRNAs (miRNA). In fact, much of the genome is transcribed, but at widely different rates.
- Cellular levels of RNA are not solely dependent upon and directly related to transcription rates; posttranscriptional mechanisms involving RNA-binding proteins also determine the level of individual RNAs.
- Dynamic changes in the transcriptome (the subject matter of this chapter) are a general feature of biological systems, and an important aspect of neuroendocrine regulation. Physiological responses such as increases in hormone secretion are accompanied by profound changes in the cellular transcriptome; different sets of genes have distinct dynamics (see section 3.3).

Transcriptome dynamics concerns the changes in transcriptome content over time. This should not be confused with the dynamics of transcription *per se* which concerns, in part, dynamic changes in the temporal association of transcription factors with chromatin (e.g. Hager *et al.*, 2009). The transcriptome of cells/tissues is inherently dynamic with overlying and interdependent 'ground' states that have two major determining factors:

Box 3.1   How many transcripts in a transcriptome?

There are some 300,000–400,000 messenger (m)RNA molecules in a somatic mammalian cell but many of these are replicates, with around 12,000 distinct mRNAs in any particular cell. Some mRNAs are in high abundance (up to ~1% of total pool or ~3000 molecules or copies per cell) whereas others are considerably more rare ('low copy number'; 1–5 molecules per cell); the median cellular copy number in one study was determined as 17 mRNA copies (Schwanhäusser *et al.*, 2011). In addition to mRNAs, there are many other forms of cellular RNA including non-coding (nc)RNAs that have diverse regulatory function. Transcriptome content is controlled, in part, by transcription factors (TFs). Of the ~1750 TFs in the human genome, a large proportion are expressed in all (somatic) cells (determined as 306–722 different TFs mRNAs at >3 copies per cell; FANTOM, 2014). However, within defined cells, particular 'cell-specific' TFs are highly enriched above low-copy values (≥100-fold; FANTOM, 2014).

---

developmental age and environment. Both of these major factors include multiple different aspects; developmental age can be considered to include a programmatic continuum of differentiation–maturation–senescence, and environment includes both stochastic (e.g. acute behavioral stress) and predictable (e.g. circadian and seasonal rhythm) components. The relative contribution and interaction of different factors in determining transcriptome dynamics is an important consideration for experimental design and data interpretation.

The interaction of multiple determining factors in driving transcriptome dynamics is nicely illustrated in one recent study from outside the neuroendocrine field (Nagano *et al.*, 2012). In this study, conducted in a 'natural' environment, the authors showed how the transcriptome of rice plant leaves (*Oryza sativa*) is determined by the combined influence of endogenous daily rhythms, environment (e.g. solar radiation), and also plant age. Importantly, they demonstrated how the influence of one determining factor (environment) on transcriptome content is time-restricted by endogenous gating mechanisms.

Transcriptome analysis (cf. individual gene measurements) advances research because it provides a comprehensive view of the set of gene transcripts that are expressed in a particular cell type/tissue at a particular moment in time. Moving to the next level of enquiry, transcriptome dynamics is then required to fully understand the changes in gene expression that mediate developmental processes and physiological responses within cells/tissues. A massively parallel approach is particularly important for the study of dynamics because there are multiple, successive, often subtle changes, involving both up- and downregulation, within the set of cellular transcripts that collectively mediate a particular biological response (Figure 3.1). Following the identification of these dynamics, it is then possible to construct testable models of the molecular interactions/pathways (e.g. transcription factor target sets) that underlie physiological responses.

In order to study transcriptome dynamics, it is necessary to consider two aspects of sampling.

3.1.1.1   Spatial

Neuroendocrine systems have a number of mechanisms for coordinating cellular responses. Consequently, discrete sampling of a specific cell population will in many cases provide a representative measure of transcriptome dynamics, although some degree of individual cellular variation will be diluted out. However, the presence of heterogeneous cell populations (e.g. neurons and glia, intermingled hypophysiotropic neurons) will result in a non-biological background that may be unacceptable in some studies. For these cases, a number of cell enrichment approaches (see Carter, 2006) have been devised and refined over the years. A caveat to cell enrichment, in the context of transcriptome dynamics, is the potential for (differential) transcript loss/gain during the enrichment procedure. **Single cell**

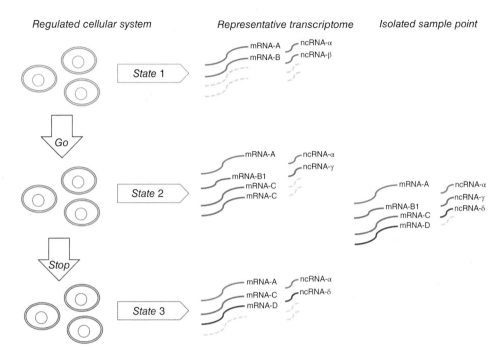

Figure 3.1 The cellular transcriptome includes many thousands of distinct transcripts, including both mRNAs and non-coding RNAs (ncRNA) of diverse types. Dynamic changes in the transcriptome reflect different physiological states – the levels of individual transcripts are differentially affected, some remaining stable, others being either up- or downregulated. Insufficient sampling will diminish the value of transcriptome dynamics analysis and generate false-positive candidate genes.

sequencing ('Method of the Year', 2013; Wu *et al.*, 2014) is now available and has been applied to hypothalamic neurons (Eberwine and Bartfai, 2011) but, as indicated above, this may be of limited value where a 'consensus' physiological dynamic is required. Stochastic patterns of gene expression in both neurons (e.g. protocadherin isoforms; Hirayama *et al.*, 2012) and pituitary cells (e.g. prolactin; Harper *et al.*, 2010) would also confound single cell approaches for some studies.

Subcellular transcript localization must also be considered; recent studies (Khaladkar *et al.*, 2013) have provided evidence of variant RNAs that are enriched in particular cell compartments and may underlie some physiological responses.

A final point to be made in the context of spatial sampling is that certain limitations can be overcome at the level of candidate gene validation, for example, by using *in situ* hybridization (ISH) to localize cellular transcript expression in downstream validation experiments (see section 3.2.1).

3.1.1.2   Temporal

Physiological changes in gene expression within a cell/tissue must be considered as a continuum. Multiple individual experimental samples across this continuum will provide a 'representative' measure of transcriptome content, and ultimately a measure of the transcriptome dynamic (see Figure 3.1). In contrast, a single sample within a physiological response (see Figure 3.1) will provide an unrepresentative measure of transcriptome content. Hence, the selection of sampling points must be carefully considered and possibly refined by single gene pilot experiments. Another approach for the refinement of sampling involves the capture of a subset of mRNAs (note, mRNAs) that are selectively associated with phosphorylated ribosomes in populations of activated neurons (Knight *et al.*, 2012).

3.1.2   Measurement and analysis of transcriptome dynamics

The field of transcriptomics has evolved remarkably in recent years from a large dependence on **microarray** technology to the wide availability of several whole transcriptome sequencing methods. Microarrays remain a valid approach for many studies and do not require the same degree of skilled interpretation (see section 3.4). However, next-generation sequencing (NGS) that enables whole transcriptome sequencing (**RNA-Seq**) delivers unbiased analysis of non-annotated sequences. This technology can therefore provide expression data on novel transcripts, for example splice variants, that may be crucial for understanding particular cellular responses. This may be particularly important in the comparative analysis of different research models because alternative splicing, as an example, is frequently species specific (Merkin *et al.*, 2012). RNA-Seq can also be used to specify strand-specific transcription. Additionally, the contribution of posttranscriptional mechanisms to transcriptome dynamics can be assessed using a new variant of RNA-Seq, **Nascent-Seq**, that provides a direct measure of transcription as opposed to simply transcript levels. This technique has been used to reveal extensive posttranscriptional regulation within circadian rhythms of gene expression (Menet *et al.*, 2012).

Perhaps the most valuable feature of RNA-Seq is that the data can be used in the future for further investigations. This is especially valuable as knowledge of genomes grows – currently the best annotated genomes are human and mouse; other genomes are less well annotated. Hence, current analysis of RNA-Seq data is limited by the version of genome used for bioinfomatics. An important aspect of this research field is therefore centralized curation of raw data that can be reanalyzed by research groups around the world. RNA-Seq (and microarray) data sets are available at ArrayExpress (www.ebi.ac.uk/arrayexpress/) and Gene Expression Omnibus (www.ncbi.nlm.nih.gov/geo/). There are a number of different computer tools for extracting and analyzing raw data from microarray and RNA-seq analysis (see section 3.4). Beyond this initial stage of data processing, transcriptome data can be used for:

- candidate gene(s) validation using QPCR, ISH, etc.
- gene network analysis using a variety of custom tools (Cytoscape, www.cytoscape.org/; Ingenuity Pathway Analysis, www.ingenuity.com/products/ipa).

3.1.3   Bioinformatic and statistical analysis of transcriptome dynamics

The analysis of transcriptome data is now an extensive research specialty. Here, this topic is limited to current insights into the analysis of NGS data that forms the basis of RNA-Seq. Some of the methods described below were used in the analysis of lncRNA dynamics in the pineal gland (see section 3.2.2; Coon *et al.*, 2012) and additional procedures used for zebrafish RNA-Seq are described in section 3.2.3. The analysis of microarray data is more standard and facilitated by commercial tools (e.g. GeneSpring; www.genomics.agilent.com). The statistical analysis of temporal dynamics in transcriptome data must be carefully considered at the experimental design stage, particularly in view of the relative fiscal expense of NGS. The requirement for sample point replicates is discussed below and also in a recent review of transcriptome dynamic analysis by Oh *et al.* (2014).

3.1.3.1   Mapping NGS data to the genome

The analysis of NGS RNA-Seq data is subject to a number of unique statistical and technical issues, and in recent years a number of specialized toolkits have evolved to perform analysis on such data. The tools described here are easily accessible via standard web browser searches. Pertinent background reading is in Steijger *et al.* (2013).

Before differential regulation or transcript quantification analyses can be performed, NGS data must first be mapped to the genome. As mRNA sequence contains numerous splices, aligner software designed for use on DNA sequencing data, such as BWA (Burrows-Wheeler Alignment Tool) or Bowtie (part of the Tuxedo suite), will generally not provide satisfactory mapping. Consequently, specialized aligners have been developed specifically for use with RNA-Seq. Such aligners can usually function with or without gene annotation, and can also align novel splice junctions and transcripts (that is, transcripts or splice junctions not found in the gene annotation). Gene annotation should be supplied to the aligner whenever it is available for the genome, as aligners will almost always perform better when annotation is provided. Popular and effective splice-aware alignment programs include GSNAP, MapSplice, TopHat2, and RNA-STAR.

Using aligned RNA-Seq data, there are a number of distinct study questions that can be addressed. In general, most studies attempt to identify features that are differentially expressed between two or more studied conditions. This differential regulation can take the form of either gene-level or isoform-level differential expression/regulation. Differential isoform usage can include either preferential expression of transcripts with alternate start sites, alternate stop sites, cassette exon skips, intron retentions, truncations or any number of similar variations to the transcripts belonging to a single gene. Regardless of the type of differential regulation that is being analyzed, all such analysis confronts certain common obstacles.

The issue of library size normalization has seen considerable development in recent years. Due to various technical limitations, different libraries are never sampled at the exact same depth. In order to compare samples to one another, one must find adjustment factors through which all samples can be measured on a common scale. Previously, the standard method of adjusting for read depth was to transform read-counts into RPKM (Reads Per Kilobase of transcript per Million mapped reads) or FPKM (Fragments PKM), a variation of 'total count' normalization, in which the implicit normalization factor is equal to the total number of reads (or read-pairs) mapped for each sample. In recent years, however, this normalization method has been shown to be inadequate and has been discarded in favor of newer normalization methods such as TMM (Trimmed Mean of M-values) or geometric normalization. Many tools also allow external normalization factors to be supplied, if needed. Of particular note, normalization via synthetic spike-ins (Jiang *et al.*, 2011) performed well in situations in which geometric or TMM normalization is not appropriate. In particular, both TMM and geometric normalization methods rely strongly on the assumption that the majority of genes are not differentially expressed. In some studies, this assumption may be violated, as in single-cell RNA-Seq or when comparing extremely different or extremely variable tissue types.

3.1.3.2 Statistical analysis and sample number

Previously, many tools assumed that RNA-Seq read-counts followed a Poisson distribution, meaning that the variance was equal to the mean. In recent years, this has been demonstrated to be inappropriate, as RNA-Seq suffers from 'overdispersion' as a consequence of biological and library-prep variation. Most tools now model read-counts via a negative binomial or β-negative-binomial distribution.

In part, as a result of this advance in understanding, running experiments without multiple replicates for each experimental condition is no longer recommended for most applications. Without replicates, it is impossible to accurately estimate the sample dispersion, making statistical analysis of small changes impossible. The number of replicates used will influence the number of differentially expressed transcripts detected, as will other factors including the abundance of the transcript and the magnitude of the changes that occur.

However, there is also value in designing RNA-Seq experiments with single samples in each condition (n = 1). The advantage of this is practical; it reduces the costs of analysis.

Rigorous statistical analysis is not important when using an opportunistic approach in which one searches only for very large differential effects involving moderately abundant transcripts, especially when such changes are limited to only a few of the thousands of transcripts in an RNA-Seq data set. This approach can identify candidate transcripts for further investigation and validation via other methods. An example of the success of this approach is a study of GnRH effects on the rat pituitary transcriptome (see section 3.2.2; Kucka *et al.,* 2013). In this study, RNA-Seq revealed a 600-fold increase in the abundance of the *Dmp1* transcript. This was confirmed by qRT-PCR, which led to an extensive analysis of the control of *Dmp1* transcription. Here, the n = 1 approach worked.

3.1.3.3   Detection of differentially expressed transcripts

The relative advantages and disadvantages of the various differential regulation tools remain a subject of much debate, and no toolset has demonstrated uniform superiority over the others. DESeq2 and EdgeR use very similar methodologies, and accordingly they generally return very similar results. It should be noted that neither tool is designed to account for variation in read-count level caused by differential isoform switching. They may detect differential expression in genes that are actually exhibiting isoform switching, or miss genes in which isoform switching obfuscates actual differential expression. In order to detect and distinguish such phenomena, different tools are required. DEXSeq uses methods similar to those used by DESeq2, but is designed to test for differential exon usage rather than gene-wide differential expression. All of these tools compare each individual feature with itself between conditions, and thus avoid the problems of sequence-specific or GC-content bias. Alternatively, both EdgeR and DESeq2 can be combined with transcript isoform quantification tools such as RSEM or eXpress.

CuffDiff and the Tuxedo suite use an entirely different approach. CuffDiff attempts to compare abundances of different features to one another across different study conditions. However, NGS RNA-Seq data are subject to numerous biases related to transcript sequence and GC content, making it difficult to directly compare different isoforms. Additionally, many reads may be mapped to locations shared by multiple isoforms. In order to estimate isoform abundances under these conditions, it performs multiple sequential passes of adjustment and likelihood maximization. However, each of these steps involves a number of implicit assumptions, and the large number of steps involved may obfuscate any violation of such assumptions. Issues regarding the analysis of RNA-Seq data are widely discussed in online forums (http://seqanswers.com/).

In some cases, identification of differentially expressed transcripts requires novel approaches that use both computational analysis and visual examination of the data, especially when dealing with a partially or poorly annotated genome. This approach was used in the effort that revealed differential expression of lncRNAs in the rat pineal gland (see section 3.2.2; Coon *et al.*, 2012). The strategy used in this study was to divide the entire genome into 1000 bp bins and to determine how many reads mapped to these bins. Bins were then filtered to select those that (1) did not align with known genes, (2) exhibited differential expression, (3) were greater than background, and (4) were enriched in the pineal gland relative to other tissues. Selected bins were then combined and the read coverage plots of the candidate lncRNAs were curated manually to identify contiguous clusters of reads that stood out from the surrounding genomic background and were not obvious extensions of known genes. This lead to the identification of over 100 lncRNAs (0.3 to >50 kb) with differential night/day expression that are relatively selectively expressed in the pineal gland. lncRNAs are poorly annotated in many tissues and the approach described above may be of value in the pursuit of differentially expressed tissue-specific lncRNAs.

3.2   Transcriptome dynamics in neuroendocrine systems

Within the neuroendocrine field, some of the most detailed analysis of transcriptome dynamics has been conducted in the pineal gland, involving both physiological and developmental studies (see section 3.3). An update on transcriptome dynamics research within the hypothalamopituitary axis is provided here.

3.2.1   Hypothalamus

With multiple subregions and intermingled neuronal phenotypes, the hypothalamus demands a skilled and considered sampling approach for effective analysis of transcriptome dynamics. Developmental dynamics of the hypothalamic transcriptome has been analyzed comprehensively by Shimogori *et al.* (2010). This focused analysis of hypothalamic transcriptome dynamics across mouse development (E10–E18, P0, P21, P42) revealed more than 1000 gene transcripts that were dynamically regulated over this developmental timecourse and also identified novel marker genes for individual hypothalamic nuclei. In this study, the authors conducted extensive ISH validation to overcome spatial limitations inherent in the initial sampling procedure. The effect of physiological stimuli on hypothalamic transcriptome dynamics has been documented in many microarray studies over the last 10–15 years but there has been limited application of RNA-Seq to neuroendocrine regulation in the hypothalamus.

3.2.2   Pituitary

Studies conducted on the hypothalamo-neurohypophysial system by the editors of the current volume have extensively documented dynamic changes in the transcriptome of this system following physiological stimuli. In contrast, there have been relatively few studies on anterior pituitary transcriptome dynamics. Naturally, such studies are complicated by the presence of multiple intermingled cell types in this tissue (see section 3.1.1). One (direct) approach that generates meaningful data in this context is to measure transcriptome dynamics after treatment with specific hypophysiotropic factors such as gonadotropin-releasing hormone (GnRH), thereby generating relative, cell-specific response profiles. Recent work in the Klein laboratory used RNA-Seq analysis to document transcriptome dynamics following GnRH stimulation of rat pituitary glands (Kucka *et al.*, 2013). Other approaches have involved the use of pituitary cell type-specific cell lines that can be used to provide data that are either complementary or hypothesis generating with respect to whole animal physiology. One study has used this approach to reveal response-specific regulatory RNA (miRNA) dynamics following GnRH stimulation of LβT2 gonadotrope cells (Yuen *et al.*, 2009).

3.3   Transcriptome dynamics in the pineal gland: lessons from different approaches

The pineal gland is a central component of a neuroendocrine system that controls nocturnal secretion of the hormone melatonin. Pineal transcriptome dynamics have been studied extensively in rodent models (primarily rat [*Rattus norvegicus*] but also mouse [*Mus musculus*]), progressing from microarray platforms (see section 3.3.1) to RNA-Seq (see section 3.3.2). Studies have also extended to other species including zebrafish (*Danio rerio*; see section 3.3.3) and chicken (*Gallus gallus*; see section 3.3.4); these nonmammalian species offer additional comparative insights because in these species, the pineal is directly photoreceptive and has intrinsic circadian clock function. Research on the pineal transcriptome has involved extensive consideration of data handling and bioinformatics.

3.3.1 Microarray analysis in the rodent pineal gland

3.3.1.1 Experimental design

A comprehensive analysis of pineal gland transcriptome dynamics in the rat (cf. mouse; see section 3.3.1.3) was conducted by a multicenter group using Affymetrix GeneChip® microarrays (Bailey *et al.*, 2009). The main aim of the experiments was to investigate nocturnal (day versus night) changes in the transcriptome; this was conducted by direct sampling of pineal glands from adult rats. The experimental design included a number of features that both enhanced the efficiency of the study and optimized the range and accuracy of outcomes.

- Samples for RNA extraction were composed of pooled glands in order to minimize inter-animal variation. All animal studies were conducted in accordance with international standards of health and welfare in order to minimize this variable.
- Parallel microarray analysis of comparator tissues was used to provide information on tissue-specific gene expression patterns. In addition, comparative *in silico* analysis was conducted using equivalent microarray data from 23 different rat tissues and cells curated at the GNF database (migrated to BioGPS; http://biogps.org/). This web resource can provide valuable insights into relative tissue/cell specificity of expression – caution is advised when comparing different DNA probes that may differentially detect RNA variants.
- A 2-timepoint (multibiological replicate) design was used for microarray analysis followed by a 9-timepoint (multibiological replicate) validation of candidate gene temporal dynamics using QPCR analysis.
- Spatial (cell type and tissue specificity) validation was performed using radiochemical ISH.
- The *in vivo* microarray analysis was replicated with *ex vivo* organ culture experiments in which regulatory mechanisms could be directly investigated.

3.3.1.2 Experimental outcomes – predictions, surprises, and lessons

The pineal transcritome dynamics study generated a large data set (GEO series: GSE12344). Current interpretations of these data have generated the following outcomes that are of general relevance.

- The day versus night dynamic in the pineal gland transcriptome involves more than 2000 gene transcripts – therefore, a significant proportion of the transcriptome changes during this physiological rhythm.
- A subset of the rhythmic transcripts comprising 604 genes exhibits a >2-fold change in abundance; of these, 72% were upregulated at night and 28% were downregulated at night. The amplitude of these changes (mid-night versus mid-day) ranged between 100-fold to 1/20-fold.
- Tissue specificity of pineal gland gene expression was confirmed using microarray data sets from comparator tissues and a relative expression computation ('rEx' values; see Bailey *et al.*, 2009). Many high rEx values in particular transcripts were shared between the pineal gland and the retina, a related tissue.
- Validation by 9-timepoint QPCR essentially confirmed the temporal microarray results although the precise numerical amplitude of day versus night rhythms differed. The extended sampling design in the QPCR studies also revealed a diversity of transcript dynamics within the transcriptome (Figure 3.2; Bailey *et al.*, 2009). Some transcripts

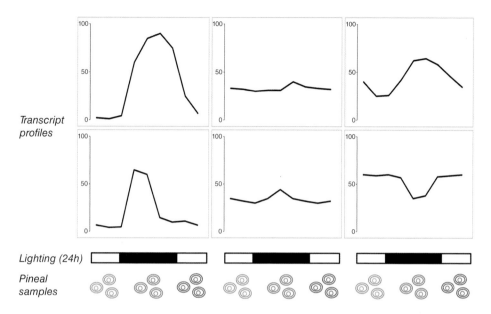

Transcript profiles

Lighting (24h)

Pineal samples

Figure 3.2 Transcript dynamics in the rat pineal gland. Examples of six different types of transcript profile from a 24-hour period of sampling (black bar represents the dark period). Profiles are derivatized from results in Bailey *et al.* (2009).

peaked relatively briefly following the onset of darkness, other transcripts exhibited an extended nocturnal peak of expression, and a large number of other, quite distinct profiles of expression were also revealed (see Figure 3.2). Presumably, these different temporal profiles accord with different functional requirements for individual transcripts.

- Validation by ISH confirmed both day versus night rhythms, and also localization (often specificity) of expression to the pineal gland. Surprisingly, some transcripts with high (pineal) rEx values were not localized to the majority (melatonin-producing) pinealocyte cell population. This result is important because it warns against the unguarded inclusion of genes into network/pathway analyses (see section 3.3.1.3) where cellular co-expression does not, in fact, exist.
- Replication of the microarray analysis in an organ culture model verified that the major pathway regulating rhythmic gene expression in the pineal is a norepinephrine-cAMP pathway. Interestingly, this experiment also revealed a subset of gene transcripts (5% of the 'day' gene set) that exhibited marked changes in expression in response to the organ culture environment (up to >30-fold higher or >30-fold lower). This result highlights the caution required in the interpretation of basal transcriptome data from *in vitro* studies.
- Many of the gene transcripts observed to be rhythmic/tissue specific had predictable associations with the key functional mechanisms in this tissue, namely adrenergic signaling or melatonin synthesis. However, a large number of rhythmic genes were not associated with these functions, indicating either ancillary roles in hormone production or, alternatively, novel biological roles for these components of the pineal gland transcriptome.
- Bioinformatic analysis of highly rhythmic gene sets demonstrated enrichment for predictable (e.g. CREB responsive) gene regulatory DNA elements. Hence, transcriptome profiling coupled with DNA data mining can reveal the genetic drivers of transcriptome dynamics. In the pineal study, however, there was a surprising absence of enrichment for some (e.g. CRX) DNA elements in tissue-specific gene sets that would be anticipated to

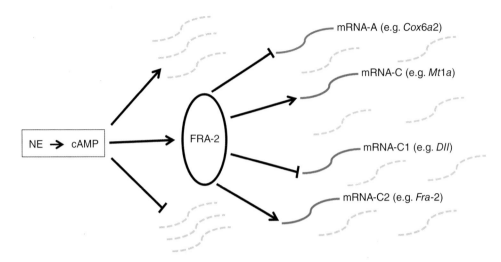

Figure 3.3 Regulation of the pineal gland transcriptome highlighting a single gene network linked to FRA-2. The schematic shows only part of the known regulatory pathways in the pineal and illustrates (i) the dominant influence of a NE->cAMP mechanism in controlling nocturnal transcription and (ii) one NE->cAMP-regulated transcription factor (FRA-2) that, in turn, modulates nocturnal dynamics of a subset of pineal transcripts (a few representative genes only are shown – see Davies *et al.*, 2011). Arrows represent activation; bars, repression.

underlie pineal/retina-specific expression. Bioinformatic analysis also provided unexpected evidence of enrichment for PAX4 elements in a 'control' subset of pineal transcripts that were neither rhythmic nor tissue specific. This result is of general interest because it suggests a potential 'hidden' level of regulation (possibly constitutive repression in this case) in some genes.

3.3.1.3   Identification of transcriptional networks

The overall transcriptome dynamic of a cellular system has multiple subcomponents that can be dissected out to reveal discrete networks (pathways) of genes that have common regulatory control. The control mechanism may or may not be transcriptional. A standard approach for establishing discrete transcriptional networks is to conduct transcriptome analysis in a TF knockout/knockdown model and thereby separate transcripts into regulated and non-regulated components. We have engineered a transgenic model that has enabled this approach for rat pineal gland transcriptomics (see Davies *et al.*, 2011). Important characteristics of our transgenic model that should be included in experimental design (where possible) include both cell type and temporal specificity of TF knockdown, thus ensuring that experimental data are not compromised by either developmental or cell-extraneous factors. Using our model combined with microarray analysis, we showed that the bZIP factor FRA-2 controls a subcomponent of the nocturnal transcriptome dynamic in the pineal (Figure 3.3). It is important to note that in addition to modulating the levels of dynamically regulated transcripts, FRA-2 also controls the levels of transcripts which are temporally stable in this system (e.g. *Cox6a2*; see Figure 3.3). Data from this type of experiment can be further analyzed using custom network/pathway tools (see section 3.1.2). Future use of this approach targeted at different genes will define additional subcomponent networks and build up a picture of the overall dynamic.

3.3.1.4   Microarray analysis of the mouse pineal – lessons from another model

The rat has been the experimental model of choice for mammalian pineal biology but the abundance of genetic mouse models has demanded transcriptome analysis in this alternative species. Experimental analysis of the mouse transcriptome, again using the Affymetrix GeneChip® platform, has provided important comparative data (Rovsing *et al.*, 2011) because the dynamics of gene expression was found to be distinct from the rat. Compared with the group of ~600 genes that exhibited a >2-fold day versus night rhythm in the rat, only 51 genes were similarly classified in the mouse. Furthermore, only 35 of the rhythmic mouse genes exhibited a >4-fold amplitude in nocturnal rhythm, compared with 130 genes achieving this level of amplitude in the rat.

These results are striking and indicative of fundamental biological differences in pineal regulation in these two species; in this context, it should be noted that differences in the transcriptional dynamics of melatonin synthesis regulation are well documented in other species such as the sheep (see discussion in Rovsing *et al.*, 2011). However, at the same time, it is important to consider that the mouse strain used in the later study (129/Sv) is an inbred strain that is distinct in this respect from the outbred Sprague Dawley rats used by Bailey *et al.* (2009; see section 3.3.1.1). The need for careful consideration of experimental model is highlighted in these studies; with respect to pineal physiology, it is known that mutations within melatonin-synthesizing enzyme genes occur variably across different mouse strains. Further studies in other mouse strains are therefore required to better define transcriptome dynamics in this species (see section 3.3.2).

3.3.2   RNA-Seq analysis in the rodent pineal gland

Recent work in the Klein laboratory has used RNA-Seq to expand upon the previous microarray analysis of the rodent pineal gland. This approach has both confirmed and extended the analysis of coding transcripts and has also made it possible to examine small RNAs and lncRNAs (>200 bp). Differential expression was investigated according to time of day, development, biochemical regulation, and tissue specificity.

3.3.2.1   Coding transcript analysis

Coding transcript analysis by RNA-Seq has confirmed that a multitude of transcripts are expressed on a daily basis, with groups of genes following patterns characterized by peaks in expression at different times during the day and night. Essentially, all of these changes in gene expression have been found to be under the control of neural input and, in most cases, these changes can be reproduced in organ culture using norepinephrine or a cyclic AMP agonist (Hartley *et al.*, 2015; Figure 3.4). With respect to development, the transcript profile that characterizes the adult rat pineal gland appears to be preprogrammed to appear in a wave of expression starting after birth independently of neural input. In addition, there are marked developmental changes reflecting a rapid decrease in expression of transcripts associated with cell division, in parallel to a decrease in mitosis. Coincident with these cell division-linked changes are marked decreases in the expression of some transcription factors, presumably because they play a critical role in the initial stages of determination of the pineal transcriptome and are not required thereafter. A third group of transcripts exhibits a transient increase in expression early in postnatal development, which may reflect critical developmental events or suppression of the development of cells or pathways not required for adult pineal function.

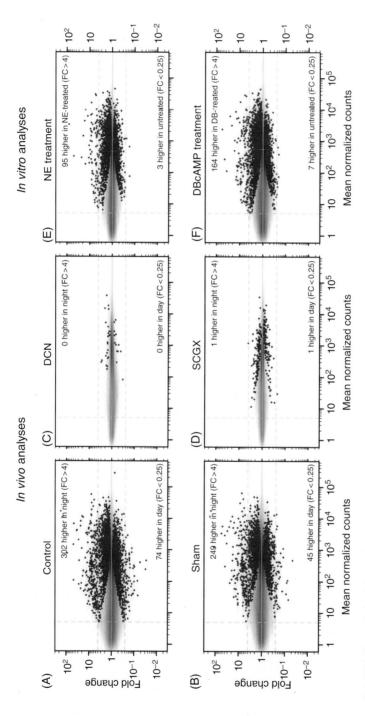

Figure 3.4 RNA-Seq analysis of the rat pineal transcriptome; neural and cAMP control. These plots display the mean normalized read-pair counts (x-axis) versus the estimated fold change (y-axis), on a log-log scale for four *in vivo* analyses and two *in vitro* analyses. The blue shading indicates the density of genes, and each red point represents a gene with statistically significant differential expression (adjusted-p < 0.001). Dashed horizontal lines mark 4-fold changes in both directions, dashed vertical line indicates minimum abundance threshold for the statistical tests. The four *in vivo* analyses compared night and day time points in adult rats for the following groups: (A) no surgery (Control); (B) neonatal sham surgery (Sham); (C) neonatal superior cervical ganglia decentralization (DCN); (D) neonatal superior cervical ganglionectomy (SCGX). The two *in vitro* analyses compared treated/untreated pineal glands: (E) norepinephrine-treated (NE) vs untreated and (F) dibutyryl-cyclic-AMP-treated (DBcAMP) vs untreated.

3.3.2.2   Non-coding transcript analysis

As noted above, a major advance provided by this technology is the detection of non- or partially annotated transcripts including lncRNAs. In the rat pineal gland, RNA-Seq revealed that over 100 lncRNAs exhibit daily changes in expression and these are relatively highly expressed in the pineal gland relative to other tissues (Coon *et al.*, 2012).

The methodology used to identify these dynamics involves computational screening combined with manual curation. The pineal lncRNAs exhibit a range of patterns of daily expression and, like coding transcripts, are, for the most part, under the control of norepinephrine acting through cyclic AMP; it has also been found that the abundance of lncRNAs changes during development. This advance has laid the foundation for future studies on the functional roles of lncRNAs in this tissue, which might include regulation of transcription including silencing and organization of complex macromolecular structures.

Micro RNAs (miRNAs) in the pineal gland have been studied by RNA-Seq using extraction methods designed specifically to recover and enrich these molecules together with software that identifies and maps the miRNAs (see Clokie *et al.*, 2012). The results of these studies have revealed that the overall profile of miRNAs is generally similar to that seen in the retina, in which a defined small group represents the majority of the miRNAs present. Amongst these are a triad including miR-182 (28% of total) and miR-183 and miR-96. In contrast to coding transcripts and lncRNAs, pineal miRNAs do not exhibit a marked night/day difference in abundance, although small changes (~2-fold) are seen in some cases. There is a global increase in the abundance of most pineal gland-enriched miRNAs during development, but some decreases. Most notably, expression of miR-483 decreases; it appears to regulate melatonin synthesis by acting on the 3'-UTR of arylalkylamine N-acetyltransferase (Aanat) mRNA, which controls the daily rhythm in melatonin production (Clokie *et al.*, 2012).

An interesting outcome of the analysis of miRNAs was the discovery of small RNAs in the pineal gland, termed pY RNA1-s1 and pY RNA1-s2 (processed Y RNA1-stem-1 and -2) that had not previously been reported (Yamazaki *et al.*, 2014). Subsequent work indicated that these were more highly expressed in the retina and that pY RNA1-s2 selectively binds the nuclear matrix protein Matrin 3 (Matr3) and to a lesser degree Hnrpull (heterogeneous nuclear ribonucleoprotein U-like protein), characteristics not shared with pY RNA1-s1. pY RNA1-s2 may act to control Matr3, which is implicated in transcriptional control and RNA editing. The retinal enrichment of pY RNA1-s2 may be a reflection of a specific role in phototransduction.

3.3.3   Transcriptome dynamics in the zebrafish pineal

In zebrafish, as in most non-mammalian vertebrates, the pineal gland is photoreceptive and contains an intrinsic circadian oscillator that drives melatonin rhythms (Falcon *et al.*, 2009). This has been known to occur even in culture, disconnected from any neuronal input (see Falcon *et al.*, 2009). In some non-mammalian vertebrates, the pineal gland is considered to be the master clock because removal of the pineal gland results in disruption of rhythmicity (see Falcon *et al.*, 2009).

Amongst non-mammalian vertebrates, the zebrafish represents a powerful model to explore the circadian timing system. It offers clear advantages for genetic analysis and easily accessible embryos, it was one of the first species for which a genome project was commenced and DNA microarray chips were commercially available. Furthermore, robust clock-controlled gene expression and melatonin production have been documented in the zebrafish embryonic pineal gland. Pineal gland markers (e.g. *aanat2*, opsins) are expressed by 22 hours post fertilization (hpf) and the circadian rhythm of melatonin production and of *aanat2* transcription begins at 2 days post fertilization (dpf), triggered by exposure to light-dark transitions (Gothilf *et al.*, 1999; Kazimi and Cahill, 1999; Vuilleumier *et al.*, 2006). With such an early development of a functional circadian clock, functional analyses of genes in intact developing zebrafish seemed conceivable.

The identification of genes that exhibit rhythmic expression in the zebrafish pineal gland has been undertaken using microarray and RNA-Seq. One hurdle was the size of the gland that enabled the isolation of minute amounts of RNA. This was handled by pooling together 12 pineal glands in each sample and by using a transgenic fish line that expresses the enhanced green fluorescent protein (EGFP) exclusively in the pineal gland. This enabled the isolation of pineal glands with the minimal amount of contaminating brain tissue. Pineal glands were collected from adult fish over two cycles at 4-hour intervals. To discern circadian clock-driven expression, fish were kept under constant darkness throughout the experiment. Initially, commercially available DNA microarray chips (Affymetrix), representing almost 15,000 genes, were utilized. In subsequent experiments, RNA-Seq was used as this methodology became available, providing information on the entire transcriptome (Tovin *et al.*, 2012).

Fast Fourier transform (FFT) analysis was used to search for periodic time-dependence followed by stringent statistics, leading to the identification of 303 rhythmic genes in the zebrafish pineal gland, which are ~3% of all the detectable genes in this tissue. This is a much lower number of rhythmic genes than previously reported in mice, in which ~10% of all detectable genes were found to be rhythmic in the suprachiasmatic nucleus (SCN) and liver (Lowrey and Takahashi, 2004). However, it is a very reliable list of genes with a very low percentage of false positives. The validity of this list is reflected by the presence of *aanat2* and the presence of 15 out of 16 known core clock genes and 12 out of 14 known accessory loop genes (Tovin *et al.*, 2012). In addition, a few genes were selected and their rhythmic expression patterns in the pineal gland were validated by real-time quantitative polymerase chain reaction (PCR) and whole mount *in situ* hybridization (Tovin *et al.*, 2012). In addition to known clock components, this list includes genes that encode for transcription factors, proteins of the phototransduction pathway and melatonin synthesis, and proteins involved in various cellular processes such as cell cycle regulation, ubiquitination, and RNA splicing. Thus, in addition to providing a comprehensive view of the expression pattern of known clock components within this master clock tissue, this endeavor revealed novel potential elements of the zebrafish circadian timing system.

One of the rhythmic genes in this list, *camk1gb*, was subjected to functional analysis using Morpholino-modified oligonucleotide (MO)-based knockdown, a commonly used method in zebrafish research which enables gene analysis during the first few days during and after embryogenesis. The effect of the knockdown on rhythmic gene expression and on rhythmic locomotor activity behavior in zebrafish larvae indicated that this gene is indeed important for the function of the circadian clock system (Tovin *et al.*, 2012). This serves as a proof of concept for future functional analyses of many other rhythmic genes that were identified in our transcriptome analysis. Moreover, as genomic technologies in zebrafish are advancing, the use of gene-specific endonuclease (TALEN and CRISPR)-mediated gene knockout will enable functional analysis of selected rhythmic genes in the adult animal.

The availability of a nearly complete genome together with a precise list of rhythmic genes enabled a computational search for *cis*-acting regulatory promoter elements that are important for the rhythmic expression of these genes in the pineal gland. Using a modified version of our *ab initio* bioinformatics search (Alon *et al.*, 2009), two elements were identified. One was a reidentification of the E-box, a sequence element known to drive rhythmic expression of genes. The second is a novel motif which is very similar to a CT-rich region in the mouse vasopressin promoter, previously suggested to be involved in driving rhythmicity (see Alon *et al.*, 2009), which obviously warrants further analysis. These (unpublished) results highlight another important outcome of such stringent genome-wide transcription analysis which could be utilized in other species and circumstances.

In zebrafish, most tissues and cells and even cell lines have circadian clocks that are entrainable by direct exposure to light via currently unknown photoreceptors. Consequently, cell-based assays were developed and are being used to study the mechanisms underlying light-induced gene expression (Vallone *et al.*, 2007), making the zebrafish instrumental in studying the effect of light on the molecular clock. Accordingly, we studied the light-induced pineal gland transcriptome using RNA-Seq followed by stringent analysis. An interesting finding in this study is that a 1-hour light pulse induces the expression of a large number of clock genes in the pineal gland, most of which function as negative factors in the molecular oscillator (Ben-Moshe *et al.*, 2014). Another intriguing finding is that cellular metabolic pathways are induced by a 1-hour light pulse. Particular interest is raised by the finding of light-induced hypoxia-inducible factor 1a (hif1a) which points to the possibility that the hypoxic pathway is involved in circadian clock entrainment. These findings obviously require further investigation.

The fact that all cells in the zebrafish are photosensitive and the availability of light-sensitive cell lines have promoted the identification of light-induced transcriptome in whole embryos and in cell lines (Weger *et al.*, 2011). It would now be interesting to compare the light-induced transcriptome of the central clock in the pineal gland to that of the peripheral clock cells.

3.3.4 Transcriptome dynamics in the chicken pineal

The chicken pineal gland is generally similar to that of the zebrafish, in that both tissues contain circadian clocks and are sensitive to light. Hence, they are of potential value in advancing our understanding of both the fundamental features of biological clocks and phototransduction mechanisms. The use of the chicken as a source of tissue for these purposes has two advantages. One is that the tissue is many fold larger than that of the zebrafish and rodents, making downstream biochemical studies easier. In addition, the retina of the chicken is easily removed and studied. Like the pineal gland, the chicken retina also contains a biological clock, and comparison of transcriptional dynamics of both tissues can identify genes that oscillate in both tissues and therefore might reflect conserved mechanisms of special functional importance.

The transcriptomes of the chicken pineal and retina were first profiled using a custom cDNA microarray (Bailey *et al.*, 2003; Karaganis *et al.*, 2008). These efforts identified transcriptional similarities between the two tissues in addition to identifying differences between the other tissues, broadly classified as involved in phototransduction, transcription/translation, cell signaling, and stress response.

More recently, RNA-Seq has been applied to circadian changes in the transcriptome of the chicken pineal gland and retina (Klein *et al.*, unpublished). This has significantly increased the number of rhythmically expressed genes seen in both tissues. In addition, it has been possible to group the rhythmically expressed transcripts into clusters that share a similar pattern of expression, with groups of genes peaking at discrete times of the day and night (see section 3.3.1.2). Functional classification of the rhythmically expressed genes has revealed groups that are associated with circadian function, phototransduction, and melatonin production. This work has also been expanded to include results from RNA-Seq analysis of other tissues, in an attempt to identify genes that are selectively expressed in the retina and pineal gland and are therefore potential cellular markers.

3.4   SN-NICHD transcriptome profiling web page

Much of the RNA-Seq data referred to here is available online: https://science.nichd.nih.gov/confluence/display/sne/Home. The data are in several formats, including graphical representation. These are linked to the UCSC genome browser, providing an opportunity to examine the read coverage plots.

3.5   Perspectives

The documentation of transcriptome dynamics in neuroendocrine systems has finally given us a high-definition picture of the molecular activity that underlies these systems. This achievement and the subsequent sharing of transcriptome data sets across laboratories will advance neuroendocrine research, firstly by providing a common context for gene-specific analyses, and secondly by providing insights into transcriptome-level function and possibly disease-related dysfunction. Current limitations in our understanding of transcriptome data sets at organizational and functional levels will be addressed by future developments of new bioinformatic approaches coupled with the new generation of gene targeting tools. For example, the CRISPR/Cas9 system can be used to edit specific genomic sites that are predicted to form regulatory links in the progression of transcriptional networks. In this way, comprehensive functional hierarchies of gene regulatory networks will be established. Clearly, however, these predicted advances depend upon our ability to engineer discrete genetic changes in a cell/tissue-specific manner and this will require further refinement of existing genome editing approaches.

References

Alon, S., Eisenberg, E., Jacob-Hirsch, J., *et al.* (2009) A new cis-acting regulatory element driving gene expression in the zebrafish pineal gland. *Bioinformatics*, **25**, 559–562.

Bailey, M.J., Beremand, P.D., Hammer, R., Bell-Pedersen, D., Thomas, T.L. and Cassone, V.M. (2003) Transcriptional profiling of the chick pineal gland, a photoreceptive circadian oscillator and pacemaker. *Molecular Endocrinology*, **17**, 2084–2095.

Bailey, M.J., Coon, S.L., Carter, D.A., *et al.* (2009) Night/day changes in pineal expression of >600 genes: central role of adrenergic/cAMP signaling. *Journal of Biological Chemistry*, **284**, 7606–7622. [*Comprehensive microarray analysis of mRNAs in a neuroendocrine transcriptome that highlights the variation in temporal dynamics of individual mRNAs.*]

Ben-Moshe, Z., Alon, S., Mracek, P., *et al.* (2014) The light-induced transcriptome of the zebrafish pineal gland reveals complex regulation of the circadian clockwork by light. *Nucleic Acids Research*, **42**, 3750–3767.

Carter, D.A. (2006) Cellular transcriptomics: the next phase of endocrine expression profiling. *Trends in Endocrinology and Metabolism*, **17**, 192–198.

Clokie, S.J., Lau, P., Kim, H.H., Coon, S.L. and Klein, D.C. (2012) MicroRNAs in the pineal gland: miR-483 regulates melatonin synthesis by targeting arylalkylamine N-acetyltransferase. *Journal of Biological Chemistry*, **287**, 25312–25324. [*Dynamics of miRNA expression revealed by miRNA-Seq analysis.*]

Coon, S.L., Munson, P.J., Cherukuri, P.F., *et al.* (2012) Circadian changes in long noncoding RNAs in the pineal gland. *Proceedings of the National Academy of Sciences of the United States of America*, **109**, 13319–13324. [*Dynamics of lncRNA expression revealed by RNA-Seq analysis.*]

Davies, J.S., Klein, D.C. and Carter, D.A. (2011) Selective genomic targeting by FRA-2/FOSL2 transcription factor: regulation of the Rgs4 gene is mediated by a variant activator protein 1 (AP-1) promoter sequence/CREB-binding protein (CBP) mechanism. *Journal of Biological Chemistry*, **286**, 15227–15239. [*Details modulation of transcriptome dynamics by a single transcription factor.*]

Eberwine, J. and Bartfai, T. (2011) Single cell transcriptomics of hypothalamic warm sensitive neurons that control core body temperature and fever response signaling asymmetry and an extension of chemical neuroanatomy. *Pharmacology & Therapeutics*, **129**, 241–259.

Falcon, J., Besseau, L., Fuentes, M., Sauzet, S., Magnanou, E. and Boeuf, G. (2009) Structural and functional evolution of the pineal melatonin system in vertebrates. *Annals of the New York Academy of Sciences*, **1163**, 101–111.

FANTOM Consortium and the RIKEN PMI and CLST (DGT) (2014) A promoter-level mammalian expression atlas. *Nature*, **507**, 462–470.

Gothilf, Y., Coon, S.L., Toyama, R., Chitnis, A., Namboodiri, M.A. and Klein, D.C. (1999) Zebrafish serotonin N-acetyltransferase-2: marker for development of pineal photoreceptors and circadian clock function. *Endocrinology*, **140**, 4895–4903.

Hager, G.L., McNally, J.G. and Misteli, T. (2009) Transcription dynamics. *Molecular Cell*, **35**, 741–753.

Harper, C.V., Featherstone, K., Semprini, S., *et al.* (2010) Dynamic organisation of prolactin gene expression in living pituitary tissue. *Journal of Cell Science*, **123**, 424–430.

Hartley, S.W., Coon, S.W., Savastano, L.E., Mullikin, J.C., NISC Comparative Sequencing Program, Fu, C. and Klein, D.C. (2015) Neurotranscriptomics: The effects of neonatal stimulus deprivation on the rat pineal transcriptome. *Plos One*, **10**, e0137548.

Hirayama, T., Tarusawa, E., Yoshimura, Y., Galjart, N. and Yagi, T. (2012) CTCF is required for neural development and stochastic expression of clustered Pcdh genes in neurons. *Cell Reports*, **2**, 345–357.

Jiang, L., Schlesinger, F., Davis, C.A., *et al.* (2011) Synthetic spike-in standards for RNA-seq experiments. *Genome Research*, **21**, 1543–1551.

Karaganis, S.P., Kumar, V., Beremand, P.D., Bailey, M.J., Thomas, T.L. and Cassone, V.M. (2008) Circadian genomics of the chick pineal gland in vitro. *BMC Genomics*, **9**, 206.

Kazimi, N. and Cahill, G.M. (1999) Development of a circadian melatonin rhythm in embryonic zebrafish. *Brain Research. Developmental Brain Research*, **117**, 47–52.

Khaladkar, M., Buckley, P.T., Lee, M.T., *et al.* (2013) Subcellular RNA sequencing reveals broad presence of cytoplasmic intron-sequence retaining transcripts in mouse and rat neurons. *PLoS One*, **8**, e76194.

Knight, Z.A., Tan, K., Birsoy, K., *et al.* (2012) Molecular profiling of activated neurons by phosphorylated ribosome capture. *Cell*, **151**, 1126–1137.

Kucka, M., Bjelobaba, I., Clokie, S.J., Klein, D.C. and Stojilkovic, S.S. (2013) Female-specific induction of rat pituitary dentin matrix protein-1 by GnRH. *Molecular Endocrinology*, **27**, 1840–1855.

Lowrey, P.L. and Takahashi, J.S. (2004) Mammalian circadian biology: elucidating genome-wide levels of temporal organization. *Annual Review of Genomics and Human Genetics*, **5**, 407–411.

Menet, J.S., Rodriguez, J., Abruzzi, K.C. and Rosbash, M. (2012) Nascent-Seq reveals novel features of mouse circadian transcriptional regulation. *ELife*, **1**, e00011.

Merkin, J., Russell, C., Chen, P. and Burge, C.B. (2012) Evolutionary dynamics of gene and isoform regulation in mammalian tissues. *Science*, **338**, 1593–1599.

Nagano, A.J., Sato, Y., Mihara, M., *et al.* (2012) Deciphering and prediction of transcriptome dynamics under fluctuating field conditions. *Cell*, **151**, 1358–1369.

Oh, S., Song, S., Dasgupta, N. and Grabowski, G. (2014) The analytical landscape of static and temporal dynamics in transcriptome data. *Frontiers in Genetics*, **5**, 35. **[*A recent review of transcriptome analysis methods that focuses on transcriptome dynamics.***]

Rovsing, L., Clokie, S., Bustos, D.M., *et al.* (2011) Crx broadly modulates the pineal transcriptome. *Journal of Neurochemistry*, **119**, 262–274.

Schwanhäusser, B., Busse, D., Li, N., *et al.* (2011) Global quantification of mammalian gene expression control. *Nature*, **473**, 337–342.

Shimogori, T., Lee, D.A., Miranda-Angulo, A., *et al.* (2010) A genomic atlas of mouse hypothalamic development. *Nature Neuroscience*, **13**, 767–775.

Steijger, T., Abril, J.F., Engström, P.G., *et al.* (2013) Assessment of transcript reconstruction methods for RNA-seq. *Nature Methods*, **10**, 1177–1184.

Tovin, A., Alon, S., Ben-Moshe, Z., *et al.* (2012) Systematic identification of rhythmic genes reveals camk1gb as a new element in the circadian clockwork. *PLoS Genetics*, **8**, e1003116.

Vallone, D., Santoriello, C., Gondi, S.B. and Foulkes, N.S. (2007) Basic protocols for zebrafish cell lines: maintenance and transfection. *Methods in Molecular Biology*, **362**, 429–441.

Vuilleumier, R., Besseau, L., Boeuf, G., *et al.* (2006). Starting the zebrafish pineal circadian clock with a single photic transition. *Endocrinology*, **147**, 2273–2279.

Weger., B.D., Sahinbas, M., Otto, G.W., *et al.* (2011) The light responsive transcriptome of the zebrafish: function and regulation. *PLoS One*, **6**, e17080.

Wu, A.R., Neff, N.F., Kalisky, T., *et al.* (2014) Quantitative assessment of single-cell RNA-sequencing methods. *Nature Methods*, **111**, 41–46. [*Considerations for single-cell transcriptome analysis.*]

Yamazaki, F., Kim, H.H., Lau, P., *et al.* (2014) pY RNA1-s2: a highly retina-enriched small RNA that selectively binds to Matrin 3 (Matr3). *PLoS One*, **9**, e88217.

Yuen, T., Ruf, F., Chu, T. and Sealfon, S.C. (2009) Microtranscriptome regulation by gonadotropin-releasing hormone. *Molecular and Cellular Endocrinology*, **302**, 12–17.

CHAPTER 4

New Players in the Neuroendocrine System: A Journey Through the Non-coding RNA World

Yongping Wang, Edward A. Mead, Austin P. Thekkumthala, and Andrzej Z. Pietrzykowski

Department of Animal Sciences, Department of Genetics, Rutgers University, New Brunswick, New Jersey, USA

Neuroendocrine integration is one of the tenets of organismal **homeostasis**. A pivotal role in this integration is played by three neuroendocrine organs located in the skull (Figure 4.1): the hypothalamus, which provides a fundamental link between the central nervous system (CNS) and the pituitary; the pituitary, a key organ controlling function of peripheral endocrine organs related to many processes including metabolism, growth, water balance, reproduction, and stress response; and the pineal gland, which connects external signals to the hypothalamus, allowing for adaptation of an organism to daily environmental changes as well as to long-term cyclical seasonality of the ecosystem driven by 'cosmic' (planetary system) events. Each of these organs contains highly specialized neuroendocrine cells, which are either central neurons secreting hormones (the hypothalamic neurons, the neuronal posterior pituitary terminals) or central endocrine cells (various anterior pituitary cells, the pinealocytes of the pineal gland).

Regulation of gene expression is one of the most fundamental biological processes controlling cell identity and function, and the neuroendocrine cells are no exception to this. In recent years, due to simultaneous advancement of several key areas of a bioscientific field (Pietrzykowski, 2010), we have witnessed an unprecedented flood of information. This contributed to the discovery of several species of non-protein-coding RNA molecules (Table 4.1) and revolutionized our view of gene expression mechanisms. Non-coding RNAs seem to be integrally interwoven in the gene expression regulatory fabric (Figure 4.2), either directly through interactions with protein-coding transcripts or indirectly, by determining protein translation efficiency.

In this chapter, we will describe our current understanding of the contribution of various non-coding RNAs to the functioning of these three main neuroendocrine organs. We will also point to specific features of regulation of neuroendocrine cells by non-coding RNAs, which deserve particular scientific attention.

4.1 Non-coding RNA contribution to gene regulation

The sequencing of the human (Venter *et al.*, 2001) and other species genomes, followed by the **ENCODE Project** that focused on the identification of all functional elements of a genome (ENCODE Project Consortium, 2012), revealed that most of the transcribed genome is non-coding. Importantly, non-coding RNA is a very diverse group (see Table 4.1).

Molecular Neuroendocrinology: From Genome to Physiology, First Edition. Edited by David Murphy and Harold Gainer.
© 2016 John Wiley & Sons, Ltd. Published 2016 by John Wiley & Sons, Ltd.
Companion website: www.wiley.com/go/murphy/neuroendocrinology

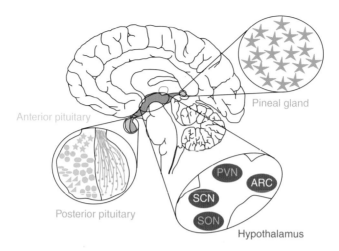

Figure 4.1 A schematic of the neuroendocrine structures discussed in this chapter. The pituitary comprises two parts: the anterior pituitary consisting of distinct groups of endocrine cells (depicted by symbols of different shapes), each releasing a different hormone; the posterior pituitary containing axons and their terminals originating in neurons located in the supraoptic nucleus (SON) and the paraventricular nucleus (PVN) of the hypothalamus. The hypothalamus contains additionally two other nuclei (clusters of neurons) discussed here: the suprachiasmatic nucleus (SCN) and the arcuate nucleus (ARC). The pineal gland contains mainly one type of cells – pinealocytes. Note that the pituitary, the hypothalamus, and the pineal gland are interconnected.

Among short (less than 200 nucleotides in length) non-coding RNAs, microRNAs are the most recognized and studied. These microRNAs serve mainly as suppressors of gene products, interfering with protein synthesis by binding, through complementarity to **3′ untranslated regions** (3′UTRs), to messenger RNAs (mRNAs). Upon microRNA binding, mRNA undergoes fast decay and/or protein synthesis is halted (Alonso, 2012). Each microRNA can regulate several targets either simultaneously or at different stages of development or aging. On the flip side, every mRNA can be targeted by different microRNAs. Rules of microRNA:mRNA interactions are the subject of intensive experimental and computational research. New microRNAs are still being discovered and validated (Martin *et al.*, 2014), as well as exceptions to the microRNA:mRNA interaction rules described (Thomson *et al.*, 2011). As mighty as microRNAs seem to be, they are just a small tip of the iceberg. Short, non-coding RNAs are a very divergent group (see Table 4.1) and many of its members still need functional recognition.

Research on **long non-coding RNA** (lncRNA, more than 200 nucleotides in length) has a much shorter history (Figure 4.3). However, this area is picking up momentum and recent years have witnessed an explosion of research geared toward understanding the role of many of the hundreds of lncRNA transcripts (Kung *et al.*, 2013).

However, that is not all. There are new, non-coding players on the block. A few years ago, microRNA sponges were designed, with a circular shape and several microRNA binding sites, as a tool to tame microRNA expression (Ebert *et al.*, 2007). It became evident shortly thereafter that circular RNAs actually exist in nature, and work in a strikingly similar way to the sponges (Hentze and Preiss, 2013; Memczak *et al.*, 2013). The discovery of different types of non-coding RNA in different parts of the neuroendocrine system adds to the complexity of homeostatic and adaptation processes.

Table 4.1 The classification of non-coding RNAs and their roles in neuroendocrine diseases.

Major class	Classes	Symbol	Characteristics	Biological functions	Role in neuroendocrine diseases
Small non-coding RNA	microRNA	miRNA	19–25 nucleotides (nt); represent 4% of the genes in the human genome; multi-step biogenesis involves Drosha and Dicer; control over 60% of protein-coding genes; typically operate in cytoplasm by binding to mRNA 3′UTR; interacts with RISC complex; degraded by 5′-to-3′ exoribonucleases 1 and 2 (XRN1 and XRN2)	Regulatory role in almost all cellular processes; typically suppresses translation; one microRNA can simultaneously target several transcripts	Several miRNAs specified here are important in neuroendocrine functions and disease
	Piwi-interacting RNA	piRNA	26–30 nt; binds Piwi proteins; biogenesis is Dicer independent; up to recently thought to be restricted to the germline; mainly operates in nucleus	Diverse functions by repressing target gene expression, involved in germ cell development, self-renewal and silencing of **retrotransposons**	Awaits discovery
	Small nucleolar RNA	snoRNA	60–300 nt; enriched in the nucleolus; in vertebrates is excised from pre-mRNA introns; always associated with a set of specific proteins – small nuclear ribonucleoproteins (snRNP)	Important for processing of pre-mRNA in the nucleus, maturation of other non-coding RNAs	Associated with two human neuroendocrine syndromes: Prader–Willi and Angelman
	Promoter-associated small RNA	PASR	20–200 nt; overlaps transcriptional start sites of protein- and non-coding genes; made by transcription of short capped transcripts	Modulates transcription of primary RNA transcripts	Awaits discovery
	Transcription initiation RNA	tiRNA	~18 nt; originates from regions just downstream of transcriptional start sites; associates with RNA Pol II binding; preferentially located in GC-rich promoters	Possibly involved in initiation of transcription and splice sites	Awaits discovery
	Centromere repeat associated small RNA	crasiRNA	34–42 nt; processed from long double-stranded RNAs (dsRNAs).	Putative role in heterochromatin formation	Awaits discovery
	Telomere-specific small RNA	Tel-sRNA	~24 nt; Dicer independent; 2′-O-methylated at the 3′ terminus; evolutionarily conserved	May play important roles in telomere structure and/or functions	Awaits discovery

(Continued)

Table 4.1 (Continued)

Major class	Classes	Symbol	Characteristics	Biological functions	Role in neuroendocrine diseases
Long non-coding RNA	Long intergenic non-coding RNA	lincRNA	Ranges from several hundreds to tens of thousands of nucleotides; lies within the genomic intervals between two genes	Involved in many biological processes occurring in the nucleus and cytoplasm; transcriptional *cis*-regulation of neighboring genes; putative interactions with microRNAs	Plays a role in some pituitary cancers
	Long intronic non-coding RNA	N/A	Over 200 nt; lies within the introns; evolutionarily conserved; tissue and subcellular expression specified	Possible link with posttranscriptional gene silencing, as precursor of shorter RNAs	Awaits discovery
	Telomere-associated ncRNA	TERRA	100 bp ->9 kb; conserved among eukaryotes; synthesized from C-rich strand; human telomeric RNA and DNA oligonucleotides form a DNA-RNA G-quadruplex	Negative regulation of telomere length and activity through inhibition of telomerase	Roles in telomere-associated diseases
	Long non-coding RNA with dual function	N/A	Over 200 bp; possesses both protein-coding capacity and non-coding regulatory function	Modulates gene expression	Awaits discovery
	Pseudogene RNA	N/A	Transcripts made from copies of coding genes through retrotransposon relocations; lost the ability to code for proteins; potential to regulate their protein coding counterparts	Regulation of tumor suppressors and oncogenes by acting as microRNA decoy	Awaits discovery
	Transcribed-ultraconserved region RNA	T-UCR	Over 200 bp; very conserved between orthologous regions of human, rat, and mouse genome; located in both intra- and intergenic regions	Antisense inhibitor for protein-coding genes or ncRNAs	Awaits discovery
Circular RNA	Circular RNA	circRNA	Forms a covalently closed continuous loop; evolutionarily conserved; abundant in humans; particularly expressed in the CNS; resembles microRNA sponge	Functions as posttranscriptional regulator of gene expression, possibly as 'microRNA sponge'; competes with other forms of RNA as decoy for microRNAs; implicated in atherosclerosis and Parkinson disease processes and aging	Awaits discovery
	Circular intronic long non-coding RNA	ciRNA	Prominently found in the nucleus, contains few (if any) miRNA binding sites, accumulates in human cells	Regulates the expression of the parent genes, function largely unknown	Awaits discovery

Non-coding RNAs are functional RNA molecules transcribed from DNA but not translated into proteins. A non-coding RNA belongs to one of the three major classes according to the transcript size or shape: small RNAs, long non-coding RNAs, and circular RNAs. The classifications of each group, their characteristics, biological functions, and role in neuroendocrine diseases are listed based on several papers: Cao et al., 2009; Carone et al., 2009; Cusanelli and Chartrand, 2015; Eddy, 2001; Fatica and Bozzoni, 2014; Kiss, 2001; Louro et al., 2009; Maicher et al., 2014; Nicholls and Knepper, 2001; Qureshi and Mehler, 2012; Rother and Meister, 2011; Sana et al., 2012; Taft et al., 2010; van Wolfswinkel and Ketting, 2010; Xu et al., 2012; Yan et al., 2011; Zhang et al., 2013.

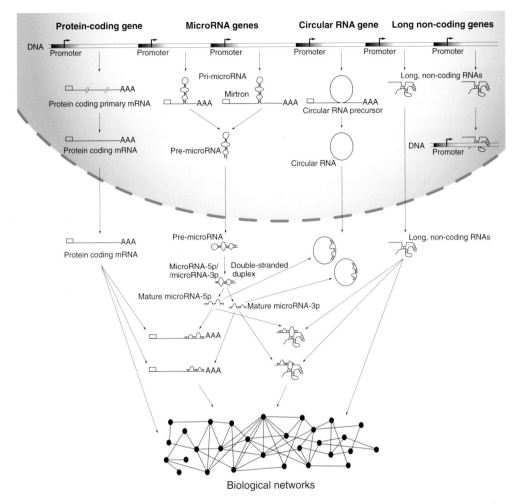

Figure 4.2 A schematic of biogenesis pathways and interactivity of various types of RNA transcripts. A protein-coding gene and three main types of the non-coding genes are shown. A microRNA primary precursor (pri-microRNA) can be produced directly from a microRNA gene or indirectly from a larger, host gene (e.g. a mirtron – a microRNA derived from an intron of a host gene). In the cytoplasm, two mature microRNAs can be produced from the complementary arms of a single pre-microRNA precursor. Mature microRNAs can interact with mRNAs, long non-coding RNAs, and circular RNAs. These reciprocal interactions contribute to the function of biological networks.

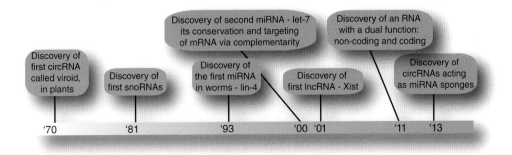

Figure 4.3 A timeline of landmark discoveries in the non-coding RNA field. Key dates pertaining to the discovery of the main types of non-coding RNA: circular RNAs (circRNAs), small nucleolar RNAs (snoRNAs), microRNAs (miRNAs), and long non-coding RNAs (lncRNAs).

4.2 Central role of the hypothalamus as a neuroendocrine organ

The hypothalamus is a key brain region coordinating centrally diverse endocrine and nervous system homeostatic functions such as body weight and metabolism (see Chapter 16), osmotic and water balance (see Chapter 15), parenting and bonding (see Chapters 18 and 19), circadian clock and sleep, and the response to stress (see Chapter 17) (Meister *et al.*, 2013). These functions are anatomically segregated amongst several discrete neuronal clusters called nuclei (see Figure 4.1), although functions of nuclei may partially overlap. The arcuate nucleus (ARC) is primarily involved in the control of appetite and feeding, to which also the paraventricular nucleus (PVN) contributes. The PVN is also involved in general energy metabolism, procreation, and water balance. The supraoptic nucleus (SON) is a primary control center of water and osmotic balance, but it also contributes to reproduction, bonding, and social identity. The suprachiasmatic nucleus (SCN) serves as the central clock of the body. Hypothalamic nuclei are anatomically and functionally linked to the pituitary, which sometimes is described as the hypothalamopituitary unit. The hypothalamus (particularly the SCN) is also directly connected to the pineal gland. We will discuss the role of non-coding RNA in the pituitary and the pineal gland later in this chapter.

Here, we focus on the hypothalamic ARC, SCN, PVN, and SON, because of recent, very interesting findings indicating a critical role of non-coding RNAs in the functions of these hypothalamic centers.

4.2.1 miRNA expression in the hypothalamus

MicroRNA profiling reveals that a large number of miRNAs are highly expressed and enriched in the developing (Smirnova *et al.*, 2005) and adult brain (He *et al.*, 2012; Krichevsky *et al.*, 2003), and that this expression is regionally diversified (Krichevsky *et al.*, 2003). Recently, a **next-generation sequencing** (NGS) technique called CLIP-Seq revealed that, in human cortex, out of hundreds of mature microRNAs expressed in the brain, five types of microRNAs are responsible for over half of all microRNA interactions with the **transcriptome** (Boudreau *et al.*, 2014): let-7/98 family, miR-9-5p, -29-3p, -124-3p, and -125-5p. Interestingly, as shown by animal studies, some of these microRNAs are also highly expressed in the hypothalamus and its specific neuronal nuclei. In mouse hypothalamic tissue, let-7c, miR-7a, -7b, -9, -124a, -125a, -136, -138, -212, -338, and -451 are the most highly expressed (Meister *et al.*, 2013). MicroRNA expression profiling of the rat ARC and PVN revealed that three microRNAs are particularly highly expressed in these nuclei: let-7c, miR-7a, and miR-9 (Amar *et al.*, 2012). Notably, in the ARC, the orexigenic neurons (known as AgRP/ NPY neurons due to their expression of two neuropeptides: agouti-related protein [AgRP] and neuropeptide Y) rather than the anorexigenic neurons (expressing pro-opiomelanocortin [POMC] and the cocaine and amphetamine-related transcript [CART]) seem to be particularly miR-7a enriched (Herzer *et al.*, 2012).

Cell-specific resolution of microRNA expression has not yet been achieved for other hypothalamic nuclei, but high expression of miR-7a was detected in the SCN and SON (Herzer *et al.*, 2012), and miR-9 in the SON (Pietrzykowski *et al.*, 2008).

4.2.2 MicroRNA role in the hypothalamic control of body weight and metabolism

The hypothalamus plays a central role in the regulation of whole-body energy homeostasis and appetite, by integration of nutrient-related signals and coordination of the neuroendocrinological response (Fliers *et al.*, 2006). Particularly important for this function is a balanced, reciprocal interaction of the orexigenic and anorexigenic neurons, located in the arcuate nucleus.

MicroRNAs seem to actively control the function of these neurons. **Dicer**, a key endoribonuclease governing one of the last steps of maturation of almost all microRNAs, is expressed in the medial hypothalamus, which contains the ARC (Schneeberger *et al.*, 2012). Co-localization studies showed that Dicer is present in both the orexigenic and the anorexigenic neurons, thus indirectly suggesting that microRNA plays a role in central energy homeostasis (Schneeberger *et al.*, 2012). Indeed, two of the main opposite factors affecting energy homeostasis, calorie restriction (fasting) and calorie excess (high-fat diet), caused opposite changes in Dicer expression, specifically in the hypothalamus. Since these experiments have not provided cell-specific results, Schneeberger and colleagues created a conditional Dicer knockout in anorexigenic, POMC-positive neurons in mice. These animals were prone to **hyperphagia** and adult-onset **metabolic syndrome**, and they displayed degeneration of POMC neurons and increased levels of circulating leptin. Although particular microRNAs were not specified in this study, it provided convincing, if indirect, evidence that Dicer-dependent microRNAs are essential for maintenance and function of ARC neurons involved in energy homeostasis. A study by Sangiao-Alvarellos and colleagues went one step further, and observed that perturbation in caloric intake leads to specific changes in a subset of microRNAs in the hypothalamus: let-7a, miR-9-3p, -30e, -132, -145, -200, and -218 (Sangiao-Alvarellos *et al.*, 2014).

Amongst these microRNAs, miR-200 could play a particularly important role in the control of metabolism homeostasis, both centrally and peripherally. It is one of the most abundant microRNAs expressed in peripheral tissues of metabolic importance (liver, pancreatic islets), and is associated with diabetes mellitus and obesity (Choi and Friso, 2010). Moreover, expression of miR-200 in adipocytes seems to be regulated by leptin, a hormone released by adipocytes proportionally to the fat content in these cells, and targeting hypothalamic ARC neurons: leptin stimulates anorexigenic, whilst inhibiting orexigenic neurons (Schwartz *et al.*, 2000). Interference with leptin signaling pathways changed expression of several microRNAs in these neurons: 34 microRNAs were upregulated, whilst four were downregulated (Benoit *et al.*, 2013). Among these, miR-200 showed the most persistent changes. Importantly, some of the miR-200 targets included elements of the leptin signaling pathway (Sangiao-Alvarellos *et al.*, 2014). Thus, miR-200 might be an attractive target of the metabolic syndrome, providing a multi-pronged therapeutic approach simultaneously aimed at central and peripheral tissues. A recent review details the role of microRNAs and other epigenetic mechanisms in metabolically critical peripheral organs (Pietrzykowski, 2014).

4.2.3 MicroRNA role in regulation of osmotic homeostasis and reproduction via vasopressin and oxytocin

The SON and PVN of the hypothalamus play important roles in the regulation of osmotic homeostasis, water and mineral balance, and some reproductive behaviors. For these functions, the SON and PVN are tightly linked to the posterior pituitary, creating one **hypothalamo-neurohypophysial system** (HNS), containing large magnocellular neurons (MCNs) producing arginine vasopressin (AVP) and oxytocin (OXT). Both hormones have similar peptide structures, which may cause some functional overlap (Gainer and Wray, 1992). Nevertheless, AVP is the main osmoregulatory hormone, whilst the primary function of OXT is to control various reproductive behaviors, including courtship, parturition, lactation, and maternal bonding (Hyodo and Urano, 1991). Cell bodies of HNS magnocellular neurons are located in the SON and PVN, and they send axonal projections to the posterior pituitary – a site of AVP and OXT release to the blood from the axonal terminals (Brownstein *et al.*, 1980).

The role of microRNAs in the regulation of osmotic homeostasis was assessed by chronic exposure (10 days) of male C57BL/6J mice to hyperosmotic drinking solution (2% w/v saline) (Choi *et al.*, 2013) and subsequent microRNA profiling of combined SON and PVN nuclei. Hypertonic saline treatment caused significant overexpression of miR-7b, -9, -29c, -137, and -451 in these nuclei. The authors focused on the potential suppression by elevated levels of miR-7b of *c-fos*, a marker of neuronal activity (Sagar *et al.*, 1988). Interestingly, as mentioned above, the miR-7 microRNA family is one of the most abundantly expressed in the hypothalamus. Changes in amount of the transcript encoding the main osmotic hormone released by the SON and PVN, AVP, were also observed (see Chapter 15). Maybe miR-7 could potentially regulate AVP expression, based on the observation of Tessmar-Raible and colleagues that miR-7 controls expression of vasotocin, a non-mammalian homolog of AVP found in annelids, gastropods, and cephalopods (Tessmar-Raible *et al.*, 2007). This study also indicates the importance of miR-7 in the creation and evolution of the neurosecretory brain centers. Further studies determining regulation of AVP transcripts by microRNA will be of great interest.

Although the AVP gene is located in the genome relatively close to its 'sister' gene encoding OXT (Gainer and Wray, 1992), separate populations of magnocellular neurons in the SON express each gene and release each hormone (Gainer, 1998, 2012). This suggests a tight, cell-specific control of divergent expression of these two hormones. Since miR-7 might be involved in regulation of AVP transcripts, can a different microRNA target OXT, adding to the specificity of the SON magnocellular neurons? Work by Choi and colleagues shows that, indeed, this may be the case. Using miRNA expression profiling, and aided by a 3′UTR luciferase assay, they determined that, in hypothalamus of C57B1/6J mice, the 3′UTR region of the OXT transcript contains a binding site for miR-24, whilst an miR-7 site seems to be absent. Interestingly, miR-24 is present in the hypothalamus and can directly bind to and trigger decay of the OXT transcripts and subsequently decrease OXT peptide production (Choi *et al.*, 2013).

Notably, hyperosmotic challenge also increases expression of miR-9 in these nuclei (Lee *et al.*, 2006). Although miR-9 seems not to target AVP or OXT transcripts, thus presumably having no direct effect on the amount of these hormones, it could affect release of AVP or OXT from neuronal terminals by targeting the hormonal release machinery. In the SON, miR-9 targets subunits of specific potassium and calcium ion channels (Pietrzykowski *et al.*, 2008), which are critical elements of the neurohormonal release machinery (Dopico *et al.*, 1996). Alcohol affects AVP or OXT release (Wang *et al.*, 1991) and elements of the release machinery (Pietrzykowski *et al.*, 2004). This effect takes place at least partially through the miR-9-dependent mechanism (Pietrzykowski *et al.*, 2008). It is tempting to speculate that miR-9 is also engaged in the regulation of AVP or OXT release by osmotic stress.

4.2.4   MicroRNA winding of the circadian clock and the control of sleep

Neurons located in the hypothalamic SCN display precisely synchronized rhythmic activity, providing a pacemaker for whole-body physiological processes (Welsh *et al.*, 2010). It is thought that the SCN clock center evolved as an adaptation to the cyclic, circadian changes in the light-dark cycle due to the rotation of the Earth around its axis (Hansen *et al.*, 2011). MicroRNAs appear to play specific roles in clock function, modulating its physiological rhythm as well as contributing to pathological processes (Mehta and Cheng, 2013).

In the mammalian SCN, two microRNAs, miR-219 and miR-132, deserve special attention. They both show oscillatory, circadian expression and seem to affect clock function. miR-219 is under direct control of the elements of the molecular clock (Cheng *et al.*, 2007). Periodic binding of the CLOCK/BMAL1 transcription factor complex to promoters of the period (*per1* and *per2*) and cryptochrome (*cry1* and *cry2*) genes allows their cyclic expression – a crux of the molecular clock rhythmicity (Gekakis *et al.*, 1998). The same complex binds to the promoter of the mir-219 gene. Moreover, *in vivo* knockdown of miR-219 lengthens the circadian day. In contrast, expression of miR-132 in the SCN is not regulated by clock genes, but is inducible by light utilizing the ERK/MAPK secondary messenger cascade and the CREB transcription factor. Direct brain injections of a miR-132 **antagomir** showed that miR-132 serves as a negative regulator of the photic clock resetting. Importantly, miR-132 serves as an orchestrator of chromatin remodeling, targeting the *per1* gene (Alvarez-Saavedra *et al.*, 2010), the overexpression of which presents in various sleep disorders (Hansen *et al.*, 2011).

Sleep deprivation can also alter miRNA expression within the hypothalamus. Davis *et al.* (2012) subjected Sprague Dawley rats to sleep deprivation. miRNA expression profiling in the hypothalamus revealed that let-7e, miR-30d, -103, -107, and -181a were significantly upregulated, whilst let-7b, miR-125a, and -128b were significantly downregulated. Notably, let-7e and let-7b are members of the let-7 family which, together with miR-125a, are primary hypothalamic microRNAs, expressed in several nuclei (Davis *et al.*, 2012; Meister *et al.*, 2013). Interestingly, neither any member of the let-7 family nor miR-125a were implicated as regulators of the circadian clock in the SCN, a pacemaker of the circadian rhythm disrupted during sleep deprivation. It would be of great interest to determine exactly which hypothalamic nuclei are exhibiting sleep deprivation-related changes in microRNA expression.

4.2.5   MicroRNA involvement in transgenerational stress

The hypothalamus plays a key role in response to stress in an adult as a part of the **hypothalamo-pituitary-adrenal (HPA) axis** (see Chapter 17). The HPA axis is sensitive to environmental changes and its dysregulation in offspring has been shown after maternal or paternal exposure to alcohol, lead or stress (Rodgers *et al.*, 2013). A dysregulated HPA is not only an element of endocrine illnesses but is also a common feature of several neuropsychiatric diseases (Borges *et al.*, 2013; Hildebrandt and Greif, 2013; Koob, 2010; Lopez-Duran *et al.*, 2009; Papadopoulos and Cleare, 2011).

Prenatal factors such as maternal nutrition and stress exposure have been shown to play a key role in fetal development, shaping of its HPA axis and subsequent stress response during adulthood. Recent research indicates that paternal stress can also affect the HPA axis in offspring (Rodgers *et al.*, 2013), and that microRNAs play a key role in that process. Rodgers and colleagues exposed F1 generation male C57BL/6:129 hybrid mice (fathers) to chronic stress for 42 weeks (duration selected to encompass a full round of spermatogenesis) and found in the F2 generation (offspring) blunted corticosterone responses to acute restraint and increased expression of glucocorticoid receptor-responsive genes in the PVN. They reported change in the expression of only one microRNA in the PVN of offspring following paternal stress, namely miR-522. In humans, miR-522 seems to putatively target elements of the calcium signaling pathway (α11 subunit of G protein, calmodulin, and calmodulin-dependent phosphodiesterase) (Pietrzykowski AZ, unpublished, Diana-microT-CDS analysis; Paraskevopoulou *et al.*, 2013) essential for neuronal function. It would be of interest to determine the persistence of changes in the expression of miR-522 and its targets in the F3 generation in this model of stress.

4.2.6   Long non-coding RNA

Recent experiments have shed some light on the expression of another group of non-coding RNA in the hypothalamus called long non-coding RNA (lncRNA). Human accelerated regions (HARs) are portions of the human genome that are typically highly conserved among vertebrates, but unique to human biology. It appeared that out of 49 HARs, 96% are located in the non-coding segments of the human genome. HAR1 is the region showing the most accelerated change. It encompasses two divergently transcribed and overlapping lncRNA genes: HAR1F and HAR1R (Pollard *et al.*, 2006). HAR1F is thought to play an important role in the developing cortex; however, Northern blot and quantitative real-time PCR (qPCR) revealed that HAR1F and HAR1R are prominent throughout adult brain regions, including the hypothalamus. Although the role of HAR1s as well as other products of HARs is thus far not well understood, they are an exciting opportunity to determine a unique aspect of the human neuroendocrine system.

4.2.7 Other non-coding RNAs

Several other types of non-coding RNAs have been discovered (see Table 4.1), including circular RNA (circRNA), which seems to act as a natural microRNA sponge and modulate microRNA expression (Hansen *et al.*, 2013). The rapidly growing inventory of non-coding RNA species clearly shows that RNA regulation of gene expression is much more complex than anyone could have imagined. It is a multiplex process involving various different RNA species and several checkpoints and feedback loops. The reciprocal relationships of different non-coding RNA species were recently put together as a **complementary endogenous RNA** (ceRNA) hypothesis (Salmena *et al.*, 2011). Exploration of the involvement of various non-coding RNAs in hypothalamic gene expression and function should help us better understand their complex roles as integrators of the neuronal and endocrine systems and drivers of homeostatic processes.

4.3 The pituitary gland and its central control of the peripheral endocrine system

The pituitary gland is a small endocrine organ located at the base of the brain and is composed of two subregions of different origin and function called the anterior (adenohypophysis) and posterior (neurohypophysis) pituitary. Both subregions are intimately linked to the hypothalamus. However, each subregion is linked in its own, unique way. The anterior pituitary contains endocrine cells which produce several hormones (adrenocorticotropic hormone [ACTH], thyroid-stimulating hormone [TSH], follicle-stimulating hormone [FSH], luteinizing hormone [LH[, growth hormone [GH], prolactin [PRL]) under tight control of hormones produced by the neurons in hypothalamic nuclei, and delivered to the anterior pituitary via the portal system, 'skipping' the general circulation. The posterior pituitary, on the other hand, is a release site of two hormones (AVP, OXT) produced by neurons located in the hypothalamic SON and PVN and delivered to the posterior pituitary via axonal processes of these neurons (the HNS). All hormones produced in the pituitary are important in the regulation of various aspects of reproduction, development, and homeostasis.

It is worth mentioning that, in contrast to the hypothalamus, where tumors are very rare, **pituitary adenomas** are relatively common, representing roughly 10% of brain tumors (Kovacs *et al.*, 1985). They are usually monoclonal in origin and arise from particular adenohypophysial cells (d'Angelo *et al.*, 2012). Pituitary tumors are typically benign and develop slowly, but due to uncontrolled production of hormones they can lead to serious disorders.

Given the fact that at least a third of protein-coding transcripts in humans may be regulated by miRNAs (He and Hannon, 2004), it should come as little surprise that non-coding RNAs were shown recently to play a prominent role in many physiological and pathological processes, including those happening in the pituitary.

4.3.1 miRNA expression in the pituitary gland

Bak and colleagues used a set of complementary next-generation techniques to determine spatial expression of microRNAs in mouse brain, including the pituitary (Bak *et al.*, 2008). They observed brain enrichment of expression of over 40 microRNAs, in agreement with other studies (Sempere *et al.*, 2004). In the pituitary, a threefold enrichment above the average expression across all brain regions was observed for the miR-7 family members miR-25, -141, -152, -200a, -200c, and -375 (Bak *et al.*, 2008). On the other hand, expression of two of the most highly expressed microRNAs in the brain, miR-9 and miR-124, was lower than average. Expression of the rest of the microRNAs was detected at a level similar to other brain regions. Together, Bak and colleagues provided a comprehensive catalog of microRNAs associated with this neuroendocrine organ (Bak *et al.*, 2008). It would be advantageous, considering that the pituitary consists of several distinct endocrine cells, to establish the microRNA profile of each pituitary cell type.

4.3.2   miRNA function in the pituitary gland physiological processes

To gain general insight into miRNA function, knockout (KO) studies of Dicer, a key enzyme involved in processing of almost all microRNAs, are often performed. Zhang and colleagues used *Pitx2-Cre* mice to conditionally knockout Dicer in anterior pituitary cells in order to assess the role of microRNA in pituitary development (Zhang Z. *et al.*, 2010). They observed that the KO had a profound effect on the pituitary, causing its hypoplasia, and an associated decrease in the levels of several hypophysial hormones: GH, PRL, and TSH.

They further showed that a specific microRNA, miR-26b, plays a critical role in pituitary developmental processes: upregulation of miR-26b downregulates the *lef-1* transcript, which encodes Lef-1, a transcriptional repressor, which targets Pit-1 (Pou1f1) – a transcriptional factor promoting differentiation of somatotropes, lactotropes, and thyrotropes. This suppression of a suppressor allows enhanced expression of Pit-1 and differentiation of these cells. Moreover, expression of the main hormones produced respectively by these cells, GH, PRL, and the β subunit of the TSH (TSHβ), was decreased in the Dicer KO. It would be of great interest to determine whether production of these hormones is directly regulated by miR-26b interactions with mRNA transcripts encoding these hormones, or rather indirectly by the Lef-1/Pit-1 pathway and binding of Pit-1 to the promoters of genes encoding these hormones. Nevertheless, this elegant study indicates that miR-26b contributes to differentiation of at least three specific anterior pituitary lineages. Are microRNAs also involved in synthesis of ACTH and FSH/LH in corticotropes and gonadotropes, respectively? Hopefully this exciting question will soon be answered.

4.3.3   miRNA role in the development of pituitary gland tumors

A pituitary tumor, an abnormal growth of the pituitary gland, is fairly common and often asymptomatic. When signs and symptoms occur, they are typically related to excess and/or dysregulation of production of one or more hormones by the tumor cells. Abnormally expressed miRNAs have been reported in pituitary tumors, implying that microRNAs may play a role in pathogenesis or development of pituitary tumors (Bottoni *et al.*, 2007).

Bottoni and colleagues identified almost two dozen microRNAs (miR-16-1, -23a, -23b, -24-2, -26a, -26b, -30a, -30b, -30c, -30d, -107, -127, -128a, -129, -203, -132, -134, -137, -140, and let-7), the expression levels of which could differentiate normal pituitary from pituitary adenomas (benign tumors) (Bottoni *et al.*, 2007). A number of miRNAs discovered in these adenomas have been implicated in regulation of cellular growth, proliferation, and apoptosis, suggesting that their dysregulation may play a critical role in the development, maintenance, and/or proliferation of pituitary neoplasms. Importantly, the authors also characterized a distinct microRNA signature of four pituitary adenomas (somatotropic, corticotropic, lactotropic, and NFA, the latter being a non-functioning adenoma). Finally, they showed that microRNA signatures of NFA tumors can separate NFA micro- from macroadenomas, or provide information about NFA tumor pharmacological treatment. All together, different combinations of microRNAs can provide valuable information about pituitary tumor type, thus serving as an important diagnostic tool. Interestingly, completely different microRNA signatures of somatotropic (downregulated miR-34b, -326, -432, -548c, -570, and -603) and corticotropic adenoma (downregulated miR-145, -21, -141, -150, -15a, -16, -143, and let-7a) have been published (Amaral *et al.*, 2009; d'Angelo *et al.*, 2012). Further work is needed to establish microRNA panels associated with each pituitary tumor.

To better understand the role of individual microRNAs in the development of pituitary tumors, it is important to also characterize their targets. Although only a few targets of pituitary tumor-relevant microRNAs have been characterized thus far, they are providing interesting insights into the pathogenesis of pituitary tumors. miR-26b regulation of the Lef-1/Pit-1 pathway described above has not been explored in the tumors, but lower levels of miR-26b were detected in GH- and PRL-secreting adenomas. This is consistent with the specific role of mR-26b and the Lef-1/Pit-1 pathway in the regulation of biosynthesis of these two hormones (Bottoni *et al.*, 2007).

A very closely related (90% sequence homology) microRNA, miR-26a, shares the same seed sequence with miR-26b, and it could also regulate the Lef-1/Pit-1 pathway (Pietrzykowski AZ, unpublished). Moreover, miR-26a (and thus miR-26b) seems to target a few additional transcripts, the products of which, if expressed at aberrant levels, can contribute to tumorigenesis: transcription factor homeobox Hox-A5 (Henderson *et al.*, 2006), protooncogene pleomorphic adenoma gene 1 (PLAG-1) (Volinia *et al.*, 2006), and protein kinase Cδ (PRKCD) (Gentilin *et al.*, 2013; Lee *et al.*, 2013). Interestingly, they can also target the checkpoint with forkhead and ring finger domains protein (Chfr), and this putative regulation is conserved between mice and human (TargetScan; Pietrzykowski AZ, unpublished). Chfr expression is lost in half of epithelial tumors and tumor cell lines (Yu *et al.*, 2005).

Putative targets of another microRNA, miR-24, also include several tumorigenic factors: oncogene kinase (Pim-1), protooncogene guanine nucleotide exchange factors for Rho family GTPases (Vav-1), vascular endothelial growth factor receptor 1 (VEGFR-1), and transcription factor caudal-type homeobox protein 2 (CDX-2). Interestingly, the latter target is expressed in 80% of neuroendocrine carcinomas of the gastrointestinal system (Barbareschi *et al.*, 2004).

Higher expression of miR-107 in both GH and NFA pituitary adenomas has been confirmed by Trivellin and colleagues (2012), who also showed experimentally that miR-107 negatively regulates the pituitary tumor suppressor aryl hydrocarbon receptor interacting protein (AIP). They also confirmed miR-107 downregulation of the hypoxia-inducible factor-1β (HIF-1β, also known as ARNT), which is a member of the AIP pathway. Thus, a single microRNA can regulate a tumorigenic biological pathway by simultaneously targeting multiple elements of that pathway.

Can several aberrantly expressed microRNAs converge on a single target during tumorigenesis in the pituitary? D'Angelo and colleagues (2012) showed both *in silico* and *in vitro* that, indeed, this can be true for pituitary somatotropic adenomas. Remarkably, miR-34b, -548c, -326, -432, and -570 can all bind to the *HMGA2* transcript and decrease levels of produced HMGA2 protein. All of these microRNAs are downregulated in somatotropic adenoma, which is also characterized by overexpression of HMGA2, a member of the high-mobility group A proteins and a regulator of gene expression (Thanos and Maniatis, 1992). In the same adenoma, HMGA2 stimulates overexpression of cyclin B2 (CCNB2) – an essential component of the cell cycle machinery (de Martino *et al.*, 2009), which may contribute to uncontrolled cell division, a characteristic feature of a tumor tissue.

When pituitary neoplasms develop, they typically overproduce a given pituitary hormone, which is a basis for classification of pituitary adenoma subtypes (Kovacs and Horvath, 1986; Sivapragasam *et al.*, 2011). However, pituitary tumors often secrete more than one hormone and therefore are of a 'mixed' hormonal subtype. For example, up to half of somatotropic tumors can co-secrete prolactin (thus they are also lactotropic tumors) (d'Angelo *et al.*, 2012; Lopes, 2010). Co-secretion of these hormones probably stems from the common origin of these two cell types from the same stem cells (Asa and Ezzat, 1998). Distinguishing different subtypes is critical for proper diagnosis and treatment. The differential expression of miRNAs in these tumors can help with that task. Indeed, d'Angelo and colleagues found that, in humans, miR-320 was differentially expressed between somatotropic (upregulated) and lactotropic (downregulated) adenomas (d'Angelo *et al.*, 2012).

Although malignant pituitary tumors (carcinomas) are very rare (Asa and Ezzat, 2009), they have a poor prognosis and microRNAs could help to determine the stage of malignancy of these tumors. Stilling and colleagues established that two microRNAs can function as markers of pituitary corticotropic carcinomas: miR-493 and miR-122. Levels of miR-493 differed between normal corticotropes and tumor tissues – it was downregulated in adenoma and upregulated in carcinoma. Interestingly, miR-493 may be involved in regulating proliferation and apoptosis of pituitary cells through targeting LGALS3. LGALS3 is expressed mainly in corticotropic tumors and produces galectin-3, a regulator of cell proliferation and apoptosis (Griffiths-Jones *et al.*, 2006; Jin *et al.*, 2005; Sivapragasam *et al.*, 2011; Stilling *et al.*, 2010). miR-122 levels were not changed between normal corticotropes and corticotropic adenoma, but significantly upregulated in corticotropic carcinoma. Interestingly, one putative miR-122 target is R-spondin 1 (Pietrzykowski AZ, unpublished), a growth factor probably involved in the development of skin carcinoma (Parma *et al.*, 2006).

4.3.4 Long non-coding RNAs in pituitary function

Long non-coding RNAs (lncRNAs) are non-protein-coding transcripts usually 200 nt or longer (Perkel, 2013), which have recently been recognized as important players in gene regulation. Two lncRNAs, MEG3 and snaR, have been recognized as important in normal pituitary function thus far. They may also play a role in some pituitary cancers.

The MEG3 gene (*Gtl2*) encodes several Meg3 isoforms of varying lengths (mean length ~1700 nt) (Baldassarre and Masotti, 2012; Zhang Z. *et al.*, 2010). Meg3 is expressed in many organs, including the pituitary gland (Baldassarre and Masotti, 2012; Schmidt *et al.*, 2000). Interestingly, it is also absent from many tumors derived from these tissues, including NFA pituitary adenomas (Baldassarre and Masotti, 2012; Zhang *et al.*, 2003; Zhang X. *et al.*, 2010), suggesting that it may function as a tumor suppressor (Baldassarre and Masotti, 2012; Zhou *et al.*, 2012). Meg3 inhibits cell proliferation by activating the cell cycle regulator p53, the dysfunction of which plays a prominent role in many cancers and inhibits cell proliferation (Baldassarre and Masotti, 2012; Zhang Z. *et al.*, 2010).

Small NF90-associated RNAs (snaRs) are lncRNAs of approximately 117 nt in length (Parrott and Mathews, 2007). They are unique to African great apes (human and chimpanzee) (Parrott *et al.*, 2011). 28 snaR genes have been identified thus far, and the expression of snaRs varies widely, with highest expression in the testis. Importantly, expression of snaR-A is highest in the pituitary compared to other brain regions – over five times more (Parrott *et al.*, 2011). Function of snaRs is not well established yet. However, association of snaR with polysomes (Parrott *et al.*, 2011) and NF90, a protein implicated in transcriptional and translational control, indicates that snaRs may participate in regulation of gene expression.

4.3.5 Circular RNAs in pituitary development

Circular RNAs (circRNAs) are the newest class of ncRNAs. The 3' and 5' ends of their spliced transcripts are joined in a head-to-tail manner. Evidence from several studies suggests that they may play roles in the expression of miRNAs and possibly also mRNAs (Hansen *et al.*, 2013; Memczak *et al.*, 2013). Interestingly, a circRNA called CDR1 is antagonistic to miR-7 and could function as an miR-7 sponge (Hansen *et al.*, 2013; Memczak *et al.*, 2013). CDR1 is expressed in human, mouse, and zebrafish brains (Memczak *et al.*, 2013) with significant differences amongst different brain regions. Interestingly, expressions of CDR1 and miR-7 seem to be negatively correlated, as one would expect from a suppressor and its target. Indeed, in the pituitary, where expression of miR-7 is high, CDR1 expression is very low compared to other brain regions (Memczak *et al.*, 2013).

4.4 The pineal gland – a connector between external environment and internal homeostasis

The pineal gland is a small endocrine organ located in the center of the brain. Pinealocytes vastly outnumber all other cell types present in the gland. The main hormone released by the pineal gland is called melatonin which is produced by pinealocytes. Stimulation of the photoreceptors in the retina by daylight is conveyed to the pineal gland by the photic conversion pathway and ultimately inhibits melatonin release, while lack of light stimulation during the night allows for melatonin release. Due to a peak of melatonin concentration during the night, melatonin is also referred to as the 'hormone of darkness.' Information from the retina on its way to the pineal gland passes through the SCN of the hypothalamus, which serves as an organismal master clock (see section 4.2.4). Due to the pineal gland's ability to detect changes in the length of the day, and its connection to the SCN, melatonin is involved in both short-term, daily adjustments of bodily functions into circadian day-and-night cycles, as well as the long-term, yearly adaptations of an organism to environmental seasonality.

These functions position the pineal gland in a central place in sexual development, seasonal breeding, hibernation, sleep pattern and alertness, contributing to homeostatic control over the endocrine and other systems (Carrillo-Vico *et al.*, 2005, 2006). Indeed, aberrant production and release of melatonin are thought to contribute to several psychiatric and endocrine disorders as well as cancer (Carrillo-Vico *et al.*, 2006; Davanipour *et al.*, 2009; Pacchierotti *et al.*, 2001).

4.4.1 MicroRNAs are important for pineal gland function

MicroRNAs have not been studied in the pineal gland until recently. Next-generation sequencing (**microRNA-Seq**) has provided a comprehensive list of microRNAs expressed in the rat (*Rattus norvegicus*) pineal gland (Clokie *et al.*, 2012). 376 miRNAs were detected with sequencing depth sufficient for accurate quantification. The vast majority of reads corresponded to known microRNAs, with less than 1% constituting possible novel microRNA species. Based on the number of reads, 98% of the total miRNA population corresponded to 75 microRNAs, with a single microRNA, miR-182, representing about one-third of all microRNA reads. miR-182 is a part of the polycistronic microRNA cluster also containing miR-183 and miR-96 (Xu *et al.*, 2007). These microRNAs not only are transcribed together but also typically function synergistically. Together, the miR-182/183/96 cluster represents over 42% of the total pineal miRNA population.

Recent studies on neuropathic pain (Lin *et al.*, 2014) determined that the miR-182/183/96 cluster is expressed in the adult dorsal root ganglion neurons. The photic conversion pathway goes from the retina to the pineal gland through the superior cervical ganglia. It would be of great interest to establish expression of these microRNAs in this intermediary checkpoint of the retina–pineal gland path.

The exact role of this microRNA cluster is not yet known. However, the miR-182/183/96 may contribute to pineal physiological processes, considering a circadian expression pattern of these microRNAs in the pineal gland, their 'co-expression' in the retina, and regulation of that expression by light in photosensitive tissues in mouse (*Mus musculus*) (Krol *et al.*,

2010) and lancelets (*Amphioxus* sp., lower *Chordata*) (Candiani *et al.*, 2011). If, indeed, these microRNAs play a fundamental role in regulation of the circadian cellular process, their overexpression in several cancers (Bandres *et al.*, 2006; Li *et al.*, 2013; Navon *et al.*, 2009; Tang *et al.*, 2013) may indicate the importance of destabilization of a circadian expression pattern in tumor progression. Interestingly, although miR-182, miR-183, and miR-96 expression in the pineal gland is high, and closely follows the circadian rhythm, the daily amplitude change does not exceed twofold.

Recently, Ben-Moshe and colleagues (2014) used similar methodology (microRNA-Seq) to identify miRNAs whose expression is induced by light in the pineal gland of zebrafish (*Danio rerio*). They observed that 14 microRNAs exhibited expression induction by light. Remarkably, these microRNAs included the miR-182/miR-183/miR-96 cluster. Additionally, they determined that miR-183 targets two important transcripts: *e4bp4-6* mRNA, which encodes a transcription factor regulating expression of many genes, and *aanat2* mRNA, which encodes arylalkylamine *N*-acetyltransferase, the penultimate enzyme of the melatonin synthesis pathway responsible for melatonin production from *N*-acetylserotonin in pinealocytes.

miR-483 is an interesting pineal microRNA. It is highly expressed in the rat (*Rattus norvegicus*) pineal gland during development but is virtually absent in the adult (50-fold decrease) (Clokie *et al.*, 2012). Remarkably, miR-483, similar to miR-183, also targets the *aanat2* mRNA. It is tempting to speculate that developmentally high levels of miR-483 suppress melatonin synthesis until pineal gland innervation is in place later in development (Clokie *et al.*, 2012), while miR-183 controls *aanat2* levels during adulthood.

Clokie and colleagues also reported 34 novel candidate microRNAs expressed in the rat pineal gland. Twelve of these candidate microRNAs were homologous to mouse and human microRNAs but not reported before in *Rattus norvegicus*. The remaining 22 candidate microRNAs were validated as novel microRNAs.

4.4.2 Long non-coding RNAs are important for pineal gland function

Long non-coding RNAs are also involved in the regulation of pineal gland function. Coon and colleagues (2012) showed that several lncRNAs are expressed in the pineal gland of rats and undergo circadian changes. In total, 112 lncRNAs were identified in the rat pineal gland with more than twofold differential night/day expression. Each was given a unique ID, consisting of the common identifier 'lncSN' (long non-coding RNA, Section on Neuroendocrinology) followed by a specific number.

Further in-depth studies of eight lncRNAs (lncSN001, lncSN004, lncSN012, lncSN016, lncSN056, lncSN081, lncSN134, lncSN215), characterized by high nocturnal expression and more than twofold differential night/day expression, indicated that the circadian change in expression of pineal lncRNAs resulted from indirect neural stimulation of pinealocytes by the central circadian clock located in the SCN. This stimulation involved norepinephrine-dependent activation of the cAMP secondary messenger system, phosphorylation of cAMP-response element binding protein (CREB), binding of CREB to lncRNA promoters either solo or as CREB/Ets/Bsx modules, and activation of lncRNA genes. Interestingly, the same system is also engaged in activation of melatonin synthesis. Although the circadian pattern of lncRNA expression regulated by light in SCN has been established, a captivating question remains: what is the function(s) of these lncRNAs in the pineal gland?

It should be noted that, in the liver, an antisense lncRNA named asPer2 controls expression of the circadian protein homolog 2 (Per2) (Kornfeld and Brüning, 2014; Vollmers *et al.*, 2012). Transcription of Per2 in the SCN is light dependent and it plays an important role in the clock entrainment by light.

Another noteworthy lncRNA is encoded by the *116HG* gene, located on human chromosome 15:q11-13 and associated with Prader–Willi syndrome. Prader–Willi syndrome is caused by loss of this non-coding RNA on the paternal chromosome. Interestingly, *116HG* long non-coding transcript is processed into multiple SNORD116 – small, nucleolar RNAs (snoRNAs), another type of non-coding RNA. Mice with *116HG* deficiency exhibit dysregulation of several circadian genes, including *Clock*, *Cry1*, and *Per2* (Powell *et al.*, 2013).

Further exploration of the roles of lncRNAs in the pineal gland is undoubtedly critical for understanding the function of this important neuroendocrine organ in both physiological and pathological processes.

4.4.3 Other non-coding RNAs important for pineal gland function

We have discussed here our current knowledge regarding mainly two types of non-coding RNAs: microRNAs and lncRNAs. This is simply because these two non-coding RNAs have been studied the most thus far. However, it doesn't mean that the other types of non-coding RNA are less important. This is true regarding all three neuroendocrine organs discussed here. Non-coding RNAs are a rich family with many divergent members (see Table 4.1). It took some time to appreciate the true meaning of the discovery of microRNA by Victor Ambros' group (Lee *et al.*, 1993), but after an initial delay, research on microRNA gained a lot of momentum (see Figure 4.3). The lag time in the lncRNA research narrative was even shorter and regarding 'rediscovery' of circular RNAs and their function as microRNA sponges, almost non-existent.

At the beginning of the 21st century, the astonishing realization started to dawn on us that regulation of gene expression is more extensive and complex than we have ever imagined. An emerging picture shows that non-coding RNAs play a central role in gene expression regulation. We are convinced that exciting, paradigm-shifting discoveries will soon unravel the fundamental roles of each type of non-coding RNAs in neuroendocrine system functions, and their contribution to diverse homeostatic processes governed by this system.

4.5 Perspectives

The recent discovery of the regulation of gene expression by non-coding RNA in neuroendocrine cells has changed our understanding of the functioning of these cells in response to internal and external environmental stimuli. The emerging image indicates that certain specific non-coding RNAs in each group (microRNA, lncRNA, and circRNA) are particularly important in neuroendocrine function. More work is now needed to determine the causal relationship between each non-coding RNA species and physiological or pathological processes. Importantly, reciprocity should be taken into consideration, as various non-coding RNAs can affect each other, with ramifications spreading across the entire transcriptome (Poliseno *et al.*, 2010; Salmena *et al.*, 2011; Tay *et al.*, 2014). Understanding the role of non-coding RNAs can also create completely novel therapeutic strategies for neuroendocrine diseases.

References

Alonso, C.R. (2012) A complex 'mRNA degradation code' controls gene expression during animal development. *Trends in Genetics*, **28**, 7888.

Alvarez-Saavedra, M., Antoun, G., Yanagiya, A., *et al.* (2010) miRNA-132 orchestrates chromatin remodeling and translational control of the circadian clock. *Human Molecular Genetics*, **20**, 731–751. [*A very elegant study describing coordinated regulation of epigenetic mechanisms by miR-132 in fine-tuning of the master circadian pacemaker.*]

Amar, L., Benoit, C., Beaumont, G., *et al.* (2012) MicroRNA expression profiling of hypothalamic arcuate and paraventricular nuclei from single rats using Illumina sequencing technology. *Journal of Neuroscience Methods*, **209**, 134–143. [*A thorough characterization of microRNAs in specific hypothalamic nuclei.*]

Amaral, F.C., Torres, N., Saggioro, F., *et al.* (2009) MicroRNAs differentially expressed in ACTH-secreting pituitary tumors. *Journal of Clinical Endocrinology and Metabolism*, **94**, 320–323.

Asa, S.L. and Ezzat, S. (1998) The cytogenesis and pathogenesis of pituitary adenomas. *Endocrine Reviews*, **19**, 798–827. [*Classification of pituitary neoplasms based on abberant mechanisms, cytoarchitecture and function.*]

Asa, S.L. and Ezzat, S. (2009) The pathogenesis of pituitary tumors. *Annual Review of Pathology*, **4**, 97–126.

Bak, M., Silahtaroglu, A., Moller, M., *et al.* (2008) MicroRNA expression in the adult mouse central nervous system. *RNA*, **14**, 432–444. [*An early inventory of microRNA expression profiles in several mouse brain regions, including hypothalamus and pituitary gland.*]

Baldassarre, A. and Masotti, A. (2012) Long non-coding RNAs and p53 regulation. *International Journal of Molecular Sciences*, **13**, 16708–16717.

Bandres, E., Cubedo, E., Agirre, X., *et al.* (2006) Identification by Real-time PCR of 13 mature microRNAs differentially expressed in colorectal cancer and non-tumoral tissues. *Molecular Cancer*, **5**, 29.

Barbareschi, M., Roldo, C., Zamboni, G., *et al.* (2004) CDX-2 homeobox gene product expression in neuroendocrine tumors: its role as a marker of intestinal neuroendocrine tumors. *American Journal of Surgical Pathology*, **28**,1169–1176.

Ben-Moshe, Z., Alon, S., Mracek, P., *et al.* (2014) The light-induced transcriptome of the zebrafish pineal gland reveals complex regulation of the circadian clockwork by light. *Nucleic Acids Research*, **42**, 3750–3767. [*A description of the microRNA role in regulation of melatonin by light.*]

Benoit, C., Ould-Hamouda, H., Crepin, D., Gertler, A., Amar, L. and Taouis, M. (2013) Early leptin blockade predisposes fat-fed rats to overweight and modifies hypothalamic microRNAs. *Journal of Endocrinology*, **218**, 35–47. [*Evidence that early life events affect hypothalamic microRNA expression in adulthood, contributing to the metabolic syndrome.*]

Borges, S., Gayer-Anderson, C. and Mondelli, V. (2013) A systematic review of the activity of the hypothalamic-pituitary-adrenal axis in first episode psychosis. *Psychoneuroendocrinology*, **38**, 603–611.

Bottoni, A., Zatelli, M.C., Ferracin, M., *et al.* (2007) Identification of differentially expressed microRNAs by microarray: a possible role for microRNA genes in pituitary adenomas. *Journal of Cellular Physiology*, **210**, 370–377. [*One of the first and most thorough assessments of the role of microRNAs in pituitary tumors.*]

Boudreau, R.L., Jiang, P., Gilmore, B.L., *et al.* (2014) Transcriptome-wide discovery of microRNA binding sites in human brain. *Neuron*, **81**, 294–305. [*A landmark paper describing a transcriptome-wide map of over 7000 microRNA binding sites in human cortex.*]

Brownstein, M.J., Russell, J.T. and Gainer, H. (1980) Synthesis, transport, and release of posterior pituitary hormones. *Science*, **207**, 373–378. [*A fundamental paper describing transcription, translation and transportation of vasopressin and oxytocin in the hypothalamo-neurohypophysial system.*]

Candiani, S., Moronti, L., de Pietri Tonelli, D., Garbarino, G. and Pestarino, M. (2011) A study of neural-related microRNAs in the developing amphioxus. *EvoDevo*, **2**, 15.

Cao, F., Li, X., Hiew, S., Brady, H., Liu, Y. and Dou, Y. (2009) Dicer independent small RNAs associate with telomeric heterochromatin. *RNA*, **15**, 1274–1281. [*Important finding of a class of small RNAs associated with telomeres.*]

Carone, D.M., Longo, M.S., Ferreri, G.C., *et al.* (2009) A new class of retroviral and satellite encoded small RNAs emanates from mammalian centromeres. *Chromosoma*, **118**, 113–125.

Carrillo-Vico, A., Guerrero, J.M., Lardone, P.J. and Reiter, R.J. (2005) A review of the multiple actions of melatonin on the immune system. *Endocrine*, **27**, 189–200.

Carrillo-Vico, A., Reiter, R.J., Lardone, P.J., *et al.* (2006) The modulatory role of melatonin on immune responsiveness. *Current Opinion in Investigational Drugs*, **7**, 423–431.

Cheng, H.Y., Papp, J.W., Varlamova, O., *et al.* (2007) MicroRNA modulation of circadian-clock period and entrainment. *Neuron*, **54**, 813–829. [*Elegant work uncovering the mechanism of light-dependent regulation of the circadian clock by miR-132 and miR-219.*]

Choi, J.W., Kang, S.M., Lee Y, *et al.* (2013) MicroRNA profiling in the mouse hypothalamus reveals oxytocin-regulating microRNA. *Journal of Neurochemistry*, **126**, 331–337.

Choi, S.W. and Friso, S. (2010) Epigenetics: a new bridge between nutrition and health. *Advances in Nutrition*, **1**, 8–16.

Clokie, S.J., Lau, P., Kim, H.H., Coon, S.L. and Klein, D.C. (2012) MicroRNAs in the pineal gland: miR-483 regulates melatonin synthesis by targeting arylalkylamine N-acetyltransferase. *Journal of Biological Chemistry*, **287**, 25312–25324. [*First microRNA profiling describing known and new microRNAs present in the pineal gland and their role in production of melatonin.*]

Coon, S.L., Munson, P.J., Cherukuri, P.F., *et al.* (2012) Circadian changes in long noncoding RNAs in the pineal gland. *Proceedings of the National Academy of Sciences of the United States of America*, **109**, 13319–13324. **[*First evidence that in the pineal gland over 100 lncRNAs are expressed, and that their expression follows daily rhythmicity.*]**

Cusanelli, E. and Chartrand, P. (2015) Telomeric noncoding RNA: telomeric repeat-containing RNA in telomere biology. *Frontiers in Genetics*, **6**, 143.

D'Angelo, D., Palmieri, D., Mussnich, P., *et al.* (2012) Altered microRNA expression profile in human pituitary GH adenomas: down-regulation of miRNA targeting HMGA1, HMGA2, and E2F1. *Journal of Clinical Endocrinology and Metabolism*, **97**, E1128–1138.

Davanipour, Z., Poulsen, H.E., Weimann, A. and Sobel, E. (2009) Endogenous melatonin and oxidatively damaged guanine in DNA. *BMC Endocrine Disorders*, **9**, 22.

Davis, C.J., Clinton, J.M. and Krueger, J.M. (2012) MicroRNA 138, let-7b, and 125a inhibitors differentially alter sleep and EEG delta-wave activity in rats. *Journal of Applied Physiology*, **113**, 756–1762.

De Martino, I., Visone, R., Wierinckx, A., *et al.* (2009) HMGA proteins up-regulate CCNB2 gene in mouse and human pituitary adenomas. *Cancer Research*, **69**, 1844–1850.

Dopico, A.M., Lemos, J.R. and Treistman, S.N. (1996) Ethanol increases the activity of large conductance, Ca(2+)-activated K+ channels in isolated neurohypophysial terminals. *Molecular Pharmacology*, **49**, 40–48. **[*One of the first indications that alcohol affects the function of specific isoforms of the BK ion channels.*]**

Ebert, M.S., Neilson, J.R. and Sharp, P.A. (2007) MicroRNA sponges: competitive inhibitors of small RNAs in mammalian cells. *Nature Methods*, **9**, 721–726. **[*A fundamental paper in the microRNA field, describing, for the first time, the concept and construction of microRNA sponges.*]**

Eddy, S.R. (2001) Non-coding RNA genes and the modern RNA world. *Nature Reviews. Genetics*, **2**, 919–929.

ENCODE Project Consortium (2012) An integrated encyclopedia of DNA elements in the human genome. *Nature*, **489**, 57–74. **[*A milestone in biological sciences – a multicentered, transcontinental project established to delineate all functional elements of the human genome.*]**

Fatica, A. and Bozzoni, I. (2014) Long non-coding RNAs: new players in cell differentiation and development. *Nature Reviews. Genetics*, **15**, 7–21. **[*A very good review on long non-coding RNA.*]**

Fliers, E., Unmehopa, U.A. and Alkemade, A. (2006) Functional neuroanatomy of thyroid hormone feedback in the human hypothalamus and pituitary gland. *Molecular and Cellular Endocrinology*, **251**, 1–8.

Gainer, H. (1998) Cell-specific gene expression in oxytocin and vasopressin magnocellular neurons. *Advances in Experimental Medicine and Biology*, **449**, 15–27.

Gainer, H. (2012) Cell-type specific expression of oxytocin and vasopressin genes: an experimental odyssey. *Journal of Neuroendocrinology*, **24**, 528–538. **[*Identification of vasopressin and oxytocin promoter regions responsible for conferring cell type specificity.*]**

Gainer, H. and Wray, S. (1992) Oxytocin and vasopressin. From genes to peptides. *Annals of the New York Academy of Sciences*, **652**, 14–28.

Gekakis, N., Staknis, D., Nguyen, H.B., *et al.* (1998) Role of the CLOCK protein in the mammalian circadian mechanism. *Science*, **280**, 1564–1569. **[*A fundamental paper describing the role of the CLOCK protein in the circadian clock.*]**

Gentilin, E., Tagliati, F., Filieri, C., *et al.* (2013) miR-26a plays an important role in cell cycle regulation in ACTH-secreting pituitary adenomas by modulating protein kinase Cdelta. *Endocrinology*, **154**, 1690–1700. **[*Translational study describing miR-26a-dependent mechanism involved in the pathogenesis of corticotropic pituitary adenomas.*]**

Griffiths-Jones, S., Grocock, R.J., van Dongen, S., Bateman, A. and Enright, A.J. (2006) miRBase: microRNA sequences, targets and gene nomenclature. *Nucleic Acids Research*, **34**, D140–144. **[*An essential paper describing the creation and function of a comprehensive microRNA registry.*]**

Hansen, K.F., Sakamoto, K. and Obrietan, K. (2011) MicroRNAs: a potential interface between the circadian clock and human health. *Genome Medicine*, **3**, 10.

Hansen, T.B., Jensen, T.I., Clausen, B.H., *et al.* (2013) Natural RNA circles function as efficient microRNA sponges. *Nature*, **495**, 384–388. **[*A great paper describing circular RNA CiRS-7 acting as a miR-7 sponge in the brain, and a product of the sex-determining region Y (Sry) gene as a circRNA miR-138 sponge in the testis.*]**

He, L. and Hannon, G.J. (2004) MicroRNAs: small RNAs with a big role in gene regulation. *Nature Reviews. Genetics*, **5**, 522–531.

He, M., Liu, Y., Wang, X., Zhang, M.Q., Hannon, G.J. and Huang, Z.J. (2012) Cell-type-based analysis of microRNA profiles in the mouse brain. *Neuron*, **73**, 35–48.

Henderson, G.S., van Diest, P.J., Burger, H., Russo, J. and Raman, V. (2006) Expression pattern of a homeotic gene, HOXA5, in normal breast and in breast tumors. *Cellular Oncology*, **28**, 305–313.

Hentze, M.W. and Preiss, T. (2013) Circular RNAs: splicing's enigma variations. *EMBO Journal*, **32**, 923–925. [*This review article provides an excellent overview of the current state of knowledge of the biogenesis and functions of circular RNAs.*]

Herzer, S., Silahtaroglu, A. and Meister, B. (2012) Locked nucleic acid-based in situ hybridisation reveals miR-7a as a hypothalamus-enriched microRNA with a distinct expression pattern. *Journal of Neuroendocrinology*, **24**, 1492–1504.

Hildebrandt, T. and Greif, R. (2013) Stress and addiction. *Psychoneuroendocrinology*, **38**, 1923–1927.

Hyodo, S. and Urano, A. (1991) Expression of neurohypophyseal hormone precursor genes in the mammalian hypothalamus. *Zoological Science*, **8**, 1005–1022.

Jin, L., Riss, D., Ruebel, K., *et al.* (2005) Galectin-3 expression in functioning and silent ACTH-producing adenomas. *Endocrine Pathology*, **16**, 107–114.

Kiss, T. (2001) Small nucleolar RNA-guided post-transcriptional modification of cellular RNAs. *EMBO Journal*, **20**, 3617–3622.

Koob, G.F. (2010) The role of CRF and CRF-related peptides in the dark side of addiction. *Brain Research*, **1314**, 3–14. [*An excellent review highlighting the role of the upper level brain stress centers in the neurobiology of addiction.*]

Kornfeld, J. and Brüning, J.C. (2014) Regulation of metabolism by long, non-coding RNAs. *Frontiers in Genetics*, **5**, 57.

Kovacs, K. and Horvath, E. (1986) Tumors of the Pituitary Gland. Atlas of Tumor Pathology fascicle 21, II series, AFPI, Washington DC.

Kovacs, K., Horvath, E. and Asa, S.L. (1985) Classification and pathology of pituitary tumors, in Neurosurgery (eds R.H. Wilkins and S. Rengachary), McGraw-Hill, New York, pp. 834–842.

Krichevsky, A.M., King, K.S., Donahue, C.P., Khrapko, K. and Kosik, K.S. (2003) A microRNA array reveals extensive regulation of microRNAs during brain development. *RNA*, **9**, 1274–1281. [*One of the first characterizations of microRNA expression in the brain.*]

Krol, J., Busskamp, V., Markiewicz, I., *et al.* (2010) Characterizing light-regulated retinal microRNAs reveals rapid turnover as a common property of neuronal microRNAs. *Cell*, **141**, 618–631. [*A landmark paper showing regulation of microRNA by light in the retina and demonstrating microRNA metabolism in neurons.*]

Kung, J.T., Colognori, D. and Lee, J.T. (2013) Long noncoding RNAs: past, present, and future. *Genetics*, **193**, 651–669.

Lee, H.J., Palkovits, M. and Young, W.S. III (2006) miR-7b, a microRNA up-regulated in the hypothalamus after chronic hyperosmolar stimulation, inhibits Fos translation. *Proceedings of the National Academy of Sciences of the United States of America*, **103**, 15669–15674.

Lee, R.C., Feinbaum, R.L. and Ambros, V. (1993) The C. elegans heterochronic gene lin-4 encodes small RNAs with antisense complementarity to lin-14. *Cell*, **75**, 843–854. [*A landmark paper describing, for the first time, a microRNA.*]

Lee, S.J., Kang, J.H., Choi, S.Y. and Kwon, O.S. (2013) PKCdelta as a regulator for TGF-beta-stimulated connective tissue growth factor production in human hepatocarcinoma (HepG2) cells. *Biochemical Journal*, **456**, 109–118.

Li, S.C., Essaghir, A., Martijn, C., *et al.* (2013) Global microRNA profiling of well-differentiated small intestinal neuroendocrine tumors. *Modern Pathology*, **26**, 685–696.

Lin, C.R., Chen, K.H., Yang, C.H., Huang, H.W. and Sheen-Chen, S.M. (2014) Intrathecal miR-183 delivery suppresses mechanical allodynia in mononeuropathic rats. *European Journal of Neuroscience*, **39**, 1682–1689.

Lopes, M.B. (2010) Growth hormone-secreting adenomas: pathology and cell biology. *Neurosurgical Focus*, **29**, E2.

Lopez-Duran, N.L., Kovacs, M. and George, C.J. (2009) Hypothalamic-pituitary-adrenal axis dysregulation in depressed children and adolescents: a meta-analysis. *Psychoneuroendocrinology*, **34**, 1272–1283.

Louro, R., Smirnova, A.S. and Verjovski-Almeida, S. (2009) Long intronic noncoding RNA transcription: expression noise or expression choice? *Genomics*, **93**, 291–298.

Maicher, A., Lockhart, A. and Luke, B. (2014) Breaking new ground: digging into TERRA function. *Biochimica et Biophysica Acta*, **1839**, 387–394.

Martin, H.C., Wani, S., Steptoe, A.L., *et al.* (2014) Imperfect centered miRNA binding sites are common and can mediate repression of target mRNAs. *Genome Biology*, **15**, R51.

Mehta, N. and Cheng, H.Y.M. (2013) Micro-managing the circadian clock: the role of microRNAs in biological timekeeping. *Journal of Molecular Biology*, **425**, 3609–3624.

Meister, B., Herzer, S. and Silahtaroglu, A. (2013) MicroRNAs in the hypothalamus. *Neuroendocrinology*, **98**, 243–253.

Memczak, S., Jens, M., Elefsinioti, A., *et al*. (2013) Circular RNAs are a large class of animal RNAs with regulatory potency. *Nature*, **495**, 333–338. [*A foundational study in the circular RNA field describing the expression of 2000 circRNAs in mammals, 700 in nematodes, and providing functional analysis of some of them.*]

Navon, R., Wang, H., Steinfeld, I., Tsalenko, A., Ben-Dor, A. and Yakhini, Z. (2009) Novel rank-based statistical methods reveal microRNAs with differential expression in multiple cancer types. *PLoS One*, **4**, e8003.

Nicholls, R.D. and Knepper, J.L. (2001) Genome organization, function, and imprinting in Prader-Willi and Angelman syndromes. *Annual Review of Genomics and Human Genetics*, **2**, 153–175.

Pacchierotti, C., Iapichino, S., Bossini, L., Pieraccini, F. and Castrogiovanni, P. (2001) Melatonin in psychiatric disorders: a review on the melatonin involvement in psychiatry. *Frontiers in Neuroendocrinology*, **22**, 18–32.

Papadopoulos, A.S. and Cleare, A.J. (2011) Hypothalamic-pituitary-adrenal axis dysfunction in chronic fatigue syndrome. *Nature Reviews. Endocrinology*, **8**, 22–32.

Paraskevopoulou, M.D., Georgakilas, G., Kostoulas, N., *et al*. (2013) DIANA-microT web server v5.0: service integration into miRNA functional analysis workflows. *Nucleic Acids Research*, **41**, W169–173. [*A paper describing a very useful online tool in microRNA research – DIANA-microT web server allowing for microRNA target prediction and functional analysis.*]

Parma, P., Radi, O., Vidal, V., *et al*. (2006) R-spondin1 is essential in sex determination, skin differentiation and malignancy. *Nature Genetics*, **38**, 1304–1309.

Parrott, A.M. and Mathews, M.B. (2007) Novel rapidly evolving hominid RNAs bind nuclear factor 90 and display tissue-restricted distribution. *Nucleic Acids Research*, **35**, 6249–6258.

Parrott, A.M., Tsai, M., Batchu, P., *et al*. (2011) The evolution and expression of the snaR family of small non-coding RNAs. *Nucleic Acids Research*, **39**, 1485–1500.

Perkel, J.M. (2013) Visiting "noncodarnia". *Biotechniques*, **54**, 301, 303–304.

Pietrzykowski, A.Z. (2010) The role of microRNAs in drug addiction: a big lesson from tiny molecules. *International Review of Neurobiology*, **91**, 1–24.

Pietrzykowski, A.Z. (2014) Epigenetics: an intermediary of environmental regulation of metabolism. *Journal of Human Nutrition & Food Science*, **2**, 1041. [*This review describes how nutrients can regulate gene expression through nuclear and cytoplasmic epigenetic mechanisms and delineates future challenges in the field.*]

Pietrzykowski, A.Z., Martin, G.E., Puig, S.I., Knott, T.K., Lemos, J.R. and Treistman, S.N. (2004) Alcohol tolerance in large-conductance, calcium-activated potassium channels of CNS terminals is intrinsic and includes two components: decreased ethanol potentiation and decreased channel density. *Journal of Neuroscience*, **24**, 8322–8332.

Pietrzykowski, A.Z., Friesen, R.M., Martin, G.E., *et al*. (2008) Posttranscriptional regulation of BK channel splice variant stability by miR-9 underlies neuroadaptation to alcohol. *Neuron*, **59**, 274–287. [*A fundamental paper establishing the contribution of microRNA to the addiction-related processes in the brain tissue.*]

Poliseno, L., Salmena, L., Zhang, J., Carver, B., Haveman, W.J. and Pandolfi, P.P. (2010) A coding-independent function of gene and pseudogene mRNAs regulates tumour biology. *Nature*, **465**, 1033–1038.

Pollard, K.S., Salama, S.R., Lambert, N., *et al*. (2006) An RNA gene expressed during cortical development evolved rapidly in humans. *Nature*, **443**, 167–172. [*A very important paper discovering regions in the human genome uniquely linked to human brain biology, and characterizing the non-coding nature of the products of these regions.*]

Powell, W.T., Coulson, R.L., Crary, F.K., *et al*. (2013) A Prader-Willi locus lncRNA cloud modulates diurnal genes and energy expenditure. *Human Molecular Genetics*, **22**, 4318–4328.

Qureshi, I.A. and Mehler, M.F. (2012) Emerging roles of non-coding RNAs in brain evolution, development, plasticity and disease. *Nature Reviews. Neuroscience*, **13**, 528–541. [*A very good summary of ncRNA related to the brain.*]

Rodgers, A.B., Morgan, C.P., Bronson, S.L., Revello, S. and Bale, T.L. (2013) Paternal stress exposure alters sperm microRNA content and reprograms offspring HPA stress axis regulation. *Journal of Neuroscience*, **33**, 9003–9012. [*A description of the link between paternal stress and abberant function of the stress axis in an offspring.*]

Rother, S. and Meister, G. (2011) Small RNAs derived from longer non-coding RNAs. *Biochimie*, **93**, 1905–1915. [*Two different classes of ncRNA are related to each other. A situation which may not be so rare.*]

Sagar, S.M., Sharp, F.R. and Curran, T. (1988) Expression of c-fos protein in brain: metabolic mapping at the cellular level. *Science*, **240**, 1328–1331.

Salmena, L., Poliseno, L., Tay, Y., Kats, L. and Pandolfi, P.P. (2011) A ceRNA hypothesis: the Rosetta Stone of a hidden RNA language? *Cell*, **146**, 353–358. [*A landmark paper providing a description of the competing endogenous RNA (ceRNA) hypothesis, unifying interactions of various elements of the transcriptome.*]

Sana, J., Faltejskova, P., Svoboda, M. and Slaby, O. (2012) Novel classes of non-coding RNAs and cancer. *Journal of Translational Medicine*, **10**, 103.

Sangiao-Alvarellos, S., Pena-Bello, L., Manfredi-Lozano, M., Tena-Sempere, M. and Cordido, F. (2014) Perturbation of hypothalamic microRNA expression patterns in male rats following metabolic distress: impact of obesity and conditions of negative energy balance. *Endocrinology*, **155**, 1838–1850.

Schmidt, J.V., Matteson, P.G., Jones, B.K., Guan, X.J. and Tilghman, S.M. (2000) The Dlk1 and Gtl2 genes are linked and reciprocally imprinted. *Genes & Development*, **14**, 1997–2002.

Schneeberger, M., Altirriba, J., García, A., *et al.* (2012) Deletion of miRNA processing enzyme Dicer in POMC-expressing cells leads to pituitary dysfunction, neurodegeneration and development of obesity. *Molecular Metabolism*, **2**, 74–85. **[Strong evidence linking microRNA dysfunction in the hypothalamus to obesity.]**

Schwartz, M.W., Woods, S.C., Porte, D. Jr, Seeley, R.J. and Baskin, D.G. (2000) Central nervous system control of food intake. *Nature*, **404**, 661–671.

Sempere, L.F., Freemantle, S., Pitha-Rowe, I., Moss, E., Dmitrovsky, E. and Ambros, V. (2004) Expression profiling of mammalian microRNAs uncovers a subset of brain-expressed microRNAs with possible roles in murine and human neuronal differentiation. *Genome Biology*, **5**, R13. **[One of the first papers describing a role for mammalian microRNAs in brain development.]**

Sivapragasam, M., Rotondo, F., Lloyd, R.V., *et al.* (2011) MicroRNAs in the human pituitary. *Endocrine Pathology*, **22**, 134–143.

Smirnova, L., Grafe, A., Seiler, A., Schumacher, S., Nitsch, R. and Wulczyn, F.G. (2005) Regulation of miRNA expression during neural cell specification. *European Journal of Neuroscience*, **21**, 1469–1477. **[An early attempt to distinguish different roles of microRNAs in neurons vs astrocytes.]**

Stilling, G., Sun, Z., Zhang, S., *et al.* (2010) MicroRNA expression in ACTH-producing pituitary tumors: up-regulation of microRNA-122 and -493 in pituitary carcinomas. *Endocrine*, **38**, 67–75.

Taft, R.J., Simons, C., Nahkuri, S., *et al.* (2010) Nuclear-localized tiny RNAs are associated with transcription initiation and splice sites in metazoans. *Nature Structural & Molecular Biology*, **17**, 1030–1034. **[The very important discovery of small ncRNAs associated with transcription start sites.]**

Tang, H., Bian, Y., Tu, C., *et al.* (2013) The miR-183/96/182 cluster regulates oxidative apoptosis and sensitizes cells to chemotherapy in gliomas. *Current Cancer Drug Targets*, **13**, 221–231.

Tay, Y., Rinn, J. and Pandolfi, P.P. (2014) The multilayered complexity of ceRNA crosstalk and competition. *Nature*, **505**, 344–352. **[An elaboration on the competing endogenous RNA (ceRNA) hypotheses.]**

Tessmar-Raible, K., Raible, F., Christodoulou, F., *et al.* (2007) Conserved sensory-neurosecretory cell types in annelid and fish forebrain: insights into hypothalamus evolution. *Cell*, **129**, 1389–1400. **[A unique and important paper describing the starting point of the evolution of the neuroendocrine brain centres, and a possible role for miR-7 in that process.]**

Thanos, D. and Maniatis, T. (1992) The high mobility group protein HMG I(Y) is required for NF-kappa B-dependent virus induction of the human IFN-beta gene. *Cell*, **71**, 777–789.

Thomson, D.W., Bracken, C.P. and Goodall, G.J. (2011) Experimental strategies for microRNA target identification. *Nucleic Acids Research*, **39**, 6845–6853.

Trivellin, G., Butz, H., Delhove, J., *et al.* (2012) MicroRNA miR-107 is overexpressed in pituitary adenomas and inhibits the expression of aryl hydrocarbon receptor-interacting protein in vitro. *American Journal of Physiology. Endocrinology and Metabolism*, **303**, E708–719.

Van Wolfswinkel, J.C. and Ketting, R.F. (2010) The role of small non-coding RNAs in genome stability and chromatin organization. *Journal of Cell Science*, **123**, 1825–1839.

Venter, J.C., Adams, M.D., Myers, E.W., *et al.* (2001) The sequence of the human genome. *Science*, **291**, 1304–1351. **[A paper which made history by describing, for the first time, results of sequencing of the human genome.]**

Volinia, S., Calin, G.A., Liu, C.G., *et al.* (2006) A microRNA expression signature of human solid tumors defines cancer gene targets. *Proceedings of the National Academy of Sciences of the United States of America*, **103**, 2257–2261.

Vollmers, C., Schmitz, R.J., Nathanson, J., Yeo, G., Ecker, J.R. and Panda, S. (2012) Circadian oscillations of protein-coding and regulatory RNAs in a highly dynamic mammalian liver epigenome. *Cell Metabolism*, **16**, 833–845.

Wang, X.M., Lemos, J.R., Dayanithi, G., Nordmann, J.J. and Treistman, S.N. (1991) Ethanol reduces vaso-pressin release by inhibiting calcium currents in nerve terminals. *Brain Research*, **551**, 338–341. **[Elegant work describing the role of ion channels in the release of neurohypophysial hormones.]**

Welsh, D.K., Takahashi, J.S. and Kay, S.A. (2010) Suprachiasmatic nucleus: cell autonomy and network properties. *Annual Review of Physiology*, **72**, 551–577.

Xu, S., Witmer, P.D., Lumayag, S., Kovacs, B. and Valle, D. (2007) MicroRNA (miRNA) transcriptome of mouse retina and identification of a sensory organ-specific miRNA cluster. *Journal of Biological Chemistry*, **282**, 25053–25066.

Xu, Y., Ishizuka, T., Yang, J., *et al.* (2012) Oligonucleotide models of telomeric DNA and RNA form a Hybrid G-quadruplex structure as a potential component of telomeres. *Journal of Biological Chemistry*, **287**, 41787–41796.

Yan, Z., Hu, H.Y., Jiang, X., *et al.* (2011) Widespread expression of piRNA-like molecules in somatic tissues. *Nucleic Acids Research*, **39**, 6596–6607. **[The surprising discovery of germline-specific piRNAs expression in somatic tissues.]**

Yu, X., Minter-Dykhouse, K., Malureanu, L., *et al.* (2005) Chfr is required for tumor suppression and Aurora A regulation. *Nature Genetics*, **37**, 401–406.

Zhang, X., Zhou, Y., Mehta, K.R., *et al.* (2003) A pituitary-derived MEG3 isoform functions as a growth suppressor in tumor cells. *Journal of Clinical Endocrinology and Metabolism*, **88**, 5119–5126.

Zhang, X., Rice, K., Wang, Y., *et al.* (2010) Maternally expressed gene 3 (MEG3) noncoding ribonucleic acid: isoform structure, expression, and functions. *Endocrinology*, **151**, 939–947.

Zhang, Y., Zhang, X.O., Chen, T., *et al.* (2013) Circular intronic long noncoding RNAs. *Molecular Cell*, **51**, 792–806. **[One of four papers published in 2013, almost in the same time, discovering circRNA in animals and placing circRNA at the center of the non-coding RNA stage.]**

Zhang, Z., Florez, S., Gutierrez-Hartmann, A., Martin, J.F. and Amendt, B.A. (2010) MicroRNAs regulate pituitary development, and microRNA 26b specifically targets lymphoid enhancer factor 1 (Lef-1), which modulates pituitary transcription factor 1 (Pit-1) expression. *Journal of Biological Chemistry*, **285**, 34718–34728. **[An important study describing pathways regulated by specific microRNAs in cellular lineage differentiation during pituitary development.]**

Zhou, Y., Zhang, X. and Klibanski, A. (2012) MEG3 noncoding RNA: a tumor suppressor. *Journal of Molecular Endocrinology*, **48**, R45–R53. **[The demonstration that MEG3 is a lncRNA acting as a tumor suppressor.]**

CHAPTER 5

Transcription Factors Regulating Neuroendocrine Development, Function, and Oncogenesis

Judy M. Coulson and Matthew Concannon
Cellular and Molecular Physiology, Institute of Translational Medicine, University of Liverpool, Liverpool, UK

5.1 The key players in transcriptional regulation

5.1.1 Core transcriptional complexes

Mammalian transcription relies on three multi-subunit core RNA polymerases. RNA pol I and III synthesize ribosomal and transfer RNAs respectively. Most pertinent here, RNA pol II is responsible for the synthesis of messenger RNA (mRNA) coding for proteins, microRNA (miRNA), and long non-coding RNA (lncRNA). The TATA binding protein (TBP) and a host of other general transcription factors are required to correctly position RNA pol II on a **gene promoter** and to support efficient transcriptional initiation. In addition, another multi-protein complex called Mediator is universally required to function as an adapter between these general transcription factors, RNA pol II, and the sequence-specific transcription factors (TFs) that ultimately determine transcriptional output.

5.1.2 Sequence-specific transcription factors (TFs)

Recent attempts to comprehensively catalog all the human or murine sequence-specific TFs estimate the total number at between 850 and 1900 (Fulton *et al.*, 2009; Vaquerizas *et al.*, 2009). A significant proportion of these TFs are at present completely uncharacterized. Many different TFs have been implicated in regulating the expression of neuropeptides and their cognate receptors, or in driving neuroendocrine development or carcinogenesis; some key examples are listed in Table 5.1. A global survey of sequence-specific TF mRNA expression shows they comprise approximately 6% of the expressed genes in all tissues, with between 150 and 300 different TFs expressed in any individual tissue. These TFs fall into two general categories: those that are expressed ubiquitously throughout the tissues of the body, and those that are restricted to one or two specific tissues (Vaquerizas *et al.*, 2009). Examples of TFs involved in neuroendocrine processes fall into both of these expression categories (see extended Table 5.1 on associated website).

All sequence-specific TFs have two major types of domain, which act independently. The first is a DNA-binding domain (DBD) that mediates direct binding of TFs to specific DNA regulatory elements. TFs are classified into more than 70 families on the basis of the specialized DBDs they utilize (Fulton *et al.*, 2009; Luscombe *et al.*, 2000). TFs use these structured domains to probe the topography of the DNA double helix, most commonly the major

Molecular Neuroendocrinology: From Genome to Physiology, First Edition. Edited by David Murphy and Harold Gainer.
© 2016 John Wiley & Sons, Ltd. Published 2016 by John Wiley & Sons, Ltd.
Companion website: www.wiley.com/go/murphy/neuroendocrinology

Table 5.1 Examples of sequence-specific transcription factors associated with regulation of neuroendocrine phenotype. HGNC human names are listed, with common names in brackets. *See associated website for extended table with further details and external links.*

Zinc finger	Basic leucine zipper
GATA2	ATF1
IKZF1	CREB1 (CREB)
INSM1	CREB3L
REST	FOS
SCRT1	JUN
SP1	
Homeobox	**bHLH**
LIM	ASCL1 (HASH1)
LHX3	NEUROD1
LHX4	USF1
	USF2
NKL	
HMX2	**bHLH-PAS**
HMX3	ARNT2
NKX2-1 (TTF1)	CLOCK
	HIF1A
POU	SIM1
POU1F1 (PIT1)	
POU3F2 (BRN2)	**Nuclear receptor**
	NR3C1 (GR)
PRD	NR5A1 (SF1)
HESX1	
OTP	**HMG-box**
PAX4	SOX3
PAX6	SOX10
PITX1	
PITX2	**Forkhead**
PROP1	FOXA2
	T-box
	TBX19 (TPIT)

groove, until they recognize the specific pattern of bases that represents their preferred binding motif. Most TFs employ an α-helix within their DBD for this purpose, and variation in the amino acid (aa) sequence of the DBD generates the differential base specificity of individual TFs. A single DBD element typically recognizes only a very short motif of several base pairs (bp) in length, which would be inadequate to provide suitable specificity within a mammalian genome. However, the scope of these recognition motifs is often extended, either by employing multimerized arrays of binding domains within the TF or by dimerization between two TFs, allowing them to recognize a longer, often inverted repeat.

Historically, the gene-by-gene empirical determination of DNA sequences bound by a TF was used to define their canonical binding motifs. This employed techniques such as **DNA footprinting**, electrophoretic mobility shift assay (**EMSA**), and reporter gene assays. However, technological advances now enable us to map the binding sites for a given TF across an entire mammalian genome. Methods include (i) computational predictions using motif searches, (ii) protein binding **microarrays**, (iii) systematic evolution of ligands by exponential enrichment (**SELEX**), or (iv) **chromatin** immunoprecipitation (ChIP), followed by either microarray analysis (**ChIP-chip**) or next-generation sequencing (**ChIP-Seq**).

SELEX is an *in vitro* method used to identify the DNA sequence binding preferences for a TF within a pool of random oligonucleotides, whilst ChIP enables a snapshot of *in vivo* binding by cross-linking TFs at their physiological binding sites within the cellular chromatin environment. High-throughput SELEX has now defined binding motifs for over 200 human TFs; although the structural families of TFs as classified by their DBDs do have distinct binding preferences, more precise binding profiles can be used to subclassify families (Jolma *et al.*, 2013). These approaches often reveal surprisingly wide-scale and diverse binding sites for TFs, and may uncover novel physiological roles through **gene ontology** and pathway profiling of the data sets. There are caveats to these studies though. Firstly, binding is highly dependent on the physiological context; computation predictions and SELEX cannot account for the cellular environment, whilst ChIP data are specific to the cell line used. Secondly, *in vivo* occupancy of sites does not always equate to transcriptional output. Some estimates suggest that only 25% of TF binding sites identified by ChIP-Seq in mammalian cells are linked to transcriptional activity (Spitz and Furlong, 2012). Expression profiling of the TF-responsive **transcriptome** by microarray or next-generation sequencing of RNA (**RNA-Seq**) therefore remains key to understanding the physiological relevance of TF binding.

The major functions of a TF are often conserved across species. However, TF binding motifs within DNA evolve rapidly, and the overlap between the human and mouse genomes may be as low as 10% (Vaquerizas *et al.*, 2009). This chapter predominantly discusses TFs with human and rodent **orthologs**, likely to perform similar functions within neuroendocrine systems. However, their genome-wide binding profiles may vary significantly between species. The DBD targets a TF in a promoter-dependent fashion. This enables precise control of the expression of individual genes, as TFs have a second class of domain that mediates protein–protein interactions to recruit transcriptional co-factors to the gene promoter. Many co-factors bear functional domains that can enhance or repress the activity of the core transcriptional complexes (Figure 5.1).

5.1.3 Transcriptional co-factors

Genomic DNA is packaged by **nucleosomes**, composed of the core **histone** proteins H2A, H2B, H3, and H4. Less structured 'tails' of each histone protrude from the complex, and aa-residues within the tails are targeted for **posttranslational modification** (PTM). These PTMs form a complex histone code that alters the dynamic chromatin environment, rendering it more or less accessible to general and sequence-specific TFs. The code is determined by many variables including the type, number, and location of the PTMs, and the position of the nucleosome within the gene architecture (Li *et al.*, 2007). The writers, readers, and erasers of the histone code are transcriptional co-factors. Most sequence-specific TFs can interact with a wide range of co-factors, either simultaneously recruiting a large complex with multiple activities towards chromatin or using alternative co-factors to provide spatial or temporal context to their activity.

Many co-factors possess enzymatic activities that add or remove histone PTMs. For example, addition of acetyl groups by histone acetyl transferases (HATs) opens up the chromatin structure and is associated with transcriptional activation, whilst their removal by histone deacetylases (HDACs) leads to chromatin condensation and transcriptional repression. In contrast, methylation presents a more complex code: increasing methylation of histone H3 on lysine 4 (H3K4) is associated with transcriptional activation, whilst H3K9 or H3K27 methylation is repressive. Families of methyl transferases and demethylases mediate these reversible modifications. Histone readers are recruited to the modified histone residues to act as scaffolds that bring in additional co-factors, ensuring an orchestrated progression of modifications to determine whether a gene is transcribed or repressed.

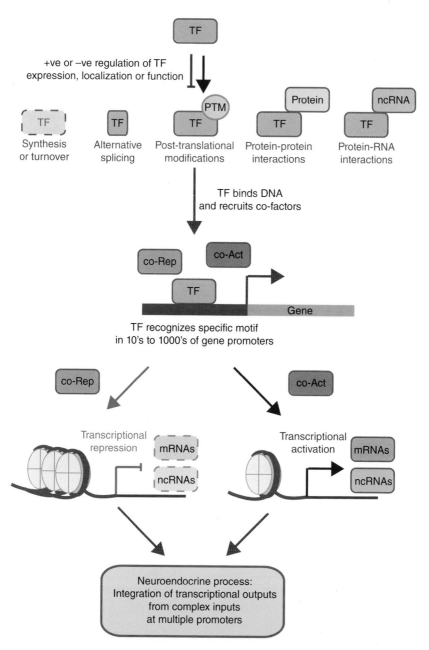

Figure 5.1 Key concepts: generalized pathway by which sequence-specific transcription factors direct physiological processes. The expression, cellular localization, and activity of transcription factors (TFs) are tightly controlled. When in an active state, TFs are targeted to bind certain gene promoters through recognition of specific DNA motifs. TFs recruit a variety of co-factor complexes, which alter the chromatin environment around the target gene to activate or repress basal transcription. TFs direct expression of both messenger RNAs (mRNA) that encode proteins and non-coding RNAs (ncRNAs) that modulate protein expression through different mechanisms. Integration of signals at a promoter determines whether the target gene is expressed, and this feeds into larger expression networks.

The **ATP-dependent chromatin remodeling** SWI/SNF complexes also modulate transcription. As nucleosome positioning influences TF occupancy at enhancers, nucleosome displacement may be required to expose low-affinity TF binding sites. Components of the remodeling complexes, such as BRG1 or BAF, may be recruited as transcriptional co-factors by **pioneer** TFs, which prime the promoter for binding of other TFs (Spitz and Furlong, 2012).

Intriguingly, some non-coding RNAs act as novel classes of transcriptional co-factor. Mechanistically, lncRNAs may act as signals that mimic TFs, decoys that titrate TFs away from DNA, guides that recruit co-factors in the absence of TFs, or scaffolds that bring together multiple TFs and/or co-factors at chromatin (Wang and Chang, 2011). Small modulatory double-stranded RNAs (smRNAs) of around 20 bp in length can also act as TF decoys by mimicking binding motifs, or may modulate interaction between the TF and its co-factors.

5.2   Classes of neuroendocrine-associated TFs

As summarized in Table 5.1, TFs that regulate the neuroendocrine phenotype fall into many different classes based on their DBDs. Full names and further details for all of these TFs can be accessed through extended Table 5.1 on the associated website. Here we briefly overview selected TFs, highlighting their DNA binding preferences and roles in neuroendocrine physiology.

5.2.1   Basic leucine zipper (bZIP)

The bZIP domain forms a long continuous α-helix consisting of two functional halves. The first is a basic region that makes contact with the DNA, typically recognizing a short sequence of 4 bp to 5 bp. The second mediates dimerization through formation of a coiled-coil structure. Homodimerization dictates that the active TF recognizes an inverted repeat, but heterodimerization generates alternate factors that recognize distinct asymmetrical binding sites. Key examples of neuroendocrine-associated bZIP TFs are the FOS and JUN family, which heterodimerize to constitute the AP1 transcription factor, and the CREB/ATF family.

CREB1 binds as a homodimer to an 8 bp palindrome known as the cAMP response element (CRE) and is the textbook example of a TF whose activity is controlled by phosphorylation. In response to cAMP signaling, protein kinase A (PKA) is activated, phosphorylating CREB1 on serine-133. CREB1 then translocates into the nucleus and interacts with its co-factor CREBBP to activate target gene transcription. CREB1 is co-activated by a family of TORCs, with TORC1 and TORC2 most highly expressed in the parvocellular and magnocellular neuroendocrine hypothalamus. TORCs are phosphorylated and held in an inactivated state in the cytoplasm by 14-3-3 proteins; when dephosphorylated, they move into the nucleus to interact with CREB1, facilitating its interaction with the transcriptional complex. This may be a requirement for CREB-dependent activation – for example, corticotropin-releasing hormone (CRH) transcription requires both phosphorylation of CREB1 and nuclear translocation of TORC2 (Aguilera and Liu, 2012). In contrast, CREB3L1 is normally sequestered in the endoplasmic reticulum membrane, from where it is cleaved in response to inducing stresses, allowing translocation into the nucleus to activate transcription. CREB3L1 was recently shown to play a pivotal role in osmotic induction of arginine vasopressin (AVP) expression (see section 5.4).

5.2.2 Basic helix-loop-helix (bHLH)

The bHLH factors also utilize a basic α-helix to contact DNA, typically binding a 6 bp E-box motif (CANNTG), the canonical form of which is the palindromic sequence CACGTG. The bHLH factors are obligate dimers, and a flexible loop region connects their DNA-binding helix to a second α-helix that enables dimerization. Although homodimerization does occur, heterodimerization is more common and interaction with different dimerization partners provides diversity in sequence recognition and co-factor recruitment.

The transcriptional activator ASCL1 has roles in neural and neuroendocrine progenitor development. ASCL1 is expressed at high levels in human neuroendocrine cancers and forced overexpression of Ascl1 is sufficient to drive development of neuroendocrine lung cancers in mice (Linnoila *et al.*, 2000). In mouse embryonic brain (E12.5) or cultured neural stem cells, genome-wide ChIP-chip identified binding sites for Ascl1 in ~1200 proximal promoters. Enriched amongst these were genes controlling the neurotransmitter biosynthetic process. Combining these data with expression profiling revealed transcriptional targets that both drive neuronal differentiation and promote cell cycle progression (Castro *et al.*, 2011). Like ASCL1, NEUROD1 and USF2 are also expressed in neuroendocrine cancers. Physiologically, NEUROD1 is required for specification of pituitary corticotropes; pathologically, it is implicated in a positive feedback loop in small cell lung cancer (SCLC), as it is upregulated in response to nicotine and increases nicotinic acetylcholine receptor subunit transcription. USF1 and USF2 predominantly heterodimerize, but also form homodimers with distinct binding specificities (Rada-Iglesias *et al.*, 2008). USF2 is overexpressed in SCLC and promotes proliferation (Ocejo-Garcia *et al.*, 2005), whilst USF1/USF2 regulates expression of neuropeptides including AVP, calcitonin gene-related peptide (CGRP), and prepro-tachykinin (PPT-A) (Coulson *et al.*, 1999a, 2003; Paterson *et al.*, 1995; Viney *et al.*, 2004).

A subfamily of bHLH-PAS factors combines this bHLH domain with PAS (Per/Arnt/Sim) domains that can bind small molecules or other proteins to sense and respond to environmental signals. A heterodimer of two bHLH-PAS factors ARNT2/SIM1 plays key roles in hypothalamic development, whilst CLOCK/BMAL1 and HIF1A contribute to regulation of the AVP promoter (see section 5.4).

5.2.3 Forkhead (FOX)

The forkhead or winged-helix domain is a distinct DBD of around 100-aa, and FOX factors bind to DNA as monomers. The hepatic factor FOXA2 plays roles in developmental systems and is implicated in regulation of neuropeptide gene expression. FOXA2 is a pioneer factor that opens up compacted chromatin to enable binding of other TFs, including nuclear receptors (Kaestner, 2010). It also works in a cooperative fashion with USF factors to activate transcription of CGRP (Viney *et al.*, 2004).

5.2.4 Homeoboxes

There are more than 300 homeobox genes of different subclasses encoded by the human genome, many of which are associated with developmental processes. They are characterized by a helical DBD, which is essential for function, and are divided into further subclasses according to their other protein domains. Functions of these TFs in the neuroendocrine hypothalamic-pituitary axis are described in section 5.4.

5.2.4.1 POU homeoboxes

Fifteen homeoboxes belong to the POU (Pit1/Oct/Unc86) subclass. They utilize two DBDs, an N-terminal POU-specific domain (~75-aa) that is separated from the C-terminal homeobox domain (~60-aa) by a non-conserved region (5-aa to 20-aa). Each domain uses a helix-turn-helix motif to contact 5 bp to 6 bp of DNA, and both are required for high-affinity DNA binding. Many of these factors have roles in neuroendocrine systems, in particular the class I factor POU1F1 (PIT1) that binds the motif TAAAT, and the class III factor POU3F2 (BRN2) (Prince *et al.*, 2011).

5.2.4.2 PRD homeoboxes

The PRD class is characterized by a serine residue at position 50 that dictates binding specificity and a second conserved PAX DBD. The PRD-like factors have a very similar homeobox but lack these two key features. A number of PRDs (e.g. PAX4, PAX6) and PRD-like factors (e.g. HESX1, OTP, PITX1, PITX2, PROP1) are involved in neuroendocrine development.

5.2.4.3 NKL homeoboxes

The NKL class of genes are related to the NK homeobox cluster in *Drosophila* and often contain an upstream TN motif. HMX2, HMX3, and NKX2-1, which participate in hypothalamic development, serve as examples of this class.

5.2.4.4 LIM homeoboxes

LIM homeodomain factors contain, in addition to a central homeobox, two N-terminal cysteine-rich LIM domains that mediate protein–protein interactions. Examples include LHX3 and LHX4 that participate in pituitary development.

5.2.5 T-box (TBX)

The TBX domain is quite large, at around 20 kDa, and is structurally distinct from other DBDs. TFs of this family bind to the DNA consensus sequence TCACACCT. These TFs are mainly involved in developmental processes and TBX19 is required for differentiation of pituitary corticotropes (see section 5.4).

5.2.6 High-mobility group box (HMG-box)

The HMG-box domain contains three α-helices, separated by loops, that make contact with DNA in the minor groove. High-affinity HMG-box binding is restricted to unwound DNA conformations. SOX3 acts as a developmental switch, counteracting the activity of proneural factors to suppress neuronal differentiation. It is required for formation of the hypothalamic-pituitary axis (see section 5.4). SOX10 is also associated with neuroendocrine tissues; it is expressed in pulmonary neuroendocrine carcinoids and is implicated in development of gonadotropin-releasing hormone (GnRH) cells in zebrafish (Whitlock *et al.*, 2005).

5.2.7 Nuclear hormone receptor (NR)

These TF sensors of steroids and other hormones typically have a C-terminal ligand-binding domain and an N-terminal activation domain, which is ligand dependent. The central DBD is composed of two zinc fingers (ZFs) and binds to the hormone response element (HRE). NRs are held in an inactive chaperone-bound state in the cytosol and, on ligand sensing, move into the nucleus and bind directly to DNA, either as monomers or as dimers. For example, the glucocorticoid receptor NR3C1 recognizes inverted repeats of a 6 bp DNA motif that are separated by a 3 bp spacer. NR3C1 requires chromatin remodeling by BRG1, a component of the SWI/SNF complex, to access many of its binding sites. In this context, FOXA2 or AP1 may act as pioneer factors to enable chromatin remodeling on which NR3C1 recruitment is dependent (Spitz and Furlong, 2012). The lncRNA GAS5 acts as a decoy for N3RC1 as its stem-loop structure mimics the glucocorticoid response element to which N3RC1 normally binds (Kino *et al.*, 2010). N3RC1 has pervasive roles in neuroendocrinology and may also interact with other transcription factors, altering expression of their responsive genes. Another NR factor, NR5A1 (SF1), is required for development of the adrenal gland, gonads, and pituitary gonadotropes. Intriguingly, NR5A1 not only binds its own canonical motif, CAAGGHCA, but can also occupy the RE1 motif used by the ZF repressor REST (Doghman *et al.*, 2013).

5.2.8   Zinc finger (ZF)

Zinc fingers are composed of around 30-aa and coordinate a single zinc ion at the base of the finger through pairs of conserved cysteine and histidine residues. Each ZF typically recognizes only 3 bp of DNA, and so they are commonly strung together in sequence to produce larger DBDs. Over 600 human TFs use ZFs to bind DNA. Many examples associated with neuroendocrine regulation primarily act as transcriptional repressors. These either silence neuroendocrine gene expression in non-neuroendocrine tissues or promote differentiation by switching off expression of genes that suppress neuroendocrine gene expression. Consideration of the preferred DNA binding motifs for some specific ZF factors illustrates that this prevalent DBD can provide diverse recognition profiles for individual TFs within the human genome (Figure 5.2).

INSM1 is a ZF repressor whose expression is tightly restricted to endocrine tissues. It has a C-terminal DBD composed of five ZFs, which recognize a 12 bp consensus motif (see Figure 5.2). INSM1 is transiently expressed during neuroendocrine differentiation and regulates development of the endocrine pancreas, as well as the noradrenergic sympathetic neurons and chromaffin cells of the sympathoadrenal gland. INSM1 is also highly overexpressed in most neuroendocrine cancers (Lan and Breslin, 2009). IKZF1 was originally described as a lymphocyte differentiation factor, although it also influences hypothalamic-pituitary cell development, differentiation, proliferation, and transformation (see

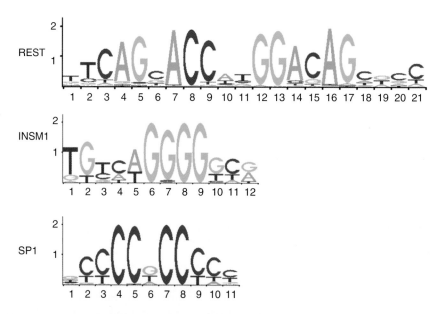

Figure 5.2 Examples of binding motifs for neuroendocrine-associated zinc finger TFs. The position weight matrices derived by ChIP-Seq (JASPAR, http://jaspar.genereg.net) are shown for three TFs that use DNA-binding domains with different configurations of zinc fingers to determine their binding specificities: REST (8 ZF), INSM1 (5 ZF), and SP1 (3 ZF).

section 5.4). IKZF1 has a C-terminal interaction domain involved in dimerization and an N-terminal DBD composed of five ZFs, although its preferential DNA recognition motif is not well established. Interestingly, IKZF1 exists as several alternatively spliced isoforms, most of which lack sufficient ZFs to bind DNA efficiently, and act in a dominant negative fashion. IKZF1 isoforms are expressed in pituitary adenomas and act as transcriptional activators or repressors for a variety of hormones, such as pro-opiomelanocortin (POMC), growth hormone (GH), prolactin (PRL), and GH-releasing hormone (GHRH) (Ezzat and Asa, 2008).

SCRT1 is a transcriptional repressor that utilizes five ZFs to bind DNA at E-box motifs, competing with bHLH factors. It is a neural-specific repressor, expressed in newly differentiated postmitotic neurons, and may mediate a switch to migratory neurons (Itoh *et al.*, 2013). SCRT1 is expressed in neuroendocrine cancers, where it antagonizes the proneural bHLH factors ASCL1 and E12 (Nakakura *et al.*, 2001). In contrast, SP1 is widely expressed with numerous physiological roles. SP1 has three ZFs that bind GC-rich DNA motifs (see Figure 5.2), and it may act as either a transcriptional repressor or an activator. SP1 is associated with transcriptional activation of POMC and GnRH. GATA2, involved in specification of pituitary gonadotropes and thyrotropes, is also quite widely expressed. It possesses a different class of GATA-type ZF, in which four cysteine residues coordinate the zinc ion. These highly conserved DBDs bind to the motif (A/T)GATA(A/G).

An example of a TF that prevents neuroendocrine expression in inappropriate tissues is the RE1-silencing transcription factor (REST), also known as NRSF. The central DBD of REST consists of eight ZFs, which bind a 21 bp consensus RE1 motif (see Figure 5.2). However, as discussed below, intensive study of genome-wide REST occupancy finds this motif to be highly divergent and surprisingly prevalent. REST is widely expressed outside the nervous system and was first described as a **silencer** of neuronal genes in non-neuronal cells (Chong *et al.*, 1995; Schoenherr and Anderson, 1995). However, REST is now known to dynamically regulate a broad spectrum of target genes and is implicated in many facets of the neuroendocrine phenotype (see section 5.3).

5.3 REST: a zinc finger TF with complex regulation and diverse function

REST controls transcription of a vast repertoire of target genes that play key roles in development and normal physiology. REST dysregulation is associated with diseases as diverse as Down syndrome, epilepsy, neurodegeneration, and cancer, in which it acts in a context-dependent fashion as either an oncoprotein or a tumor suppressor (Coulson, 2005; Negrini *et al.*, 2013). Importantly, the loss of REST in neuroendocrine lung cancers licenses inappropriate expression of neuropeptides, neurosecretory pathway components and neurotransmitter receptors, which can convey growth advantages (Coulson *et al.*, 1999b, 2000; Gurrola-Diaz *et al.*, 2003; Moss *et al.*, 2009). REST is a bipartite repressor, which recruits a variety of co-factors through N-terminal (RD1) and C-terminal (RD2) repression domains (Figure 5.3). It is part of the pluripotency network in embryonic stem cells and decreases as progenitors differentiate along a neuronal program, permitting expression of neural-specific transcripts (Ballas *et al.*, 2005). However, REST also controls expression of many other protein-coding mRNAs, as well as regulatory non-coding RNAs, which may act in feedback loops. Perhaps unsurprisingly, its own expression and function are tightly regulated. Here we use REST as a paradigm to illustrate the complexity of TF regulation and function (Figure 5.4).

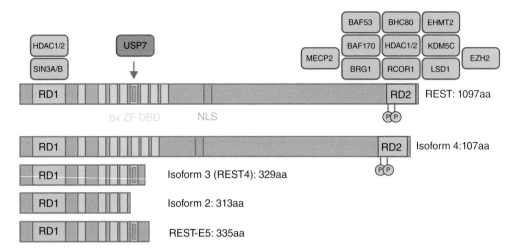

Figure 5.3 REST isoforms and co-factors. The major REST isoforms have two repression domains, RD1 and RD2, that recruit differential transcriptional co-factor complexes (green, shown above). Alternative splicing potentially generates multiple REST isoforms lacking key domains, which may antagonize REST function. Examples shown are numbered according to Uniprot (http://www.uniprot.org/uniprot/Q13127). Truncated isoforms are generated by inclusion of a neural-specific exon between exons 3 and 4 (isoforms 2 and 3) or the use of an alternative 3′ exon 5 (Chen and Miller, 2013); these lack several ZFs of the DBD, RD2 and the phosphodegron (P). ZF5 of the DBD domain, which recruits USP7 and mediates nuclear localization, is deleted in isoforms 2 and 4. NLS: nuclear localisation signal.

5.3.1   Transcription and alternative splicing of REST

REST function is regulated in many ways, including altering its transcription or by alternative splicing that generates isoforms lacking key domains (see Figure 5.3). During neurogenesis, the reduction in REST is partly attributed to abrogation of REST transcription, and this may also be downregulated in SCLC by promoter methylation (Kreisler *et al.*, 2010). However, alternative splicing in neurons, neuroblastoma, and SCLC also alters REST function (Coulson *et al.*, 2000; Palm *et al.*, 1998, 1999). A common splice variant retains an internal neural-specific exon and encodes a truncated isoform, known as REST4 or sNRSF, lacking the C-terminal repression domain. Intriguingly, the splicing regulator SRRM4 (nSR100), expressed in both neurons and SCLC, promotes inclusion of this exon and is itself a REST target gene (Raj *et al.*, 2011; Shimojo *et al.*, 2013). Other splice variants skip a domain required for nuclear translocation (Shimojo *et al.*, 2001), or truncate REST by using an alternative 3′ exon (Chen and Miller, 2013).

Although REST4 retains only five of the eight ZFs in the DBD, reducing its binding affinity, it may compete with REST at a subset of RE1 motifs. The prevalence and functional roles of the different REST isoforms remain under debate. However, the absence of RD2 in REST4 may mitigate repression of target genes. For example, REST4 induction is seen on differentiation of human embryonic stem cells into neural progenitor cells where neuronal gene expression is activated (Ovando-Roche *et al.*, 2014) and in epilepsy models, REST4 induction corresponds with that of the neuropeptide PPT-A (Spencer *et al.*, 2006). Further physiological evidence comes from a rodent study into the effect of early life stress on subsequent chronic stress. In this model, as the hypothalamic-pituitary-adrenal axis response increases, both REST4 expression and the transcription of REST target genes are upregulated in the prefrontal cortex (Uchida *et al.*, 2010).

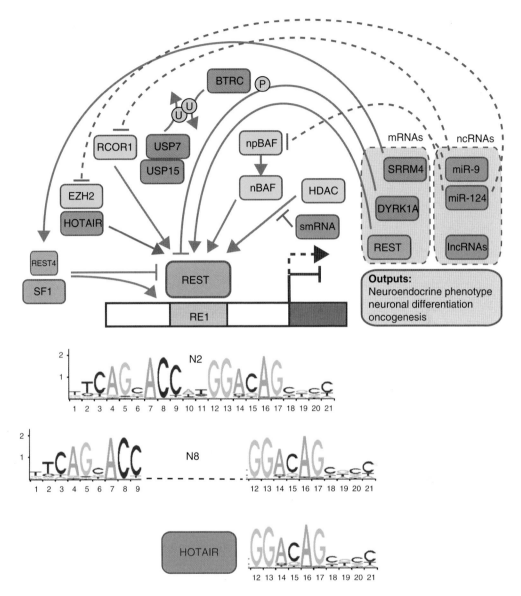

Figure 5.4 REST as a paradigm for diversity and feedback in transcription factor regulation and function. REST binds to a diverse array of RE1 motifs and recruits co-repressors (*green*) to switch off transcription. In the absence of REST, transcription is enabled that promotes the neuronal/neuroendocrine phenotype. Target mRNAs include regulatory proteins (*orange*) and miRNAs (*blue*) that establish feedback loops with REST. Other TFs (*purple*) may compete for RE1, and ncRNAs modulate REST interactions with the RE1 and protein co-factors. Gray lines show protein interactions and blue lines show ncRNA interactions.

5.3.2 Posttranslational modification, stability, and cellular localization

In common with many TFs, the functionality, localization, and stability of REST are controlled by reversible PTMs and protein interactions. Mature REST is glycosylated (Lee *et al.*, 2000; Pance *et al.*, 2006) which, although still poorly characterized, is associated with nuclear localization. The targeting of REST to the nucleus has also been associated with the fifth ZF that is spliced out in some variants (Shimojo *et al.*, 2001) or by the interacting proteins RILP (PRICKLE1), p150-glued (DCTN1), and huntingtin (HTT) (Shimojo, 2011; Shimojo and Hersh, 2003). In addition to relocalization, REST activity is also controlled by acute ubiquitin-mediated proteasomal degradation.

REST becomes acutely phosphorylated during neural differentiation, cell division, and adenoviral infection. Using mass spectrometry, this has been mapped to two independent phosphodegrons close to the C-terminal repression domain. Several candidate kinases have been suggested. The Down syndrome-associated kinase DYRK1A, a transcriptional target of REST, interacts with the REST-SWI/SNF complex, potentially establishing a negative feedback loop (Lu *et al.*, 2011), whilst casein kinase (CK1) phosphorylates REST in adult neurons (Kaneko *et al.*, 2014). Activation of REST phosphodegrons triggers acute polyubiquitylation of REST by the E3 ligase SCFβ^{TrCP} (BTRC), leading to its degradation (Guan and Ricciardi, 2012; Guardavaccaro *et al.*, 2008; Westbrook *et al.*, 2008). In the case of neural differentiation, REST degradation is antagonized by the deubiquitylase USP7 (Huang *et al.*, 2011). Interestingly, different REST isoforms lack residues required for either phosphorylation or interaction with USP7 (see Figure 5.3). REST protein abundance changes during the cell cycle, notably at the G2/M and M/G1 transitions; REST degrades as cells enter mitosis but rapidly recovers at mitotic exit. We recently identified the deubiquitylase USP15 as a regulator of REST stability by siRNA screening. Using mitotic and translational inhibitors, we demonstrated that USP15 specifically promotes new REST synthesis (Faronato *et al.*, 2013). Intriguingly, USP15 expression is relatively low in postmitotic neurons, but is amplified in glioblastoma (Eichhorn *et al.*, 2012) where REST has oncogenic function (Kamal *et al.*, 2012).

Another player in the regulation of REST activity is the telomere repeat protein TRF2. In pluripotent cells, TRF2-REST complexes are sequestered in aggregated nuclear PML bodies and protected from proteasomal degradation. However, during development, there is a switch in TRF2 isoforms, which now sequester REST in the cytoplasm, leading to derepression of target gene expression and promoting acquisition of the neuronal phenotype. Intriguingly, TRF2 also binds to the C-terminal of the REST4 isoform, protecting its stability in neural progenitor cells (Ovando-Roche *et al.*, 2014; Zhang *et al.*, 2008, 2011).

5.3.3   REST transcriptional co-factors

REST recruits a diverse cohort of transcriptional co-factors (see Figure 5.3). For an extensive discussion of this topic and full referencing, we refer the reader to two comprehensive reviews (Bithell, 2011; Ooi and Wood, 2007). Here we focus on the emerging understanding of their cooperative functions and the significance of alternative REST co-factor complexes.

5.3.3.1   Protein co-factors

Yeast two-hybrid screening has identified two major REST co-repressor complexes: SIN3A/B, that binds RD1 serving as a docking site for HDAC1/2 (Grimes *et al.*, 2000; Huang *et al.*, 1999), and RCOR1 (coREST) that binds RD2 (Andres *et al.*, 1999). RCOR1 was subsequently shown to recruit many histone modifiers that contribute to the repressive chromatin environment. These include HDAC1/2 and BHC80, the demethylases LSD1 (H3K4me/me2) and KDM5C (JARID1C or SMCX, H3K4me2/me3), the methyl transferases EHMT2 (G9a, H3K9me2) and EZH2, a component of the polycomb repressive complex PRC2 (H3K9 and H3K27). Intriguingly, whilst both RD1 and RD2 must be retained for full repression of some target genes, a single repression domain is sufficient to repress others; this is important when considering the activity of isoforms like REST4. Both full-length REST and RCOR1 can also interact with components of the ATP-dependent chromatin-remodeling complex, including BRG1 (SMARCA4), BAF53 (ACTL6A), and BAF170 (SMARCC2), and with the methyl-binding protein MECP2. In addition, REST can block the basal transcription machinery: it binds to TBP, inhibiting formation of the preinitiation complex, and to SCP1, inhibiting RNA pol II activity.

It is suggested that step-wise recruitment of these co-factors coordinates progressive chromatin changes that ultimately switch off expression of target genes. The nucleosome remodeling activity of BRG1 may be an early requirement, to provide better access and stabilize REST binding at RE1 sites. Profiling of nucleosome positioning and of 38 histone modifications by ChIP-Seq analysis revealed the complexity of the chromatin landscape remodeled by REST (Zheng *et al.*, 2009). This study provides good evidence for coordination of histone modifications, as REST binding is often correlated with decreased acetylation (H3K4ac, H4K8ac) and active methylation marks (H3K4me3), but increased repressive methylation (H3K27me3, H3K9me2). However, not all co-factors are recruited to each REST locus concomitantly, and this may vary according to the cellular context (Greenway *et al.*, 2007; Hohl and Thiel, 2005). Thus target genes may acquire different chromatin modifications as a consequence of REST binding. The selective engagement of co-factors may be linked to the strength and dynamics of binding and repression, such that alternative co-factor complexes may distinguish between transient repression and long-term silencing mechanisms. In this context, MECP2 recognizes repressive methylation marks within CpG islands and can retain repression at promoters once REST is no longer bound (Ballas *et al.*, 2005).

5.3.3.2 Non-coding RNA co-factors

To date, two ncRNAs have been shown to modulate transcriptional repression by REST through contrasting mechanisms (see Figure 5.4). HOTAIR, a lncRNA transcribed from within the HOXC cluster, acts as both a guide and a scaffold, to repress transcription of the HOXD cluster. HOTAIR recruits PRC2/EZH2 through binding to its 5' sequence, and the LSD1/RCOR1/REST complex at its 3' sequence; this molecular bridge coordinates H3K27 methylation by EZH2 with H3K4 demethylation by LSD1. Interestingly, this HOTAIR-REST complex now uses the right-hand RE1 half-site to bind DNA, potentially altering its profile of target genes (Tsai *et al.*, 2010). In contrast, a double-stranded smRNA found in neurons mimics the RE1 binding site for REST and results in transcriptional activation of REST target genes, specifying the fate of adult neural stem cells. However, this smRNA does not act as a decoy, as ChIP analysis shows that REST still binds to target gene promoters, but without recruitment of its usual co-repressors. The smRNA was therefore suggested to switch the function of chromatin-associated REST from that of a repressor to a transcriptional activator (Kuwabara *et al.*, 2004).

5.3.4   Diversity of transcriptional targets

5.3.4.1   Genome-wide RE1 identification

REST has proved of particular interest for genome-wide profiling, due to the long recognition motif for its DBD (see Figure 5.2). Numerous studies attempted to predict RE1 binding sites (reviewed in Bithell (2011) and Ooi and Wood (2007)). However, early empirical global studies revealed many more binding sites than expected. One used serial analysis of chromatin occupancy (SACO) in human lymphocytes; the other, in mouse kidney cells, was the first published ChIP-Seq study (Johnson *et al.*, 2007; Otto *et al.*, 2007). The increase in binding sites was partly due to the discovery that the RE1 motif functions as two half sites separated by a spacer, which varies in length from 2 bp, found in the most common canonical sequence, up to at least 8 bp (see Figure 5.4). Intriguingly, RE1 motifs were later divided into subgroups that are human, primate or mammal specific, and a small group that are deeply conserved across reptiles, amphibians, and fish (Johnson *et al.*, 2009). On comparison with the mouse genome, human RE1 motifs fell into three equal groups that aligned to mouse RE1, aligned with the mouse genome despite the lack of a murine RE1, or failed to align with the mouse genome at all. The most recent compilation across global occupancy studies suggests up to 21,000 REST binding sites within the human genome (Rockowitz *et al.*, 2014).

Broadly speaking, REST binding at both canonical and expanded RE1 motifs correlates with loss of transcription and occurrence of the expected histone marks (Zheng *et al.*, 2009). However, some studies suggest that only half of REST occupancy sites recruit co-factors (Yu *et al.*, 2011). The sequence context around an RE1 influences co-factor recruitment, and specific co-factors mark higher (SIN3A) or lower (EZH2) expressed targets (Rockowitz *et al.*, 2014). It is clear that REST occupancy is dynamic and depends on the cellular context. For example, tumor suppressors are identified as targets in cancer cells, but a very different profile of REST targets is seen in neurons compared to non-neuronal cells (Rockowitz *et al.*, 2014). Intriguingly, whilst a number of ChIP-validated occupancy sites are not RE1 (Johnson *et al.*, 2008), other TFs may also compete for binding at RE1 motifs. ChIP-Seq for SF1 in adrenocortical cells shows enriched occupancy at RE1 in addition to the SF1 consensus site. Indeed, SF1 could relieve REST repression of key steroidogenic genes (Doghman *et al.*, 2013). From a physiological perspective, genome-wide occupancy and transcription analyses concur that REST controls diverse processes, regulating expression of neuropeptides, neurotransmitter receptors, synaptic signaling and neuroendocrine secretion, as well as other TFs that drive neuronal and endocrine differentiation.

5.3.4.2 Transcriptional targets: mRNAs and non-coding RNAs

REST, via its myriad binding sites, regulates both mRNA and ncRNA expression. REST targets of both classes operate in feedback loops that influence protein expression of target genes, and directly impact on REST function. The contribution of such mechanisms to REST-dependent expression networks is highlighted in Figure 5.4.

Our interest in REST arose from the discovery that it was a negative regulator of neuropeptides, including PPT-A (Quinn *et al.*, 2002) and AVP (see section 5.4). Other neuropeptides and hypophysiotropic hormones are also REST target genes, including CRH (Korosi *et al.*, 2010), establishing REST as a neuroendocrine-associated TF. Indeed, gene ontology analysis from the first global ChIP study identified a role for REST in coordinating neuroendocrine pancreatic development (Johnson *et al.*, 2007). Recently, IL6 was found to induce neuroendocrine differentiation of prostate cancer cells through downregulating USP7 and accelerating REST turnover (Zhu *et al.*, 2014). Targeted transcript analysis and DNA microarray studies of the REST-dependent transcriptome, conducted in REST-deficient PC12 cells, on dominant negative REST expression in neuronal cells, or following siRNA depletion of REST in lung cancer cells, have all highlighted a role for REST in regulating the neurosecretory phenotype (d'Alessandro *et al.*, 2008, 2009; Hohl and Thiel, 2005; Moss *et al.*, 2009; Pance *et al.*, 2006). Target genes include many synaptic and dense-core vesicle proteins, as well as the chromogranin and prohormone convertase families.

Non-coding RNA is diverse in form and function (see Chapter 4) and lncRNA and miRNA targets of REST were identified through genome-wide occupancy and microarray studies (Conaco *et al.*, 2006; Gao *et al.*, 2012; Ng *et al.*, 2012; Rockowitz *et al.*, 2014). Most recent data suggest REST occupancy at 14% of currently annotated human miRNAs, with 4.2% of these differentially expressed in neurons (Rockowitz *et al.*, 2014). Currently, only a handful of these have been extensively investigated, most notably miR-9 and miR-124. These

REST-regulated miRNAs often exert feedback on REST function by targeting REST expression, or its co-factors including SCP1, RCOR1, MECP2 and EZH2, as well as switching neural progenitor BAF53a for neural BAF53b in the chromatin remodeling complex (Packer *et al.*, 2008; Rockowitz *et al.*, 2014; Visvanathan *et al.*, 2007; Wu and Xie, 2006; Yoo *et al.*, 2009). Intriguingly, several mRNAs that are normally repressed by REST also feed back to regulate REST function, including the splicing factor SRRM4 (Raj *et al.*, 2011) and the kinase DYRK1A (Lu *et al.*, 2011). Developmentally, miRNAs expressed as a consequence of REST downregulation contribute to establishing neuronal phenotype. For example, in combination with the TFs POU3F2 and MYTL1, miR-124 expression is sufficient to induce conversion of fibroblasts into neurons (Ambasudhan *et al.*, 2011). Cross-regulation of these miRNAs also integrates REST into networks with other neuronal and neuroendocrine TFs such as POU3F2, NEUROD1, and CREB1 (Rockowitz *et al.*, 2014; Wu and Xie, 2006). The context-specific studies published to date provide a glimpse into the extensive feedback between REST and ncRNAs that is proposed to govern maintenance and renewal of neuronal stem cells, differentiation, and establishment of neural identity (Qureshi and Mehler, 2012).

5.4 Cooperation of TFs in neuroendocrine phenotype and function

5.4.1 Transcriptional networks in neuroendocrine development

Neuronal differentiation is a highly coordinated process during which cells commit to a neuronal fate, acquire positional identities, exit the cell cycle, migrate, and terminally differentiate. Key to these processes are cascades of TFs that establish gene expression programs to develop, define and maintain the correct phenotypes. Here, we overview the TFs implicated in the development and physiological function of specific cells within the neuroendocrine hypothalamus and the anterior pituitary gland.

5.4.1.1 Magnocellular and parvocellular neurons of the hypothalamus

The hypothalamus sits below the thalamus and above the pituitary gland, to which it is connected; together they play a major role in homeostasis. Two classes of hypothalamic neurons form functional nuclei. Magnocellular neurons originate in the paraventricular nucleus (PVN) and the supraoptic nucleus (SON) of the hypothalamus and extend their axons into the posterior pituitary. In response to physiological stimuli, they release the neuropeptides oxytocin (OXT) and AVP directly into the circulation. In contrast, parvocellular neurons project from hypothalamic nuclei to the median eminence, where they secrete hypophysiotropic hormones. From here, the hypophysial portal system runs down the pituitary stalk into the anterior lobe, where the hormones act on specialized pituitary cells. The parvocellular neurons are classified according to the hormones they produce: CRH and thyroid-releasing hormone (TRH) neurons are found in the PVN; somatostatin (SS) neurons in the anterior periventricular (aPeV) nucleus; SS, GHRH, and dopamine (DA) neurons in the arcuate (ARC) nucleus; GnRH neurons in the preoptic area (POA); and gonadotropin-inhibiting hormone (GnIH) neurons in the dorsal-medial nucleus (DMN). A number of TFs expressed in the developing hypothalamus were mapped to progressive definition of these neuroendocrine lineages using human disease mutations and rodent models (Figure 5.5).

Otp is expressed from E10 in the mouse diencephalon, and by E17 is restricted to the regions from which the hypothalamic neuroendocrine nuclei originate (Simeone *et al.*, 1994). Otp is required at multiple stages of development, from the initial proliferation and migration of progenitor cells, through neuroendocrine differentiation, and during hormone

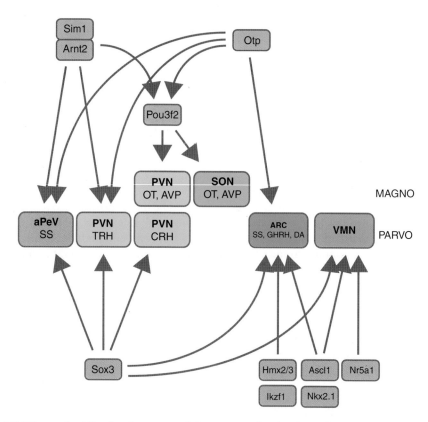

Figure 5.5 TFs required for development of the neuroendocrine hypothalamus. Examples of TFs that promote early commitment and later differentiation of the hypothalamic magnocellular and parvocellular neurons.

expression from established nuclei. These pervasive and essential roles of Otp are apparent in knockout mice, which fail to form both the magnocellular and parvocellular neurons of the aPeV, ARC, PVN or SON, and lack hypothalamic expression of the neuropeptides CRH, TRH, AVP, OT, SS, and DA (reviewed in del Giacco *et al.*, 2008). Sim1/Arnt2 act in parallel with Otp and, although not required in progenitor cells, Sim1/Arnt2 mutant mice have a reduced number of hypothalamic cells. These mice fail to establish the SON, lack parvocellular and magnocellular neurons of the PVN and SS neurons of the aPeV, and ultimately lose production of all these neuroendocrine hormones. Downstream of both Otp and Sim1/Arnt2 is Pou3f2 (also known as Brn2), which is normally expressed in the SON and much of the PVN. Pou3f2 knockout mice do not express CRH, OT or AVP, as they fail to establish the requisite neurons of the SON and PVN, although they do retain expression of TRH and SS (reviewed in Prince *et al.*, 2011; Szarek *et al.*, 2010).

Sox3 may be required for proper development of most parvocellular neurons. Sox3 null mice, and human patients with SOX3-linked hypopituitarism disorder, have multiple pituitary hormone deficiencies (Szarek *et al.*, 2010). However, this may also be linked to additional roles for Sox3 in the anterior pituitary itself, where it is required for development but not normal function. Other TFs implicated in development of specific parvocellular nuclei include Ascl1, Ikzf1, Nkx2.1, Hmx2/Hmx3, and Nr5a1. Proneural Ascl1 (also known as Mash1) is broadly required for neurogenesis throughout the central nervous system, and

Ascl1 null mice fail to develop the ARC and ventromedial nucleus (VMN) nuclei. Ascl1 is linked to neuronal subtype specification and, in the context of the hypothalamus, is required for establishment of GHRH-expressing neurons. Ikzf1 is also expressed in the developing GHRH neurons, and Ikzf1 knockout mice display severe neuroendocrine phenotypes, including dwarfism (Ezzat and Asa, 2008). Nkx2.1 (also known as Ttf1 or T/ebp) was originally described as a thyroid-specific TF, but is also expressed in developing lung and the presumptive hypothalamus. Nkx2.1 mutant mice die at birth, exhibiting lung, thyroid, and ventral hypothalamus defects, specifically in the ARC and VMN. Two closely related TFs, Hmx2 and Hmx3, may have redundant functions in hypothalamic development. Mice that are null for both Hmx2 and Hmx3 have a severe deficiency of GHRH neurons in the ARC, but not the VMN, and exhibit dwarfism. Lastly, Nr5a1 (also known as SF1) is required for development of the adrenals, gonads, and pituitary gonatotropes. Within the hypothalamus, Nr5a1 expression is restricted to the VMN, and is broadly required from the initial growth and migration of VMN precursors, to their terminal differentiation (Szarek *et al.*, 2010).

The downstream transcriptional pathways for many developmentally important TFs remain incompletely characterized. However, these and other TFs directly regulate transcription of neuropeptides or hypophysiotropic hormones. For example, the CRH promoter is directly repressed by REST (Korosi *et al.*, 2010) and activated by POU3F2 and CREB1 (Aguilera and Liu, 2012), IKZF1 induces GHRH transcription (Ezzat and Asa, 2008) and NKX2-1 is a transcriptional regulator of GnRH.

5.4.1.2 TFs that specify the anterior pituitary

In contrast to the posterior pituitary, the anterior pituitary is a true gland. Cells of the anterior pituitary fall into five distinct subtypes: gonadotropes that produce follicle-stimulating hormone (FSH) and luteinizing hormone (LH); thyrotropes that produce thyroid-stimulating hormone (TSH); lactotropes that produce PRL; somatotropes that produce GH; and corticotropes that synthesize POMC which is processed into adrenocorticotropic hormone (ACTH). These pituitary hormones are released under the control of the hypothalamic parvocellular neurons, as receptors on the pituitary cell surface recognize the appropriate hypophysiotropic hormone, which increases or decreases their hormone secretion into the general circulation. Mutations in several TFs result in impaired pituitary function, and transgenic models have clarified the cascades of TFs that specify development of anterior pituitary cells. The development of the human pituitary follows a similar, although not identical program, and disease-associated mutations in human patients suggest that the TF orthologs play similar roles. A simplified overview highlighting some key TFs in this developmental network is shown in Figure 5.6.

The anterior pituitary is derived from Rathke's pouch, an invagination of the oral ectoderm, under the control of a series of signaling pathways (reviewed in de Moraes *et al.*, 2012). Several homeobox TFs are required early in this process. Pitx1 (also called Tpit) and Pitx2 are expressed in Rathke's pouch and persist in the gonadotropes and thyrotropes of the adult pituitary. Lhx3, and the related Lhx4, are key regulators of anterior pituitary cell commitment and differentiation, being required for early development of Rathke's pouch. Experiments in knockout mice show that pituitary expression of Lhx3 is dependent on both Pitx1 and Pitx2. Lhx3 also persists in the adult pituitary, where it directly activates transcription of various pituitary hormones and other regulatory TFs.

HesX1 is present before Rathke's pouch forms and its downregulation is required for anterior pituitary cell differentiation. HesX1 negatively regulates pituitary-specific Prop1, first expressed at E10.5. Prop1 mutation is responsible for the hypoplastic pituitary phenotype of the Ames dwarf mouse, which lacks expression of GH, TSH, PRL, LH, and FSH. Prop1 regulates downstream expression of another pituitary-specific TF, Pou1f1 (Pit1), which is

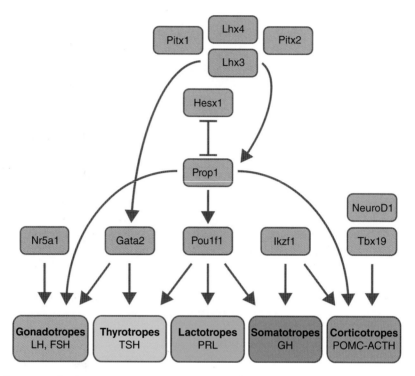

Figure 5.6 TF cascades in anterior pituitary development. Examples of TFs that promote early commitment and later differentiation of anterior pituitary cells.

expressed in mice from E13.5 through to adulthood. Pou1f1 specifies thyrotropes, lacto-tropes and somatotropes, all of which are lacking in dwarf mice with Pou1f1 mutation. It regulates transcription of many genes within these lineages, including GH, PRL, and TSHβ; human patients with Pou1f1 mutations are deficient in these same neurohormones (Prince *et al.*, 2011). Additional TFs including Nr5a1, Gata2, Izkf1, Tbx19, and NeuroD1 are required later in differentiation to specify hormone-secretory pituitary cell types (see Figure 5.6). For example, Ikzf1 expression is important for anterior pituitary cell growth, differentiation, and survival. Ikzf1 directly regulates POMC expression in cooperation with PitX1 by recruit-ing the co-activator SRC/P160, and increases PRL but decreases GH expression (Ezzat and Asa, 2008). POMC processing relies on a number of convertases, including PCSK1, which is transcriptionally repressed by REST (Moss *et al.*, 2009).

5.4.2   Context-dependent regulation of the AVP promoter

To conclude this chapter, we will briefly consider the context-dependent expression of the neuropeptide AVP (see Chapter 15). In a normal physiological context, AVP is transcribed in and released from magnocellular neurons of the SON and PVN in response to changes in osmolality, and acts on AVP receptors in the kidneys and blood vessels to maintain homeo-stasis. AVP is also transcribed in the suprachiasmatic nucleus (SCN) in response to circadian cues. However, the AVP gene was first cloned and sequenced from a SCLC cell line (Sausville *et al.*, 1985) and is commonly overexpressed in these neuroendocrine tumors, where it can lead to the syndrome of inappropriate secretion of antidiuretic hormone (SIADH) and dilu-tional hyponatremia (Johnson *et al.*, 1997).

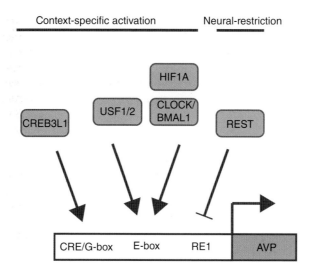

Figure 5.7 Context-dependent TF regulation of the AVP promoter. Examples of TFs that activate or repress transcription through the AVP proximal promoter in response to osmotic, circadian or pathological cues.

In 2002, we reviewed the binding motifs and TFs that regulated pathological expression of the AVP gene promoter in SCLC, highlighting roles for loss of repression by REST through an RE1 motif at the transcriptional start site, and activation by USF1/USF2 through proximal E-box motifs (Coulson, 2002). Interestingly, whilst USF1/USF2 bind the major E-box of the AVP promoter in SCLC, the bHLH-PAS heterodimer CLOCK/BMAL1 (Jin *et al.*, 1999) utilizes this same E-box during circadian regulation of AVP transcription in the SCN. Another bHLH-PAS factor, HIF1A, mediates cross-talk between hypoxic and circadian signaling by promoting BMAL1 recruitment (Ghorbel *et al.*, 2003). Although the physiological transcription of AVP in the magnocellular neurons is induced by hyperosmotic stress and cAMP signaling, until recently it remained unclear which TFs mediated this response. New studies found no direct role for CREB1, but instead show a key role for CREB3L1 (see Chapter 15). Both transcriptional induction and cellular relocalization of CREB3L1 are seen in response to osmotic challenge, and CREB3L1 can bind and activate the AVP promoter (Greenwood *et al.*, 2014). The TFs that have been physically mapped to the AVP promoter are summarized in Figure 5.7.

5.5   Perspectives

The complex networks that regulate transcription of physiological processes are slowly being uncovered. Systems biology approaches are required to understand how these transcriptional networks are integrated, but we do not yet know the full complement of TFs encoded by the human or murine genomes. Study of even a single TF reveals unexpected complexity, with multiple levels of regulation that contribute to contextual differences in their transcriptional activity. Considerable advances in the techniques available to study TFs are enabling their roles to be established in different tissues, through development, and in response to specific stimuli. Genome-wide maps of TF occupancy are helping to build networks, but this is hampered by the interspecies evolution of binding sites, and the incomplete correlation of binding with TF activity. Given these limitations, mapping transcriptional regulation of the neuroendocrine phenotype remains a work in progress.

References

Aguilera, G. and Liu, Y. (2012) The molecular physiology of CRH neurons. *Frontiers in Neuroendocrinology*, **33**, 67–84.

Ambasudhan, R., Talantova, M., Coleman, R., *et al.* (2011) Direct reprogramming of adult human fibroblasts to functional neurons under defined conditions. *Cell Stem Cell*, **9**, 113–118.

Andres, M.E., Burger, C., Peral-Rubio, M.J., *et al.* (1999) CoREST: a functional corepressor required for regulation of neural-specific gene expression. *Proceedings of the National Academy of Sciences of the United States of America*, **96**, 9873–9878.

Ballas, N., Grunseich, C., Lu, D.D., Speh, J.C. and Mandel, G. (2005) REST and its corepressors mediate plasticity of neuronal gene chromatin throughout neurogenesis. *Cell*, **121**, 645–657.

Bithell, A. (2011) REST: transcriptional and epigenetic regulator. *Epigenomics*, **3**, 47–58. [**In-depth review on REST function and roles in disease.**]

Castro, D.S., Martynoga, B., Parras, C., *et al.* (2011) A novel function of the proneural factor Ascl1 in progenitor proliferation identified by genome-wide characterization of its targets. *Genes & Development*, **25**, 930–945.

Chen, G.L. and Miller, G.M. (2013) Extensive alternative splicing of the repressor element silencing transcription factor linked to cancer. *PLoS One*, **8**, e62217.

Chong, J.A., Tapia-Ramirez, J., Kim, S., *et al.* (1995) REST: a mammalian silencer protein that restricts sodium channel gene expression to neurons. *Cell*, **80**, 949–957.

Conaco, C., Otto, S., Han, J.J. and Mandel, G. (2006) Reciprocal actions of REST and a microRNA promote neuronal identity. *Proceedings of the National Academy of Sciences of the United States of America*, **103**, 2422–2427.

Coulson, J.M. (2002) Positive and negative regulators of the vasopressin gene promoter in small cell lung cancer. *Progress in Brain Research*, **139**, 329–343.

Coulson, J.M. (2005) Transcriptional regulation: cancer, neurons and the REST. *Current Biology*, **15**, R665–668.

Coulson, J.M., Fiskerstrand, C.E., Woll, P.J. and Quinn, J.P. (1999a) E-box motifs within the human vasopressin gene promoter contribute to a major enhancer in small-cell lung cancer. *Biochemical Journal*, **344**, 961–970.

Coulson, J.M., Fiskerstrand, C.E., Woll, P.J. and Quinn, J.P. (1999b) Arginine vasopressin promoter regulation is mediated by a neuron-restrictive silencer element in small cell lung cancer. *Cancer Research*, **59**, 5123–5127.

Coulson, J.M., Edgson, J.L., Woll, P.J. and Quinn, J.P. (2000) A splice variant of the neuron-restrictive silencer factor repressor is expressed in small cell lung cancer: a potential role in derepression of neuroendocrine genes and a useful clinical marker. *Cancer Research*, **60**, 1840–1844.

Coulson, J.M., Edgson, J.L., Marshall-Jones, Z.V., Mulgrew, R., Quinn, J.P. and Woll, P.J. (2003) Upstream stimulatory factor activates the vasopressin promoter via multiple motifs, including a non-canonical E-box. *Biochemical Journal*, **369**, 549–561.

D'Alessandro, R., Klajn, A., Stucchi, L., Podini, P., Malosio, M.L. and Meldolesi, J. (2008) Expression of the neurosecretory process in PC12 cells is governed by REST. *Journal of Neurochemistry*, **105**, 1369–1383.

D'Alessandro, R., Klajn, A. and Meldolesi, J. (2009) Expression of dense-core vesicles and of their exocytosis are governed by the repressive transcription factor NRSF/REST. *Annals of the New York Academy of Sciences*, **1152**, 194–200.

Del Giacco, L., Pistocchi, A., Cotelli, F., Fortunato, A.E. and Sordino, P. (2008) A peek inside the neurosecretory brain through Orthopedia lenses. *Developmental Dynamics*, **237**, 2295–2303. [**Detailed overview of developmental roles for OTP.**]

De Moraes, D.C., Vaisman, M., Conceicao, F.L. and Ortiga-Carvalho, T.M. (2012) Pituitary development: a complex, temporal regulated process dependent on specific transcription factors. *Journal of Endocrinology*, **215**, 239–245. [**Concise review of transcriptional pathways in pituitary development.**]

Doghman, M., Figueiredo, B.C., Volante, M., Papotti, M. and Lalli, E. (2013) Integrative analysis of SF-1 transcription factor dosage impact on genome-wide binding and gene expression regulation. *Nucleic Acids Research*, **41**, 8896–8907.

Eichhorn, P.J., Rodon, L., Gonzalez-Junca, A., *et al.* (2012) USP15 stabilizes TGF-beta receptor I and promotes oncogenesis through the activation of TGF-beta signaling in glioblastoma. *Nature Medicine*, **18**, 429–435.

Ezzat, S. and Asa, S.L. (2008) The emerging role of the Ikaros stem cell factor in the neuroendocrine system. *Journal of Molecular Endocrinology*, **41**, 45–51.

Faronato, M., Patel, V., Darling, S., *et al.* (2013) The deubiquitylase USP15 stabilizes newly synthesized REST and rescues its expression at mitotic exit. *Cell Cycle*, **12**, 1964–1977.

Fulton, D.L., Sundararajan, S., Badis, G., *et al.* (2009) TFCat: the curated catalog of mouse and human transcription factors. *Genome Biology*, **10**, R29.

Gao, Z., Ding, P. and Hsieh, J. (2012) Profiling of REST-dependent microRNAs reveals dynamic modes of expression. *Frontiers of Neuroscience*, **6**, 67.

Ghorbel, M.T., Coulson, J.M. and Murphy, D. (2003) Cross-talk between hypoxic and circadian pathways: cooperative roles for hypoxia-inducible factor 1alpha and CLOCK in transcriptional activation of the vasopressin gene. *Molecular and Cellular Neursciences*, **22**, 396–404.

Greenway, D.J., Street, M., Jeffries, A. and Buckley, N.J. (2007) RE1 silencing transcription factor maintains a repressive chromatin environment in embryonic hippocampal neural stem cells. *Stem Cells*, **25**, 354–363.

Greenwood, M., Bordieri, L., Greenwood, M.P., *et al.* (2014) Transcription factor CREB3L1 regulates vasopressin gene expression in the rat hypothalamus. *Journal of Neuroscience*, **34**, 3810–3820.

Grimes, J.A., Nielsen, S.J., Battaglioli, E., *et al.* (2000) The co-repressor mSin3A is a functional component of the REST-CoREST repressor complex. *Journal of Biological Chemistry*, **275**, 9461–9467.

Guan, H. and Ricciardi, R.P. (2012) Transformation by E1A oncoprotein involves ubiquitin–mediated proteolysis of the neuronal and tumor repressor REST in the nucleus. *Journal of Virology*, **86**, 5594–5602.

Guardavaccaro, D., Frescas, D., Dorrello, N.V., *et al.* (2008) Control of chromosome stability by the beta-TrCP-REST-Mad2 axis. *Nature*, **452**, 365–369.

Gurrola-Diaz, C., Lacroix, J., Dihlmann, S., Becker, C.M. and von Knebel Doeberitz, M. (2003) Reduced expression of the neuron restrictive silencer factor permits transcription of glycine receptor alpha1 subunit in small-cell lung cancer cells. *Oncogene*, **22**, 5636–5645.

Hohl, M. and Thiel, G. (2005) Cell type-specific regulation of RE-1 silencing transcription factor (REST) target genes. *European Journal of Neuroscience*, **22**, 2216–2230.

Huang, Y., Myers, S.J. and Dingledine, R. (1999) Transcriptional repression by REST: recruitment of Sin3A and histone deacetylase to neuronal genes. *Nature Neuroscience*, **2**, 867–872.

Huang, Z., Wu, Q., Guryanova, O.A., *et al.* (2011) Deubiquitylase HAUSP stabilizes REST and promotes maintenance of neural progenitor cells. *Nature Cell Biology*, **13**, 142–152.

Itoh, Y., Moriyama, Y., Hasegawa, T., Endo, T.A., Toyoda, T. and Gotoh, Y. (2013) Scratch regulates neuronal migration onset via an epithelial-mesenchymal transition-like mechanism. *Nature Neuroscience*, **16**, 416–425.

Jin, X.W., Shearman, L.P., Weaver, D.R., Zylka, M.J., DeVries, G.J. and Reppert, S.M. (1999) A molecular mechanism regulating rhythmic output from the suprachiasmatic circadian clock. *Cell*, **96**, 57–68.

Johnson, B.E., Chute, J.P., Rushin, J., *et al.* (1997) A prospective study of patients with lung cancer and hyponatremia of malignancy. *American Journal of Respiratory and Critical Care Medicine*, **156**, 1669–1678.

Johnson, D.S., Mortazavi, A., Myers, R.M. and Wold, B. (2007) Genome-wide mapping of *in vivo* protein-DNA interactions. *Science*, **316**, 1497–1502. [**The first example of ChIP-seq to investigate genome-wide TF occupancy.**]

Johnson, R., The, C.H., Kunarso, G., *et al.* (2008) REST regulates distinct transcriptional networks in embryonic and neural stem cells. *PLoS Biology*, **6**, e256.

Johnson, R., Samuel, J, Ng, C.K., Jauch, R., Stanton, L.W. and Wood, I.C. (2009) Evolution of the vertebrate gene regulatory network controlled by the transcriptional repressor REST. *Molecular Biology and Evolution*, **26**, 1491–1507.

Jolma, A., Yan, J., Whitington, T., *et al.* (2013) DNA-binding specificities of human transcription factors. *Cell*, **152**, 327–339.

Kaestner, K.H. (2010) The FoxA factors in organogenesis and differentiation. *Current Opinion in Genetics & Development*, **20**, 527–532.

Kamal, M.M., Sathyan, P., Singh, S.K., *et al.* (2012) REST regulates oncogenic properties of glioblastoma stem cells. *Stem Cells*, **30**, 405–414.

Kaneko, N., Hwang, J.Y., Gertner, M., Pontarelli, F. and Zukin, R.S. (2014). Casein kinase 1 suppresses activation of REST in insulted hippocampal neurons and halts ischemia-induced neuronal death. *Journal of Neuroscience*, **34**, 6030–6039.

Kino, T., Hurt, D.E., Ichijo, T., Nader, N. and Chrousos, G.P. (2010) Noncoding RNA gas5 is a growth arrest- and starvation-associated repressor of the glucocorticoid receptor. *Science Signaling*, **3**, ra8.

Korosi, A., Shanabrough, M., McClelland, S., *et al.* (2010) Early-life experience reduces excitation to stress-responsive hypothalamic neurons and reprograms the expression of corticotropin-releasing hormone. *Journal of Neuroscience*, **30**, 703–713.

Kreisler, A., Strissel, P.L., Strick, R., Neumann, S.B., Schumacher, U. and Becker, C.M. (2010) Regulation of the NRSF/REST gene by methylation and CREB affects the cellular phenotype of small-cell lung cancer. *Oncogene*, **29**, 5828–5838.

Kuwabara, T., Hsieh, J., Nakashima, K., Taira, K. and Gage, F.H. (2004) A small modulatory dsRNA specifies the fate of adult neural stem cells. *Cell*, **116**, 779–793.

Lan, M.S. and Breslin, M.B. (2009) Structure, expression, and biological function of INSM1 transcription factor in neuroendocrine differentiation. *FASEB Journal*, **23**, 2024–2033.

Lee, J.H., Shimojo, M., Chai, Y.G. and Hersh, L.B. (2000) Studies on the interaction of REST4 with the cholinergic repressor element-1/neuron restrictive silencer element. *Molecular Brain Research*, **80**, 88–98.

Li, B., Carey, M. and Workman, J.L. (2007) The role of chromatin during transcription. *Cell*, **128**, 707–719.

Linnoila, R.I., Zhao, B., DeMayo, J.L., *et al.* (2000) Constitutive achaete-scute homologue-1 promotes airway dysplasia and lung neuroendocrine tumors in transgenic mice. *Cancer Research*, **60**, 4005–4009.

Lu, M., Zheng, L., Han, B., *et al.* (2011) REST regulates DYRK1A transcription in a negative feedback loop. *Journal of Biological Chemistry*, **286**, 10755–10763.

Luscombe, N.M., Austin, S.E., Berman, H.M. and Thornton, J.M. (2000) An overview of the structures of protein-DNA complexes. *Genome Biology*, **1**, REVIEWS001.

Moss, A.C., Jacobson, G.M., Walker, L.E., Blake, N.W., Marshall, E. and Coulson, J.M. (2009) SCG3 transcript in peripheral blood is a prognostic biomarker for REST-deficient small cell lung cancer. *Clinical Cancer Research*, **15**, 274–283.

Nakakura, E.K., Watkins, D.N., Schuebel, K.E., *et al.* (2001) Mammalian Scratch: a neural-specific Snail family transcriptional repressor. *Proceedings of the National Academy of Sciences of the United States of America*, **98**, 4010–4015.

Negrini, S., Prada, I., D'Alessandro, R. and Meldolesi, J. (2013) REST: an oncogene or a tumor suppressor? *Trends in Cell Biology*, **23**, 289–295.

Ng, S.Y., Johnson, R. and Stanton, L.W. (2012) Human long non-coding RNAs promote pluripotency and neuronal differentiation by association with chromatin modifiers and transcription factors. *EMBO Journal*, **31**, 522–533.

Ocejo-Garcia, M., Baobkah, T.A., Ashurst, H.L., *et al.* (2005) Roles for USF-2 in lung cancer proliferation and bronchial carcinogenesis. *Journal of Pathology*, **206**, 151–159.

Ooi, L. and Wood, I.C. (2007) Chromatin crosstalk in development and disease: lessons from REST. *Nature Reviews. Genetics*, **8**, 544–554. [*In-depth review on REST focused on molecular mechanisms.*]

Otto, S.J., McCorkle, S.R., Hover, J., *et al.* (2007) A new binding motif for the transcriptional repressor REST uncovers large gene networks devoted to neuronal functions. *Journal of Neuroscience*, **27**, 6729–6739.

Ovando-Roche, P., Yu, J.S., Testori, S., Ho, C. and Cui, W. (2014) TRF2-mediated stabilization of hREST4 is critical for the differentiation and maintenance of neural progenitors. *Stem Cells*, **32**, 2111–2122.

Packer, A.N., Xing, Y., Harper, S.Q., Jones, L. and Davidson, B.L. (2008) The bifunctional microRNA miR-9/miR-9* regulates REST and CoREST and is downregulated in Huntington's disease. *Journal of Neuroscience*, **28**, 14341–14346.

Palm, K., Belluardo, N., Metsis, M. and Timmusk, T. (1998) Neuronal expression of zinc finger transcription factor REST/NRSF/XBR gene. *Journal of Neuroscience*, **18**, 1280–1296.

Palm, K., Metsis, M. and Timmusk, T. (1999) Neuron-specific splicing of zinc finger transcription factor REST/NRSF/XBR is frequent in neuroblastomas and conserved in human, mouse and rat. *Brain Research. Molecular Brain Research*, **72**, 30–39.

Pance, A., Livesey, F.J. and Jackson, A.P. (2006) A role for the transcriptional repressor REST in maintaining the phenotype of neurosecretory-deficient PC12 cells. *Journal of Neurochemistry*, **99**, 1435–1444.

Paterson, J.M., Morrison, C.F., Mendelson, S.C., McAllister, J. and Quinn, J.P. (1995) An upstream stimulatory factor (usf) binding motif is critical for rat preprotachykinin–a promoter activity in PC12 cells. *Biochemical Journal*, **310**, 401–406.

Prince, K.L., Walvoord, E.C. and Rhodes, S.J. (2011) The role of homeodomain transcription factors in heritable pituitary disease. *Nature Reviews. Endocrinology*, **7**, 727–737. [*In-depth review of homeobox TFs in human pituitary disease.*]

Quinn, J.P., Bubb, V.J., Marshall-Jones, Z.V. and Coulson, J.M. (2002) Neuron restrictive silencer factor as a modulator of neuropeptide gene expression. *Regulatory Peptides*, **108**, 135–141.

Qureshi, I.A. and Mehler, M.F. (2012) Emerging roles of non-coding RNAs in brain evolution, development, plasticity and disease. *Nature Reviews. Neuroscience*, **13**, 528–541. [*Review of ncRNA mechanisms in neurology.*]

Rada-Iglesias, A., Ameur, A., Kapranov, P., *et al.* (2008) Whole-genome maps of USF1 and USF2 binding and histone H3 acetylation reveal new aspects of promoter structure and candidate genes for common human disorders. *Genome Research*, **18**, 380–392.

Raj, B., O'Hanlon, D., Vessey, J.P., *et al.* (2011) Cross-regulation between an alternative splicing activator and a transcription repressor controls neurogenesis. *Molecular Cell*, **43**, 843–850.

Rockowitz, S., Lien, W.H., Pedrosa, E., *et al.* (2014) Comparison of REST cistromes across human cell types reveals common and context-specific functions. *PLoS Compuational Biology*, **10**, e1003671. [*Insightful integration of multiple REST ChIP-Seq datasets.*]

Sausville, E., Carney, D. and Battey, J. (1985) The human vasopressin gene is linked to the oxytocin gene and is selectively expressed in a cultured lung cancer cell line. *Journal of Biological Chemistry*, **260**, 10236–10241.

Schoenherr, C.J. and Anderson, D.J. (1995). The neuron-restrictive silencer factor (NRSF): a coordinate repressor of multiple neuron-specific genes. *Science*, **267**, 1360–1363.

Shimojo, M. (2011) RE1-silencing transcription factor (REST) and REST-interacting LIM domain protein (RILP) affect P19CL6 differentiation. *Genes to Cells*, **16**, 90–100.

Shimojo, M. and Hersh, L.B. (2003) REST/NRSF-interacting LIM domain protein, a putative nuclear translocation receptor. *Molecular Cell Biology*, **23**, 9025–9031.

Shimojo, M., Lee, J.H. and Hersh, L.B. (2001) Role of zinc finger domains of the transcription factor neuron-restrictive silencer factor/repressor element-1 silencing transcription factor in DNA binding and nuclear localization. *Journal of Biological Chemistry*, **276**, 13121–13126.

Shimojo, M., Shudo, Y., Ikeda, M., Kobashi, T. and Ito, S. (2013) Small cell lung cancer-specific isoform of RE1-silencing transcription factor (REST) is regulated by neural-specific Ser/Arg repeat-related protein of 100 kDa (nSR100). *Molecular Cancer Research*, **11**, 1258–1268.

Simeone, A., D'Apice, M.R., Nigro, V., *et al.* (1994) Orthopedia, a novel homeobox-containing gene expressed in the developing CNS of both mouse and Drosophila. *Neuron*, **13**, 83–101.

Spencer, E.M., Chandler, K.E., Haddley, K., *et al.* (2006) Regulation and role of REST and REST4 variants in modulation of gene expression in in vivo and in vitro in epilepsy models. *Neurobiology of Disease*, **24**, 41–52.

Spitz, F. and Furlong, E.E. (2012) Transcription factors: from enhancer binding to developmental control. *Nature Reviews. Genetics*, **13**, 613–626. [*In-depth discussion of integrated transcription factor function.*]

Szarek, E., Cheah, P.S., Schwartz, J. and Thomas, P. (2010) Molecular genetics of the developing neuroendocrine hypothalamus. *Molecular Cell Endocrinology*, **323**, 115–123. [*Overview of signaling pathways and TFs required for NE hypothalamus development.*]

Tsai, M.C., Manor, O., Wan, Y., *et al.* (2010) Long noncoding RNA as modular scaffold of histone modification complexes. *Science*, **329**, 689–693.

Uchida, S., Hara, K., Kobayashi, A., *et al.* (2010) Early life stress enhances behavioral vulnerability to stress through the activation of REST4-mediated gene transcription in the medial prefrontal cortex of rodents. *Journal of Neuroscience*, **30**, 15007–15018.

Vaquerizas, J.M., Kummerfeld, S.K., Teichmann, S.A. and Luscombe, N.M. (2009) A census of human transcription factors: function, expression and evolution. *Nature Reviews. Genetics*, **10**, 252–263. [*Comprehensive high-quality census of TFs in the human genome.*]

Viney, T.J., Schmidt, T.W., Gierasch, W., *et al.* (2004) Regulation of the cell-specific calcitonin/CGRP enhancer by USF and the Foxa2 forkhead protein. *Journal of Biological Chemistry*, **279**, 49948–49955.

Visvanathan, J., Lee, S., Lee, B., Lee, J.W. and Lee, S.K. (2007) The microRNA miR-124 antagonizes the anti-neural REST/SCP1 pathway during embryonic CNS development. *Genes & Development*, **21**, 744–749.

Wang, K.C. and Chang, H.Y. (2011) Molecular mechanisms of long noncoding RNAs. *Molecular Cell*, **43**, 904–914.

Westbrook, T.F., Hu, G., Ang, X.L., *et al.* (2008) SCFbeta-TRCP controls oncogenic transformation and neural differentiation through REST degradation. *Nature*, **452**, 370–374.

Whitlock, K.E., Smith, K.M., Kim, H. and Harden, M.V. (2005) A role for foxd3 and sox10 in the differentiation of gonadotropin-releasing hormone (GnRH) cells in the zebrafish Danio rerio. *Development*, **132**, 5491–5502.

Wu, J. and Xie, X. (2006) Comparative sequence analysis reveals an intricate network among REST, CREB and miRNA in mediating neuronal gene expression. *Genome Biology*, **7**, R85.

Yoo, A.S., Staahl, B.T., Chen, L. and Crabtree, G.R. (2009) MicroRNA-mediated switching of chromatin-remodelling complexes in neural development. *Nature*, **460**, 642–646.

Yu, H.B., Johnson, R., Kunarso, G. and Stanton, L.W. (2011) Coassembly of REST and its cofactors at sites of gene repression in embryonic stem cells. *Genome Research*, **21**, 1284–1293.

Zhang, P., Pazin, M.J., Schwartz, C.M., *et al.* (2008) Nontelomeric TRF2-REST interaction modulates neuronal gene silencing and fate of tumor and stem cells. *Current Biology*, **18**, 1489–1494.

Zhang, P., Casaday-Potts, R., Precht, P., *et al.* (2011) Nontelomeric splice variant of telomere repeat-binding factor 2 maintains neuronal traits by sequestering repressor element 1-silencing transcription factor. *Proceedings of the National Academy of Sciences of the United States of America*, **108**, 16434–16439.

Zheng, D., Zhao, K. and Mehler, M.F. (2009) Profiling RE1/REST-mediated histone modifications in the human genome. *Genome Biology*, **10**, R9.

Zhu, Y., Liu, C., Cui, Y., Nadiminty, N., Lou, W. and Gao, A.C. (2014) Interleukin-6 induces neuroendocrine differentiation (NED) through suppression of RE-1 silencing transcription factor (REST). *The Prostate*, **74**, 1086–1094.

CHAPTER 6

Epigenetics

Chris Murgatroyd

School of Healthcare Science, Manchester Metropolitan University, Manchester, UK

6.1 Introduction

There is a close relationship between the quality of early life and health in later life (for review see Gluckman *et al.*, 2008). Many studies illustrate that various aspects of the early environment can lead to dramatic changes in physiological and neurological development underlying disease. For example, adverse conditions during periods of early life can shape individual differences in neuroendocrine function, programming a vulnerability to stress-related disorders throughout later life, dependent on the degree of 'match' and 'mismatch' between early and later life environments.

This, therefore, raises the question as to how environmental experiences can become integrated at the molecular and cellular level to lead to such long-term changes. Developmental plasticity defines the ability of an organism to adapt to the early environment and program long-lasting biological alterations on the assumption that the environmental conditions during this early period will persist throughout later life. Taking the example of adverse experiences during early life, this might severely influence the experience-dependent maturation of structures underlying emotional functions and neuroendocrine responses to stress, such as the hypothalamo-pituitary-adrenal (HPA) system – an integral component of the body's stress response – leading to increased stress responsivity in adulthood. Indeed, depressed patients with a history of childhood abuse or neglect are often characterized by hyperactivity of the HPA axis.

Current work suggests that so-called epigenetic mechanisms of gene regulation, which alter the activity of genes without changing the order of their DNA sequence, could explain how early life experiences can leave indelible chemical marks on the brain and influence both physical and mental health later in life, even when the initial trigger is long gone. In this chapter, we will address the different epigenetic mechanisms involved in regulating gene expression and describe some of the methods by which these processes can be measured. We will then discuss some clinical and animal studies that have investigated how biological stress systems, particularly the HPA axis, are shaped by adversity and the role of epigenetic systems in mediating transcriptional programming of key genes. The dynamic nature of epigenetic mechanisms in neuroendocrine regulation may have important implications when considering the possibility of therapeutic interventions.

Molecular Neuroendocrinology: From Genome to Physiology, First Edition. Edited by David Murphy and Harold Gainer.
© 2016 John Wiley & Sons, Ltd. Published 2016 by John Wiley & Sons, Ltd.
Companion website: www.wiley.com/go/murphy/neuroendocrinology

6.2   Early life adversity shapes the HPA axis

The body activates stress systems on exposure to a stressor; the autonomic nervous system initiates a rapid and relatively short-lived 'fight-or-flight' response, whilst the HPA axis is slower, instigating a more protracted response. Tight regulation of the HPA axis is therefore core to the long-term control of systems governing stress responsivity.

Following a stressor, the neuropeptides corticotropin-releasing hormone (CRH) and arginine vasopressin (AVP) are released from the paraventricular nucleus (PVN) of the hypothalamus into the portal vessel system. These bind specific receptors (the CRHR1 and V1b receptors for CRH and AVP, respectively) in the anterior pituitary that stimulate the release of adrenocorticotropic hormone (ACTH). This in turn acts on the adrenal cortex to synthesize and release glucocorticoid hormones, i.e. cortisol/corticosterone. These then mobilize glucose from energy stores and increase cardiovascular tone, among further widespread effects. Feedback loops, primarily mediated at the PVN and pituitary through glucocorticoid receptors (GR), restrain the responsiveness of the HPA axis to reset the system to baseline activity (Figure 6.1). In sum, the forward loop prepares the organism to anticipate and respond optimally to a threat, while the feedback loop ensures returning efficiently to a homeostatic balance when it is no longer challenged (for review see de Kloet *et al.*, 1998).

It has been shown that the negative feedback control of the HPA axis can become dysregulated, particularly following periods of chronic stress. This may critically impact the development of affective disorders; altered HPA activity is one of the most commonly

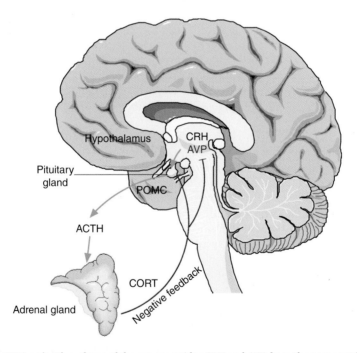

Figure 6.1 The HPA axis. The release of the neuropeptides CRH and AVP from the paraventricular nucleus of the hypothalamus into the portal blood vessels leads to stimulation of pituitary ACTH secretion and in turn of corticosterone from the adrenals. The activational effects of the HPA axis are counteracted by the negative feedback inhibitory effects of glucocorticoid receptors expressed in the hippocampus, hypothalamus, and pituitary. ACTH, adrenocorticotropic hormone; AVP, arginine vasopressin; CORT, corticosterone; CRH, corticotropin-releasing hormone; POMC, pro-opiomelanocortin.

observed neuroendocrine symptoms in major depressive disorder (MDD) and dysregulation of cortisol secretion can be found in as many as 80% of depressed patients if controlled for age ranges (for review see Holsboer, 2000). Childhood stress has also been shown to be a strong predictor of impaired inhibitory feedback regulation of the HPA axis, with evidence linking to a role of CRH and/or AVP systems. For example, postmortem studies of brain tissue reveal elevated CRH and AVP in the hypothalamus of depressed individuals. Studies in rodent models further support the concept that exposure to a chronic stressor can lead to long-term changes in HPA regulation and behavior stemming from changes in neuropeptide regulation (for review see Murgatroyd and Spengler, 2011a).

Given that early life stress may lead to enduring dysregulation of the HPA axis, and the close link between HPA dysfunction and major depression, it is widely suggested that early life stress may predispose individuals to psychiatric diseases in later life. One of the most studied models for early life stress is maternal separation in which rodents are separated for around 3 hours a day for the first 2 weeks of life. This procedure can lead to persistent increases in anxiety-related behaviors and life-long hyperactivity of the HPA axis in response to stressors (for review see Holmes *et al.*, 2005). Elucidation of neuroendocrine changes underlying the persistent effects of early life stress in rodents has become a major research area. A growing number of reports have documented permanent increases in neurotransmission and hypothalamic expression of CRH and AVP following early life stress. In addition, the ability of hippocampal GRs to attenuate the HPA axis may be persistently disrupted; rats subjected to maternal separation exhibit significantly reduced expression of forebrain GRs and an impaired negative feedback inhibition of the HPA axis in adults (for review see Murgatroyd and Spengler, 2011a).

6.3 Epigenetic mechanisms: changes in the regulation of gene activity and expression that are not dependent on gene sequence

There is a strong body of data demonstrating that various aspects of the early environment can program long-term changes in neuroendocrine regulation of the HPA axis. This begs the question as to how early life exposure to adverse environments is able to induce such long-lasting alterations in hormonal regulation and behavioral responses lasting into adulthood.

> In 1940, Conrad Waddington introduced the word '**epigenetics,**' fusing 'epigenesis' and 'genetics' together to refer to the study of the processes by which the genotype gives rise to the phenotype.

'Epigenetics' is a term used to describe the study of stable alterations in gene expression potential that arise during development, differentiation and under the influence of the environment (for review see Jaenisch and Bird, 2003). For example, cells of a multicellular organism, though genetically identical, are functionally and structurally distinct owing to the different expression patterns of a multitude of genes. Many of these gene regulatory differences arise during development and are subsequently retained through cell division. Therefore, **epigenetic marks** that underlie gene regulatory changes are tissue specific; this is in contrast to DNA sequence that is identical in all tissues. Genomes can therefore be considered to contain two layers of information: first, there is the DNA sequence that is conserved throughout life and mostly identical in all cells of a body, and second, there are the epigenetic marks that are cell specific.

Epigenetic processes therefore enable a cell and organism to integrate intrinsic and environmental signals into the genome, resulting in regulatory control of gene expression and thus facilitating adaptation. In this way, epigenetic mechanisms could be thought of as conferring plasticity to the hard-coded genome. In the context of the early life environment, epigenetic changes offer a plausible mechanism by which early experiences could be integrated into the genome as a kind of memory to program adult hormonal and behavioral responses.

The 'epigenome' refers to the collective of coordinated epigenetic marks on a genome. These marks are able to govern accessibility of the DNA to the machinery driving gene expression; inaccessible genes become silenced whereas accessible genes are actively transcribed. Although the understanding of the interplay between epigenetic modifications is still evolving, the methylation of DNA at the cytosine side chain in cytosine-guanine (CpG) dinucleotides and modification of core histones that package the DNA into **chromatin** represent the best understood epigenetic marks.

6.3.1 DNA methylation – a process by which a methyl group is added to a DNA nucleotide

DNA methylation describes the addition of a methyl group to DNA. This modification specifically occurs at the 5 position of cytosine residues within a CG sequence, known as CpG – the 'p' in CpG refers to the phosphodiester bond between the cytosine and the guanine, indicating that the C and the G are next to each other. This generally leads to transcriptional repression or silencing when located at gene regulatory regions. Since DNA methylation represents a covalent bond, it is often considered to be a relatively stable epigenetic mark.

In 1975, Riggs, Holliday and Pugh, by critically reviewing the literature, proposed that chemical modifications of the DNA (i.e. methylation) could influence gene expression, introducing hypotheses about maintenance and *de novo* methylation and the involvement of this in differentiation and X inactivation.

6.3.1.1 Occurrence of DNA methylation

CpG sequences are found at lower levels than expected in mammalian genomes. The vast majority (70–80%) of these are methylated. Around 85% of all CpGs are located in repetitive sequences such as transposons which include Alu elements, long terminal repeats (LTR)-retrotransposons and long and short interspersed nuclear elements (LINE and SINE, respectively) that together constitute around half of the human genome. These DNA elements could interfere with the regulation of gene expression and genome structure by means of insertions, deletions, inversions, and translocations. The genome limits this potential damage by ensuring that these sequences are highly methylated and effectively silenced. This explains why the vast majority of the genome is methylated.

The other 15% of CpG dinucleotides not found in repetitive elements typically cluster within GC-rich regions known as 'CpG islands' (Figure 6.2A). CpG islands are defined as regions of greater than 500 bp that have cytosine-guanine content of greater than 55%. Up to 60% of CpG islands are in the 5' regulatory (promoter) regions of genes. However, CpG islands that are not in promoter regions can also be found within coding and non-coding regions of genes, which may be targets for methylation. Accounting for ~1% of the genome, these CpG islands are largely unmethylated in somatic cells while regions showing lower CpG density are more frequently methylated (Weber *et al.*, 2007).

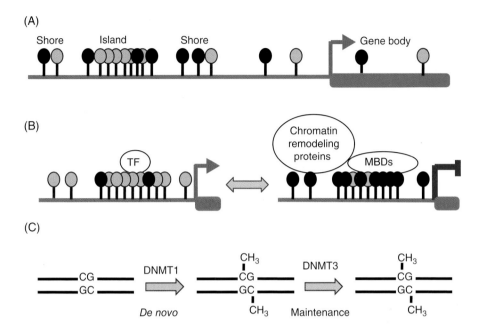

Figure 6.2 DNA methylation. (A) GC-rich regions are known as CpG islands and areas around 2 kb away, within sequences of intermediate CpG density, are referred to as CpG shores. (*Black circle*, methylated. *Blue circle*, unmethylated.) (B) DNA methylation (*black circle*) can lead to gene repression through inhibiting transcription factor (TF) binding and recruitment of methylated DNA-binding proteins (MBDs) and chromatin remodeling factors. (C) DNMT1 is considered the primary maintenance methyltransferase while DNMT3A and DNMT3B are considered *de novo* methyltransferases important in establishing methylation.

Though cytosine methylation (**5mC**) in mammalian cells occurs predominantly at CpG dinucleotides, in plant cells methylation is frequently found in both CpG and non-CpG (i.e. CHG and CHH, where H is A, C or T) sequences. However, very recent studies have also revealed the presence of non-CpG methylation (CpH) in a few mammalian cell types, including embryonic stem cells, oocytes, and brain cells. Interestingly, it appears that both methylated CpGs and mCpHs can repress transcription *in vitro* and bind methyl-CpG binding protein 2 (MeCP2) in a similar way (Guo *et al.*, 2014).

> In 1985, Adrian Bird and colleagues described a small fraction of the mouse genome that is frequently cleaved by a methylation-sensitive restriction enzyme as 'CpG islands,' i.e. CpG-rich fragments with very low levels of methylation.

6.3.1.2   Function of DNA methylation

In general, the endpoint of DNA methylation is either long-term silencing or fine-tuning of gene expression potential. The insertion of methyl groups changes the appearance and structure of DNA, which may either directly block DNA binding of transcription factors or attract factors that preferentially bind to methylated or unmethylated DNA to interfere with transcription factor accessibility (Figure 6.2B). Three families of proteins that bind methylated DNA have been identified: methyl-CpG-binding domain (MBD) proteins, Kaiso and Kaiso-like proteins, and SRA domain proteins. By recruiting these proteins, DNA methylation marks can promote particular histone states, such as deacetylation, thus enabling posttranslational

histone modifications. These complexes can additionally recruit **DNA methyltransferases** (DNMTs) to promote, maintain and enforce gene repression, compatible with a reciprocal cross-talk between DNA methylation and chromatin marks in the regulation of gene transcription. In general, patterns of methylation therefore tend to correlate with chromatin structure. For example, active regions of the chromatin associate with hypomethylated DNA whereas hypermethylated DNA is found with inactive chromatin. MBD protein 2 (MeCP2), a member of the MBD family, binds to methyl-CpG and recruits HDACs and DNMTs, which promote chromosome condensation, DNA methylation, and transcriptional repression.

The conventional view is therefore of DNA methylation as a silencing mark. Hence, promoters and enhancer elements of transcriptionally active genes are usually methylated but may become silenced once targeted by DNA methylation. Although CpG islands, promoters, and enhancer regions appear to attract most attention in regard to DNA methylation and gene transcription, it is becoming increasingly clear that methylation at other gene elements plays important regulatory roles. Indeed, genome-wide sequencing of DNA methylation patterns among different tissues is revealing the presence and locations of tissue-specific, differentially methylated regions. Strikingly, many of these do not seem to occur in CpG islands themselves but tend to locate in regions next to CpG islands around 2 kb away within sequences of intermediate CpG density. These are now referred to as 'CpG shores' (Irizarry *et al.*, 2009) (see Figure 6.2A). Further regions can be divided up as 'CpG shelves' as 2 kb regions extending from the shores while CpGs located yet further away from CpG islands are defined as being in 'open sea.' In contrast to repression-associated DNA methylation in promoters and enhancers, higher CpG methylation within gene bodies is being frequently observed in transcriptionally active genes, perhaps suggesting a further functional role of DNA methylation in transcriptional regulation.

6.3.1.3 Establishment and maintenance of DNA methylation – DNMTs

When cells divide, DNA is replicated. However, the DNA replication machinery is unable to copy DNA methylation and retain such marks after cell division. As such, cells contain a set of specific methylating enzymes, firstly, to establish new cell type-specific DNA methylation patterns and, secondly, to maintain methylation marks once cells undergo division and replication of their genome. This is performed by a family of enzymes known as DNA methyltransferases (DNMTs) that use S-adenosylmethionine as a methyl donor. DNMT1 acts as the maintenance DNMT by recognizing hemimethylated DNA and methylating appropriate cytosines in newly synthesized daughter strands formed during replication. As such, DNA methylation patterns established during embryonic development are copied through somatic cell divisions in order to maintain a gene in a transcriptionally active or inactive state. In contrast, DNMT3A and DNMT3B can methylate unmethylated DNA which is supportive of their roles as *de novo* methylases (see Figure 6.2C). They also contain similar domains for binding to chromatin.

Obviously, the activity of DNMTs is crucial to the vast array of developmental programs in mammals and is, therefore, tightly regulated. During early embryogenesis, epigenetic silencing of genes from either maternal or paternal origins occurs predominantly through the DNA methylation activity of the 'maintenance' methyltransferase DNMT1 while further tissue-specific gene expression and DNA methylation during later postnatal development require also the activity of the *'de novo'* methyltransferases DNMT3a and DNMT3b (for review see Turek-Plewa and Jagodziński, 2005).

6.3.1.4 Removal of DNA methylation

As mentioned, 5mC can be a relatively stable epigenetic mark. However, DNA methylation can be a dynamic modification. For example, during development, there are increases and decreases in amounts of methylation occurring both at the global and gene-specific levels. Such DNA 'demethylation' can be achieved either passively, through simply not methylating the newly replicated DNA strand, or actively by a replication-independent process. One striking example of active demethylation in mammalian cells is of the global changes in DNA methylation taking place in fertilized oocytes, where methylation is rapidly lost from the paternal genome.

Though the active removal of DNA methylation has been observed for a relatively long time, the actual identity of the players involved had remained elusive. However, recently, a new epigenetic mark, 5-hydroxymethylcytosine (5hmC), was discovered. Hailed as the 'sixth base,' this modified 5mC appears to be particularly abundant in human and mouse brains. Considering that cytosine residues can be hydroxymethylated to 5hmC, an active mechanism for demethylation has been hypothesized in which this modified 5hmC serves as an intermediate in the removal of methylated cytosines. A family of enzymes known as TET are thought to drive active demethylation through hydroxylating and then further oxidizing methylated cytosines to catalyze their conversion to 5hmC. Therefore, such modifications might regulate dynamic DNA methylation and gene expression patterns.

6.3.2   Histone modifications: changes that associate with distinct gene regulatory states

Chromatin is a complex of DNA wrapped around histone proteins. Discovered in avian red blood cell nuclei (unlike mammalian erythrocytes which are anucleate) by Albrecht Kossel, histones were dismissed, until relatively recently, as simply serving as inert packing material for DNA. Since Vincent Allfrey and colleagues' pioneering studies in the early 1960s, we know that these protein components of chromatin are posttranslationally modified, it becoming further apparent that these modified histones can repress gene transcription. Later research has now allowed us to understand that histones can play both positive and negative roles in gene expression, through various modifications forming the basis of a 'histone code' for gene regulation (Jenuwein and Allis, 2001).

> In 1882, Walther Flemming first identified and named chromatin due to its affinity for the red aniline dye he was using. He was also the first to discover that chromosomes split into two identical halves and segregate during mitosis, providing the basis for Mendel's rule for heredity.

Histones can be modified at their N-terminal tails by methylation, phosphorylation, acetylation, and ubiquitination as part of a histone signature serving to define accessibility to the DNA, i.e. densely packaged 'closed' chromatin (heterochromatin) in contrast to accessible 'open' chromatin (euchromatin). Histone acetylation is known to be a predominant signal for active chromatin configurations while some specific histone methylation reactions are associated with either gene silencing or activation.

> In 1869, Hoppe-Seyler and student Friedrich Miescher first isolated DNA. Fifteen years later, another of Hoppe-Seyler's students, Albrecht Kossel, isolated a series of acid-soluble proteins which he called histones. He received the Nobel Prize in Physiology and Medicine in 1910.

6.3.2.1   Occurrence of histone marks

In regard to chromatin, genomes can generally be divided into two distinct regions that characteristically differ in histone modifications, though there appear to be no simple rules. Euchromatin contains most of the active genes while heterochromatin regions are relatively compact structures containing mostly inactive genes. Two distinct heterochromatic regions have been defined: facultative heterochromatin consists of genomic regions containing genes that are differentially expressed through development and/or differentiation which then become silenced. A classic example of this type of heterochromatin is the inactive X-chromosome present within mammalian female cells, which is heavily marked by H3K27me3. Constitutive heterochromatin contains permanently silenced genes in genomic regions such as the centromeres and telomeres, generally characterized by high levels of H3K9me3, for example.

Euchromatin is also heterogeneous in histone mark distributions in which certain regions are enriched with specific histone modifications. For instance, in a transcriptionally active gene, enhancers may contain relatively high levels of H3K4me1, the transcriptional start site may have a high enrichment of H3K4me3, and H3K36me3 can be highly enriched throughout the transcribed region.

6.3.2.2   Regulation of histone marks

Though histone acetylation was first discovered in the 1960s, the enzymes responsible, histone acetyltransferases (HATs) and deacetylases (HDACs), were not identified until the mid-1990s. There is now an array of known histone-modifying enzymes that catalyze the addition or removal of an array of covalent modifications in histone and non-histone proteins, such as HATs that acetylate and HDACs which deacetylate histone tails. Further enzymes have been identified for methylation, demethylation, phosphorylation, ubiquitination, sumoylation, ADP-ribosylation, deimination, and proline isomerization. These enzymes are generally recruited through interactions with specific transcription factors that recognize and bind to certain *cis*-acting sequences in genes.

> In 1964, Allfrey, Faulkner and Mirsky demonstrated that acetylation of histones can reduce their efficacy as inhibitors of transcription, and suggested that this implied a dynamic mechanism for activation as well as repression of transcription.

6.3.2.3   Functions of histone marks

Chromatin remodeling plays a central role in gene regulation by providing the transcription machinery with dynamic access to an otherwise tightly packaged genome. Although histone modifications were initially thought to primarily affect gene expression through a direct effect on histone–DNA interaction, these marks also play an important role in recruiting proteins which can themselves influence transcription and other chromatin-templated processes. The previously mentioned 'histone code' can therefore be 'read' by specific proteins, which have particular domains that recognize certain histone modifications to produce an effect on gene expression. For example, acetylated histones are typically recognized by factors containing a bromodomain, whereas methylated histones are recognized by various domains, including the chromo- and PHD finger domains.

6.4   Methods of epigenetic analysis

No single method of epigenetic analysis will be appropriate for every application and experiment. Therefore, it is important to understand the type of information and limitation associated with each method to allow the selection of a particular method.

6.4.1   Techniques for measuring DNA methylation

There are many methods available to study DNA methylation and to quantify 5mC levels. The choice of technique will depend upon whether global, genome-wide or gene-specific analysis information is required.

> McGhee and Ginder published the first DNA methylation technique in 1979. Using restriction enzymes distinguishing between methylated and unmethylated DNA, they demonstrated that the β-globin locus was unmethylated in cells that expressed β-globin but methylated in other cell types that did not express this gene.

6.4.1.1   Global quantification of DNA methylation

Several methods can be used to detect the total overall (i.e. global) levels of 5mC in a genome. Genomic DNA can be digested into single nucleotides and total genomic 5mC can be quantified by high-performance liquid chromatography (HPLC) or liquid chromatography-mass spectroscopy (LC-MS). Though highly sensitive, such methods also require large amounts of DNA and specialized equipment and can be relatively time-consuming. As stated, DNA methylation predominantly occurs in repetitive elements such as transposons that constitute 45% of the human genome. Polymerase chain reaction (PCR) methods can therefore be used to measure approximate global DNA methylation by assessing methylation of the aforementioned repetitive DNA elements, such as LINE-1 methylation, for example.

As global methylation analyses provide no information on the genomic positions at which methylation is altered, it is difficult to link such changes to functional outcomes. However, it may be important to quantify global patterns of DNA methylation so as to evaluate the effect of certain diets or drugs, for example. Indeed, it is also becoming clear that a growing list of environmental factors and pollutants such as polycyclic aromatic hydrocarbons have global effects on DNA methylation.

6.4.1.2   Gene-specific quantification of DNA methylation

As global methylation analysis does not allow the determination of specific genomic positions of methylation, other high-resolution approaches are needed. Techniques for measuring locus or site-specific 5mC are generally based on one of three main pretreatments: sodium bisulfite, restriction enzyme digestion, and affinity capture.

Most gene-specific methylation analysis methods are primarily PCR based and combined with one of the three pretreatment approaches. Two different strategies have been used in the design of primers for such reactions. The first approach is based on design of primers that specifically amplify methylated or unmethylated templates, and is adopted by methylation-specific PCR (MSP) and quantitative MSP. The second approach is based on primers that amplify a region of the desired template including CpG islands, no matter what its methylation status is. In this case, methylation-independent PCR (MIP) is first performed and information on the methylation status of that region is obtained through post-PCR analysis techniques like bisulfite sequencing, restriction digestion, single-strand conformation analysis, and high-resolution melting.

6.4.1.3   Sodium bisulfite treatment deaminates unmethylated cytosines

During PCR amplification, DNA methylation information is lost as DNA polymerase cannot distinguish between methylated and unmethylated cytosines. Sodium bisulfite treatment modifies DNA prior to PCR, so enabling the methylation information to be preserved. This chemical deaminates cytosine to uracil – importantly, the rate of deamination of 5mC to thymine is

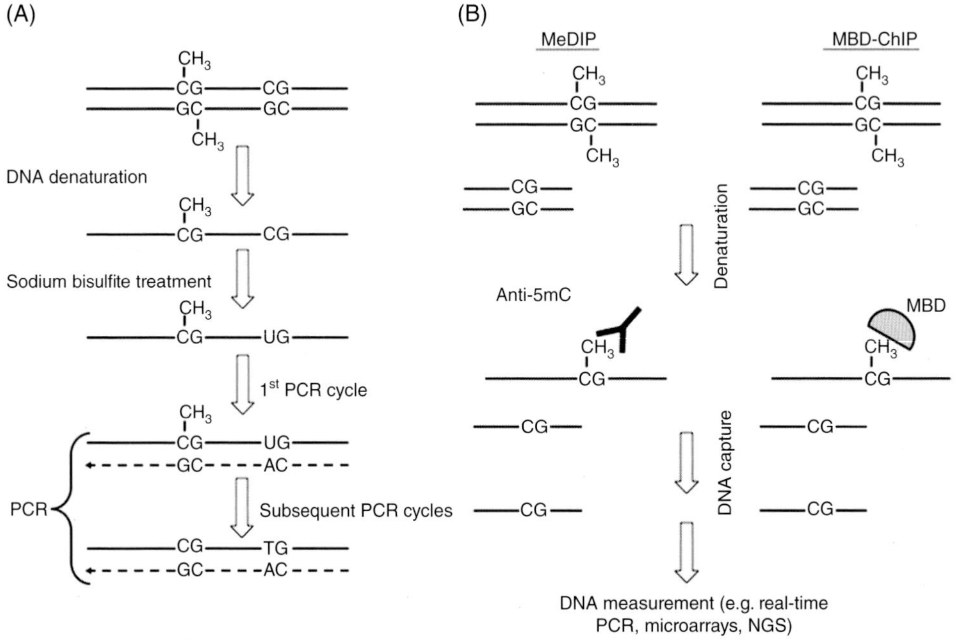

(A)

(B)

Figure 6.3 Methods for DNA methylation analysis. (A) Bisulfite conversion. DNA is denatured and then treated with sodium bisulfite to convert unmethylated cytosine (C) bases to uracil (U). This is then converted to thymine (T) by PCR. (B) MeDIP and MBD-ChIP analysis of methylation. Genomic DNA is denatured and then affinity purified with either an antibody (MeDIP) or a methyl-binding protein (MBD) (MBD-ChIP) that can be attached to a column, agarose beads or magnetic beads. NGS, next-generation sequencing; PCR, polymerase chain reaction.

much slower than the conversion of cytosine to uracil. Therefore, one assumes that the only cytosines remaining after treatment are those that were methylated. When performing PCR, uracil residues are replicated as thymine residues and 5mC residues are replicated as cytosine moieties (Figure 6.3A). Sodium bisulfite treatment remains one of the most widely used epigenetic techniques and forms the backbone of a variety of commercial kits for this purpose.

In 1976, Hikoya Hayatsu described that sodium bisulfite causes the deamination of cytosine. In 1980, Wang and colleagues further demonstrated that 5mC is resistant to this modification, allowing Frommer and colleagues in 1992 to devise a method to analyze 5mC in DNA – 'bisulfite genomic sequencing.'

6.4.1.4   Methylation-sensitive restriction enzymes differentially digest methylated DNA

Some restriction enzymes such as HpaI, HpaII, and MspI are 5mC sensitive and so will not cut sequences in which a cytosine is methylated. Using corresponding 5mC-insensitive restriction enzymes, that have identical recognition sequences (isoschizomers) though cut indifferently of methylation, it is possible to distinguish methylated from unmethylated cytosines.

6.4.1.5   Specific antibodies can be used to purify regions of methylated DNA

The most commonly used affinity-based assay for measuring levels of DNA methylation is the methylated DNA immunoprecipitation (Me-DIP). This uses a 5mC-specific antibody or an MBD protein. DNA is fragmented to an average size of around 400 bp using either sonication or restriction enzymes. This is then bound to anti-5mC or anti-MBD, combined with Sepharose or magnetic beads, and then purified from uncaptured DNA. Real-time PCR can be used to test for methylation status at a single locus or the pulled-down DNA may be used for microarray analyses to test numerous positions across the genome (MeDIP-ChIP). Newer methods, known as MeDIP-Seq, combine this technique with next-generation sequencing (NGS) technologies; DNA enriched following immunoprecipitation is high-throughput sequenced to provide total genome-wide coverage (see Figure 6.3B). This strategy was used to generate the first whole-genome methylation profile of a mammalian genome in 2008 by Down and colleagues and has since been successfully used to provide methylation profiles of several tissues, including brain.

6.4.1.6 Genome-wide DNA methylation analysis

Rapidly developing sequencing technologies are increasing the capability to analyze DNA sequences throughout a genome. Such approaches, in combination with microarrays or NGS, hold the promise of providing increasingly greater density genome-wide epigenetic studies.

Microarray-based methylation approaches

The main classes of microarray-based methods developed to quantify DNA methylation are those using sodium bisulfite, methylation-sensitive restriction enzymes or methylated DNA enrichment. Advances in microarray designs have seen the development of high-quality commercial oligonucleotide arrays such as bead arrays made by Illumina, lithographic arrays made by Affymetrix, and inkjet arrays manufactured by Agilent.

The differing technologies and chemistries behind arrays influence the methylation assays that these can be adapted to. Oligonucleotide-based microarrays are designed using oligonu-cleotide hybridization probes targeting the CpG sites of interest. With the Infinium methylation assay (Illumina), researchers can measure methylation sites at single-nucleotide resolution, covering over 450,000 methylation sites spanning 99% of genes, with an average of 17 CpG sites per gene region distributed across the promoter, 5' and 3' untranslated regions and exons. It also covers 96% of CpG islands, with additional coverage in island shores and the regions flanking them. Such approaches are therefore ideal for screening epigenome-wide association study (GWAS) populations.

Next-generation sequencing-based methylation analysis

New high-throughput NGS technology platforms are opening up a wide range of applications and are particularly adapted to epigenomic research for profiling DNA methylation and chro-matin at high resolution and relatively low cost. Sodium bisulfite conversion followed by NGS (known as either BS-Seq or MethylC-Seq) has become an increasingly used technique. Providing sequencing coverage at a single-base pair resolution, it is able to provide a compre-hensive coverage of >90% of cytosines in the human genome. Lister and colleagues (2013) used this method to discover that during mammalian brain development, there is extensive global DNA methylation reconfiguration, further revealing high levels of cell type-specific intragenic methylation patterns and over 100,000 developmental and cell type-specific differ-entially CG-methylated regions. Importantly, it also enabled the detection of highly conserved non-CG methylation (mCH) in neurons.

6.4.2 Chromatin

The field of neuroscience is fast realizing the importance of histone modifications in both devel-opment and brain function. A critical role for chromatin regulation in behavior has now been demonstrated in several rodent models in different brain regions. Moreover, chromatin modi-fications play an important role in psychiatric disorders such as schizophrenia, depression, and drug addiction.

6.4.2.1 Global histone analyses

Using antibodies raised against specific histone modifications in Western blot analysis on nuclear extracts is a relatively simple technique able to reveal the type, tissue location, and degree of histone posttranslational modifications. For example, in 2013, Bousiges and colleagues, measuring global histone acetylation levels in the rat hippocampus, revealed differential acetylation of H3, H4, and H2B at early stages of spatial or fear memory formation. H3K9K14 acetylation was mostly responsive to any experimental conditions compared to naive animals, whereas H2B N-terminus and H4K12 acetylations were mostly associated with memory for either spatial or fear learning.

6.4.2.2 Sequence-specific and genome-wide histone analyses

Incredible progress has been seen in characterizing histone modifications on genome-wide scales, since the development of the chromatin immunoprecipitation (ChIP) method. The technique has now developed into a powerful and widely applied tool used to analyze the spatial and temporal association of specific proteins (e.g. modified histones, DNA-binding proteins, transcription factors, etc.) within native cells *in vivo*. It further forms the basis of numerous related adapted approaches. The basic principle of this method is that living cells are treated with formaldehyde to cross-link proteins and DNA within ~2 Å of each other. As a result, proteins are covalently bound to their target sequences on the DNA and provide a snapshot of transcription factor occupancy. Next, the cross-linked DNA is fragmented into pieces of 300–600 bp and protein-DNA complexes are immunoprecipitated by antibodies against the protein of interest (Figure 6.4). This allows the enrichment of specific DNA sequences which can then be identified through amplification by PCR, hybridization to microarrays (ChIP-on chip), cloning or NGS (ChIP-Seq) (for review see Murgatroyd *et al.*, 2012).

In 1984, John Lis and David Gilmour developed the chromatin immunoprecipitation (ChIP) method, first using UV irradiation to covalently cross-link proteins in contact with neighboring DNA followed by immunoprecipitation with an antibody to bacterial RNA polymerase. A year later, they used the same methodology to study distribution of eukaryotic RNA polymerase II on fruit fly heat shock genes. Two months later, Mark Solomon and Alexander Varshavsky developed the use of formaldehyde as a cross-linking reagent to examine the gene locations of histone H4.

6.5 Alterations in epigenetic processes

The ~85 billion cells constituting a brain have the same DNA sequence yet are differentiated for their diverse functions through epigenetic programming during pre- and postnatal development and possibly throughout life. Epigenetic mechanisms are therefore gatekeepers to brain development, differentiation and maturation – alterations in the molecular machinery regulating these processes can associate with neurological diseases.

6.5.1 Mutations in the epigenetic machinery

Deregulation of epigenetic pathways can lead to either silencing or inappropriate expression of specific sets of genes manifesting with diseases. There are a large number of genes encoding epigenetic regulators that, when mutated, can give rise to mental retardation (for review see Murgatroyd and Spengler, 2012). Though one might assume that different epigenetic factors would orchestrate the expression of a large number of potentially unrelated genes, disruptions in distinct epigenetic regulators seemingly lead to symptomatically similar mental retardation syndromes. Ergo, it is conceivable that mental retardation does not take root in changes of

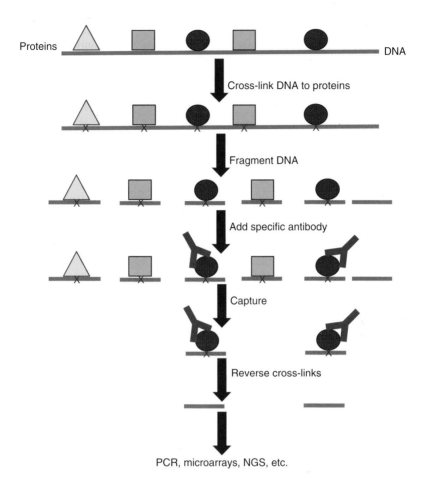

Figure 6.4 Chromatin immunoprecipitation (ChIP) procedure. Cells are initially treated with a cross-linking agent (e.g. formaldehyde) that covalently links DNA-interacting proteins to the DNA. The genomic DNA is then isolated and sheared (e.g. sonication) into a suitable fragment size distribution (usually around 300 bp). An antibody that specifically recognizes the protein of interest is then added and immunoprecipitation used to isolate appropriate protein-DNA complexes. The cross-links are then reversed and the DNA fragments purified and measured. NGS, next-generation sequencing; PCR, polymerase chain reaction.

specific target gene(s) but by the inability of concerned neurons to respond adequately to environmental signals under conditions of greatly distorted transcriptional homeostasis. One of the most common causes of mental retardation in females is Rett syndrome, a progressive neurodevelopmental disorder resulting from mutations in the methyl-CpG binding protein MeCP2 located on the X chromosome. Interestingly, less disruptive gene mutations or alterations in expression of this gene also appear to underlie some cases of autism, along with further members of the MBD family in a small number of cases.

6.5.2   Alterations in epigenetic regulation

There are an increasing number of studies linking abnormal epigenetic marks with the development of mental pathologies later in life. However, the question remains whether these alterations originate during early development or as responses to environmental exposure(s) during later life. For example, DNA **hypomethylation** at the promoter of the gene for catechol-*O*-methyltransferase (COMT), an enzyme regulating the level of dopamine, has been found to associate with schizophrenia and bipolar disorder. At the promoter region of the RELN gene, encoding a protein implicated in long-term memory, a number of studies have evidenced **hypermethylation**, correlating with reduced expression, in schizophrenia (for review see Maric and Svrakic, 2012).

6.5.3   Epigenetics and aging

Starting from the pioneering salmon sperm studies of Berdyshev, epigenetic alterations have been increasingly recognized as part of aging and aging-related diseases. A number of important studies analyzing global and gene-specific methylation levels in mono- (MZ) and dizygotic (DZ) twins have allowed the elucidation of age-related divergence of epigenetic marks over time. These firstly revealed lower epigenetic differences between MZ than DZ twins, supporting the influence of genetics. In addition, although MZ twins are epigenetically indistinguishable during the early years of life, older individuals exhibited significant tissue-specific differences in their overall content and genomic distribution of 5mC, affecting global gene expression. Furthermore, these differences already occur during the course of childhood development, suggesting that early environmental factors can establish long-lasting epigenetic changes (for review see d'Aquila *et al.*, 2013).

A gradual loss of total 5mC content with age occurs in most mammalian tissues which predominantly affects non-island CpGs and interspersed repetitive sequences, though further studies indicate age-associated hypomethylation also at promoters, exonic, intronic, and intergenic regions. However, some specific promoters do show progressive elevations in DNA methylation levels across the lifespan, suggesting a more complex picture with both increases and decreases in intraindividual global methylation levels over time. Importantly, dietary deficiencies in nutrients, including exposures to various metals, have been demonstrated to affect DNA methylation status in a tissue- and age-dependent manner. It should be further mentioned that these age-related epigenetic factors are also characterized by the occurrence of different types or combination of histone modifications (for review see d'Aquila *et al.*, 2013).

6.6   The epigenome and early life adversity

Aside from controlling constitutive gene expression, epigenetic mechanisms can also serve to fine-tune gene expression potential in response to environmental cues. It has therefore been proposed that conditions of early life environment can evoke changes in DNA methylation, facilitating epigenetic programming of critical genes involved in regulating stress responsivity that may in turn manifest with neuroendocrine and behavioral symptoms in adulthood (Figure 6.5).

6.6.1   Maternal care and epigenetic programming of GR

One well-characterized model established to study the effects of early life environment on stress programming examines variations in the quality of early postnatal maternal care, as measured by levels of licking and grooming. Rats who received high levels of maternal care during early life developed sustained elevations in GR expression within the hippocampus and reduced HPA axis responses to stress. Meany and colleagues found an important role for epigenetic regulation, revealing that the enhanced GR expression was associated with a persistent DNA hypomethylation at specific CpG dinucleotides within the hippocampal GR exon 1_7 promoter and increased histone acetylation. The lower CpG methylation facilitated binding of the transcriptional activator nerve growth factor-inducible protein A (NGF1a) to this region (Weaver *et al.*, 2004).

Interestingly, some further studies using a different early life stressor in different species of rats do not find the same changes in DNA methylation at the same hippocampal GR promoter,

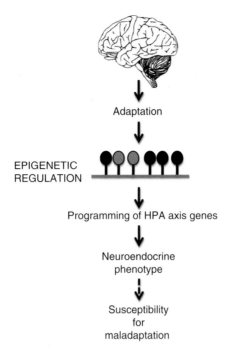

Figure 6.5 **Programming**. Early life experience can persistently alter expression levels of key genes through epigenetic marking which can underpin changes in behavior, neuroendocrine and stress responsivity throughout later life.

supporting again the idea that variations in genetic make-up, and environmental disparities, are critical determinants in organization of the epigenome. In support of this idea, subsequent studies detected altered GR promoter methylation in human postmortem hippocampal tissue of depressed suicide patients who had a history of early life abuse and neglect (McGowan *et al.*, 2009). In contrast, suicide patients who were not exposed to early life adversity, or patients suffering from major depression only, revealed no epigenetic marking of hippocampal GR (Alt *et al.*, 2010).

6.6.2   Maternal separation stress and epigenetic programming of AVP
Through the study of maternal separation in mice, early life adversity has been found to induce sustained expression of hypothalamic *Avp* mRNA, a peptide specific to the parvocellular subpopulation of the paraventricular nucleus (PVN). This underlined elevated corticosterone secretion, heightened endocrine responsiveness to subsequent stressors and altered HPA axis feedback inhibition. Importantly, this altered expression was associated with reduced levels of DNA methylation in the PVN at particular CpG dinucleotides within an enhancer region important for *Avp* gene activity. Further investigations revealed that hypomethylation at this region reduced the ability of MeCP2 to bind and recruit further epigenetic machinery such as HDACs and DNMTs, supporting previous evidence for a role of MeCP2 as an epigenetic platform upon which histone and DNA modifications are carried out to confer transcriptional repression. Signals controlling MeCP2 occupancy at this early step were then explored, revealing that neuronal depolarization is able to trigger Ca^{2+}-dependent phosphorylation of MeCP2. This causes MeCP2 to dissociate from the AVP enhancer, resulting in loss of both its repressing activity and its organization of DNA methylation and histone marks (for review see Murgatroyd and Spengler, 2011b).

6.6.3   Early life stress and epigenetic programming of CRH

A growing number of studies support the role of epigenetic modifications in the control of further neuroendocrine genes regulating the HPA axis. Epigenetic programming of hypothalamic CRH expression in response to stress in the prenatal period, early life or during adulthood has been demonstrated. Mueller and Bale reported that pregnant mice subjected to chronic variable stressors during early gestation revealed a hypomethylation at the CRH promoter in the hypothalamus of the offspring in adulthood while Chen and colleagues demonstrated that maternal separation in rat pups also induced hypomethylation at the CRH promoter. Again, similar CRH hypomethylation was found by Elliot and colleagues in chronic social stress in adult mice in the PVN of a subset of animals displaying subsequent social avoidance (for review see Murgatroyd and Spengler, 2011a.)

6.6.4   Clinical studies of epigenetic programming by early life adversity

A number of key studies demonstrate that the epigenetic programming seen in the early life stress animal models may extrapolate to human studies. For example, McGowan and colleagues (2009) found hypermethylation of the GR gene (NR3C1) promoter among suicide victims with a history of abuse in childhood, but not among controls or suicide victims who did not suffer such early life stress. These data appear consistent with previous studies demonstrating that the epigenetic status of the homologous GR gene promoter is regulated by parental care during early postnatal development in rats (Weaver *et al.*, 2004). These studies therefore suggest that epigenetic processes in the brain could mediate the effects of the early environment on gene expression, and that stable epigenetic marks such as DNA methylation might then persist into adulthood to influence HPA axis activity.

Might epigenetic changes in blood provide a clinically valuable surrogate for what is happening in the brain? Some early studies have shown partial correlations in gene expression between various brain regions and blood, which have been supported by studies showing epigenetic differences in lymphocytes associating with the brain in Rett syndrome and Alzheimer disease. However, might DNA methylation associated with exposure to early life stress also demonstrate levels of conservation between the brain and periphery? Initial studies in rhesus macaques by Provencal *et al.* (2012), examining DNA methylation in the prefrontal cortex and peripheral T cells in parallel in response to differential rearing, did reveal differences in common between both tissues. However, results as a whole tended to suggest the T cells were not a direct surrogate marker of brain tissue but probably reflected the response of the immune system to early life stress.

The idea that the epigenomic response to early life stress is not limited to the brain and could be studied in peripheral lymphocytes has gained support from several other tissues. For example, a number of studies have correlated DNA methylation of the GR promoter in DNA from neonatal cord blood mononuclear cells with maternal depression and childhood adversity and maltreatment (e.g. Oberlander *et al.*, 2008). Longitudinal studies investigating these changes would be important, especially if this could highlight possible environmental or pharmacological interventions.

6.7   Perspectives

With the advent of new high-throughput sequencing technologies, research is now scaling up to epigenome-wide analyses of all epigenetic marks throughout a genome in a specific tissue. Given that epigenetic modifications are sensitive to changes in the environment, it might be anticipated that these efforts will identify epigenomic signatures for mental disorders and molecular dysregulations resulting from early life stress (Albert, 2010). Whilst such a strategy appears highly attractive in the field of cancer research addressing clonally expanded cell populations, rather than heterogeneous tissues, the field of epigenetic association studies in neuroendocrine disorders and psychiatry might be more challenging. Epigenetic modifications reported to date in animal models and postmortem brain tissues are seldom 'all or none' but gradual, and seem to occur in a highly cell type-specific manner. In this view, the extraordinary complexity and heterogeneity of neural tissues pose a major hurdle to the derivation of epigenetic biomarkers in psychiatric disease.

- Taken together, we have covered the importance of epigenetic processes in programming long-term neuroendrocrine regulation in response to the early life environment. Adversity during this early period is able to shape this towards dysregulation of the HPA axis.
- Epigenetic mechanisms appear key to allowing an organism to respond to the environment through changes in gene expression.
- DNA methylation and chromatin marks can be measured through a number of techniques, and the advent of next-generation sequencing technologies has heralded epigenome-wide studies.
- Numerous studies of early life stress in animal models are revealing the importance of the epigenetic regulation of key neuroendocrine genes. Some clinical studies demonstrate translational aspects of this research. The role of epigenetic marks in peripheral tissues as a convenient surrogate marker for the brain needs to be further explored.

In sum, understanding how early life experiences can give rise to lasting epigenetic memories conferring increased risk for mental disorders is emerging at the epicenter of modern psychiatry. Whether suitable social or pharmacological interventions could reverse deleterious epigenetic programming triggered by adverse conditions during early life should receive highest priority on future research agendas. Progress in this field will further garner public interest, a general understanding and appreciation of the consequences of childhood abuse and neglect for victims in later life.

References

Albert, P.R. (2010) Epigenetics in mental illness: hope or hype? *Journal of Psychiatry & Neuroscience*, **35**, 366–368.

Alt, S.R., Turner, J.D., Klok, M.D., *et al.* (2010) Differential expression of glucocorticoid receptor transcripts in major depressive disorder is not epigenetically programmed. *Psychoneuroendocrinology*, **35**, 544–556.

Bousiges, O., Neidl, R., Majchrzak, M., *et al.* (2013) Detection of histone acetylation levels in the dorsal hippocampus reveals early tagging on specific residues of H2B and H4 histones in response to learning. *PLoS One*, **8**, e57816.

D'Aquila, P., Rose, G., Bellizzi, D. and Passarino, G. (2013) Epigenetics and aging. *Maturitas*, **74**, 130–136.

De Kloet, E.R., Vreugdenhil, E., Oitzl, M.S. and Joëls, M. (1998) Brain corticosteroid receptor balance in health and disease. *Endocrine Reviews*, **19**, 269–301.

Down, T.A., Rakyan, V.K., Turner, D.J., *et al.* (2008) A Bayesian deconvolution strategy for immunoprecipitation-based DNA methylome analysis. *Nature Biotechnology*, **26**, 779–785.

Frommer, M., McDonald, L.E., Millar, D.S., *et al.* (1992) A genomic sequencing protocol that yields a positive display of 5-methylcytosine residues in individual DNA strands. *Proceedings of the National Academy of Sciences of the United States of America*, **89**, 1827–1831. **[This landmark paper describes the sodium bisulfite sequencing method most commonly used for DNA methylation analysis today.]**

Gluckman, P.D., Hanson, M.A., Cooper, C. and Thornburg, K.L. (2008) Effect of in utero and early-life conditions on adult health and disease. *New England Journal of Medicine*, **359**, 61–73.

Guo, J.U., Su, Y., Shin, J.H., *et al.* (2014) Distribution, recognition and regulation of non-CpG methylation in the adult mammalian brain. *Nature Neuroscience*, **17**, 215–222.

Holmes, A., Le Guisquet, A.M., Vogel, E., Millstein, R.A., Leman, S. and Belzung, C. (2005) Early life genetic, epigenetic and environmental factors shaping emotionality in rodents. *Neuroscience and Biobehavioral Reviews*, **29**, 1335–1346.

Holsboer, F. (2000) The corticosteroid receptor hypothesis of depression. *Neuropsychopharmacology*, **23**, 477–501.

Irizarry, R.A., Ladd-Acosta, C., Wen, B., *et al.* (2009) The human colon cancer methylome shows similar hypo- and hypermethylation at conserved tissue-specific CpG island shores. *Nature Genetics*, **41**, 178–186.

Jaenisch, R. and Bird, A. (2003) Epigenetic regulation of gene expression: how the genome integrates intrinsic and environmental signals? *Nature Genetics*, **33**, 245–254.

Jenuwein, T. and Allis, C.D. (2001) Translating the histone code. *Science*, **293**, 1074–1080. [*This paper outlines a model, termed the histone code hypothesis, to predict that a combination of covalent modifications of histone tails functions as a target for the specific binding of effector proteins.*]

Lister, R., Mukamel, E.A., Nery, J.R., *et al.* (2013) Global epigenomic reconfiguration during mammalian brain development. *Science*, **341**, 1237905. [*Whole genomic methylation analysis of 5-methyl and 5-hydroxymethylcytosine and non-CpG methylation in a comparison between human and mouse brain that revealed patterning, cell specificity and dynamics of DNA methylation at single-base resolution.*]

Maric, N.P. and Svrakic, D.M. (2012) Why schizophrenia genetics needs epigenetics: a review. *Psychiatria Danubina*, **24**, 2–18.

McGowan, P.O., Sasaki, A., D'Alessio, A.C., *et al.* (2009) Epigenetic regulation of the glucocorticoid receptor in human brain associates with childhood abuse. *Nature Neuroscience*, **12**, 342–348.

Murgatroyd, C. and Spengler, D. (2011a) Epigenetics of early child development. *Frontiers in Psychiatry*, **2**, 16.

Murgatroyd, C. and Spengler, D. (2011b) Epigenetic programming of the HPA axis: early life decides. *Stress*, **14**, 581–559.

Murgatroyd, C. and Spengler, D. (2012) Genetic variation in the epigenetic machinery and mental health. *Current Psychiatry Reports*, **14**, 138–149.

Murgatroyd, C., Hoffmann, A. and Spengler, D. (2012) In vivo ChIP for the analysis of microdissected tissue samples. *Methods in Molecular Biology*, **809**, 135–148.

Oberlander, T.F., Weinberg, J., Papsdorf, M., Grunau, R., Misri, S. and Devlin, A.M. (2008) Prenatal exposure to maternal depression, neonatal methylation of human glucocorticoid receptor gene (NR3C1) and infant cortisol stress responses. *Epigenetics*, **3**, 97–106.

Provençal, N., Suderman, M.J., Guillemin, C., *et al.* (2012) The signature of maternal rearing in the methylome in rhesus macaque prefrontal cortex and T cells. *Journal of Neuroscience*, **32**, 15626–15642.

Turek-Plewa, J. and Jagodziński, P.P. (2005) The role of mammalian DNA methyltransferases in the regulation of gene expression. *Cellular & Molecular Biology Letters*, **10**, 631–647.

Weaver, I.C., Cervoni, N., Champagne, F.A., *et al.* (2004) Epigenetic programming by maternal behaviour. *Nature Neuroscience*, **7**, 847–854. [*A landmark paper that demonstrates that changes in early life maternal care in a rat model induced long-lasting epigenetic programming of a key gene important in behavior.*]

Weber, M., Hellmann, I., Stadler, M.B., *et al.* (2007) Distribution, silencing potential and evolutionary impact of promoter DNA methylation in the human genome. *Nature Genetics*, **39**, 457–466.

Further reading

McGhee, J.D. and Ginder, G.D. (1979) Specific DNA methylation sites in the vicinity of the chicken beta-globin genes. *Nature*, **280**, 419–420. [*First observation of the presence of DNA methylation differences in the vicinity of β-globin genes. This lead to a role of this mark in gene repression.*]

Zhang, T.Y., Labonté, B., Wen, X.L., Turecki, G. and Meaney, M.J. (2013) Epigenetic mechanisms for the early environmental regulation of hippocampal glucocorticoid receptor gene expression in rodents and humans. *Neuropsychopharmacology*, **38**, 111–123.

Proteins, Posttranslational Mechanisms, and Receptors

CHAPTER 7

Proteome and Peptidome Dynamics

Lloyd D. Fricker

Department of Molecular Pharmacology, Albert Einstein College of Medicine, Bronx, New York, USA

7.1 Introduction

The endocrine system involves a large number of intercellular signaling molecules that communicate between various components of the system. Many of these signaling molecules are peptides and proteins. In many cases, the same molecules that communicate within the endocrine system also signal the brain. For example, insulin is secreted from the pancreas in response to elevated blood sugar, and signals to peripheral tissues and specific populations of neurons that express the insulin receptor. Insulin, as well as other peptide hormones and neuropeptides, is secreted in a transient manner. Many aspects of the production, secretion, and degradation of endocrine peptides are regulated, adding to the complexity of the system. Classic endocrine peptides such as insulin are produced within the regulated secretory pathway, stored within secretory vesicles, and released upon appropriate stimulation. After secretion, some peptides are processed into forms that retain biological activity, often with properties that are different from the form of the peptide that was secreted. Eventually, the peptide is degraded into inactive fragments by one or more extracellular peptidases. This process is also highly dynamic and can be regulated by drugs and/or endogenous molecules. The first section of this review discusses the proteins and peptides in the classic regulated secretory pathway (RSP), with an emphasis on the dynamics of the various steps involved in the production, secretion, and degradation of the bioactive moieties.

Neuropeptide: a peptide secreted from a neuron that influences a nearby cell, functioning as a neurotransmitter.

In addition to the classic endocrine peptides, some intercellular signaling molecules arise from other pathways. For example, the bioactive peptide angiotensin II is not stored within secretory vesicles; instead, this peptide is produced from angiotensinogen by a series of enzymes that convert the protein into a peptide (angiotensin I) and then remove two C-terminal residues from the peptide to generate the bioactive form (angiotensin II). These processing steps occur in plasma in a highly dynamic process. The importance of the protein and peptide cleavages is evident from the powerful effects that inhibitors of angiotensin-converting enzyme have on the levels of angiotensin II and the subsequent biological effect of this peptide. Using tools developed to study neuropeptides (see Chapter 8), novel peptides have been detected in brain and neuroendocrine tissues. Some of these peptides are

---

Molecular Neuroendocrinology: From Genome to Physiology, First Edition. Edited by David Murphy and Harold Gainer.
© 2016 John Wiley & Sons, Ltd. Published 2016 by John Wiley & Sons, Ltd.
Companion website: www.wiley.com/go/murphy/neuroendocrinology

similar to the classic neuropeptides, being produced from relatively small proteins within the RSP. Other peptides found in tissue extracts during peptidomic analyses are derived from proteins present within the cytosol or other intracellular compartments (nuclei, mitochondria). Some of these peptides are secreted from cells and influence extracellular receptors; this group of peptides has been named **non-classic neuropeptides**, by analogy to non-classic neurotransmitters such as anandamide, which are produced transiently upon demand. Although the biological relevance of non-classic neuropeptides is not yet well known, this potentially important emerging concept is included in this review.

The final section of this review discusses the dynamics of intracellular proteins and peptides. Proteins within a cell are never static, but are continually produced and degraded. The ubiquitin-proteasome system is an important mediator of protein degradation. The proteasome converts proteins into peptides, most of which are rapidly degraded by cytosolic peptidases. However, some peptides that arise from intracellular proteins were detected in the peptidomic analyses, and only a subset of these is secreted. The non-secreted intracellular peptides may have functional roles, much like microRNA and other small RNAs have functions. While some evidence supports the idea that intracellular peptides are functional, this remains fairly speculative but is included briefly in this review along with a discussion of the dynamics of intracellular proteins.

7.2   Classic neuropeptides and proteins in the RSP

Although discovered in the early 1900s by Bayliss and Starling, it took several decades before the sequence of the first known peptide signaling molecule was determined (Mutt *et al.*, 1970). The peptide was named secretin, because it influenced the secretion of acid from the stomach, and the term 'hormone' was coined to describe chemical messengers such as secretin. In the century subsequent to the discovery of secretin, dozens of bioactive peptide hormones were discovered throughout the endocrine system (Fricker, 2012; Strand, 2003). Some of these peptide hormones have also been found in brain where they function as **classic neuropeptides** or **direct neuropeptides**. In addition, other neuropeptides have been identified only in brain and not in the endocrine system.

The vast majority of neuropeptides and peptide hormones are produced from precursors by selective cleavages at specific sites (Eipper and Mains, 1980; Fricker, 2012; Kemmler *et al.*, 1971). In most cases, these sites contain the basic amino acids Lys and/or Arg, usually as pairs of basic amino acids (Figure 7.1). Of the four possible combinations, Lys-Arg and Arg-Arg are the most common, with fewer examples of Lys-Lys and Arg-Lys serving as cleavage sites (Lindberg and Hutton, 1991). Some cleavage sites contain a pair of Arg residues separated by two, four or six other residues (i.e. Arg-Xaa-Xaa-Arg, etc.). In the early days of the field, these were often considered 'monobasic' cleavages, although technically they are 'dibasic' cleavage sites and cleaved by the same enzymes that cleave at Arg-Arg sites (described later in this chapter). Some neuropeptides and peptide hormones are produced by cleavages at true monobasic sites, lacking an upstream basic residue, and others require cleavage at non-basic sites (although these are relatively uncommon).

The conversion of the precursors into the bioactive peptides requires a series of steps that occur in the various intracellular compartments (see Figure 7.1). Initially, the signal peptide on the N-terminus of the precursor is removed by the signal peptidase (Tuteja, 2005). The signal peptide is necessary for targeting the protein into the lumen of the endoplasmic reticulum. The term 'signal peptide' should not be confused with the term 'signaling peptide,' which is occasionally used to refer to peptides that signal between cells (i.e. peptide hormones and neuropeptides). Most proteins that enter into the lumen of the endoplasmic

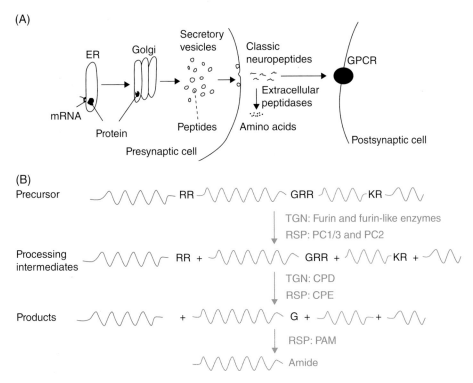

Figure 7.1 Schematic diagram of the production and degradation of classic neuropeptides. (A) Cellular compartments that neuropeptides traverse during their synthesis and secretion. (B) Enzymatic steps of the endopeptidases, carboxypeptidases, and amidating enzyme for a typical precursor containing three cleavage sites (note that the number of cleavage sites varies among precursors, ranging from one to over 20). Neuropeptide precursors are translocated into the lumen of the endoplasmic reticulum (ER) during protein synthesis. Once inside the ER, the protein is transported via a vesicle-mediated process into the lumen of the Golgi apparatus. In the trans-Golgi network (TGN), the protein can be initially processed by furin and furin-like endopeptidases if consensus sites for these enzymes are present in the precursor. After the precursor is sorted into immature secretory vesicles and the pH becomes slightly acidic, the precursor is processed by the endopeptidase prohormone convertases 1/3 (PC1/3) and 2 (PC2). Following the action of the endopeptidases, all of the products (except for the C-terminal peptide) require a carboxypeptidase to remove the C-terminal basic residues. In the TGN, this enzyme is carboxypeptidase D (CPD) while in the secretory granules this step is mediated by carboxypeptidase E (CPE). Peptides that contain a Gly on the C-terminus following the action of CPE or CPD are further cleaved in a two-step process to generate peptides with a C-terminal amide group in place of the Gly residue. The enzymes that perform this reaction are collectively referred to as peptidyl glycine α-amidating monooxygenase. Secretion of the peptides and other soluble contents of the vesicles occurs when the vesicle fuses with the plasma membrane; this step is regulated, and the pathway is referred to as the regulatory secretory pathway (RSP). Following secretion, the peptides can bind to receptors; typical neuropeptide receptors are in the G protein-coupled receptor (GPCR) family. Extracellular peptidases cleave the neuropeptide into smaller peptides, and eventually into amino acids. Some of the smaller peptides can be biologically active (not shown). Source: © Lloyd Fricker.

reticulum have a signal peptide on the N-terminus, and this step is not specific for neuropeptides and peptide hormones.

Once the signal peptide is removed and the protein enters into the lumen of the endoplasmic reticulum, the protein can undergo posttranslational modifications such as glycosylation (Bennett, 1991). Only a subset of the precursors of neuropeptides and peptide hormones are glycosylated, although most of the enzymes that process the precursors are glycosylated.

Folding and disulfide bond formation also occurs in the endoplasmic reticulum (Braakman and Hebert, 2013). Disulfide bond formation is very important for most peptide-processing enzymes but is only required for a subset of the precursors of neuropeptides and peptide hormones (examples include oxytocin, vasopressin, and insulin). After the protein is folded, it is transported to the Golgi where it can undergo further posttranslational modifications such as sulfation and phosphorylation (Bennett, 1991; Huttner, 1988; Tagliabracci et al., 2013). If an N-linked carbohydrate was added in the endoplasmic reticulum, the sugar side chains are further modified in the Golgi.

For some peptide precursors, cleavage begins in the latter part of the Golgi, especially the trans-Golgi network. Endopeptidases that are active in the trans-Golgi network include furin and related enzymes (Nakayama, 1997; Steiner, 1998). These enzymes cleave peptides with the consensus site Arg-Xaa-Lys/Arg-Arg, where Xaa is any amino acid other than Cys. Cleavage occurs to the C-terminal side of the consensus sequence, producing peptide intermediates that contain C-terminal basic residues. These basic residues are subsequently removed by carboxypeptidase D (CPD) which is present and active in the environment of the Golgi and trans-Golgi network (Fricker, 2002; Song and Fricker, 1995). As a side point, although both furin and CPD are transmembrane-bound enzymes that are predominantly located in the trans-Golgi network, they cycle to the cell surface and back to the trans-Golgi network (Thomas, 2002; Varlamov and Fricker, 1998).

The majority of the proteolytic processing of prohormones into bioactive peptides occurs after the molecules are sorted into immature secretory granules (see Figure 7.1). These cleavages are mediated by the prohormone convertases (PCs) and carboxypeptidase E (CPE). There are two major peptide-processing PCs, named prohormone convertase 1 (PC1) and prohormone convertase 2 (PC2) (Hoshino and Lindberg, 2012; Seidah and Chretien, 2004a, b; Zhou et al., 1999). PC1 was identified simultaneously by two different groups, one of which used the name prohormone convertase 3 (see Chapter 9) (Seidah et al., 1990; Smeekens and Steiner, 1990). As a result, this enzyme is commonly referred to as PC1/3. The PCs are evolutionarily related to furin and to the yeast endopeptidase Kex2 (Fuller et al., 1989). Similarly, CPE is related to CPD and the two enzymes have nearly identical substrate specificities, cleaving C-terminal basic residues from a wide range of peptides (Fricker, 2004a, b). In contrast to CPE and CPD, PC1/3 and PC2 have distinct specificities and both are much different from furin in that the PCs require only two basic residues for optimal activity (i.e. Arg-Arg or Arg-Xaa-Xaa-Arg, etc.) whereas the optimal furin consensus site includes a third basic residue, as described above. Because most prohormones have PC consensus sites and not furin consensus sites, processing of the precursor is primarily mediated by the PCs and then CPE to remove the basic C-terminal residues.

Following the action of the PCs and CPE, if the peptide contains a C-terminal Gly residue this is converted into a C-terminal amide group by a two-step process collectively known as peptidyl-glycine α-amidating monooxygenase (PAM) (Prigge et al., 2000). Interestingly, PAM is composed of two distinct enzyme activities. The first is named peptidyl-glycine α-hydroxylating monooxygenase (PHM); this enzyme oxidizes the Gly residue to introduce an oxygen atom. The second enzyme, peptidyl-α-hydroxyglycine α-amidating lyase (PAL), removes the two-carbon piece of Gly, leaving behind the amine group as the amide.

In addition to neuropeptide processing steps that occur within the secretory pathway, neuropeptides undergo further cleavages after secretion. A number of different extracellular aminopeptidases, carboxypeptidases, and endopeptidases have been identified and found to play a role in the cleavage of various peptides (for review, see Fricker, 2012). In some cases, these cleavages inactivate the peptide, much as acetylcholine is inactivated by acetylcholine esterase located outside neurons. However, some of the extracellular peptidases play a more complex role in modulating neuropeptide activity (described in the next section). Peptide processing can also occur after the peptide-receptor complex is internalized into the endocytic pathway. For example, endothelin-converting enzyme-1 plays a role in degrading calcitonin gene-related peptide after the internalization of this peptide together with its receptor into early endosomes (Padilla *et al.*, 2007). Similarly, endothelin-converting enzyme-1 also cleaves substance P after internalization along with its receptor (Pelayo *et al.*, 2011). Finally, there are cytosolic peptidases that have been implicated in the degradation of peptides (for review, see Fricker, 2012). For example, insulin-degrading enzyme is an oligopeptidase which is primarily located in the cytosol. It is not known how extracellular insulin is transported into the cytosol. Alternatively, it is possible the insulin-degrading enzyme is secreted by a non-conventional mechanism and interacts with insulin in the extracellular environment.

7.2.1 Dynamics of peptide processing

The processing of neuroendocrine peptide precursors into the bioactive forms is not a static process, but one that is highly dynamic. Some peptide precursors are converted into a variety of peptides, depending on the presence of various processing enzymes, and the resulting peptides have substantially different biological properties. A classic example of this is the processing of pro-opiomelanocortin (POMC) (see Figure 8.1, Chapter 8). In the anterior pituitary, POMC is primarily cleaved into larger peptides such as adrenocorticotropic hormone (ACTH). In intermediate pituitary and brain hypothalamus, the ACTH is further cleaved into two peptides: α-melanocyte-stimulating peptide (α-MSH) and corticotropin-like intermediate lobe peptide. The properties of ACTH and α-MSH are much different; both bind to various melanocortin receptors but with greatly different affinities, and so the differential processing of POMC into these distinct peptides plays an important role in controlling the physiological activity of the peptides (Bicknell, 2008). Similarly, the POMC-derived peptide β-endorphin 1-31 is an agonist at the μ opioid receptor while the C-terminally truncated β-endorphin 1-27 is an antagonist at this receptor (Hammonds *et al.*, 1984). β-Endorphin 1-27 is formed from 1-31 by PC2 cleavage at the Lys-Lys (positions 28–29) and then removal of the C-terminal Lys residues by CPE (Allen *et al.*, 2001). ACTH is converted into α-MSH and CLIP by PC2 followed by the removal of basic residues by CPE (Zhou *et al.*, 1993). The differential expression of PC1/3 and PC2 plays a critical role in the formation of specific sets of peptides.

   Extracellular peptide processing is also a highly dynamic process that regulates the activity of the peptide. For example, the enkephalin heptapeptide (YGGFMRF) is converted by extracellular angiotensin-converting enzyme into Met-enkephalin (YGGFM) and the dipeptide Arg-Phe (Norman *et al.*, 1985). The heptapeptide has comparable subnanomolar affinity for all three opioid receptors, with Ki values of 0.37, 0.57, and 0.34 nM for the μ, δ and κ opioid receptors, respectively (Mansour *et al.*, 1995). However, Met-enkephalin has highest affinity for the δ opioid receptor (Ki = 0.45 nM), and lower affinity for the μ and κ opioid receptors (1.8 and 47 nM, respectively) (Mansour *et al.*, 1995). Therefore, angiotensin-converting enzyme changes a non-selective opioid agonist into a δ opioid receptor-selective agonist. There are other cases where bioactive peptides are activated by extracellular peptidases. For example, bradykinin 1-9 binds preferentially to the B2 bradykinin receptor and has weak affinity for the B1 bradykinin receptor, while bradykinin 1-8 is a potent agonist of the B1 receptor (Zhang *et al.*, 2008). Extracellular processing of bradykinin 1-9 into bradykinin 1-8 by carboxypeptidase M or N converts the peptide from a B2 receptor agonist into a B1 receptor agonist (Zhang *et al.*, 2008). Thus, the extracellular processing of peptides is not always a degradative step, as it is for acetylcholine, and the particular peptidases that are present in the extracellular environment can modulate the physiological action of peptides in complex ways.

In addition to cleavages within the secretory pathway and outside the cell, endosomal cleavages can also have a profound effect on the signaling of the peptide-receptor complex. Enzymatic cleavages mediated by endothelin-converting enzyme-1 have been found to increase recycling of the receptors to the cell surface, presumably by degrading the peptide within the early endosome and allowing the dissociation of the receptor-arrestin complex (Padilla *et al.*, 2007; Pelayo *et al.*, 2011). It is likely that intracellular processing within the endocytic system occurs for a number of different peptides, and this is a relatively unexplored area.

7.2.2 Regulation of peptide processing

A large number of studies have examined the regulation of gene expression of peptide precursors and/or peptide-processing enzymes. Some of these studies focused on seasonal changes in food intake in rodents. Many genes are up- and downregulated in ground squirrels during the seasonal changes that precede or follow hibernation; some of these genes encode prohormones that are involved in food intake and energy balance (Schwartz *et al.*, 2013). In Siberian hamsters, differences in prohormone expression levels were found between the short and long photoperiods, although some of the changes did not seem to correlate with feeding behavior (Helwig *et al.*, 2006). For example, POMC levels were decreased during the short-day cycle when the animals were eating relatively low amounts of food. But because POMC is the precursor of the anorexic peptide α-MSH, it was predicted that POMC levels would be higher during this period, not lower (Helwig *et al.*, 2006). The resolution of this apparent paradox is presumably due to the extent of POMC processing, and not the absolute levels of POMC gene expression. The expression levels of PC1/3 and PC2 were found to change with the photoperiod (Helwig *et al.*, 2006). The variation in enzyme levels caused changes in the processing of POMC such that greater amounts of the anorexogenic α-MSH were produced during the short-day period than the long-day period. Similarly, levels of CPE were also found to be regulated by the change in photoperiod (Helwig *et al.*, 2013). However, because CPE is not a rate-limiting enzyme in the production of α-MSH, it is not clear if the change in CPE is the driving force for the altered peptide levels.

Another interesting mechanism of regulation of peptide processing involves inhibition of the PCs and CPE by catecholamines. Many decades ago, it was noted that treatment of cultured bovine adrenal chromaffin cells with reserpine resulted in a large increase in levels of enkephalin. Reserpine blocks catecholamine transport into secretory vesicles, thereby lowering the levels of dopamine, norepinephrine, and epinephrine inside these vesicles. Some of the catecholamine-containing vesicles also contain proenkephalin and the various processing enzymes (PCs and CPE). But the mechanism of the reserpine effect on enkephalin levels was a mystery for many years. An early report found that CPE activity was elevated by the reserpine treatment (Hook *et al.*, 1985), but it was not clear how this alone would raise enkephalin levels because CPE is not a rate-limiting enzyme. Recently, it was found that catecholamines directly inhibit PC1/3, PC2, and CPE (Helwig *et al.*, 2011). By lowering catecholamine levels inside the vesicles, reserpine activates PC and CPE, thereby increasing the conversion of proenkephalin into enkephalin. This is an example of cross-talk between catecholamines and neuropeptides, and shows that levels of peptides can be regulated by changes in activity of the processing enzymes caused by naturally occurring compounds.

Another way that neuropeptides can be regulated is by inhibition of the extracellular peptidases that cleave the neuropeptides (Figure 7.2). For example, the peptide opiorphin was found to be a potent analgesic, but surprisingly did not directly bind to the opioid receptors. Instead, opiorphin inhibits the breakdown of enkephalin by extracellular peptidases, thus prolonging the analgesic activity of enkephalin. I have proposed the term '**indirect neuropeptide**' to describe opiorphin and other peptides that produce biological effects by

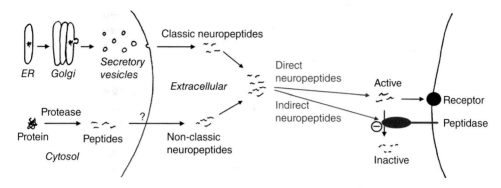

Figure 7.2 Comparison of classic neuropeptides with other types of peptides that can function like neuropeptides. Classic neuropeptides are produced in the regulated secretory pathway, as described in Figure 7.1. After secretion, classic neuropeptides can directly bind to neuropeptide receptors or be cleaved by extracellular peptidases into inactive peptides. In some cases, neuropeptides are cleaved into biologically active peptides that have different receptor-binding properties from the initial neuropeptide. Peptides are also produced within the cytosol by the action of proteases that cleave cellular proteins into peptides. Several families of proteases function within the cytosol, including proteasomes, calpains, and caspases. All of these proteases convert proteins into peptides. Some of the cytosolic peptides are secreted via an unknown mechanism. Peptides produced from cytosolic proteins that bind to cell surface receptors are collectively referred to as 'non-classic' neuropeptides. Non-classic neuropeptides are an emerging concept in the field; their role has not been fully established. In addition to neuropeptides (classic or non-classic) which bind directly to cell surface receptors, other biologically active peptides appear to exert their effects by inhibiting extracellular peptidases, thereby influencing levels of direct-acting neuropeptides. This idea of 'indirect neuropeptides' is also an emerging concept in the field, and further studies are needed to prove that they are physiologically important. ER, endoplasmic reticulum. Source: © Lloyd Fricker.

inhibition of the extracellular peptidases (see Figure 7.2); this term is analogous to the term 'indirect agonist' used to refer to compounds such as acetylcholine esterase inhibitors, which elevate neurotransmitters by blocking their degradation. Although only a small number of indirect neuropeptides are known, it is likely that many peptides influence the activity of the extracellular peptidases simply by competing for the active site on the peptidase. In considering this issue, it is important to keep in mind that peptides are always secreted as mixtures consisting of all peptides derived from the precursor. Although these other peptides are often ignored in studies that focus on a single neuropeptide, it is certainly possible that the mixture of peptides alters the extracellular cleavage of the neuropeptide, thereby acting as indirect neuropeptides.

7.3   Techniques used to study the rate of peptide biosynthesis

Many previous studies examined the rate of neuroendocrine peptide biosynthesis by labeling cells with radioactive amino acids and then following the incorporation of the label into peptides, which were usually isolated using immunoprecipitation (Gold *et al.*, 1982a, b; Howell *et al.*, 1965; Mains and May, 1988; Sando *et al.*, 1972). Modern approaches use non-radioactive amino acids containing stable isotopes such as deuterium (^{2}H), ^{13}C, and/or ^{15}N instead of radioactive amino acids, and then follow the incorporation of the label into the peptide using mass spectrometry (Che *et al.*, 2004). The latter approach using mass spectrometry is better for detecting small changes in the incorporation of the labeled amino acid. Using this approach, it was found that treating cells with chloroquine (which disrupts the acidification of secretory granules) slows peptide biosynthesis (Che *et al.*, 2004), as expected based on the acidic pH optima of the various peptide-processing enzymes. However, a previous study using radioactive amino acids came to the opposite conclusion, that chloroquine had no effect on the rate of peptide biosynthesis (Mains and May, 1988); this other study could not detect small changes due to the limitations of the technique.

The basic approach of pulse-chase labeling (with either radioactive or stable isotopic forms of amino acids) has provided useful information about the time-course of neuroendocrine peptide processing, such as the concept that the majority of the proteolytic cleavages occur in the late secretory pathway once the precursor and enzymes are packaged into immature secretory granules. Another finding is that the newly made neuropeptide is secreted preferentially over the older material during basal conditions. This result was found using primary cultures of rat islets, primary cultures of intermediate pituitary and cell lines derived from islets and anterior pituitary (Gold *et al.*, 1982a, b; Howell *et al.*, 1965; Sando *et al.*, 1972). Stimulating the cells with secretagogues caused the secretion of the older material, along with some of the newly made peptide. The finding that newly produced peptide is preferentially secreted under basal conditions was unexpected as it was initially thought that the newly made secretory granules would be slowly transported to the cell surface and secreted in the order produced, much like a conveyor belt moves products from a factory to the shipping dock to be loaded on a truck. Instead, it appears as if the old vesicles are stored in a manner that reduces their secretion under basal conditions. This is supported by studies examining the spatial segregation of old and new granules in bovine adrenal chromaffin cells using a fluorescent cargo protein that changes color over time (Duncan *et al.*, 2003). Under basal conditions, the newly synthesized vesicles were found to be docked at the plasma membrane while the older granules were found within the cell, away from the plasma membrane. Because the secretory granules contain active peptidases along with the peptides, the peptidases can continue to cleave the peptides over the hours/days that some vesicles are stored. The sites within the precursors that are not efficiently cleaved by the processing enzymes (such as some Lys-Lys, Arg-Lys, and single basic sites) can be cleaved as the granules age, changing the composition of peptides which in turn can influence the signaling properties of the peptides.

7.4   Dynamics of intracellular proteins and peptides

The preceding section focused on proteins/peptides within the secretory pathway. A distinct set of proteins and peptides are found in compartments such as the cytosol, nucleus, and mitochondria. These proteins/peptides are collectively referred to as intracellular, to distinguish them from proteins/peptides in the secretory pathway that are destined for secretion, although this term is not perfect in that some proteins in the secretory pathway are also largely intracellular (i.e. furin, CPD), owing to the presence of a transmembrane domain. In this review, the term 'intracellular' will be used to refer to proteins/peptides that are initially synthesized within the cytosol and not imported into the endoplasmic reticulum. This includes proteins that remain in the cytosol as well as proteins that are transported into the nucleus or mitochondria.

Many of the mass spectrometry-based techniques used to study proteins and peptides do not distinguish between secretory pathway and intracellular molecules. For example, pulse-chase labeling techniques used to examine the turnover of proteins in brain detected mainly intracellular proteins and a very small number of secretory pathway proteins (Price *et al.*, 2010). Interestingly, the turnover of proteins in mouse brain was found to be slower than that of proteins in liver or blood, with average lifetimes of 9.0, 3.0, and 3.5 days, respectively (Price *et al.*, 2010). Even when similar proteins were compared (for example, mitochondrial proteins), the turnover rate was slower in brain than other tissues. A small subset of brain proteins, such as myelin basic protein, had a half-life of up to 1 year. Unfortunately, the only neuropeptide precursor detected in this study (proSAAS) could not be quantified, possibly because the turnover rate was too fast and the proSAAS-derived peptides were never detected in a form with both ^{14}N and ^{15}N at the same time-point. Other studies have used similar isotopic labeling approaches to study protein turnover in rat brain, with generally similar results (Savas *et al.*, 2012). Collectively, these studies show that most brain proteins are highly dynamic and are turned over within days or weeks.

Many fragments of intracellular proteins were detected when peptidomic approaches were used to look for neuropeptides in extracts of mouse brain (Che *et al.*, 2005; Fricker, 2010; Rioli *et al.*, 2003; Skold *et al.*, 2002; Svensson *et al.*, 2003). The presence of intracellular protein fragments in brain was somewhat surprising. Even though intracellular proteins are cleaved by the proteasome complex into peptides, these peptides are thought to be rapidly degraded by intracellular peptidases (Herberts *et al.*, 2003; Reits *et al.*, 2004). It appears that hundreds of peptides escape this rapid degradation in brain and other mouse tissues, as well as all cell lines examined (Fricker, 2010). Interestingly, most of these peptides are not derived from highly abundant or unstable proteins. Recent studies on cell lines suggest that the proteasome is involved in the production of the intracellular peptides, based on the finding that levels of many peptides were reduced by treatment of cells with a proteasome inhibitor (Fricker *et al.*, 2012; Gelman *et al.*, 2011). An intriguing possibility is that intracellular peptides are functional; there is a variety of possible roles for these peptides.

One potential role for peptides derived from intracellular proteins is as cell-to-cell signaling molecules, much like the classic neuropeptides produced in the secretory pathway (see Figure 7.2). Some of the peptides derived from intracellular proteins have been found to be secreted from cultured brain slices (Gelman *et al.*, 2013). Although the mechanism of peptide secretion is not known, a number of cytosolic proteins are secreted by non-conventional mechanisms (Nickel and Rabouille, 2009). Proteins secreted from the cytosol include some interleukins, growth factors, and other signaling molecules. One of the peptides secreted from brain slices (named RVD-hemopressin) was found to bind to and activate the CB1 cannabinoid receptor (Gomes *et al.*, 2009). This concept of non-classic neuropeptides (i.e. neuropeptides derived from cytosolic proteins) is not new; the yeast *Saccharomyces cerevisiae* produces two mating factors, named α-factor and a-factor. While α-factor is produced in the secretory pathway by Kex2 (a furin-like endopeptidase) along with other enzymes, a-factor is produced in the cytosol and secreted via a transporter protein (Fuller *et al.*, 1988). Thus, there is a precedent for peptides generated in the cytosol to be secreted and function in cell-to-cell signaling, although more studies are needed to establish the physiological importance of this in brain and other neuroendocrine tissues.

Another potential role for peptides derived from intracellular proteins is as signaling molecules within the cell. This idea is analogous to the role of microRNA and other small non-coding RNAs in regulating mRNA levels and/or function. Small synthetic peptides of 10–20 amino acids are commonly used by scientists to influence protein-protein dimers, or otherwise interact with cellular proteins and alter their function (Arkin and Whitty, 2009; Churchill *et al.*, 2009; Rubinstein and Niv, 2009). However, even though peptides are clearly capable of interacting and modulating proteins, it remains to be demonstrated that this occurs under physiological conditions. Several studies have tested synthetic peptides that correspond to naturally occurring peptides. Interestingly, many intracellular processes are altered by these peptides (Berti *et al.*, 2012; Cunha *et al.*, 2008; Ferro *et al.*, 2004). In *C. elegans*, peptides derived from mitochondrial proteins were found to signal the cell's nucleus (Haynes *et al.*, 2010). In addition, there are examples of peptides which are cleaved from transmembrane proteins and enter the nucleus where they play a role in gene transcription (Brown and Goldstein, 1999; Francone *et al.*, 2010; Ye *et al.*, 2000). Thus, peptides produced from intracellular proteins can perform important roles in signaling.

The dynamics of protein processing into peptides is very important for the intracellular peptides. Unlike peptides in the secretory pathway, which are made in advance and stored within vesicles until being released by a stimulus, the peptides within the cytosolic and other non-secretory pathway compartments are generated on demand. The regulation of protein cleavage into peptides therefore plays a critical role in the generation of the signal. In some cases, the regulatory process is well characterized, such as the cholesterol-mediated generation of the sterol regulatory element-binding protein (Brown and Goldstein, 1999). In other cases, such as the generation of RVD-hemopressin (the peptide that activates the CB1 cannabinoid receptor), the levels of peptide are known to be regulated but the mechanism has not been determined (Gelman *et al.*, 2010).

7.5 Perspectives

Many years ago, the dogma was one gene, one mRNA, one protein, and one function. This is clearly wrong at all levels. First, multiple versions of mRNA are often produced from each gene due to differential splicing of the exons. If the differential splicing occurs within the coding region, these RNAs produce different proteins. Proteolytic processing can also generate distinct forms of each protein, including smaller peptides that may be functional. In most cases, the functions of the different proteins/peptides produced from a particular gene are distinct. The production and processing of proteins and peptides are highly dynamic within the cell, and to fully understand the system it is necessary to know how the synthesis, processing, and secretion of the molecules are regulated.

References

Allen, R.G., Peng, B., Pellegrino, M.J., *et al.* (2001) Altered processing of pro-orphanin FQ/nociceptin and pro-opiomelanocortin-derived peptides in the brains of mice expressing defective prohormone convertase 2. *Journal of Neuroscience*, **21**, 5864–5870.

Arkin, M.R. and Whitty, A. (2009) The road less traveled: modulating signal transduction enzymes by inhibiting their protein-protein interactions. *Current Opinion in Chemical Biology*, **13**, 284–290.

Bennett, H.P.J. (1991) Glycosylation, phosphorylation, and sulfation of peptide hormones and their precursors, in *Peptide Biosynthesis and Processing* (ed L.D. Fricker), CRC Press, Boca Raton, pp. 111–140.

Berti, D.A., Russo, L.C,. Castro, L.M., *et al.* (2012) Identification of intracellular peptides in rat adipose tissue: insights into insulin resistance. *Proteomics*, **12**, 2668–2681.

Bicknell, A.B. (2008) The tissue-specific processing of pro-opiomelanocortin. *Journal of Neuroendocrinology*, **20**, 692–699.

Braakman, I. and Hebert, D.N. (2013) Protein folding in the endoplasmic reticulum. *Cold Spring Harbor Perspectives in Biology*, **5**, a013201.

Brown, M.S. and Goldstein, J.L. (1999) A proteolytic pathway that controls the cholesterol content of membranes, cells, and blood. *Proceedings of the National Academy of Sciences of the United States of America*, **96**, 11041–11048.

Che, F.Y., Yuan, Q., Kalinina, E. and Fricker, L.D. (2004) Examination of the rate of peptide biosynthesis in neuroendocrine cell lines using a stable isotopic label and mass spectrometry. *Journal of Neurochemistry*, **90**, 585–594.

Che, F.Y., Lim, J., Biswas, R., Pan, H. and Fricker, L.D. (2005) Quantitative neuropeptidomics of microwave-irradiated mouse brain and pituitary. *Molecular & Cellular Proteomics*, **4**, 1391–1405.

Churchill, E.N., Qvit, N. and Mochly-Rosen, D. (2009) Rationally designed peptide regulators of protein kinase C. *Trends in Endocrinology and Metabolism*, **20**, 25–33.

Cunha, F.M., Berti, D.A., Ferreira, Z.S., Klitzke, C.F., Markus, R.P. and Ferro, E.S. (2008) Intracellular peptides as natural regulators of cell signaling. *Journal of Biological Chemistry*, **283**, 24448–24459.

Duncan, R.R., Greaves, J., Wiegand, U.K., *et al.* (2003) Functional and spatial segregation of secretory vesicle pools according to vesicle age. *Nature*, **422**, 176–180. [*This paper shows that all secretory vesicles are not the same, and supports previous studies which suggested that newly synthesized vesicles were secreted before older vesicles. Duncan and colleagues used a fluorescent cargo protein that changed color with time and found that newly assembled vesicles are docked at the plasma membrane, whereas older vesicles are located inside the cell, away from the plasma membrane. They also showed that the different types of vesicles could be preferentially released by different secretagogues.*]

Eipper, B.A. and Mains, R.E. (1980) Structure and biosynthesis of proACTH/endorphin and related peptides. *Endocrine Reviews*, **1**, 1–27.

Ferro, E.S., Hyslop, S. and Camargo, A.C. (2004) Intracellullar peptides as putative natural regulators of protein interactions. *Jorunal of Neurochemistry*, **91**, 769–777. [*This paper describes the novel hypothesis that intracellular peptides can affect cellular activity by influencing protein interactions. Subsequent studies have found additional evidence in support of this hypothesis.*]

Francone, V.P., Ifrim, M.F., Rajagopal, C., *et al.* (2010) Signaling from the secretory granule to the nucleus: Uhmk1 and PAM. *Mol Endocrinol* **24**: 1543–1558.

Fricker, L.D. (2002) Carboxypeptidases E and D, in *The Enzymes, Volume 23: Co- and Post-translational Proteolysis of Proteins* (eds R.E. Dalbey and D.S. Sigman), Academic Press, San Diego, pp. 421–452.

Fricker, L.D. (2004a) Carboxypeptidase E, in *Handbook of Proteolytic Enzymes* (eds A.J. Barrett, N.D. Rawlings and J.F. Woessner), Academic Press, San Diego, pp. 840–844.

Fricker, L.D. (2004b) Metallocarboxypeptidase D, in *Handbook of Proteolytic Enzymes* (eds A.J. Barrett, N.D. Rawlings and J.F. Woessner), Academic Press, San Diego, pp. 848–851.

Fricker, L.D. (2010) Analysis of mouse brain peptides using mass spectrometry-based peptidomics: implications for novel functions ranging from non-classical neuropeptides to microproteins. *Molecular Biosystems*, **6**, 1355–1365.

Fricker, L.D. (2012) *Neuropeptides and Other Bioactive Peptides,* Morgan & Claypool Life Sciences, Charleston.

Fricker, L.D., Gelman, J.S., Castro, L.M., Gozzo, F.C. and Ferro, E.S. (2012) Peptidomic analysis of HEK293T cells: effect of the proteasome inhibitor epoxomicin on intracellular peptides. *Journal of Proteome Research*, **11**, 1981–1990.

Fuller, R.S., Sterne, R.E. and Thorner, J. (1988) Enzymes required for yeast prohormone processing. *Annual Review of Physiology*, **50**, 345–362.

Fuller, R.S., Brake, A.J. and Thorner, J. (1989) Intracellular targeting and structural conservation of a prohormone-processing endoprotease. *Science*, **246**, 482–486.

Gelman, J.S., Sironi, J., Castro, L.M., Ferro, E.S. and Fricker, L.D. (2010) Hemopressins and other hemoglobin-derived peptides in mouse brain: comparison between brain, blood, and heart peptidome and regulation in Cpefat/fat mice. *Journal of Neurochemistry*, **113**, 871–880.

Gelman, J.S., Sironi, J., Castro, L.M., Ferro, E.S. and Fricker, L.D. (2011) Peptidomic analysis of human cell lines. *Journal of Proteome Research*, **10**, 1583–1592.

Gelman, J.S., Dasgupta, S., Berezniuk, I. and Fricker, L.D. (2013) Analysis of peptides secreted from cultured mouse brain tissue. *Biochimica et Biophysica Acta*, **1834**, 2408–2417.

Gold, G., Gishizky, M.L. and Grodsky, G.M. (1982a) Evidence that glucose "marks" beta cells resulting in preferential release of newly synthesized insulin. *Science*, **218**, 56–58.

Gold, G., Landahl, H.D., Gishizky, M.L. and Grodsky, G.M. (1982b) Heterogeneity and compartmental properties of insulin storage and secretion in rat islets. *Journal of Clinical Investigation*, **69**, 554–563.

Gomes, I., Grushko, J.S., Golebiewska, U., *et al.* (2009) Novel endogenous peptide agonists of cannabinoid receptors. *FASEB Journal*, **23**, 3020–3029.

Hammonds, R.G. Jr., Nicolas, P. and Li, C.H. (1984) beta-endorphin-(1-27) is an antagonist of beta-endorphin analgesia. *Proceedings of the National Academy of Sciences of the United States of America*, **81**, 1389–1390.

Haynes, C.M., Yang, Y., Blais, S.P., Neubert, T.A. and Ron, D. (2010) The matrix peptide exporter HAF-1 signals a mitochondrial UPR by activating the transcription factor ZC376.7 in C. elegans. *Molecular Cell*, **37**, 529–540.

Helwig, M., Khorooshi, R.M., Tups, A., *et al.* (2006) PC1/3 and PC2 gene expression and post-translational endoproteolytic pro-opiomelanocortin processing is regulated by photoperiod in the seasonal Siberian hamster (Phodopus sungorus). *Journal of Neuroendocrinology*, **18**, 413–425.

Helwig, M., Vivoli, M., Fricker, L.D. and Lindberg, I. (2011) Regulation of neuropeptide processing enzymes by catecholamines in endocrine cells. *Molecular Pharmacology*, **80**, 304–313.

Helwig, M., Herwig, A., Heldmaier, G., Barrett, P., Mercer, J.G. and Klingenspor, M. (2013) Photoperiod-dependent regulation of carboxypeptidase E affects the selective processing of neuropeptides in the seasonal Siberian hamster (Phodopus sungorus). *Journal of Neuroendocrinology*, **25**, 190–197. [**The two papers by Helwig et al. (2006, 2013) demonstrate that the peptide-processing enzymes are dynamically regulated in animals, and this regulation alters the specific peptides that are produced.**]

Herberts, C., Reits, E. and Neefjes, J. (2003) Proteases, proteases and proteases for presentation. *Nature Immunology*, **4**, 306–308.

Hook, V.Y.H., Eiden, L.E. and Pruss, R.M. (1985) Selective regulation of carboxypeptidase peptide hormone-processing enzyme during enkephalin biosynthesis in cultured bovine adrenomedullary chromaffin cells. *Journal of Biological Chemistry*, **260**, 5991–5997.

Hoshino, A.L. and Lindberg, I. (2012) Peptide Biosynthesis: Prohormone Convertases 1/3 and 2, Morgan & Claypool Life Sciences, New Jersey.

Howell, S.L., Parry, D.G. and Taylor, K.W. (1965) Secretion of newly synthesized insulin in vitro. *Nature*, **208**, 487.

Huttner, W.B. (1988) Tyrosine sulfation and the secretory pathway. *Annual Review of Physiology*, **50**, 363–376.

Kemmler, W., Peterson, J.D. and Steiner, D.F. (1971) Studies on the conversion of proinsulin to insulin. *Journal of Biological Chemistry*, **246**, 6786–6791.

Lindberg, I. and Hutton, J.C. (1991) Peptide processing proteinases with selectivity for paired basic residues, in Peptide Biosynthesis and Processing (eds L.D. Fricker), CRC Press, Boca Raton, pp. 141–174.

Mains, R.E. and May, V. (1988) The role of a low pH intracellular compartment in the processing, storage, and secretion of ACTH and endorphin. *Journal of Biological Chemistry*, **263**, 7887–7894.

Mansour, A., Hoversten, M.T., Taylor, L.P., Watson, S.J. and Akil H (1995) The cloned mu, delta and kappa receptors and their endogenous ligands: evidence for two opioid peptide recognition cores. *Brain Research*, **700**, 89–98.

Mutt, V., Jorpes, J.E. and Magnusson, S. (1970) Structure of porcine secretin. *The amino acid sequence. European Journal of Biochemistry*, **15**, 513–519.

Nakayama, K. (1997) Furin: a mammalian subtilisin/Kex2p-like endoprotease involved in processing of a wide variety of precursor proteins. *Biochemical Journal*, **327**, 625–635.

Nickel, W. and Rabouille, C. (2009) Mechanisms of regulated unconventional protein secretion. *Nature Reviews. Molecular Cell Biology*, **10**, 148–155.

Norman, J.A., Autry, W.L. and Barbaz, B.S. (1985) Angiotensin-converting enzyme inhibitors potentiate the analgesic activity of [Met]-enkephalin-Arg6-Phe7 by inhibiting its degradation in mouse brain. *Molecular Pharmacology*, **28**, 521–526.

Padilla, B.E., Cottrell, G.S., Roosterman, D., *et al.* (2007) Endothelin-converting enzyme-1 regulates endosomal sorting of calcitonin receptor-like receptor and beta-arrestins. *Journal of Cell Biology*, **179**, 981–997.

Pelayo, J.C., Poole, D.P., Steinhoff, M., Cottrell, G.S. and Bunnett, N.W. (2011) Endothelin-converting enzyme-1 regulates trafficking and signalling of the neurokinin 1 receptor in endosomes of myenteric neurones. *Journal of Physiology*, **589**, 5213–5230.

Price, J.C., Guan, S., Burlingame, A., Prusiner, S.B. and Ghaemmaghami, S. (2010) Analysis of proteome dynamics in the mouse brain. *Proceedings of the National Academy of Sciences of the United States of America*, **107**, 14508–14513.

Prigge, S.T., Mains, R.E., Eipper, B.A. and Amzel, L.M. (2000) New insights into copper monooxygenases and peptide amidation: structure, mechanism and function. *Cellar and Molecular Life Sciences*, **57**, 1236–1259.

Reits, E., Neijssen, J., Herberts, C., *et al.* (2004) A major role for TPPII in trimming proteasomal degradation products for MHC class I antigen presentation. *Immunity*, **20**, 495–506.

Rioli, V., Gozzo, F.C., Heimann, A.S., *et al.* (2003) Novel natural peptide substrates for endopeptidase 24.15, neurolysin, and angiotensin-converting enzyme. *Journal of Biological Chemistry*, **278**, 8547–8555.

Rubinstein, M. and Niv, M.Y. (2009) Peptidic modulators of protein-protein interactions: progress and challenges in computational design. *Biopolymers*, **91**, 505–513.

Sando, H., Borg, J. and Steiner, D.F. (1972) Studies on the secretion of newly synthesized proinsulin and insulin from isolated rat islets of Langerhans. *Journal of Clinical Investigation*, **51**, 1476–1485.

Savas, J.N., Toyama, B.H., Xu, T., Yates, J.R. III and Hetzer, M.W. (2012) Extremely long-lived nuclear pore proteins in the rat brain. *Science*, **335**, 942.

Schwartz, C., Hampton, M. and Andrews, M.T. (2013) Seasonal and regional differences in gene expression in the brain of a hibernating mammal. *PLoS One*, **8**, e58427.

Seidah, N.G. and Chretien, M. (2004a) Proprotein convertase 2, in *Handbook of Proteolytic Enzymes* (eds A.J. Barrett, N.D. Rawlings and J.F. Woessner), Academic Press, San Diego, pp. 1865–1868.

Seidah, N.G. and Chretien, M. (2004b) Proprotein convertase I, in *Handbook of Proteolytic Enzymes* (eds A.J. Barrett, N.D. Rawlings and J.F. Woessner), Academic Press, San Diego, pp. 1861–1864.

Seidah, N.G., Gaspar, L., Mion, P., Marcinkiewicz, M., Mbikay, M. and Chretien, M. (1990) cDNA sequence of two distinct pituitary proteins homologous to Kex2 and furin gene products: tissue-specific mRNAs encoding candidates for prohormone processing proteinases. *DNA Cell Biology*, **9**, 415–424.

Skold, K., Svensson, M., Kaplan, A., Bjorkesten, L., Astrom, J. and Andren, P.E. (2002) A neuroproteomic approach to targeting neuropeptides in the brain. *Proteomics*, **2**, 447–454.

Smeekens, S.P. and Steiner, D.F. (1990) Identification of a human insulinoma cDNA encoding a novel mammalian protein structurally related to the yeast dibasic processing protease Kex2. *Journal of Biological Chemistry*, **265**, 2997–3000.

Song, L. and Fricker, L.D. (1995) Purification and characterization of carboxypeptidase D, a novel carboxypeptidase E-like enzyme, from bovine pituitary. *Journal of Biological Chemistry*, **270**, 25007–25013.

Steiner, D.F. (1998) The proprotein convertases. *Current Opinion in Chemical Biology*, **2**, 31–39.

Strand, F.L. (2003) Neuropeptides: general characteristics and neuropharmaceutical potential in treating CNS disorders. *Progress in Drug Research*, **61**, 1–37.

Svensson, M., Skold, K., Svenningsson, P. and Andren, P.E. (2003) Peptidomics-based discovery of novel neuropeptides. *Journal of Proteome Research*, **2**, 213–219. [**This paper describes a breakthrough in the development of peptidomic techniques to identify peptides in mouse brain. Using focused microwave irradiation to rapidly inactivate proteases in mouse brain, the authors were able to detect neuropeptides above the background of protein degradation fragments.**]

Tagliabracci, V.S., Xiao, J. and Dixon, J.E. (2013) Phosphorylation of substrates destined for secretion by the Fam20 kinases. *Biochemical Society Transactions*, **41**, 1061–1065.

Thomas, G. (2002) Furin at the cutting edge: from protein traffic to embryogenesis and disease. *Nature Reviews. Molecular Cell Biology*, **3**, 753–766.

Tuteja, R. (2005) Type I signal peptidase: an overview. *Archives of Biochemistry & Biophysics*, **441**, 107–111.

Varlamov, O. and Fricker, L.D. (1998) Intracellular trafficking of metallocarboxypeptidase D in AtT-20 cells: localization to the trans-Golgi network and recycling from the cell surface. *Journal of Cell Science*, **111**, 877–885.

Ye, J., Rawson, R.B., Komuro, R., *et al.* (2000) ER stress induces cleavage of membrane-bound ATF6 by the same proteases that process SREBPs. *Molecular Cell*, **6**, 1355–1364.

Zhang, X., Tan, F., Zhang, Y. and Skidgel, R.A. (2008) Carboxypeptidase M and kinin B1 receptors interact to facilitate efficient b1 signaling from B2 agonists. *Journal of Biological Chemistry*, **283**, 7994–8004.

Zhou, A., Bloomquist, B.T. and Mains, R.E. (1993) The prohormone convertases PC1 and PC2 mediate distinct endoproteolytic cleavages in a strict temporal order during proopiomelanocortin biosynthetic processing. *Journal of Biological Chemistry*, **268**, 1763–1769.

Zhou, A., Webb, G., Zhu, X. and Steiner, D.F. (1999) Proteolytic processing in the secretory pathway. *Journal of Biological Chemistry*, **274**, 20745–20748.

CHAPTER 8

Neuropeptidomics: The Characterization of Neuropeptides and Hormones in the Nervous and Neuroendocrine Systems

Ning Yang,[1] Samuel J. Irving,[2] Elena V. Romanova,[1] Jennifer W. Mitchell,[3] Martha U. Gillette,[3] and Jonathan V. Sweedler[1]

[1] Department of Chemistry and the Beckman Institute, University of Illinois at Urbana-Champaign, Urbana, Illinois, USA
[2] Department of Molecular and Integrative Physiology, University of Illinois at Urbana-Champaign, Urbana, Illinois, USA
[3] Department of Cell and Developmental Biology and the Beckman Institute, University of Illinois at Urbana-Champaign, Urbana, Illinois, USA

8.1 Neuropeptides – one gene, multiple products

The completion of the sequencing of the genomes of many model animals, especially the human genome (International Human Genome Sequencing Consortium, 2004), has illuminated the functional significance of numerous genes and hinted at the benefits of personalized genetic medicine. However, turning this genomics vision into reality requires dedicated effort to understand the molecular architecture encoded by the genome. A key challenge is unraveling the network of regulatory mechanisms that govern the coordinated actions of the myriad cells in the body.

Neuropeptides are a diverse group of molecules involved in both local and long-distance cell-to-cell communication that play an important role in the central and peripheral nervous systems (Jekely, 2013). They participate in various physiological processes and endocrine regulation such as reproduction, growth, food intake, memory, pain sensation, diseases, and circadian rhythm control, to name but a few. Neuropeptides are often grouped into families based on their sequence similarities. Yet even when neuropeptide families have considerable sequence homology, they can have diverse structural identities and functions, partly due to differential gene expression and posttranslational modifications (PTMs). Alternative splicing of genes contributes to the diversity of neuropeptides before the translation; alternative splicing of mRNA from prohormone genes also is known. As one example, it was found that alternative splicing in calcitonin gene expression led to two different polypeptide products, designated as calcitonin peptide and calcitonin gene-related peptide (Amara *et al.*, 1982). Other prohormones, such as gastrin-releasing peptide prohormone and protachykinin, follow this same process (Amara *et al.*, 1982; Spindel *et al.*, 1986).

Neuropeptide diversity can result from intrinsic factors affecting the regulation of transcription and translation, as well as from environmental stimuli, cellular catalytic activity, and many others. Neuropeptide processing is complex. For example, the predicted protein encoded by a neuropeptide gene, a preprohormone, is cleaved before it is entirely translated; during ribosomal synthesis, the N-terminus comprising the signal sequence of the

Molecular Neuroendocrinology: From Genome to Physiology, First Edition. Edited by David Murphy and Harold Gainer.
© 2016 John Wiley & Sons, Ltd. Published 2016 by John Wiley & Sons, Ltd.
Companion website: www.wiley.com/go/murphy/neuroendocrinology

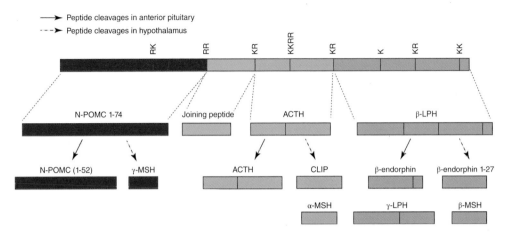

Figure 8.1 While a preprohormone is a predictable gene product, the prohormone undergoes additional processing to form the final bioactive peptides, including a series of well-orchestrated cleavages at specific locations; it is these processing steps that result in the production of a number of bioactive neuropeptides. The final peptide forms can vary in a tissue-specific and even physiological state-dependent manner, making their prediction from a gene sequence difficult. As in the example shown here, pro-opiomelanocortin (POMC) prohormone is cleaved into a number of peptides, including N-POMC (1-52), adrenocorticotropic hormone (ACTH), β-endorphin and γ-lipotropin (LPH) formed in the anterior pituitary and γ-melanotropin (MSH), corticotropin-like intermediary peptide (CLIP), α-MSH, β-endorphin 1-27, and β-MSH formed in the hypothalamus.

forming preprohormone translocates into the endoplasmic reticulum, and its signal peptide is cleaved off, even before the rest of the prohormone is translated, so that the intact preprohormone is not detected. The prohormone (minus its signal peptide) is transported along the secretory pathway where many other processing steps occur (Figure 8.1). Many of the enzymatic cleavages and other PTMs take place in the Golgi apparatus and secretory vesicles. The myriad processing steps result in the production of a number of peptides, some of which are bioactive neuropeptides, from a single prohormone. The final products may vary in a tissue-specific manner, and peptides can even be released in an inactive form and then become bioactive after additional PTMs take place after release. Neuropeptides constitute arguably the largest and most diverse class of signaling molecules, making their wholesale characterization difficult.

8.2   Mining the neuropeptidome 21st-century style using mass spectrometry-based 'omics approaches

When investigating the function of a specific brain region, it is useful to know the complement of the possible cell-to-cell signaling molecules used by that region. What peptides are present in a brain region? While the question seems simple, it can be difficult to answer without considerable effort. In the past, this challenge would have been confronted one peptide at a time, using specific tests based on molecular-selective probes. In recent years, **mass spectrometry** (MS) has emerged as an effective analytical tool for the global characterization of a diverse range of molecules. The terminology for MS-based 'omics approaches is based on the class of molecule under investigation: metabolomics (small molecules), **peptidomics** (peptides and smaller protein products), and proteomics (proteins). Today, MS-based peptidomics, the global and comprehensive characterization of the peptides in a selected tissue, can provide a library of peptide forms that are present in cells or tissues. Two of the commonly used MS-based methods for peptide discovery are **liquid chromatography**/tandem mass spectrometry (LC)-MS/MS and **matrix-assisted laser desorption/ionization** (MALDI)-MS (Li and Sweedler, 2008; Romanova *et al.*, 2013) (Figure 8.2).

8.2.1 Mass spectrometry-based peptidomics

Distinct from traditional methods used to study neuropeptides, such as enzyme-linked immunosorbent assay (ELISA) and radioimmunoassay (RIA), MS does not require anti-bodies, which must be produced only for known peptides, and if many peptides are to be characterized, can be time-consuming to develop. While a global peptidomics measurement does not require analyte preselection, it only detects peptides if they:

- are present in the sample in detectable amounts
- are stable throughout the sampling and analysis processes
- have physiochemical properties that allow retention and selective elution during a separation
- are structurally suitable for efficient ionization during MS analysis.

However, even the long lists of peptides characterized from a brain region via MS-based peptidomics are not expected to be comprehensive. With tandem MS (MS/MS) approaches, unknown and unexpected peptides can be characterized, including their PTMs; some PTMs offer insight into the bioactive forms of numerous neuropeptides, information not generally obtained via ELISA and RIA. The advantages of MS are enhanced by efficient separation techniques such as LC and new bioinformatics tools. In what follows, the fundamental requirements of an effective peptidomics experiment are discussed.

8.2.1.1 Sample preparation: the foundation for successful MS-based neuropeptide investigation

No measurement approach can recover analytes lost during the sampling process; therefore, sample preparation is crucial. One of the most critical issues that can arise in mammalian neuropeptide sampling is rapid protein degradation. This causes a large increase in the number of protein fragments, which can significantly overshadow signals from less abundant endogenous peptides. A number of methods are available to reduce postmortem degradation. Sacrificing animals with focused microwave irradiation or microwave-irradiating tissues after animal sacrifice (Mathe *et al.*, 1990) increases the temperature of tissues rapidly and uniformly denatures the proteases. Cryostat dissection followed by boiling buffer treatment also blocks non-specific protein degradation effectively (Taban *et al.*, 2007). Some tissues of sufficiently small size, such as defined brain nuclei, can be frozen immediately after dissection. Instead of requiring a slow thawing process during which degradation occurs, treating them with boiling water during thawing almost instantly equilibrates the temperature of the tissue with the surrounding environment. Working with fresh tissues that have not been heated (or frozen) can have advantages when isolating small samples or measuring neuropeptide release (Romanova *et al.*, 2014).

In the next step, the peptides – together with other intracellular contents – are extracted from the samples. Several groups have compared the extraction performance of different solutions; ice-cold diluted acid solution has been found to work well in extracting peptides. One group found that more peptides can be extracted when the same samples are treated sequentially with several extraction solutions such as ice-cold acid, acidified acetone, and acidified methanol (Dowell *et al.*, 2006).

In addition to peptides, initial tissue homogenates also contain proteins, lipids, salts, and other molecules. This complexity may affect the separation of peptides via LC, and even

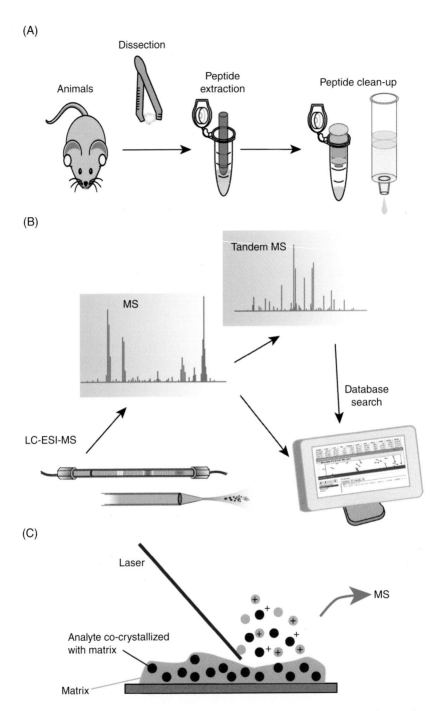

Figure 8.2 Schema showing the major steps of a typical peptidomics experiment. (A) The initial steps include isolation of tissue, extraction of peptides from the tissue, and sample clean-up and preconcentration. (B) One of the more common approaches involves LC separation and MS analysis of peptides, database searches, and other bioinformatics manipulations. (C) Another approach involves the direct measurement of a sample via a MALDI process. The sample containing neuropeptides is mixed with matrix that co-crystallizes with it. Upon laser irradiation, the matrix molecules absorb the energy and, along with the analyte, are vaporized and ionized. The resultant ions are characterized via an appropriate mass analyzer, often a time of flight instrument. The data manipulations after the MS experiment are similar to those shown in (B).

hinder their ionization during MS. To remove large molecules, molecular weight filters are frequently used. Solid phase extraction is commonly employed to desalt the samples (Li and Sweedler, 2008). After this clean-up process, the extracted peptides are ready for MS analysis. MALDI-MS can quickly reveal peptides from minimum sample amounts and LC-MS provides unambiguous identification of peptides.

8.2.1.2 Liquid chromatography-MS for high-throughput peptide sequencing and identification

The efficiency of using MS for peptide discovery is boosted by coupling the mass spectrometer to a separation modality such as LC. Figure 8.2 shows a typical LC-MS peptidomics workflow. After sample preparation and clean-up (Figure 8.2A), peptides are injected into LC columns where the analytes are separated into discrete bands (Figure 8.2B), thereby reducing the number of peptides entering the mass spectrometer at a specific time. Separation conditions can be optimized to allow peptides to be concentrated on-column prior to MS detection. The scale of the analysis can be adjusted to the amount of sample and peptide concentration; capillary- or nano-scale reversed phase (RP)-LC are suitable approaches for separating endogenous peptides. Multidimensional high-performance LC separation provides improvements in separation efficiency and resolution, and is often used to study neuroendocrine samples. A number of other separation methods, such as ion exchange chromatography, hydrophilic interaction chromatography, and size exclusion chromatography, can be coupled to subsequent RPLC-MS, with such sequential separations being well suited for complex samples. For peptides, strong cation exchange chromatography is frequently used as the first dimension of separation. Use of RP-RP-LC separations has increased the number of identified peptides (Dowell *et al.*, 2006).

Implementing a combination of complementary MS platforms in one experiment also increases peptidome coverage as each type of mass analyzer has distinct strengths and advantages. For example, analyses with Fourier transform (FT) ion cyclotron MS yield high mass resolution and the most accurate mass measurements, thus aiding peptide identification. In contrast, quadrupole and time-of-flight mass analyzers are faster and more sensitive, and thus are often more suitable for the detection and quantitation of low-abundance analytes.

Liquid chromatography-MS/MS offers an unsurpassed capability to measure hundreds of molecules in a single experiment. To aid in the MS/MS data interpretation, bioinformatics tools with suitable mathematical algorithms are required. A number of open source web-based tools are available to accomplish peptide identification, including Mascot (www.matrixscience.com/search_form_select.html), SeQuest (http://fields.scripps.edu/sequest/), and OMSSA (www.chem.wisc.edu/~coon/software.php), which compare the acquired MS/MS spectra with the predicted fragmentation patterns of proteins and peptides contained in protein databases such as SwissProt and NCBInr. Peptide databases can also be built manually using open source tools such as NeuroPred (http://neuroproteomics.scs.illinois.edu/neuropred.html), which predicts expected peptides from a prohormone (Southey *et al.*, 2006). Another MS-based peptide sequencing option is *de novo* sequencing, accomplished using commercially available software such as PEAKS and PepNovo. This approach deduces peptide structure by calculating mass shifts among the fragment ions and comparing these to ion series formed from amino acids under known rules, without reference to a database. This independence from a sequence database makes *de novo* sequencing an exciting option for discovering peptides in organisms having an unknown genome. Successful *de novo* sequencing depends on the efficiency of the fragmentation, intensity of fragment ions and overall quality of the MS/MS spectra since all of the sequence information is extrapolated from the data obtained.

8.2.1.3   MALDI-MS for peptide analysis in small volume samples

With the advantages of high sensitivity, fast acquisition, reasonable salt tolerance, and small sample consumption, MALDI-MS is a widely applied technique for the analysis of complex biological samples in either a solid or liquid state. To obtain an informative peptide profile, a submicroliter volume of a biological sample is mixed or coated with an energy-absorbing matrix and irradiated by short laser pulses under vacuum in a mass spectrometer (Figure 8.2C). Two commonly used matrices for analytes below 10 kDa are 2,5-dihydroxybenzoic acid (DHB) and α-cyano-4-hydroxycinnamic acid (CHCA), either of which are effective for peptide analysis. DHB solutions can be used to extract peptides with or without tissue homogenization; peptides extracted directly in DHB solution can be subjected to MALDI-MS without any other further sample pretreatment, which reduces sample losses due to handling (Romanova *et al.*, 2008). Studies using MALDI-MS have greatly broadened our knowledge on neuropeptide families and contributed to the discovery of many novel neuropeptides in a variety of model organisms, including a urotensin II-like peptide (Romanova *et al.*, 2012), feeding circuit-activating peptides in the cerebral-buccal connective of *Aplysia californica* (Sweedler *et al.*, 2002), and a novel C-terminal KFamide peptide in the hemolymph of *Cancer borealis* (Chen *et al.*, 2009). Due to the distinct ionization processes used in **electrospray ionization** and MALDI, their combination expands neuropeptide coverage. Among the 56 identified neuropeptides in rat hypothalamus and striatum punches, 14 neuropeptides, or nearly 25%, were exclusively detected with MALDI-FTMS (Dowell *et al.*, 2006).

The high sensitivity and low sample consumption of MALDI-MS make it the tool of choice for the analysis of low-abundance peptides, such as those present in releasates from a cultured neuronal network or even single neurons. We recently developed a microfluidic device off-line coupled to MALDI-MS that selectively and controllably applies different chemical stimulants to a network of cultured neurons, and collects the peptides released from the neurons in response to the stimulation (Croushore and Sweedler, 2013). MALDI-MS analysis of the collected releasates provided information on the limited temporal dynamics of the cells under investigation. We have also used a dual capillary releasate collection device and interfaced it to MALDI-MS to profile the peptides released at defined regions of the central nervous system, with the detection levels ranging from neuron clusters down to individual cells (Fan *et al.*, 2013).

8.2.1.4   MALDI-MS for direct tissue profiling and imaging

A key advantage of MS is its ability to probe biological tissue to determine the localization and distribution of hundreds of peptides simultaneously in one experiment. Spatial information can be obtained by isolating a selected area of a tissue sample and assaying it directly with MALDI-MS to generate a profile of the peptides present, or by scanning the entire sample to obtain a chemical image. This direct profiling approach has been used successfully to characterize the neuropeptide complement in identified neurons, as well as elucidating the proteolytic processing of newly discovered prohormones (Jespersen *et al.*, 1999; Rubakhin *et al.*, 2006; Weaver and Audsley, 2010).

Mass spectrometry imaging (MSI) produces ion maps that are inherently registered to each other and provides an optical image of the analyzed tissue, thereby making interpretation of the data straightforward. Figure 8.3 illustrates the process of constructing a peptide distribution map using MSI. Preserving sample integrity, and the stability and spatial arrangement of the peptides, requires special considerations during sampling. A thorough discussion of the MSI sampling process is available in two recent reviews (Lanni *et al.*, 2013; Tucker *et al.*, 2011).

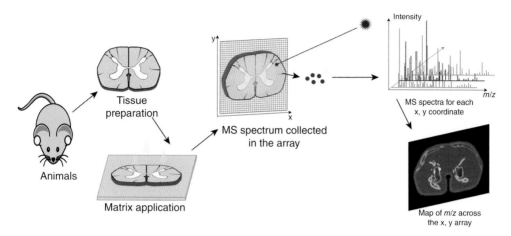

Figure 8.3 Mass spectrometry imaging (MSI) provides the chemical information content of MS with the addition of spatial information on the localization of the peptides from a tissue section. Most commonly, the tissue is sampled using MALDI-MS in an ordered array of locations. Any *m/z* can be selectively extracted from the data set and replotted according to its spatial coordinates, creating an image. The remapped image of selected ions shows their distributions across the tissue section. While the chemical information content is high, typical spatial resolutions are of the orders of tens of microns when imaging peptides.

When is it appropriate to use imaging and direct tissue profiling? Both approaches have several advantages. First, peptide extraction and desalting/fractionation are not needed, thereby minimizing sample loss and the chance that contaminants will be introduced into the sample (Rubakhin *et al.*, 2006). Also, the integrity of the tissue is preserved, allowing one to connect peptide distribution information from MSI to specific anatomical or morphological structures (Cornett *et al.*, 2007). One important benefit of MSI is that it can be used to compare and track the distribution and expression-level changes of peptides, including neuropeptides, in response to certain functional or behavioral disorders (Minerva *et al.*, 2011).

8.2.2 Selecting novel peptides for functional annotation using MS approaches

Empowered by highly multiplexed, sensitive and specific mass spectrometric assays, researchers can probe and mine hundreds of potential neuropeptides without having *a priori* information on their structure or localization (Lee *et al.*, 2013). For example, Xie and colleagues (2010) identified not only the genome-predicted peptides but also previously unknown, alternatively processed forms of known functional neuropeptides in the zebrafish brain. The qualitative study of peptides through both direct tissue analysis and LC-MS can provide exciting, hypothesis-generating strategies for exploring known and new molecular players in biochemical pathways and physiological functions. However, the vast numbers of peptides typically reported in peptidomics studies must be assessed carefully to differentiate the putative bioactive peptides among the vast majority of proteinaceous compounds generally found in complex biological samples due to postmortem degradation, catabolism of ubiquitous proteins or sampling artifacts. Once putative neuropeptide candidates have been verified, their functional and physiological contexts can be characterized using a variety of methods.

To distinguish the physiologically relevant peptides from the extraneous, specific questions should be addressed. Do the peptides come from known, previously characterized prohormones or secreted proteins? Are those proteins known to be expressed in the tissue being analyzed? Are the neuropeptides cleaved from prohormones at conventional cleavage sites, or are the unconventional cleavage sites supported by *in silico* prediction? Are the putative neuropeptides actively released from a neuron? And lastly, do the neuropeptides have bioactivity under the specific experimental paradigm? Importantly, one must carefully inspect the wealth of MS data that is obtained to determine whether there are non-specific degradation products from intracellular proteins, often manifested by numerous peptide forms with single amino acid truncations.

Due to advances in genetic sequencing technologies, many organisms have had their genomes sequenced. These genomic libraries can be used to identify potential neuropeptides and their regulatory regions. Sequences can also be annotated to include details on putative functions and functional structures. Propeptide processing is known to be variable, especially with larger prepro- and propeptides (Pritchard and White, 2007), so that different organs, morphological regions, and tissues may differentially process these precursors, resulting in overlapping but not identical peptide subsets found in different locations throughout the body. One example of a differentially processed neuropeptide precursor in the brain is pro-opiomelanocortin (POMC), which is processed into a variety of peptides that regulate diverse physiological functions, ranging from metabolic balance to melanin production in the skin (Pritchard and White, 2007) (see Figure 8.1). POMC processing occurs in a tissue-specific manner in the pituitary, hypothalamus, and skin. In addition to the multitude of processing products of POMC, there is a plethora of PTMs that add to the potential products identified (Yasuda *et al.*, 2013).

8.2.2.1 Monitoring peptide release

Once identified, the next step may be to determine which newly characterized and putative bioactive peptides are actively released in a regulated fashion. For example, knowing that peptides are released in a time-of-day fashion in the appropriate brain regions suggests an activity related to circadian rhythms. Said differently, if a measured peptide is a processing intermediate and is not released from cells, it is not a cell-to-cell signaling peptide.

Investigating releasate may be hampered by the large number of distinct stimulation parameters that can be used to initiate release. Identifying a physiological system that simplifies release experiments is advantageous. To this end, we know organisms have evolved several basic mechanisms to respond to stimuli and to predict external environmental threats and resources, and organize their homeostatic systems. One well-conserved mechanism is related to circadian rhythms (roughly 24-hour cycles); these cycles synchronize the internal environment with the external environment, allowing the anticipation of environmental cues, such as light, food, reproductive partners, and predators. This type of overarching activity makes a good target for neuropeptide discovery. In mammals, syncing internal systems to external stressors is controlled on a daily basis by the central clock in the brain, the suprachiasmatic nucleus (SCN); the SCN utilizes a diverse range of neuropeptides to connect the brain and, subsequently, the whole animal to the master pacemaker. These neuropeptides have varying concentrations in the brain throughout the day, corresponding to the signal, target, and downstream effects that ensure that the organism's homeostatic processes are performed at the proper time and during the appropriate activity. Dysregulation of clock coupling signals has been implicated in sleep disorders, obesity, immune dysregulation, and even cancer (Castanon-Cervantes *et al.*, 2010; Marcheva *et al.*, 2013; Schroeder and Colwell, 2013).

Additionally, neuropeptides can be released in response to other stimuli. For example, the SCN responds to the presence of environmental light and synchronizes the internal homeostatic clock to the external environment. To illustrate this in an experimental system, hypothalamic brain slices with intact SCN networks and optic nerves have been stimulated electrically. The result is a rapid response within the SCN that stimulates neuropeptide release (Hatcher *et al.*, 2008).

Thus, after careful application of the meaningful criteria outlined above to the entirety of the neuropeptidomics results, the list of putative neuropeptides relevant to a certain biological hypothesis becomes substantially shorter, more focused, and amenable to follow-up bioactivity assays.

8.2.2.2   Monitoring peptide level changes via MS-based quantitation

Because they are involved in complex signaling pathways, often a change in biological status leads to a change in the abundance of a neuropeptide. MS can be used to compare peptide levels in samples arising from different behavioral and disease states, as well as exposure to drugs (Romanova *et al.*, 2013). Isotopic labeling and label-free methods are two major MS-based approaches being used for global peptide quantitation.

As the name suggests, isotopic labeling requires peptides from two or more sets of samples to be labeled, respectively, with light or heavy forms of chemically similar isotopic labeling tags, which create peptides with different sets of distinct masses. The differentially labeled samples are combined and analyzed using LC-MS or LC-MS/MS. Ideally, the light and heavy labeled peptides share identical LC behaviors and MS responses so that the relative levels of the same peptides in different samples can be calculated from their peak intensity (or area) ratio in light and heavy forms. Many isotopic tags for peptide labeling are designed to react with amino groups in the peptides. Acetic anhydride, succinic anhydride, methyl chloride, and 4-trimethylammoniumbutyryl-*N*-hydroxysuccinimide are commonly used isotopic labeling reagents for peptides; they are inexpensive, synthetically or commercially available, and stable after being incorporated into the peptides.

In contrast, iTRAQ-type reagents are isobaric tags for quantitation based on MS/MS. In addition to nearly identical performance in LC and MS, peptides with different labeled forms have the same mass, leading to simultaneous selection and isolation for MS/MS. Reporter groups, substructures of the iTRAQ tags with different masses, are detached from the labeled peptides and detected during fragmentation in the mass spectrometer collision cell. Similarly, the peptide amounts in different samples can be calculated from the peak intensity or area of these reporter ions. iTRAQ allows peptides from up to eight samples to be quantified in a single run; this increases the throughput of peptide quantitation since the combined use of the stable isotopes, ^{13}C and ^{15}N, produces eight reporter groups each with a distinct mass. In addition, technical issues such as variable injection volumes or retention times are eliminated when samples are combined for parallel measurements during one analysis. However, pooling of multiple samples increases the chemical complexity of the combined sample, and may result in greater ion suppression and lower peptide coverage.

Peptides in a complex environment can also be quantified using label-free approaches, which are promising alternatives to isotopic labeling quantitation. Free from the addition of any labeling reagent, label-free methods overcome hurdles from isotopic labeling, such as the increased sample loss inherent in multistep labeling protocols, incomplete labeling of peptides, and an excess of labeling materials. Therefore, label-free methods are better at quantifying peptides present in low abundance and peptides with a blocked N-terminus (so that many labels will not react). However, since peptides are not labeled in different forms, each sample group must be analyzed separately, which makes **label-free quantitation** dependent on the precision of each step in the measurement process.

Several label-free quantitation approaches are used, including spectral count and peak intensity/area-based methods. Spectral count relies on counting the number of times a specific peptide is identified with MS/MS, since a higher abundance of a peptide results in a corresponding higher probability that the peptide is selected for each MS/MS analysis. The second approach accomplishes quantitation by comparing the peak intensity or area of peptides among two or more LC-MS runs. Since it is a run-to-run comparison of peak intensity or peak area, retention time alignment, detected (mass/ charge) m/z accuracy, and signal normalization are important for an accurate and reliable label-free quantitation. In a study of the circadian-related peptides in the rat SCN, multiple neuropeptides were found to have statistically significant abundance changes at two distinct times of day using this label-free quantitation method (Lee *et al.*, 2013). Other works have used spectral count to study circadian rhythm and diet-related neuropeptides (Frese *et al.*, 2013; Southey *et al.*, 2014). Peptides found to possess meaningful abundance changes are likely to be involved in function or behavior. They can be synthesized and tested with physiological and biological methods to confirm whether they are able to influence expected responses or behaviors.

8.3   What do all these peptides do? Follow-up functional studies

Once specific peptides are discovered that change levels in a meaningful way or are released in an activity-dependent manner, what is next? When one characterizes a single peptide at a time in an experiment, the answer is straightforward. When an MS-based peptidomics experiment creates a list that often includes hundreds of peptides, how does one select which peptides should receive follow-up studies and how does one implement them? Such analyses of neuropeptide activity can be performed using a variety of techniques using fixed or live tissue samples. Both static and dynamic experimental samples provide information that complements the identification and characterization information from the MS-based neuropeptide discovery and offers benefits for functional peptide characterization.

Generally, fixed tissue methods allow for amplification of peptide levels and an increase in the localization detail. Working with live tissues provides the opportunity to look for physiological responses and cellular changes from peptides in real time within the same sample. One can study changes in morphology, transcriptional changes, translational regulation, membrane excitability, and even *in vivo* organism responses. As highlighted above, such studies can be combined with MS measurements; for example, a series of samples taken from a living system and analyzed for peptide content allows one to understand how the system responds to stimuli such as by releasing specific peptides at specific time points or under well-defined stimulation protocols (Hatcher *et al.*, 2008).

Next, we discuss follow-up approaches that have been categorized according to whether they work with static or dynamic samples.

8.3.1   Static approaches

Static approaches allow detailed measurement of the chemical or spatial content of a tissue, but require sample preservation. Often this involves a fixation step to chemically and spatially freeze the sample, such as via formaldehyde fixation or rapid freezing, followed by cellular lytic methods that can prevent neuropeptide degradation. The tissue can then be assessed using established microscopic imaging methodologies that allow precise localization via antibody labeling, and peptidergic neuronal networks can be mapped to determine connections within the brain. As one example, Patel and colleagues (2014) identified similar neuropeptides in the brain of *Rhodnius prolixus* using MS, localized the neuropeptides using immunohistochemistry, and addressed their function *in vivo*.

For static measurements, a range of transcriptional assays are available. As one example, fluorescence *in situ* hybridization can be used to detect and localize specific mRNA transcripts.

Cellular transcripts are probed through RNA-fluorescent probes complementary to a sequence of mRNA. This approach is often used to determine the localization of transcripts and can verify whether a specific neuropeptide prohormone is synthesized in the tissue of interest. Besides visualization approaches, quantitation of protein levels across a time course can use non-MS approaches such as Western blot analyses (Atkins *et al.*, 2010).

8.3.2 Dynamic approaches

Dynamic studies allow the researcher to address functionality from a single cell to a whole organism, at times in real time, frequently utilizing the same sample for control and experimental purposes, thereby providing information on bioactivity and function that is difficult to obtain in other ways. There are many assays from which to choose, depending on the systems under study.

Perhaps one of the most accepted approaches to investigating neuronal function involves electrophysiology. Analyzing membrane excitability through single and multi-electrode extracellular recording and patch-clamp methods provides insight into the function of neuropeptides. Membrane physiology changes can take place due to the expression of ion channels, PTMs that regulate channel conformations, and/or by regulating cellular concentrations of second messenger signals, which regulate channels (Rood and Beck, 2013). Neuropeptide stimulation from MS-identified neuropeptides changes the predicted peak excitability pattern, denoting plasticity and responsiveness, allowing putative peptides to be confirmed as bioactive (Atkins *et al.*, 2010).

Another approach involves monitoring the enzymatic co-factors NAD and FAD; these are intrinsically fluorescent and enable non-invasive imaging. The metabolic state of neurons directly regulates the activity of channels and pumps, as well as the transcription-translation process. If a neuropeptide changes the metabolic state of a cell, it may have a global effect on the neuron. Intrinsic co-factor ratios have been used to calculate a relative redox state within neurons of the SCN (Wang *et al.*, 2012). This redox state has been linked to the regulation of the oscillation in membrane excitability of the SCN around the clock (Belle and Piggins, 2012).

As a final example, appropriate transgenic animal models can be used effectively to delineate a peptide's function. Genetic manipulations through gain or loss of function, gene knockouts, and neuropeptide-reporter fusions are effective options that provide unmatched information, although the results obtained may need to be interpreted with caution. For example, when a peptide is knocked out, a phenotype is observed, and one often attributes the phenotype to the peptide; however, genetic manipulations frequently have broad or masking physiological phenotypes. A specific neuropeptide may have a unique function during development and a separate function in the adult. Although one may observe a peptide's effect in an animal, it may be hard to ascribe this to a specific form or developmental time point. Alternatively, Cre-lox animal models have the gene of interest selectively excised from the genome in the cells that express the neuropeptide in a specific region or time. This is an elegant and more specific way to avoid genetic lethality and minimize the tissues affected by genetic manipulation.

There are many additional types of tests that can be performed to determine the function of a peptide, with the tests dependent on the specific peptide and its function. Establishing a stimulation paradigm associated with a specific neuropeptide's function is more straightforward when the peptide is expressed in a specific, well-defined location with a known function. For example, the stimulus can be applied directly on the region of interest and a response monitored. Previous studies using cannulation and dialysis have included examining the effects of neuropeptide Y administration on a specific hypothalamic nucleus *in vivo* (Anastasiou *et al.*, 2009; McMinn *et al.*, 1998; Yavropoulou *et al.*, 2008).

In summary, the responses to a neuropeptide stimulus can be seen in the cell through changes in metabolism, membrane excitability, transcription, translation or PTMs, whereas organismal responses are typically measured through behavioral or homeostatic changes. While observations of bioactivity are exciting, caution should be exercised when addressing a lack of bioactivity of neuropeptides for several reasons:

- for receptor assays, identified neuropeptides may interact with more than one receptor or work through unknown receptors
- they exhibit substantial variability in their bioactive concentrations
- many may function only in a specific developmental state
- they may exert differential effects at different times and in different brain regions.

8.4   Perspectives

Reading the genetic blueprint of an organism has become easier, faster, and less expensive, greatly aiding research on discovering new neuropeptides and their biological roles. However, as genomics does not fully unravel the final structure of these bioactive compounds, additional experiments are necessary. Unfortunately, comprehensive neuropeptide analyses are still challenging. In addition, a mystery of the postgenomic era is the number of genes with still unknown functions, like those coding for orphan G protein-coupled receptors (GPCRs) (see Chapter 10). Among the family of about 800 GPCRs, 30% have natural ligands, with drug targets having only been identified for ~6% (Tang *et al.*, 2012). Many known GPCRs do not yet have ligands associated with them, limiting investigation of their function. As many of the characterized GPCRs bind to peptide ligands, continued discovery of new bioactive peptides may illuminate novel GPCR systems in our bodies.

Although MS-based analytical tools are becoming increasingly user friendly and affordable, high-quality peptide measurement remains a complex process. The three components of an experiment are sampling, MS-based measurement, and data analyses. One area needing improvement is handling the overwhelming amount of data obtained in a peptidomics investigation, including the selection of candidates for follow-up studies using complementary methods, a skill that does not come prepackaged with a mass spectrometer. Peptide characterization has evolved into a coordinated and synergistic amalgam of cross-training and close collaborations between analytical chemists, biologists, programmers, and statisticians. Peptidomics is an interdisciplinary pipeline that enables scientists to efficiently screen and discover markers of disease, elucidate normal physiological states or develop therapeutic drugs. We envision expanded use of MS-based analyses in clinical laboratories for the sensitive detection of hormone levels in complex matrices of biological fluids in the effort to achieve diagnosis and management of endocrine-related conditions. As the expertise to perform MS-based characterization of tissues becomes more commonplace, its use will expand to a greater range of scientific and clinical laboratories, similar to what has happened with transcriptomic measurements.

Acknowledgments

This work was supported by award numbers P30 DA018310 from the National Institute on Drug Abuse, R21 MH100704 from the National Institute of Mental Health, and IOS 08-18555 from the National Science Foundation. The content is solely the responsibility of the authors and does not necessarily represent the official views of the awarding agencies.

References

Amara, S.G., Jonas, V., Rosenfeld, M.G., Ong, E.S. and Evans, R.M. (1982) Alternative RNA processing in calcitonin gene expression generates mRNAs encoding different polypeptide products. *Nature*, **298**, 240–244.

Anastasiou, O.E., Yavropoulou, M.P., Kesisoglou, I., Kotsa, K. and Yovos, J.G. (2009) Intracerebroventricular infusion of neuropeptide Y modulates VIP secretion in the fasting conscious dog. *Neuropeptides*, **43**, 41–46.

Atkins, N. Jr., Mitchell, J.W., Romanova, E.V., *et al.* (2010) Circadian integration of glutamatergic signals by little SAAS in novel suprachiasmatic circuits. *PLoS One*, **5**, e12612.

Belle, M.D. and Piggins, H.D. (2012) Physiology. Circadian time redoxed. *Science*, **337**, 805–806.

Castanon-Cervantes, O., Wu, M., Ehlen, J.C., *et al.* (2010) Dysregulation of inflammatory responses by chronic circadian disruption. *Journal of Immunology*, **185**, 5796–5805.

Chen, R., Ma, M., Hui, L., Zhang, J. and Li, L. (2009) Measurement of neuropeptides in crustacean hemolymph via MALDI mass spectrometry. *Journal of the American Society for Mass Spectrometry*, **20**, 708–718.

Cornett, D.S., Reyzer, M.L., Chaurand, P. and Caprioli, R.M. (2007) MALDI imaging mass spectrometry: molecular snapshots of biochemical systems. *Nature Methods*, **4**, 828–833. [*The state of MALDI mass spectrometry as an information-rich imaging technique with broad current and future application possibilities.*]

Croushore, C.A. and Sweedler, J.V. (2013) Microfluidic systems for studying neurotransmitters and neurotransmission. *Lab on a Chip*, **13**, 1666–1676.

Dowell, J.A., Heyden, W.V. and Li, L. (2006) Rat neuropeptidomics by LC-MS/MS and MALDI-FTMS: enhanced dissection and extraction techniques coupled with 2d RP-RP HPLC. *Journal of Proteome Research*, **5**, 3368–3375.

Fan, Y., Lee, C.Y., Rubakhin, S.S. and Sweedler, J.V. (2013) Stimulation and release from neurons via a dual capillary collection device interfaced to mass spectrometry. *The Analyst*, **138**, 6337–6346.

Frese, C.K., Boender, A.J., Mohammed, S., Heck, A.J., Adan, R.A. and Altelaar, A.F. (2013) Profiling of diet-induced neuropeptide changes in rat brain by quantitative mass spectrometry. *Analytical Chemistry*, **85**, 4594–4604.

Hatcher, N.G., Atkins, N., Annangudi, S.P., *et al.* (2008) Mass spectrometry-based discovery of circadian peptides. *Proceedings of the National Academy of Sciences of the United States of America*, **105**, 12527–12532.

International Human Genome Sequencing Consortium (2004) Finishing the euchromatic sequence of the human genome. *Nature*, **431**, 931–945.

Jekely, G. (2013) Global view of the evolution and diversity of metazoan neuropeptide signaling. *Proceedings of the National Academy of Sciences of the United States of America*, **110**, 8702–8707. [*Experimental evidence for deep conservation of neuropeptidergic regulation between different metazoan phyla obtained through analysis of stable evolutionary association of GPCR–ligand pairs.*]

Jespersen, S., Chaurand, P., van Strien, F.J., Spengler, B. and van der Greef, J. (1999) Direct sequencing of neuropeptides in biological tissue by MALDI-PSD mass spectrometry. *Analytical Chemistry*, **71**, 660–666.

Lanni, E.J., Rubakhin, S.S. and Sweedler, J.V. (2013) Visualizing the proteome: mapping protein changes in disease states with mass spectrometry imaging. *Journal of Neurochemistry*, **124**, 581–583.

Lee, J.E., Zamdborg, L., Southey, B.R., *et al.* (2013) Quantitative peptidomics for discovery of circadian-related peptides from the rat suprachiasmatic nucleus. *Journal of Proteome Research*, **12**, 585–593.

Li, L. and Sweedler, J.V. (2008) Peptides in the brain: mass spectrometry-based measurement approaches and challenges. *Annual Review of Analytical Chemistry*, **1**, 451–483.

Marcheva, B., Ramsey, K.M., Peek, C.B., Affinati, A., Maury, E. and Bass, J. (2013) Circadian clocks and metabolism. *Handbook of Experimental Pharmacology*, **217**, 127–155. [*Comprehensive reviews of the emerging relationship between the molecular clock and human health, as well as negative effects of clock disruption.*]

Mathe, A.A., Stenfors, C., Brodin, E. and Theodorsson, E. (1990) Neuropeptides in brain: effects of microwave irradiation and decapitation. *Life Sciences*, **46**, 287–293.

McMinn, J.E., Seeley, R.J., Wilkinson, C.W., Havel, P.J., Woods, S.C. and Schwartz, M.W. (1998) NPY-induced overfeeding suppresses hypothalamic NPY mRNA expression: potential roles of plasma insulin and leptin. *Regulatory Peptides*, **75–76**, 425–431.

Minerva, L., Boonen, K., Menschaert, G., Landuyt, B., Baggerman, G. and Arckens, L. (2011) Linking mass spectrometric imaging and traditional peptidomics: a validation in the obese mouse model. *Analytical Chemistry*, **83**, 7682–7691.

Patel, H., Orchard, I., Veenstra, J.A. and Lange, A.B. (2014) The distribution and physiological effects of three evolutionarily and sequence-related neuropeptides in rhodnius prolixus: adipokinetic hormone, corazonin and adipokinetic hormone/corazonin-related peptide. *General and Comparative Endocrinology*, **195**, 1–8.

Pritchard, L.E. and White, A. (2007) Minireview: neuropeptide processing and its impact on melanocortin pathways. *Endocrinology*, **148**, 4201–4207.

Romanova, E.V., Rubakhin, S.S. and Sweedler, J.V. (2008) One-step sampling, extraction, and storage protocol for peptidomics using dihydroxybenzoic acid. *Analytical Chemistry*, **80**, 3379–3386.

Romanova, E.V., Sasaki, K., Alexeeva, V., *et al.* (2012) Urotensin II in invertebrates: from structure to function in Aplysia californica. *PLoS One*, **7**, e48764.

Romanova, E.V., Dowd, S.E. and Sweedler, J.V. (2013) Quantitation of endogenous peptides using mass spectrometry based methods. *Current Opinion in Chemical Biology*, **17**, 801–808. [***Brief overview of current approaches for quantitative peptidomics and the technical challenges that stimulate new advances in the field, all on the examples from newest literature on functional characterizations of endogenous peptides using mass spectrometry.***]

Romanova, E.V., Aerts, J.T., Croushore, C.A. and Sweedler, J.V. (2014) Small-volume analysis of cell-cell signaling molecules in the brain. *Neuropsychopharmacology*, **39**, 50–64.

Rood, B.D. and Beck, S.G. (2013) Vasopressin indirectly excites dorsal raphe serotonin neurons through activation of the vasopressin1A receptor. *Neuroscience*, **260**, 205–216.

Rubakhin, S.S., Churchill, J.D., Greenough, W.T. and Sweedler, J.V. (2006) Profiling signaling peptides in single mammalian cells using mass spectrometry. *Analytical Chemistry*, **78**, 7267–7272.

Schroeder, A.M. and Colwell, C.S. (2013) How to fix a broken clock. *Trends in Pharmacological Science*, **34**, 605–619.

Southey, B.R., Amare, A., Zimmerman, T.A., Rodriguez-Zas, S.L. and Sweedler, J.V. (2006) NeuroPred: a tool to predict cleavage sites in neuropeptide precursors and provide the masses of the resulting peptides. *Nucleic Acids Research*, **34**, W267–272.

Southey, B.R., Lee, J.E., Zamdborg, L., *et al.* (2014) Comparing label-free quantitative peptidomics approaches to characterize diurnal variation of peptides in the rat suprachiasmatic nucleus. *Analytical Chemistry*, **86**, 443–452.

Spindel, E.R., Zilberberg, M.D., Habener, J.F. and Chin, W.W. (1986) Two prohormones for gastrin-releasing peptide are encoded by two mRNAs differing by 19 nucleotides. *Proceedings of the National Academy of Sciences of the United States of America*, **83**, 19–23.

Sweedler, J.V., Li, L., Rubakhin, S.S., *et al.* (2002) Identification and characterization of the feeding circuit-activating peptides, a novel neuropeptide family of *Aplysia*. *Journal of Neurosciences*, **22**, 7797–7808.

Taban, I.M., Altelaar, A.F.M., van der Burgt, Y.E.M., *et al.* (2007) Imaging of peptides in the rat brain using MALDI-FTICR mass spectrometry. *Journal of the American Society for Mass Spectrometry*, **18**, 145–151.

Tang, X.L., Wang, Y., Li, D.L., Luo, J. and Liu, M.Y. (2012) Orphan G protein-coupled receptors (GPCRs): biological functions and potential drug targets. *Acta Pharmacologica Sinica*, **33**, 363–371.

Tucker, K.R., Lanni, E.J., Serebryannyy, L.A., Rubakhin, S.S. and Sweedler, J.V. (2011) Stretched tissue mounting for MALDI mass spectrometry imaging. *Analytical Chemistry*, **83**, 9181–9185.

Wang, T.A., Yu, Y.V., Govindaiah, G., *et al.* (2012) Circadian rhythm of redox state regulates excitability in suprachiasmatic nucleus neurons. *Science*, **337**, 839–842.

Weaver, R.J. and Audsley, N. (2010) MALDI-TOF mass spectrometry approaches to the characterisation of insect neuropeptides. *Methods in Molecular Biology*, **615**, 101–115.

Xie, F., London, S.E., Southey, B.R., *et al.* (2010) The zebra finch neuropeptidome: prediction, detection and expression. *BMC Biology*, **8**, 28.

Yasuda, A., Jones, L.S. and Shigeri, Y. (2013) The multiplicity of post-translational modifications in pro-opiomelanocortin-derived peptides. *Frontiers in Endocrinology*, **4**, 186.

Yavropoulou, M.P., Kotsa, K., Kesisoglou, I., Anastasiou, O. and Yovos, J.G. (2008) Intracerebroventricular infusion of neuropeptide Y increases glucose dependent-insulinotropic peptide secretion in the fasting conscious dog. *Peptides*, **29**, 2281–2285.

Further reading

Fricker, L.D., Lim, J., Pan, H. and Che, F.Y. (2006) Peptidomics: identification and quantification of endogenous peptides in neuroendocrine tissues. *Mass Spectrometry Reviews*, **25**, 327–344. [***A thorough introduction to peptidomics with emphasis on differences from proteomics, challenges of peptide characterization as opposed to proteins, and peptide quantification labeling methods.***]

Mitchell, J.W., Atkins, N. Jr., Sweedler, J.V. and Gillette, M.U. (2011) Direct cellular peptidomics of hypothalamic neurons. *Frontiers in Neuroendocrinology*, **32**, 377–386. [***An overview of innovating sampling approaches for collection of released neuropeptides in spatial, temporal and event-dependent patterns that*** *provide remarkable new insights into the complexity of neuropeptidergic cell-to-cell signaling central to neuroendocrine physiology.*]

Neilson, K.A., Ali, N.A., Muralidharan, S., *et al.* (2011) Less label, more free: approaches in label-free quantitative mass spectrometry. *Proteomics*, **11**, 535–553. [*Detailed discussion of workflow schemas, advantages and limitations of label-free quantitation as applied to peptides and proteins, a reliable, versatile and cost-effective alternative to stable isotopic labeling methods.*]

Pagotto, U., Fanelli, F. and Pasquali R. (2013) Insights into tandem mass spectrometry for the laboratory endocrinology. *Reviews in Endocrine and Metabolic Disorders*, **14**, 141. [*A thoughtful reflection on the strength of LC-MS/MS in terms of clinical effectiveness as well as on current progress in measurement of vitamins and hormones.*]

So, P.K., Hu, B. and Yao, Z.P. (2013) Mass spectrometry: towards in vivo analysis of biological systems. *Molecular Biosystems*, **9**, 915–929. [*A fun-to-read literature review on relatively new mass spectrometric techniques for in* in vivo *analysis of select molecules.*]

CHAPTER 9

Posttranslational Processing of Secretory Proteins

Nabil G. Seidah and Johann Guillemot

Laboratory of Biochemical Neuroendocrinology, Institut de Recherches Cliniques de Montréal, Montréal, Canada

9.1 Posttranslational modifications of secretory proteins

Of the ~20,000 genes that comprise the mammalian genome, it is estimated that ~4000 encode proteins that transit through the lumen of the endoplasmic reticulum (ER) to reach their final destination (Ladunga, 2000). These so-called membrane or secreted proteins (MSPs) are the secretory proteins that form the main subject of this chapter. Proteins destined to enter the secretory pathway typically possess N-terminal hydrophobic signal sequences, which direct their initial transport through the membrane of the ER. Once inside the ER lumen, they can traverse the secretory pathway by vesicular transport for their delivery to membrane-enclosed organelles and/or the plasma membrane or for their secretion into the extracellular matrix (ECM). Thus, transport across the ER is the only membrane translocation event that these proteins require, representing the commitment step in entry to the secretory pathway. Co-translational protein targeting to the ER represents an evolutionarily conserved mechanism regulating the cellular sorting of secretory proteins. In this targeting pathway, proteins possessing signal sequences are recognized at the ribosome by the signal recognition particle while they are still undergoing synthesis. This triggers their delivery to the ER protein translocation channel, leading to their direct translocation into the ER where chain elongation continues in the lumen. The signal sequence consists of a 15–30 amino acids (aa) targeting signal peptide or stop transfer transmembrane sequence of hydrophobic residues near the amino terminal end of the growing polypeptide. Rapidly the signal peptide is excised by the signal peptidase complex to release the secretory protein into the lumen of the ER (Nyathi *et al.*, 2013).

In the majority of secretory proteins, a series of complex N-glycosylation modifications start with the co-translational attachment of a core-N-glycosylation chain to Asn residues with a recognition motif of Asn-X-Ser/Thr, where X is any aa except Pro or Asp. The tripeptide sequence Asn-X-Ser/Thr is widely recognized as the consensus motif for attachment of high mannose N-linked oligosaccharides to nascent polypeptides of constant common carbohydrate structure $Glc_{1-3}Man_9GlcNAc_2$-(Asn), where the di-N-acetyl glucosamine chain $(GlcNAc_2)$ is attached to Asn followed by nine branched mannoses and ending with 1–3 glucose residues (Kornfeld and Kornfeld, 1985). However, in rare cases the atypical sequence Asn-X-Cys can also be N-glycosylated, for example, in human plasma protein C (Gil *et al.*, 2009). The N-glycosylated protein then undergoes a series of interactions with ER-resident chaperone proteins to ensure that it adopts a folded and stable conformation (Bedard *et al.*, 2005;

Molecular Neuroendocrinology: From Genome to Physiology, First Edition. Edited by David Murphy and Harold Gainer.
© 2016 John Wiley & Sons, Ltd. Published 2016 by John Wiley & Sons, Ltd.
Companion website: www.wiley.com/go/murphy/neuroendocrinology

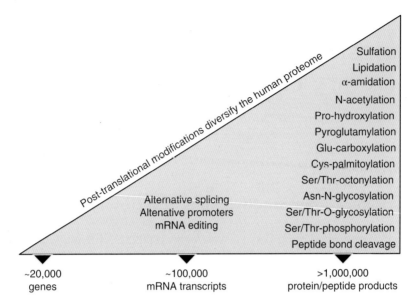

Post-translational modifications diversify the human proteome

Sulfation
Lipidation
α-amidation
N-acetylation
Pro-hydroxylation
Pyroglutamylation
Glu-carboxylation
Cys-palmitoylation
Ser/Thr-octonylation
Asn-N-glycosylation
Ser/Thr-O-glycosylation
Ser/Thr-phosphorylation
Peptide bond cleavage

Alternative splicing
Altenative promoters
mRNA editing

~20,000
genes

~100,000
mRNA transcripts

>1,000,000
protein/peptide products

Figure 9.1 Posttranslational modifications diversify the mammalian proteome. The number of genes encoded in the human proteome is greatly exceeded by the number of distinct covalent forms of existing proteins. This is in part due to posttranscriptional modifications such as mRNA editing and alternative splicing but even more to the posttranslational modifications giving rise to the large human proteome. The figure illustrates the expansion from number of genes to number of transcripts to number of proteins. A segment of the more than 200 known posttranslational modifications is listed at the far right. Adapted from Schjoldager and Clausen (2012).

Helenius *et al.*, 1992), following which it is allowed to exit the ER and continue its journey into the Golgi apparatus and up to the last cisternae called the trans-Golgi network (TGN). Along this route, some proteins may remain as residents in the *cis*-, medial- or trans-Golgi compartments or will traffic to the TGN and bifurcate from there to reach their final cellular or ECM destination.

During their transport through the ER and Golgi, the Asn-glycosylated oligosaccharide chains undergo trimming and remodeling by the removal of glucose and mannose moieties and addition of various sugar modifiers such as fucose, galactose, and terminal sialic acid (Kornfeld and Kornfeld, 1985). Aside from Asn, other residues are also modified. Some of the most common **posttranslational modifications** of secretory proteins are listed in Figure 9.1. Posttranslational modifications can lead to both reversible (for example, phosphorylation, Cys-palmitoylation) and irreversible (for example, N-glycosylation, proteolytic cleavage, C-terminal amidation, N-terminal acetylation, sulfation, pyroglutamination, Glu-carboxylation) modifications of proteins (Figures 9.1 and 9.2). Some of the major irreversible posttranslational modifications are afforded by proteases, which either cleave peptide bonds within proteins and peptides, resulting in two or more products, or trim them from the N-terminus (aminopeptidases) or C-terminus (carboxypeptidases), leading to the generation or loss of functional activity (Seidah *et al.*, 2013). Proteases comprise ~1.7% of the predicted proteins of the human genome (Puente *et al.*, 2003). Thus, it is forecast that the ~500–600 mammalian proteases expressed as active enzymes belong to one of the five classes of eukaryotic proteases, which are classified based on the required nucleophile and on the identity of the catalytic subunit between each family member (Figure 9.3).

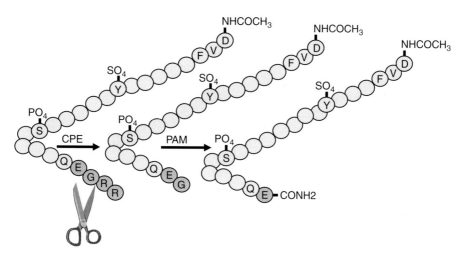

Figure 9.2 Schema of some posttranslational modifications of secretory proteins. On the left is shown part of a hypothetical secretory protein posttranslationally modified by phosphorylation at either a Ser (S) or Thr, sulfation at Tyr (Y), and N-acetylated at an Asp acid (D) residue. Cleavage of this protein at the Arg-Arg (RR) pair of basic amino acids by one of the basic-aa convertases (represented schematically by a pair of scissors) would expose a C-terminal Arg residue, which is then trimmed by a basic-aa-specific carboxypeptidase such as CPE (others include CPD and CPX), thereby exposing a C-terminal Gly (G) residue which is a substrate for the amidation enzyme PAM, which will then amidate the penultimate Glu (E) residue.

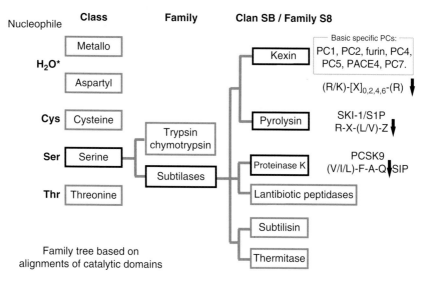

Figure 9.3 Classification of proteases with emphasis on the secretory PCs. The five classes of proteases are represented on the left with the nucleophile used with each enzyme class. The emphasis is on serine proteases of the subtilisin type, of which six subfamilies within the clan SB can be identified. The basic amino acid PCs are related to yeast kexin, while the convertases SKI-1/S1P and PCSK9 are related to pyrolysin and proteinase K, respectively. The three different motifs generally recognized by the seven basic-aa-specific convertases and the two non-basic-aa-specific convertases SKI-1/S1P and PCSK9 are also shown, with the downward arrow indicating the cleavage site. Here X = any amino acid and Z is any amino acid except Glu, Asp, Val, Pro, and Cys.

Proteases are either of a digestive nature, resulting in protein degradation via multiple cleavages of the protein/peptide substrate, or participate in the limited proteolysis of precursor proteins (proproteins) to generate diversity within the proteome by increasing the number of polypeptide products that result from a single proprotein. The cleavage process can take place intracellularly, at the cell surface or in the ECM. Within the cell, processing can occur in various subcellular locations and organelles, including the cytosol, nucleus, mitochondria, ER, various lamellae of the Golgi apparatus, within secretory granules (SG), endosomes, and lysosomes. In this review, we will concentrate only on the cleavage of proteins within the secretory pathway. This includes the processing of proteins that enter the ER (via a signal peptide or stop transfer hydrophobic sequences) and that either remain there or are transported to the Golgi apparatus and beyond, including the cell surface, dense-core SG, recycling endosomes, and lysosomes.

9.2 The family of proprotein convertases

The mammalian **proprotein convertases** (PCs) represent a family of secretory serine proteinases related to bacterial subtilisin but distinct from the trypsin/chymotrypsin subfamily. Across the phylae, we can distinguish six different subtilase clans, but so far the nine members of the secretory mammalian PC family belong to only three of these (see Figure 9.3). The seven mammalian convertases related to yeast kexin (Figures 9.3 and 9.4) cleave after basic aa (basic aa-specific PCs) and are known as PC1/3, PC2, furin, PC4, PC5/6, PACE4, and PC7. These enzymes cleave after single or paired basic residues predominantly within the motif (Arg/Lys)-$X_{0,2,4,6}$-Arg↓, where Xn are variable even numbers of aa separating the two canonical basic residues required for cleavage recognition (see Figure 9.3). The other two convertases (SKI-1/S1P and PCSK9) that are related to pyrolysin and proteinase K cleave at non-basic residues within the motifs R-X-(hydrophobic)-Z↓ and (V/I/L)FAQ↓, respectively (see Figures 9.3 and 9.4). All PCs are first synthesized in the ER as inactive zymogens (proPCs). Upon folding, most of these (except for PC2) undergo an initial autocatalytic zymogen processing in the ER at an N-terminal site in their prosegment, resulting in a tight binding complex of the inhibitory N-terminal prosegment non-covalently associated to the main protease (Figure 9.5). Except for PC4, PC7, and PCSK9, a second autocatalytic zymogen activation event is usually needed to liberate the protease from its prosegment, thus specifying the location within the cell where the protease can then cut substrates in *trans* (Seidah and Prat, 2012; Seidah *et al.*, 2013, 2014).

The zymogen activation of PCs can occur in at least six different subcellular organelles: (i) in the *cis/medial* Golgi (for SKI-1), (ii) in the early TGN as for PC5/6B, (iii) in the last lamella of the TGN that is insensitive to brefeldin A (BFA) as is the case for furin, (iv) at the cell surface as for PC5/6A and PACE4, (v) within endosomes as for PC7, or (vi) in dense-core SG as for PC1/3 and PC2 (Figures 9.5 and 9.6). PCSK9 is the only convertase that always remains associated with its inhibitory prosegment and hence never acts as a protease except on itself, resulting in an autocatalytic cleavage of its prosegment in the ER at the $VFAQ_{152}$↓ sequence (Seidah and Prat, 2012; Seidah *et al.*, 2013, 2014). The site of separation of the inhibitory N-terminal prosegment from mature PC7 (and hence activation of PC7), and possibly PC4 (Gyamera-Acheampong *et al.*, 2011), is not clear, as this is thought to occur either at the cell surface or in endosomes, without requiring a secondary cleavage. The prosegment of PC7 is secreted intact into the medium, leaving an active membrane-bound PC7 in the cell (Rousselet *et al.*, 2011a).

Analysis of the subcellular distribution and activity of the convertases (Seidah *et al.*, 2008) allowed their classification into at least six distinct groups (see Figure 9.6).

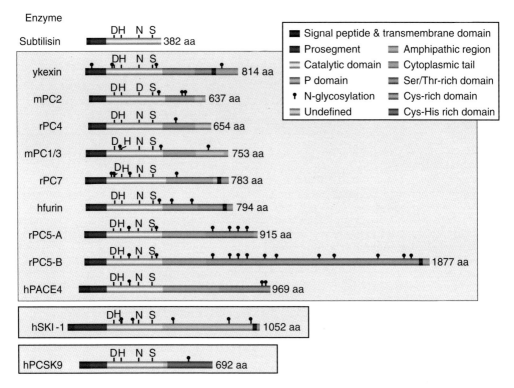

Figure 9.4 Schematic representation of the primary structure and domains of the PCs. The schematized primary structures and domains of the nine mammalian PCs (h = human, r = rat, m = mouse) are compared to the original subtilisin and yeast (y) kexin. The basic amino acid-specific PCs, together with ykexin, SKI-1, and PCSK9, are individually boxed to emphasize their distinct subclasses. Notice that only PC5 exhibits two validated alternative spliced forms, namely PC5A and PC5B. The membrane-bound PCs include furin, PC5B, PC7, and SKI-1. The presence of a signal peptide, a prosegment and catalytic domain is common to all convertases that exhibit the typical catalytic triad residues Asp, His, and Ser and the oxyanion hole Asn. Except for SKI-1 and PCSK9, all the basic-aa-specific convertases exhibit a β-barrel P-domain that apparently stabilizes the catalytic pocket. The C-terminal domain of each convertase contains unique sequences regulating their cellular localization and trafficking. PC5/6 and PACE4 contain a specific Cys-rich domain (CRD), which binds HSPGs at the cell surface. In contrast, PCSK9 exhibits a Cys-His-rich domain (CHRD) that is required for cell surface binding in an LDLR-dependent fashion.

1 The membrane-bound convertase SKI-1/S1P is mostly localized in the *cis/medial* Golgi, and is autocatalytically shed, resulting in the secretion of a soluble form of the proteinase.

2 The membrane-bound convertases furin, the isoform PC5/6-B, and PC7 that cycle from the cell surface to the TGN through the endosomal pathway. Their subcellular trafficking is regulated by signals in their cytosolic tail. These membrane-bound convertases process most precursors in the TGN, cell surface or in endosomes. Furin and PC5/6B (but not PC7) are shed as soluble secreted enzymes. Shedding/cleavage of furin and PC5/6B is performed by membrane-bound metalloproteases, including disintegrin and metalloprotease (ADAMs) enzymes.

3 The secretory convertases PC5/6A and PACE4 contain a specific Cys-rich domain (CRD) (see Figure 9.4) that binds to heparin sulfate proteoglycans (HSPGs), thereby retaining

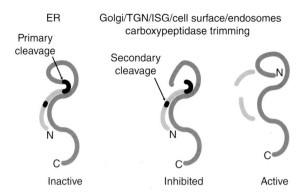

ER Golgi/TGN/ISG/cell surface/endosomes carboxypeptidase trimming

Primary cleavage

Secondary cleavage

N N N

C C C

Inactive Inhibited Active

Figure 9.5 Zymogen activation of the PCs. A PC usually undergoes a first autocatalytic zymogen cleavage in the endoplasmic reticulum (ER), generating an N-terminal inhibitory prosegment that remains non-covalently tightly bound to the enzyme. This bimolecular complex then exits the ER and is directed towards the Golgi apparatus. For SKI-1, a second autocatalytic processing event in the prosegment activates the enzyme in the *cis/medial* Golgi. Zymogen activation of furin, PC7, and PC5B occurs in the *trans* Golgi network (TGN), where the acidic (pH 6.2) and high calcium concentrations (~5–10 mM) favor a second autocatalytic cleavage of the prosegment, thereby liberating the enzyme from its inhibitory prosegment. These enzymes can then cleave *in trans* other substrates. In the case of PC5 and possibly PACE4, it seems that the second autocatalytic cleavage may either occur in the TGN or at the cell surface, where these proteinases may then cleave and sometimes inactivate specific proteins. The zymogen activation steps of the regulated endocrine and neural convertases PC1/3 and PC2 occur in the highly acidic (~pH 5.2) immature secretory granules (SG), which explains why they are maximally active only in dense-core SG. Finally, the secreted PCSK9 is complexed with its prosegment and zymogen activation seems to occur in recycling endosomes upon reinternalization of the secreted bimolecular complex.

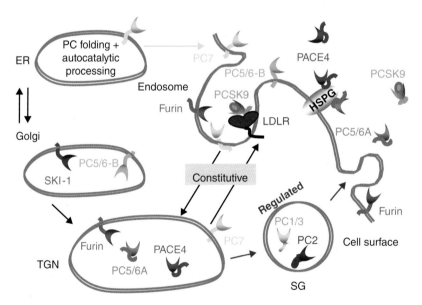

Figure 9.6 Subcellular localization of the PCs. Upon exiting the ER, most of the basic amino acid-specific PCs traverse the Golgi apparatus towards the *trans* Golgi network (TGN). At this branching point, the convertases PC1/3 and PC2 are sorted to immature secretory granules (SG) where they are activated, cleave polypeptide hormones and are stored in dense-core SG awaiting signals for regulated secretion. On the other hand, the constitutively secreted furin, PC5/6, and PACE4 reach the cell surface from the TGN. PC7 is able to reach the same destination from both the TGN and by an unconventional secretion pathway from the ER directly. Upon binding to HSPGs, PC5/6A and PACE4 are retained at the cell surface and/or ECM. The membrane-bound furin, PC5/6B, and PC7 are recycled through endosomes back to the TGN. The activated membrane-bound SKI-1/S1P is mostly concentrated in the *cis/medial* Golgi, from where it is then sent to lysosomes for degradation, and does not normally reach the cell surface. PCSK9 is secreted from the TGN directly into the medium as an enzymatically inactive non-covalent complex of the protease and its prosegment. Upon binding the LDLR at the cell surface, the PCSK9≡LDLR complex is then internalized into endosomes and then sent to lysosomes for degradation.

this complex at the cell surface, a site where processing of specific HSPG-bound precursors takes place.

4 While PC7 reaches the cell surface by the conventional ER-TGN-cell surface secretory pathway, amazingly, a fraction of PC7 (~30% in HEK293 cells) rapidly reaches the cell surface in ≤10 min after synthesis via an unconventional route, directly from the ER to the plasma membrane (see Figure 9.6) (Rousselet *et al.*, 2011a).

5 The regulated convertases PC1/3 and PC2 are sorted to dense-core SG and process therein most polypeptide hormones, which are then secreted following specific stimuli.

6 Finally, the last member of the family, PCSK9, is secreted and then reinternalized as a complex with the low density lipoprotein receptor (LDLR) into clathrin-coated endosomes and then targeted to lysosomes, where in a non-enzymatic fashion it enhances the degradation of the LDLR and/or its other target receptors (Seidah and Prat, 2012; Seidah *et al.*, 2014).

The processing of protein precursors by furin, PC5/6, PACE4, PC7, and SKI-1/S1P can be inhibited by overexpression of their prosegments, by specific chloromethylketone (CMK) containing membrane-permeable decanoylated (dec-) peptides such as dec-RVKR-CMK or dec-RRLL-CMK, by variants of the serpin α1-PDX (Seidah and Prat, 2012), or even by specific non-peptidyl inhibitors such as 2,5-dideoxystreptamine (Jiao *et al.*, 2006) and dicoumarol (Komiyama *et al.*, 2009) derivatives, a naphthofluorescein disodium salt (B3) (Coppola *et al.*, 2008). A tripeptide cell-permeable reversible basic aa-specific inhibitor that inhibits PC1/3, furin, PC4, and PACE4 with similar potency, but not PC2 or PC7, is now commercially available (www.millipore.com/catalogue/item/537076-2mg). An aminopyrrolidineamide SKI-1/S1P inhibitor, PF-429242 (Hawkins *et al.*, 2008), inhibits Lassa virus glycoprotein processing and may have applications as a novel antiviral agent (Urata *et al.*, 2011). In cell-based assays, these inhibitors were shown to block metastasis, viral infection, and tumor growth, suggesting that inhibitors/modulators of the basic aa-specific furin-like PCs (furin, PC5/6, PACE4, and PC7), and possibly other PCs, may have applications in a broad range of disorders, including cancer/metastasis, viral infections, inflammatory diseases, and hypercholesterolemia (Seidah and Prat, 2012). The use of PC inhibitors in neurological or neurodegenerative diseases is not yet clear, although a brain-penetrable PC7 inhibitor may find application as an anxiolytic agent (Besnard *et al.*, 2012), and could also be used to control other emotional disorders (Wetsel *et al.*, 2013).

9.3 The neural and endocrine functions of the proprotein convertases

Since a number of excellent reviews have been written on the activation and general functions of the PCs and their targets (Constam, 2014; Seidah and Prat, 2012; Seidah *et al.*, 2008, 2013; Steiner, 1998), we will only consider here some of the novel aspects of the implication of the PCs in various physiological functions with emphasis on their roles in the central (CNS) and peripheral (PNS) nervous systems and neural pathways.

9.3.1 The neuroendocrine convertases PC1/3 and PC2

PC1/3 and PC2 are the principal activators of prohormones and proneuropeptides within the regulated secretory pathway of neural and endocrine cells. PC1/3 and PC2 process many precursors into their functional hormones: pro-opiomelanocortin (POMC; precursor of adrenocorticotropic hormone [ACTH] and β-endorphin), proinsulin, promelanin-concentrating hormone (MCH), secretogranin II, proenkephalin, prodynorphin, proglucagon, prosomatostatin, chromogranins A and B, and PACAP (Seidah *et al.*, 2013). The functional activation of the convertases PC2 is regulated by a ~30 kDa neuroendocrine PC2-binding protein 7B2 that is co-regulated with it (Mbikay *et al.*, 2001). Both proteases are calcium-dependent enzymes that function within the confines of dense-core SG (see Figure 9.6), where they exert their maximal activity.

The physiological importance of PC1/3 and PC2 was deduced from studies of the pheno-types of their gene knockout (KO) in mice and from the discovery of some human patients with defects in PC1/3. The anticipated physiological importance of PC1/3 and PC2 has been confirmed by analysis of their KO phenotypes in mice (Furuta *et al.*, 2001; Zhu *et al.*, 2002a). Although these mice were viable, those lacking both PC1/3 and PC2 died during embryogenesis (Steiner DF, personal communication), suggesting some redundancy of these enzymes.

The disruption of the *Pcsk1* mouse gene results in a greatly reduced number of pups born. Furthermore, PC1/3 KO mice exhibit growth retardation and dwarfism (Zhu *et al.*, 2002a), probably due to a deficit in growth hormone-releasing hormone (GHRH), whose precursor is processed C-terminally by PC1/3 (Dey *et al.*, 2004; Posner *et al.*, 2004). Interestingly, insulin growth factor 1 (IGF-1) and GHRH levels were significantly reduced in KO mice along with pituitary GH mRNA levels, suggesting that this reduction contributes to the growth retardation observed in these mice. In agreement, these mice accumulate precursor and intermediate processing forms of the hypothalamic GHRH, but also of the pituitary POMC, proinsulin, and intestinal proglucagon (Wardman *et al.*, 2010; Zhu *et al.*, 2002b). However, PC1/3 KO mice can still partially process POMC to ACTH, probably due to the ability of another PC to perform such function, and have normal levels of blood corticosterone. In addition, PC1/3 KO mice exhibit marked innate immune defects and uncontrolled cytokine secretion (Refaie *et al.*, 2012).

Complete PC1/3 deficiency in a compound heterozygote patient exhibiting a splicing defect and a non-synonymous mutation resulted in neonatal massive obesity, abnormal glucose homeostasis, moderate adrenal insufficiency, and infertility of hypothalamic origin (Jackson *et al.*, 1997), in part related to low circulating levels of insulin and ACTH. Obesity was also observed in a 6-year-old boy who carried a homozygote mutation S307L associated with partial loss of PC1/3 activity (Farooqi *et al.*, 2007). Two genome-wide screens confirmed the existence of a link between the PC1/3 gene (*Pcsk1*) and monogenic obesity (Benzinou *et al.*, 2008; Corpeleijn *et al.*, 2010). In addition, it was reported that ~1% of extremely obese European patients exhibited deleterious heterozygote **loss-of-function** (LOF) point mutations in the coding region of PC1/3 (Creemers *et al.*, 2012). In another study of a pediatric cohort, it was reported that congenital human PC1/3 deficiency causes malabsorptive diarrhea and other endocrinopathies (Martin *et al.*, 2013). Curiously, although neither heterozygote nor KO mice are obese, a homozygote mutation N222D that results in a ~60% lower PC1/3 activity led to obesity, hyperphagia, and increased metabolic efficiency (Lloyd *et al.*, 2006). Because a partial loss of PC1/3 activity may not equally reduce the rate of cleavage of all its substrates, the selective loss of processing of poor PC1/3 substrate(s) may be responsible for the obesity phenotype.

PC2 KO mice exhibited retarded growth, chronic fasting hypoglycemia, and a deficiency in circulating glucagon (Furuta *et al.*, 1997, 2001). These mice also showed defects in the processing of various precursors, including those of somatostatin, neuronal cholecystokinin, neurotensin, neuromedin N, dynorphin, nociceptin, and POMC-derived peptides (Berman *et al.*, 2000; Furuta *et al.*, 1997, 1998; Zhang *et al.*, 2010). Because PC2 is involved in the generation of most opioid peptides such as β-endorphin, dynorphin or met- and leu-enkephalins, it was important to investigate the role of PC2 in pain perception. Unexpectedly, after a short forced swim in warm water, PC2-null mice were significantly less (rather than more) responsive to the stimuli than wild-type mice, an indication of increased opioid-mediated stress-induced analgesia (Croissandeau *et al.*, 2006). The enhanced analgesia in PC2-null mice may be caused by an accumulation of opioid precursor processing intermediates with potent analgesic effects, or by loss of antiopioid peptides, such as substance P (SP). Thus, the presence of abnormal cocktails of pain neuropeptides in the brain of PC2 KO mice is likely to disturb pain perception mechanisms in ways that remain to be fully elucidated.

As a representative example of a precursor processing by more than one PC, we investigated the potential cellular processing of the precursor of substance P (proSP) by the PCs. The undecapeptide C-terminally amidated SP belongs to the tachykinin family, which in humans comprises three genes coding for preprotachykinin-A (PPTA), -B, and -C. Only PPTA contains the sequence of SP, and through its differentially spliced mRNA, results in four isoforms – PPTAα, β, γ, δ. Although SP is flanked by two typical PC-cleavage sites, the convertases involved in the production of SP are still not defined. SP is widely distributed in the CNS and PNS and is in particular associated with enhanced pain transmission (Harrison and Geppetti, 2001). Recently, SP was co-localized with PC1/3, PC2, and furin in mouse duodenum and jejunum (Gagnon *et al.*, 2009).

To examine the roles of PCs in processing of PPTA, we co-expressed the longest form PPTAβ carrying a Flag-tag at its N-terminus just following an optimized signal peptide and a V5-tag at its C-terminus (Figure 9.7A) with the convertases PC1/3, furin, soluble furin, PC5/6A, PC5/6B, PACE4, PC7 or soluble PC7 in human embryonic kidney HEK293 cells. Biosynthetic analysis following a 3-hour pulse labeling with ^{35}S-Met revealed that, except for PACE4, all tested PCs can process PPTAβ in this *ex vivo* system (Figure 9.7B). Because PCs cleave substrates after basic residues, we mutated the basic aa in the proposed cleavage sites of PPTAβ that flanked the two neuropeptides SP and neurokinin A (SP and NKA; Figure 9.7A). Because furin is the most efficient convertase (Figure 9.7B), we co-expressed furin with wild-type (WT) PPTAβ or its mutants R57A, R71A, R97A or R110A (notice that we did not include the length of the tags in the nomenclature). The results revealed that all mutants were cleaved by furin (Figure 9.7C). Via Western blot analysis with a V5 mAb in the neuronal mouse Neuro2A cells, we also observed that PC1/3 and PC2 were able to process the C-terminus of SP (Figure 9.7D). In sum, our data suggest that PC7 and furin (Figure 9.7C, D) can best process PPTAβ at the N-terminal cleavage site $RIAR_{57}\downarrow RP$ of SP, whereas PC1/3, PC2, furin, PC5/6, and PC7 can all process the C-terminus of SP at $M\underline{G}KR_{71}\downarrow DA$ (Figure 9.7E). However, SP is amidated at its C-terminus, in agreement with the presence of a Gly_{69} just preceding the $LysArg_{71}\downarrow$ cleavage site, which is usually trimmed by carboxypeptidase E in SGs (Fricker, 1988). The exposure of a C-terminal Gly is an absolute requirement for the C-terminal amidation of peptides (Bradbury and Smyth, 1991). Since amidation occurs in dense-core SGs and both PC1/3 and PC2 are maximally active in SGs, it is more likely that one or both of these convertases is/are implicated in the physiological processing of PPTAβ at $M\underline{G}KR_{71}\downarrow DA$.

Analysis of mice lacking either PC1/3 or PC2 should give more insights into the physiological *in vivo* processing of proSP into SP. Indeed, quantitative neuropeptidomics of mouse brains and pituitary demonstrated that SP was reduced in PC2 KO mice (Pan *et al.*, 2006; Zhang *et al.*, 2010), but not in PC1/3 KO mice (Pan *et al.*, 2005; Wardman *et al.*, 2010). Moreover, a significant decrease (but not abolition) in the level of circulating SP was recently observed in PC2 KO mice compared to WT (Gagnon *et al.*, 2011). Altogether, these data suggest that PC2 and possibly another enzyme(s) are needed for the production of SP *in vivo*, probably through cleavage at $M\underline{G}KR_{71}\downarrow DA$, and that a decrease in SP production may explain the analgesia observed in PC2 KO mice (Croissandeau *et al.*, 2006). Thus, PC2 by cleavage of proneuropeptides participates in the balance between pro- and antinociceptive peptides and consequently in regulating analgesia and nociception phenomena. However,

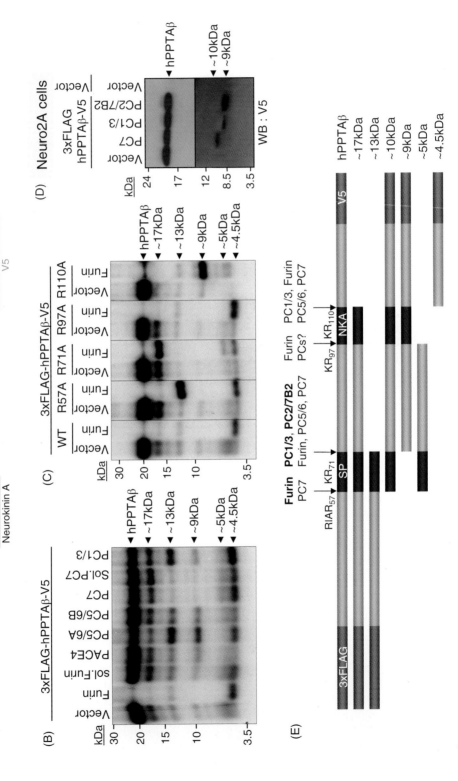

since SP is expressed in both CNS and PNS, it has yet to be determined if PC1/3 is not involved in the production of SP in other tissues than the brain. Finally, the above example of proSP processing emphasizes the need to validate processing reactions *in vivo*, and that cellular assays can only guide further analysis *in vivo*, but should not be considered as providing the final answer.

9.3.2 The ubiquitous convertases furin, PC7, and SKI-1/S1P

9.3.2.1 Furin

Furin, a ubiquitous membrane-bound protein, is initially produced as a ~104 kDa precursor rapidly converted into an active ~98 kDa form. This autocatalytic cleavage, occurring in the ER, is a prerequisite for the exit of mature furin molecules from the ER to the TGN and cell surface. Furin and PC5/6 seem to exhibit partial redundancy of their *in vitro* cleavage selectivity of a number of substrates and sensitivity to certain modified serpin inhibitors, such as α1-PDX or their prosegments. In the CNS, furin is responsible for the processing of a number of growth factors, including the neurotropins proNerve growth factor (proNGF), proBrain derived neurotropic factor (proBDNF) and neural cell adhesion and cueing proteins such as L1 CAM and semaphorins, respectively (Seidah and Prat, 2012). Recently, furin was shown to be needed for the proper development of the brain since *in vivo* evidence revealed that it is involved in axonal growth and that cleavage of the repulsive guidance molecule RGMa is essential for neogenin-mediated outgrowth inhibition (Tassew *et al.*, 2012). RGMa is a tethered membrane-bound molecule, and proteolytic processing amplifies RGMa diversity by creating soluble versions with long-range effects.

Inactivation of the fur gene (*FURIN*) causes embryonic death at about embryonic day 11 (E11), due to hemodynamic insufficiency and cardiac ventral closure defects (Roebroek *et al.*, 1998). Mutant embryos failed to develop large vessels despite the presence of endothelial cell precursors. Transforming growth factor (TGF) β1 was shown to be best processed by furin (Dubois *et al.*, 2001) and inactivation of its gene produces a phenotype similar to that of furin-null embryos (Bonyadi *et al.*, 1997). We also observed that loss of furin expression

---

Figure 9.7 *Ex vivo* **processing of preprotachykinin-A (PPTA) by the PCs.** (A) Amino acid sequence of the human tagged PPTAβ used for the *ex vivo* experiments. Depicted are the preprotrypsin signal peptide (*in italic*), the N-terminal 3xFlag-tag and the C-terminal V5-tag (*in gray*), the neuropeptides substance P (SP) and neurokinin A (NKA) are shown in black, and the potential PC-processing sites are in bold. (B,C) HEK293 cells were co-transfected with 3xFlag-hPPTAβ-V5 cDNA carrying either no mutation (WT) or R57A, R71A, R97A, and R110A mutations (nomenclature based on the complete human PPTAβ sequence), and vectors expressing no protein (vector), furin, its soluble form (sol.Furin), PACE4, PC5/6A, PC5/6B, PC7, its soluble form (sol.PC7), and PC1/3. Following overnight expression, the cells were incubated for 3 hours with ^{35}S-Met, and the resulting media were immunoprecipitated with a rabbit polyclonal antibody raised against the human full-length PPTAβ (sc-15322, Santa-Cruz). The immunoprecipitated proteins/peptides were separated by SDS-PAGE, and the dried gel was then autoradiographed. (D) Western blot analysis of cell lysates from Neuro2A cells transiently transfected with 3xFlag-hPPTAβ-V5 and PC7, PC1/3, PC2 or empty plasmid (vector). Proteins were revealed using an anti-HRP conjugated mAb-V5 (Sigma-Aldrich). (E) Schematic representation of human PPTAβ processing deduced from the analysis of the autoradiographed gels and immunoblots. The cleavage sites and the PCs implicated are indicated. Dark gray and black squares show the location of the tag (3xFlag and V5) and neuropeptides (SP and NKA), respectively. The apparent molecular weights of hPPTAβ-derived fragments are indicated. Note the high percentage of acidic aa in the N-terminal part of hPPTAβ construct depicted in **A** just preceding SP (aa 1-76 in this construct), which includes three copies of the Flag-tag. It is known that acidic proteins migrate abnormally on SDS-PAGE gels which may explain the apparent high molecular weights of the fragments ~17 kDa and ~13 kDa.

in endothelial cells results in a lethal phenotype at birth, with the KO mice exhibiting a heart ventricular/septal defect (Kim *et al.*, 2012). Conditional KO of furin in liver resulted in viable mice with mild phenotypes (Essalmani *et al.*, 2011; Roebroek *et al.*, 2004). This revealed some redundancy with other PCs, since some typical furin substrates were cleaved to a lesser extent (Roebroek *et al.*, 1998). In contrast, furin can uniquely process the Ac45 subunit of the vacuolar type H^+-ATPase in pancreatic β-cells (Louagie *et al.*, 2008) and PCSK9 (Essalmani *et al.*, 2011).

9.3.2.2 PC7

PC7 (gene *PCSK7*), the seventh member of the PC family, is widely expressed (Seidah *et al.*, 1996b), being abundant in neurons (Dong *et al.*, 1997), as well as microglia (Marcinkiewicz *et al.*, 1999). PC7 specifically enhances the processing of proepidermal growth factor (Rousselet *et al.*, 2011b) and sheds human transferrin receptor (Guillemot *et al.*, 2013), suggesting that it can play a role in cellular growth and iron metabolism. It was also shown that PC7, which is highly expressed in the immune system (Seidah *et al.*, 1996b), may be implicated in MHC class I-mediated antigen presentation (Leonhardt *et al.*, 2010). Recently, it was reported that PC7 mRNA knockdown in *Xenopus laevis* led to early embryonic death. Embryos lack eyes and brain, and exhibit abnormal anterior neural development (Senturker *et al.*, 2012). Moreover, the PC7 mRNA knockdown in zebrafish indicates that the larvae display severe developmental defects that result in 100% mortality within the first 7 days of life. The lack of functional PC7 interferes with the organogenesis of the brain, eyes, and otic vesicles (Turpeinen *et al.*, 2013). PC7 in amphibians and fish has clearly essential neuronal functions, whereas mice lacking PC7 did not exhibit overtly apparent phenotypes. However, recent data revealed that PC7 KO mice exhibit anxiolytic-like and novelty-seeking behaviors that are partly regulated by dopamine (DA), as a specific D2/D4 receptor antagonist reversed the phenotypes observed (Besnard *et al.*, 2012). Furthermore, these mice display impaired episodic and emotional memories, that are reversed by a BDNF tropomyosin-related kinase-B (TrkB) receptor agonist, 7,8 dihydroxyflavone (Wetsel *et al.*, 2013).

It was reported that mature BDNF derives from the intracellular processing of its precursor, proBDNF, by several members of the PC family, i.e. PC1/3, furin, PC5, PACE4 (Seidah *et al.*, 1996a), and PC7 (Wetsel *et al.*, 2013) in cell lines. BDNF is expressed in many tissues and in particular is highly expressed in hippocampus, amygdala, cerebral cortex, and cerebellum. BDNF belongs to the family of neurotropins and serves essential functions in the brain in synaptic plasticity and is crucial for learning and memory processes (Autry and Monteggia, 2012). Although it was reported that PC1/3 is expressed in hippocampus (Seidah *et al.*, 1991), the levels of hippocampal BDNF were similar between WT and PC1/3 KO mice (Wetsel *et al.*, 2013). In contrast, BDNF levels were ~45% lower in the hippocampus (and also amygdala) of PC7 KO compared to WT mice (Wetsel *et al.*, 2013). Interestingly, PC7 KO mice exhibit learning and memory impairments, suggesting that the reduced BDNF levels are probably responsible for these deficits, as they are rescued by an agonist of the BDNF receptor TrkB (Wetsel *et al.*, 2013). Altogether, these results suggest that in mouse hippocampus (and amygdala), ~45% of proBDNF is processed by PC7 (but not by PC1 or PC2) and that the release of mature BDNF *in vivo* occurs via the constitutive secretory pathway.

Interestingly, it was reported that iron deficiency, the most common nutrient deficiency, is related to learning and memory impairment (Fretham *et al.*, 2011). Iron deficiency may induce these deficits by decreasing the expression of growth factors in the central nervous system like IGFI/II or BDNF (Estrada *et al.*, 2013). Recently, we have observed that iron depletion by deferoxamine induced a decrease of PC7 expression (Guillemot *et al.*, 2013). So, we can postulate that a decrease of PC7 levels under iron deficiency conditions may

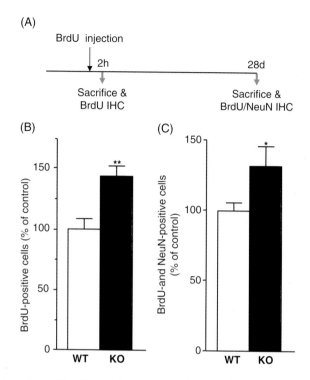

Figure 9.8 Hippocampal proliferation and neurogenesis are enhanced in PC7 KO mice.
(A) Experimental timeline. IHC, immunohistochemistry. (B) Quantification of BrdU+ cells in the dentate gyrus 2 hours after a single injection of BrdU in WT and PC7 KO mice. Data are represented by the mean ± SEM (% of WT mice) from nine WT and 11 PC7 KO mice; **P<0.01. (C) Quantification of [BrdU+/NeuN+] cells in the dentate gyrus 4 weeks after BrdU injection. Data are represented by the mean ± SEM (% of WT mice); N = 8 mice/genotype; *P<0.05.

promote a reduction of BDNF production, possibly leading to learning and memory deficits observed in such a nutritional deficiency state.

Animal models also indicated that overexpression of BDNF causes anxiety-like behaviors (Govindarajan *et al.*, 2006; Papaleo *et al.*, 2011). In particular, it was shown that BDNF over-expression leads to increased anxiety and is associated with a higher dendritic spine density in the basolateral amygdala (Govindarajan *et al.*, 2006). At the opposite end, PC7 KO mice exhibit anxiolytic-like behaviors (Besnard *et al.*, 2012), suggesting that the decrease of BDNF levels in amygdala of PC7 KO mice (Wetsel *et al.*, 2013) may be involved in the anxiolytic phenotype observed in these mice.

Another neuronal phenotype observed in PC7 KO mice is an increased proliferation and neurogenesis in the dentate gyrus of the hippocampus. Indeed, examination of cell proliferation by immunohistochemistry in dividing cells 2 hours after bromodeoxyuridine (BrdU) injection (Figure 9.8A) indicated that the number of BrdU-positive cells in the hip-pocampus was significantly higher than that in WT mice (Figure 9.8B). To determine whether newly generated BrdU-positive cells were transformed subsequently into mature neurons, double immunofluorescence staining was performed 28 days after the last BrdU injection (Figure 9.8A). The proportion of BrdU-positive cells that co-localized with the neuronal marker NeuN in PC7 KO compared to WT dentate gyrus was significantly increased by ~32% (Figure 9.8C), indicating that deletion of PC7 in mice increases hippocampal proliferation and neurogenesis. Because adult hippocampal neurogenesis is a dynamic process that is regulated, both positively and negatively, by growth factors and environmental conditions (Warner-Schmidt and Duman, 2006), it is still too early to identify which factors are involved in this process in the absence of PC7. Finally, reduced neurogenesis has been associated with increased anxiety-like behaviors in mice (Revest *et al.*, 2009). Thus, the increased neurogenesis in the hippocampus of PC7 KO mice may participate in the etiology of the marked anxiolytic phenotype reported in these mice (Besnard *et al.*, 2012). PC7 and/ or its substrates could therefore represent pharmacological targets for the development of novel anxiolytics.

9.3.2.3 SKI-1/S1P

SKI-1/S1P is a key enzyme in the regulation of lipid metabolism and cholesterol homeostasis that cleaves the transcription factors sterol regulatory element binding proteins (SREBP-1 and SREBP-2). The latter are synthesized as precursors harboring two transmembrane domains separated by a short ER luminal loop and comprising N- and C-terminal cytosolic domains. These precursors are cleaved in an SREBP-cleavage-activating protein (SCAP) and insulin-induced gene (Insig)-dependent fashion (Brown and Goldstein, 1999). When cellular cholesterol levels are high, the insulin-regulated proteins Insig-1 and/or Insig-2 bind and retain the SCAP-SREBP complex in the ER. When cells are deprived of sterols, Insigs separate, allowing the transport of the SREBP-SCAP complex to the Golgi. Therein, a two-step proteolytic process starting with SKI-1/S1P at the sequence R-X-V-L↓ (see Figure 9.2) and then site 2 protease (S2P) releases the cytosolic N-terminal segments of SREBPs from cell membranes, allowing their translocation to the nucleus (nSREBP), where they activate transcription of more than 35 mRNAs coding for proteins/enzymes required for the biosynthesis and uptake of cholesterol and unsaturated fatty acids, as well as the LDLR (Seidah and Prat, 2007).

Similar to SREBPs, the ER-anchored type II membrane-bound transcription factor ATF6 plays a major role in the unfolded protein response (UPR). Under normal conditions, it is held in the ER by the chaperone BIP, with its N-terminal DNA binding domain facing the cytosol and its COOH terminus in the ER lumen. Accumulation of improperly folded proteins in the ER, which can be induced by calcium depletion (thapsigargin) or inhibition of N-glycosylation (tunicamycin), leads to an ER stress response resulting in BIP dissociation from proATF6. The latter is then translocated in a SCAP-independent fashion to the Golgi where it is first cleaved by SKI-1/S1P and then by S2P. This releases the cytosolic N-terminal domain, which reaches the nucleus (nATF6) to activate ER stress target genes (Ye *et al.*, 2000).

Other type II membrane-bound substrates include at least six CREB-like basic leucine zipper transcription factors. BDNF is a soluble substrate and the study of its processing led to the initial cloning of SKI-1/S1P (Seidah *et al.*, 1999). We also showed that the soluble pro-somatostatin is cleaved by SKI-1/S1P to release the N-terminal peptide antrin (Mouchantaf *et al.*, 2004).

The essential roles played by SKI-1/S1P are evident from the fact that deletion of its gene results in embryonic death at the earliest stages of cell division, i.e. from one- to two-cell stage, hence preventing blastocyst formation. Early data with specific deletion of SKI-1/S1P function in the CNS revealed a major role in RGMa cleavage and neuronal development (Tassew *et al.*, 2012). When expression of SKI-1/S1P was abrogated in liver, the level of circulating total cholesterol was decreased by 50%, clearly emphasizing its critical control of cholesterol synthesis and uptake (Yang *et al.*, 2001). Thus, SKI-1/S1P is a major enzyme controlling cholesterol and fatty acid synthesis (Rawson *et al.*, 1998), that also regulates a number of homeostatic functions such as cellular stress response (Ye *et al.*, 2000), bone mineralization (Gorski *et al.*, 2011; Patra *et al.*, 2007), neuronal survival and guidance (Tassew *et al.*, 2012), and targeting proteins to lysosomes (Marschner *et al.*, 2011). SKI-1/S1P also enhances viral infections through surface glycoprotein processing of hemorrhagic fever viruses (Lenz *et al.*, 2001). So far, this gene was found not to be polymorphic and no single nucleotide polymorphic variations (SNPs) that could be genetically associated with a known pathology are known.

9.3.3 The convertases PC5/6 and PACE4

The convertases PC5/6 and PACE4 seem to form a class of their own based on their primary structures and their ability to bind the cell surface via their C-terminal Cys-rich domains which bind HSPGs, and in many cases inactivate HSPG-bound proteins such as endothelial and lipoprotein lipases (Jin *et al.*, 2005), and possibly adhesion molecules such as N-cadherin (Maret *et al.*, 2012). Since in cellular experiments and *in vitro*, many of the substrates processed by either enzyme can also be cleaved by furin and/or PC7, the complete set of specific physiological substrates of PC5/6 and PACE4 is yet to be unraveled. In the CNS, it was shown that PC5/6 can process the neural adhesion molecule L1, assisting in neuronal repair and migration (Schnegelsberg *et al.*, 2014). Ontogeny and tissue distribution analysis of these convertases showed that PC5/6 expression is detected early during embryonic development, appearing first in extraembryonic tissues. By embryonic day E9, it is also specifically expressed in cells of the maternal-embryonic junction, where no other convertase is expressed (Essalmani *et al.*, 2006). Present data strongly suggest unique tissue-specific functions of PC5/6 and PACE4. Thus, PC5/6 mRNA was detected only in neuronal cells, whereas PACE4 mRNA was expressed in both neuronal and glial cells. In areas that are rich in neuropeptides such as cortex, hippocampus, and hypothalamus, mRNA levels of PC5/6 were high but those of PACE4 were low or undetectable. In regions such as the amygdaloid body and thalamus, distinct but complementary distributions of PC5/6 and PACE4 mRNAs were observed. The medial habenular and cerebellar Purkinje cells expressed very high levels of PACE4 mRNA.

The KO of PACE4 and PC5/6 genes in mice resulted in different phenotypes. Thus, while PACE4 KO results in a 75% viable phenotype with bone morphogenesis defects, that of PC5/6 causes embryonic death at birth.

9.3.3.1 PACE4

PACE4 KO (gene Pcsk6) in mice revealed that about 25% of embryos die at E14 with severe cardiac malformations (ventriculoseptal defects), strong defects in the left/right axis (left pulmonary isomerism; dextrocardia and cyclopism), and bone morphogenesis defects (Constam and Robertson, 2000). However, these phenotypes have a relatively low penetrance that depends on the genetic background. The left/right axis defects are preceded at embryonic day 8.5 by abnormal expression of the axis determining TGFβ-like growth factors Nodal and Lefty, which are both PC substrates (Blanchet *et al.*, 2008). A recent GWAS study revealed PACE4 SNPs associated with the degree of left-handedness and/or dyslexia in some individuals (Arning *et al.*, 2013; Scerri *et al.*, 2011), as well as with relative hand skills (Brandler *et al.*, 2013), possibly related to the activity of PACE4 on the left/right axis determinants Nodal and Lefty (Constam, 2009). Very recently, it was discovered that PACE4 is the physiological protease that activates the heart enzyme corin, which is itself responsible for the generation of atrial natriuretic peptide (ANP). The lack of PACE4 results in salt-sensitive hypertension. Thus, PACE4 is the long-sought corin activator and is important for sodium homeostasis and normal blood pressure (Chen *et al.*, 2015).

9.3.3.2 PC5/6

PC5/6 KO newborns die at birth, lack kidneys and exhibit major defects in the anteroposterior axis with extrathoracic and -lumbar vertebrae, and a lack of tail (Essalmani *et al.*, 2008), a phenotype also reported for Gdf11 KO mice (McPherron *et al.*, 1999). We showed that

Gdf11 is a favorite substrate of PC5/6, in part due to an Asn residue at P1', the first position after the cleavage site $\underline{\textbf{R}}\textbf{S}\underline{\textbf{RR}}_{296}\!\downarrow\!\underline{\textbf{N}}\textbf{L}$ (Essalmani *et al.*, 2008). In addition, magnetic resonance imaging revealed severe phenotypes reminiscent of those observed in newborns exhibiting VACTERL (vertebral, anorectal, cardiac, tracheoesophageal, renal, limb) malformations (Szumska *et al.*, 2008). Finally, exon sequencing of control and VACTERL patients linked mutations in the human PC5/6 gene (*PCSK5*) to this syndrome (Szumska *et al.*, 2008). Thus, PC5/6, at least in part via Gdf11, coordinately regulates caudal Hox paralogs to control anteroposterior patterning, nephrogenesis, skeletal, and anorectal development. PC5/6 KO also exhibit Gdf11-independent phenotypes (Essalmani *et al.*, 2008), suggesting the lack of processing of other substrates that are not yet defined. Interestingly, GDF11 is now considered as a youth rejuvenating factor, as exposure of aged mice to GDF11 decreases cardiac hypertrophy which is associated with aging (Loffredo *et al.*, 2015).

9.3.4   The convertase PCSK9

PCSK9 (originally named NARC-1), first characterized by our group in 2003, is highly expressed in adult liver, gut, and kidney (Seidah *et al.*, 2003). During development, PCSK9 mRNA levels are transiently high in telencephalon (E12–E15), in the rostral extension of the olfactory peduncle and in cerebellar neurons. Under resting conditions, the expression of PCSK9 in adult mouse brain is restricted to the olfactory peduncle and cerebellum. Overexpression of PCSK9 in telencephalic progenitor cells enhanced proliferation and neuronal differentiation. Whether specific signals can regulate the levels of PCSK9 in the CNS is yet to be defined. PCSK9 expression in liver hepatocytes is now known to be upregulated by SREBP-2 and downregulated by cholesterol (Dubuc *et al.*, 2004). We established the first association between single-point mutations in PCSK9 and autosomal dominant hypercholesterolemia (ADH) in two French families (Abifadel *et al.*, 2003). Accordingly, mutations associated with hypercholesterolemia result in a **gain of function** (GOF) of PCSK9 that triggers the degradation of LDLR in acidic compartments, probably endosomes (Seidah and Prat, 2012; Seidah *et al.*, 2014). By a yet unknown mechanism, high levels of PCSK9 lead to a faster rate of lysosomal degradation of cell surface LDLR, resulting in increased circulating LDL-cholesterol, as LDL uptake in hepatocytes by LDLR is diminished (Seidah *et al.*, 2014). PCSK9 is similar to other PCs in that it autocatalytically cleaves its prosegment in the ER to exit this compartment, but then it differs from all other PCs since the inhibitory prosegment remains tightly bound to the enzyme, always keeping it as an inactive protease. Thus, all functions of PCSK9 that involve the escort of its partners to lysosomes do not implicate its enzymatic activity.

PCSK9-null mice exhibit increased LDLR protein, but not mRNA, and a two-fold drop in circulating cholesterol, whereas mice overexpressing PCSK9 in liver exhibit high levels of circulating cholesterol (Zaid *et al.*, 2008). LOF mutations protective against human disease often provide *in vivo* validation of therapeutic targets. Indeed, 2 years after the discovery of PCSK9, it was shown that nonsense PCSK9 mutations, probably resulting in LOF, are associated with hypocholesterolemia in ~2% of black subjects (Cohen *et al.*, 2005). A 32-year-old educated female aerobics instructor with two compound heterozygote mutations leading to complete loss of PCSK9 expression was recently identified. Although healthy, fertile, and normotensive, with normal liver and renal functions, her LDL-cholesterol is remarkably low (14 mg/dL). This and the observation that LOF nonsense mutations could lead to 88% reduction in the risk of development of cardiovascular artery disease (CAD) over a 15-year period indicate that inhibition of PCSK9 may represent a safe and effective strategy for the primary prevention of CAD (Horton *et al.*, 2009). Various Phase 2/3 clinical trials are now under way to test the efficacy of monoclonal antibodies that block the PCSK9≡LDLR interaction in reducing circulating LDL-cholesterol (Seidah, 2013; Seidah and Prat, 2012; Seidah *et al.*, 2014). So far, the results look very promising (Stein, 2013; Stein and Raal, 2014).

A study in zebrafish suggested that PCSK9 plays a significant role in brain development and that its absence results in extensive neuronal damage and death. Indeed, specific knockdown of zebrafish PCSK9 mRNA resulted in a general disorganization of cerebellar neurons and loss of hindbrain-midbrain boundaries, leading to embryonic death at approximately 96 hours after fertilization (Poirier *et al.*, 2006). These data support a novel role for PCSK9 in CNS development, distinct from that in cholesterogenic organs such as liver. However, the loss of expression of mammalian PCSK9 in brain does not affect embryonic development in mouse or human, as KO mice are viable and fertile, as are humans with complete LOF of PCSK9 (Seidah *et al.*, 2014). Accordingly, the role of PCSK9 in the mammalian brain is less clear, and may only become apparent under stressful or pathogenic conditions. PCSK9 and LDLR are co-expressed in mouse brain during development and at adulthood upregulation of LDLR protein levels in PCSK9 KO mice enhances apoE degradation (Rousselet *et al.*, 2011c). During embryonic development, LDLR levels increased and were accompanied by a reduction of apoE levels. Upon ischemic stroke, PCSK9 was expressed in the dentate gyrus between 24 and 72 hours following brain reperfusion. Although mouse behavior and lesion volume were similar, LDLR protein levels dropped ~2-fold in the PCSK9 KO-lesioned hippocampus, without affecting apoE levels and neurogenesis. Thus, PCSK9 downregulates LDLR levels during brain development and following transient ischemic stroke in adult mice (Rousselet *et al.*, 2011c).

The Alzheimer disease-associated aspartyl protease β-secretase BACE1 is transiently acetylated on seven Lys residues in the lumen of the ER/ERGIC. The acetylated intermediates of the nascent protein are able to reach the Golgi apparatus, whereas the non-acetylated ones are retained and degraded in a post-ER compartment. PCSK9 was reported to contribute to the disposal of non-acetylated BACE1 in the ER/ERGIC (Jonas *et al.*, 2008). This interesting observation still requires *in vivo* validation, as well as the possible regulatory role of neuronal apoptosis by PCSK9 and oxidized LDL-cholesterol that is increased under hyperlipidemic conditions (Wu *et al.*, 2014). In addition, it was recently reported that a small proportion of individuals receiving high doses (420 mg) of a monoclonal antibody to PCSK9 every 4 weeks together with a cholesterol-lowering 'statin' complained of neurological dysfunctions such as amnesia, headache, and migraine (Koren *et al.*, 2014). Thus, care must be exercised when reducing LDL-cholesterol to very low levels as it may result in unwanted side effects in humans.

9.4 Perspectives

The nine-membered family of the proprotein convertases (PCs) comprises seven basic amino acid-specific subtilisin-like serine proteinases, related to yeast kexin, known as PC1/3, PC2, furin, PC4, PC5/6, PACE4, and PC7, and two other subtilases that cleave at non-basic residues, called SKI-1/S1P and PCSK9. Except for the testicular PC4, all the other convertases are expressed in brain and play a critical role in various neuronal functions, including the production of diverse neuropeptides as well as neural growth factors and receptors, the regulation of cellular adhesion/migration, cholesterol and fatty acid homeostasis and repair mechanisms and growth/differentiation of progenitor cells. While *in vitro* work on the PCs gave us some clues as to their specificity and cellular localizations, the physiological functions of the PCs were mostly derived from *in vivo* studies in **animal models** and humans. Some of these convertases process proteins implicated in neuropathologies, such as Alzheimer disease β-secretase (BACE1), neural growth such as NGF, as well as adhesion molecules such as L1 and N-CAM, important for cancer malignancies and neuronal regeneration and repair following injury.

Taken together, the major advances in our understanding of how the PCs function in various organs, and what their targets are, have allowed us to better define their physiological roles in normal and pathological conditions. In some cases, their dysregulation results in a fundamental neural and endocrine axis dysfunction, as exemplified by PC1/3-specific LOF mutations associated with morbid obesity and increased food intake.

We hope that the topics discussed here will encourage the investigation and development of potent and effective PC inhibitors/silencers. For example, PCSK9-inhibiting monoclonal antibodies are now being evaluated in Phase 3 clinical trials as cholesterol-lowering agents, which is one of the most exciting developments in the field of PCs and their translational applications, which rapidly progressed from discovery in 2003 to clinical applications in 2012 (Seidah and Prat, 2012; Seidah *et al.*, 2013, 2014).

PC7 is the only other convertase whose absence could be favorable under certain conditions. Thus, lack of PC7 in the CNS is not lethal, but rather would result in loss of anxiety and emotional memory associated with traumatic conditions. Therefore, it is feasible that a brain-penetrable inhibitor of PC7 may find applications in some neurological and/or psychiatric disorders. Furthermore, since PC7 sheds transferrin receptor, it is possible that an inhibitor of PC7 would raise the levels of this receptor at the surface of endothelial cells of the blood–brain barrier (BBB). Thus, enhanced targeting of antitransferrin receptor antibody to brain vasculature will increase the transport across the BBB to deliver non-lipophilic compounds to the brain, for example, a BACE1-specific antibody as proposed for Alzheimer disease (Bien-Ly *et al.*, 2014).

The challenge of identification of safe, orally active small molecule inhibitors of some of the convertases, for example, for PCSK9 or PC7, is definitively a future goal that should be pursued vehemently as they may find applications in the control of specific pathologies, which would be highly beneficial in the clinical treatment of diseases in which these PCs play a key role. The future is indeed bright and awaits innovative and practical approaches to target these exciting and multifunctional secretory enzymes.

References

Abifadel, M., Varret, M., Rabes, J.P., *et al.* (2003) Mutations in PCSK9 cause autosomal dominant hypercholesterolemia. *Nature Genetics*, **34**, 154–156. [**This is the first evidence linking PCSK9 to LDL-cholesterol regulation.**]

Arning, L., Ocklenburg, S., Schulz, S., *et al.* (2013) VNTR polymorphism is associated with degree of handedness but not direction of handedness. *PLoS One*, **8**, e67251.

Autry, A.E. and Monteggia, L.M. (2012) Brain-derived neurotrophic factor and neuropsychiatric disorders. *Pharmacological Reviews*, **64**, 238–258.

Bedard, K., Szabo, E., Michalak, M. and Opas, M. (2005) Cellular functions of endoplasmic reticulum chaperones calreticulin, calnexin, and ERp57. *Intinternational Review of Cytology*, **245**, 91–121.

Benzinou, M., Creemers, J.W., Choquet, H., *et al.* (2008) Common nonsynonymous variants in PCSK1 confer risk of obesity. *Nature Genetics*, **40**, 943–945.

Berman, Y., Mzhavia, N., Polonskaia, A., *et al.* (2000) Defective prodynorphin processing in mice lacking prohormone convertase PC2. *Journal of Neurochemistry*, **75**, 1763–1770.

Besnard, J., Ruda, G.F., Setola, V., *et al.* (2012) Automated design of ligands to polypharmacological profiles. *Nature*, **492**, 215–220.

Bien-Ly, N., Yu, Y.J., Bumbaca, D., *et al.* (2014) Transferrin receptor (TfR) trafficking determines brain uptake of TfR antibody affinity variants. *Journal of Experimental Medicine*, **211**, 233–244.

Blanchet, M.H., Le Good, J.A., Mesnard, D., *et al.* (2008) Cripto recruits Furin and PACE4 and controls Nodal trafficking during proteolytic maturation. *EMBO Journal*, **27**, 2580–2591.

Bonyadi, M., Rusholme, S.A., Cousins, F.M., *et al.* (1997) Mapping of a major genetic modifier of embryonic lethality in TGF beta 1 knockout mice. *Nature Genetics*, **15**, 207–211.

Bradbury, A.F. and Smyth, D.G. (1991) Peptide amidation. *Trends in Biochemical Sciences*, **16**, 112–115.

Brandler, W.M., Morris, A.P., Evans, D.M., *et al.* (2013) Common variants in left/right asymmetry genes and pathways are associated with relative hand skill. *PLoS Genetics*, **9**, e1003751.

Brown, M.S. and Goldstein, J.L. (1999) A proteolytic pathway that controls the cholesterol content of membranes, cells, and blood. *Proceedings of the National Academy of Sciences of the United States of America*, **96**, 11041–11048.

Chen, S., Cao, P., Dong, N., *et al.* (2015) PCSK6-mediated corin activation is essential for normal blood pressure. *Nature Medicine*, doi: 10.1038/nm.3920. [Epub ahead of print].

Cohen, J., Pertsemlidis, A., Kotowski, I.K., Graham, R., Garcia, C.K. and Hobbs, H.H. (2005) Low LDL cholesterol in individuals of African descent resulting from frequent nonsense mutations in PCSK9. *Nature Genetics*, **37**, 161–165.

Constam, D.B. (2009) Running the gauntlet: an overview of the modalities of travel employed by the putative morphogen Nodal. *Current Opinion in Genetics & Development*, **19**, 302–307.

Constam, D.B. (2014) Regulation of TGFbeta and related signals by precursor processing. *Seminars in Cell & Developmental Biology*, **32C**, 85–97.

Constam, D.B. and Robertson, E.J. (2000) SPC4/PACE4 regulates a TGFbeta signaling network during axis formation. *Genes & Development*, **14**, 1146–1155. **[This is the first evidence that PACE4 is critical for body axis formation.]**

Coppola, J.M., Bhojani, M.S., Ross, B.D. and Rehemtulla, A. (2008) A small-molecule furin inhibitor inhibits cancer cell motility and invasiveness. *Neoplasia*, **10**, 363–370.

Corpeleijn, E., Petersen, L., Holst, C., *et al.* (2010) Obesity-related polymorphisms and their associations with the ability to regulate fat oxidation in obese Europeans: the NUGENOB study. *Obesity*, **18**, 1369–1377.

Creemers, J.W., Choquet, H., Stijnen, P., *et al.* (2012) Heterozygous mutations causing partial prohormone convertase 1 deficiency contribute to human obesity. *Diabetes*, **61**, 383–390.

Croissandeau, G., Wahnon, F., Yashpal, K., *et al.* (2006) Increased stress-induced analgesia in mice lacking the proneuropeptide convertase PC2. *Neuroscience Letters*, **406**, 71–75.

Dey, A., Norrbom, C., Zhu, X., *et al.* (2004) Furin and prohormone convertase 1/3 are major convertases in the processing of mouse pro-growth hormone-releasing hormone. *Endocrinology*, **145**, 1961–1971.

Dong, W., Seidel, B., Marcinkiewicz, M., Chretien, M., Seidah, N.G. and Day, R. (1997) Cellular localization of the prohormone convertases in the hypothalamic paraventricular and supraoptic nuclei: selective regulation of PC1 in corticotrophin-releasing hormone parvocellular neurons mediated by glucocorticoids. *Journal of Neuroscience*, **17**, 563–575.

Dubois, C.M., Blanchette, F., Laprise, M.H., Leduc, R., Grondin, F. and Seidah, N.G. (2001) Evidence that furin is an authentic transforming growth factor-beta1-converting enzyme. *American Journal of Pathology*, **158**, 305–316.

Dubuc, G., Chamberland, A., Wassef, H., *et al.* (2004) Statins upregulate PCSK9, the gene encoding the proprotein convertase neural apoptosis-regulated convertase-1 implicated in familial hypercholesterolemia. *Arteriosclerosis, Thrombosis & Vascular Biology*, **24**, 1454–1459.

Essalmani, R., Hamelin, J., Marcinkiewicz, J., *et al.* (2006) Deletion of the gene encoding proprotein convertase 5/6 causes early embryonic lethality in the mouse. *Molecular & Cellular Biology*, **26**, 354–361.

Essalmani, R., Zaid, A., Marcinkiewicz, J., *et al.* (2008) In vivo functions of the proprotein convertase PC5/6 during mouse development: Gdf11 is a likely substrate. *Proceedings of the Nationall Academy of Sciences of the Unied States of America*, **105**, 5750–5755. **[This is the first evidence that lack of PC5/6 results in lack of homeotic transformations.]**

Essalmani, R., Susan-Resiga, D., Chamberland, A., *et al.* (2011) In vivo evidence that furin from hepatocytes inactivates PCSK9. *Journal of Biological Chemistry*, **286**, 4257–4263.

Estrada, J.A., Contreras, I., Pliego-Rivero, F.B. and Otero, G.A. (2013) Molecular mechanisms of cognitive impairment in iron deficiency: alterations in brain-derived neurotrophic factor and insulin-like growth factor expression and function in the central nervous system. *Nutritional Neuroscience*, **17**, 193–206.

Farooqi, I.S., Volders, K., Stanhope, R., *et al.* (2007) Hyperphagia and early-onset obesity due to a novel homozygous missense mutation in prohormone convertase 1/3. *Journal of Clinical Endocrinology and Metabolism*, **92**, 3369–3373.

Fretham, S.J., Carlson, E.S. and Georgieff, M.K. (2011) The role of iron in learning and memory. *Advances in Nutrition*, **2**, 112–121.

Fricker, L.D. (1988) Carboxypeptidase E. *Annual Review of Physiology*, **50**, 309–321.

Furuta, M., Yano, H., Zhou, A., *et al.* (1997) Defective prohormone processing and altered pancreatic islet morphology in mice lacking active SPC2. *Proceedings of the National Academy of Sciences of the United States of America,* **94**, 6646–6651. **[First in vivo evidence for the endocrine role of PC2.]**

Furuta, M., Carroll, R., Martin, S., *et al.* (1998) Incomplete processing of proinsulin to insulin accompanied by elevation of Des-31,32 proinsulin intermediates in islets of mice lacking active PC2. *Journal of Biological Chemistry,* **273**, 3431–3437.

Furuta, M., Zhou, A., Webb, G., *et al.* (2001) Severe defect in proglucagon processing in islet A-cells of prohormone convertase 2 null mice. *Journal of Biological Chemistry,* **276**, 27197–27202.

Gagnon, J., Mayne, J., Mbikay, M., Woulfe, J. and Chretien, M. (2009) Expression of PCSK1 (PC1/3), PCSK2 (PC2) and PCSK3 (furin) in mouse small intestine. *Regulatory Peptides,* **152**, 54–60.

Gagnon, J., Mayne, J., Chen, A., *et al.* (2011) PCSK2-null mice exhibit delayed intestinal motility, reduced refeeding response and altered plasma levels of several regulatory peptides. *Life Sciences,* **88**, 212–217.

Gil, G.C., Velander, W.H. and van Cott, K.E. (2009) N-glycosylation microheterogeneity and site occupancy of an Asn-X-Cys sequon in plasma-derived and recombinant protein C. *Proteomics,* **9**, 2555–2567.

Gorski, J.P., Huffman, N.T., Chittur, S., *et al.* (2011) Inhibition of proprotein convertase SKI-1 blocks transcription of key extracellular matrix genes regulating osteoblastic mineralization. *Journal of Biological Chemistry,* **286**, 1836–1849.

Govindarajan, A., Rao, B.S., Nair, D., *et al.* (2006) Transgenic brain-derived neurotrophic factor expression causes both anxiogenic and antidepressant effects. *Proceedings of the National Academy of Sciences of the United States of America,* **103**, 13208–13213.

Guillemot, J., Canuel, M., Essalmani, R., Prat, A. and Seidah, N.G. (2013) Implication of the proprotein convertases in iron homeostasis: proprotein convertase 7 sheds human transferrin receptor 1 and furin activates hepcidin. *Hepatology,* **57**, 2514–2524.

Gyamera-Acheampong, C., Sirois, F., Denis, N.J., *et al.* (2011) The precursor to the germ cell-specific PCSK4 proteinase is inefficiently activated in transfected somatic cells: evidence of interaction with the BiP chaperone. *Molecular and Cellular Biochemistry,* **348**, 43–52.

Harrison, S. and Geppetti, P. (2001) Substance P. *International Journal of Biochemistry & Cell Biology,* **33**, 555–576.

Hawkins, J.L., Robbins, M.D., Warren, L.C., *et al.* (2008) Pharmacologic inhibition of site 1 protease activity inhibits sterol regulatory element-binding protein processing and reduces lipogenic enzyme gene expression and lipid synthesis in cultured cells and experimental animals. *Journal of Pharmacology and Experimental Therapeutics,* **326**, 801–808.

Helenius, A., Marquardt, T. and Braakman, I. (1992) The endoplasmic reticulum as a protein-folding compartment. *Trends in Cell Biology,* **2**, 227–231.

Horton, J.D., Cohen, J.C. and Hobbs, H.H. (2009) PCSK9: a convertase that coordinates LDL catabolism. *Journal of Lipid Research,* **50(suppl)**, S172–S177.

Jackson, R.S., Creemers, J.W., Ohagi, S., *et al.* (1997) Obesity and impaired prohormone processing associated with mutations in the human prohormone convertase 1 gene. *Nature Genetics,* **16**, 303–306.

Jiao, G.S., Cregar, L., Wang, J., *et al.* (2006) Synthetic small molecule furin inhibitors derived from 2,5-dideoxystreptamine. *Proceedings of the National Academy of Sciences of the United States of America,* **103**, 19707–19712.

Jin, W., Fuki, I.V., Seidah, N.G., Benjannet, S., Glick, J.M. and Rader, D.J. (2005) Proprotein convertases are responsible for proteolysis and inactivation of endothelial lipase. *Journal of Biological Chemistry,* **280**, 36551–36559.

Jonas, M.C., Costantini, C. and Puglielli, L. (2008) PCSK9 is required for the disposal of non-acetylated intermediates of the nascent membrane protein BACE1. *EMBO Reports,* **9**, 916–922.

Kim, W., Essalmani, R., Szumska, D., *et al.* (2012) Loss of endothelial furin leads to cardiac malformation and early postnatal death. *Molecular and Cellular Biology,* **32**, 3382–3391.

Komiyama, T., Coppola, J.M., Larsen, M.J., *et al.* (2009) Inhibition of furin/proprotein convertase-catalyzed surface and intracellular processing by small molecules. *Journal of Biological Chemistry,* **284**, 15729–15738.

Koren, M.J., Giugliano, R.P., Raal, F.J., *et al.* (2014) Efficacy and safety of longer-term administration of evolocumab (AMG 145) in patients with hypercholesterolemia: 52-week results from the Open-Label Study of Long-Term Evaluation Against LDL-C (OSLER) randomized trial. *Circulation,* **129**, 234–243.

Kornfeld, R. and Kornfeld, S. (1985) Assembly of asparagine-linked oligosaccharides. *Annual Review of Biochemistry,* **54**, 631–664.

Ladunga, I. (2000) Large-scale predictions of secretory proteins from mammalian genomic and EST sequences. *Current Opinion in Biotechnology*, **11**, 13–18.

Lenz, O., ter Meulen, J., Klenk, H.D., Seidah, N.G. and Garten, W. (2001) The Lassa virus glycoprotein precursor GP-C is proteolytically processed by subtilase SKI 1/S1P. *Proceedings of the National Academy of Sciences of the United States of America*, **98**, 12701–12705.

Leonhardt, R.M., Fiegl, D., Rufer, E., Karger, A., Bettin, B. and Knittler, M.R. (2010) Post-endoplasmic reticulum rescue of unstable MHC class I requires proprotein convertase PC7. *Journal of Immunology*, **184**, 2985–2998.

Lloyd, D.J., Bohan, S. and Gekakis, N. (2006) Obesity, hyperphagia and increased metabolic efficiency in Pc1 mutant mice. *Human Molecular Genetics*, **15**, 1884–1893.

Loffredo, F.S., Steinhauser, M.L., Jay, S.M., *et al.* (2013) Growth differentiation factor 11 is a circulating factor that reverses age-related cardiac hypertrophy. *Cell*, **153**, 828–839.

Louagie, E., Taylor, N.A., Flamez, D., *et al.* (2008) Role of furin in granular acidification in the endocrine pancreas: identification of the V-ATPase subunit Ac45 as a candidate substrate. *Proceedings of the National Academy of Sciences of the United States of America*, **105**, 12319–12324.

Marcinkiewicz, M., Marcinkiewicz, J., Chen, A., Leclaire, F., Chretien, M. and Richardson, P. (1999) Nerve growth factor and proprotein convertases furin and PC7 in transected sciatic nerves and in nerve segments cultured in conditioned media: their presence in Schwann cells, macrophages, and smooth muscle cells. *Journal of Comparative Neurology*, **403**, 471–485.

Maret, D., Sadr, M.S., Sadr, E.S., Colman, D.R., Del Maestro, R.F. and Seidah, N.G. (2012) Opposite roles of furin and PC5A in N-cadherin processing. *Neoplasia*, **14**, 880–892.

Marschner, K., Kollmann, K., Schweizer, M., Braulke, T. and Pohl, S. (2011) A key enzyme in the biogenesis of lysosomes is a protease that regulates cholesterol metabolism. *Science*, **333**, 87–90.

Martin, M.G., Lindberg, I., Solorzano-Vargas, R.S., *et al.* (2013) Congenital proprotein convertase 1/3 deficiency causes malabsorptive diarrhea and other endocrinopathies in a pediatric cohort. *Gastroenterology*, **145**, 138–148.

Mbikay, M., Seidah, N.G. and Chretien, M. (2001) Neuroendocrine secretory protein 7B2: structure, expression and functions. *Biochemical Journal*, **357**, 329–342.

McPherron, A.C., Lawler, A.M. and Lee, S.J. (1999) Regulation of anterior/posterior patterning of the axial skeleton by growth/differentiation factor 11. *Nature Genetics*, **22**, 260–264.

Mouchantaf, R., Watt, H.L., Sulea, T., *et al.* (2004) Prosomatostatin is proteolytically processed at the amino terminal segment by subtilase SKI-1. *Regulatory Peptides*, **120**, 133–140.

Nyathi, Y., Wilkinson, B.M. and Pool, M.R. (2013) Co-translational targeting and translocation of proteins to the endoplasmic reticulum. *Biochimica et Biophysica Acta*, **1833**, 2392–2402.

Pan, H., Nanno, D., Che, F.Y., et al. (2005) Neuropeptide processing profile in mice lacking prohormone convertase-1. *Biochemistry*, **44**, 4939–4948.

Pan, H., Che, F.Y., Peng, B., Steiner, D.F., Pintar, J.E. and Fricker, L.D. (2006) The role of prohormone convertase-2 in hypothalamic neuropeptide processing: a quantitative neuropeptidomic study. *Journal of Neurochemistry*, **98**, 1763–1777.

Papaleo, F., Silverman, J.L., Aney, J., *et al.* (2011) Working memory deficits, increased anxiety-like traits, and seizure susceptibility in BDNF overexpressing mice. *Learning Memory*, **18**, 534–544.

Patra, D., Xing, X., Davies, S., *et al.* (2007) Site-1 protease is essential for endochondral bone formation in mice. *Journal of Cell Biology*, **179**, 687–700.

Poirier, S., Prat, A., Marcinkiewicz, E., *et al.* (2006) Implication of the proprotein convertase NARC-1/PCSK9 in the development of the nervous system. *Journal of Neurochemistry*, **98**, 838–850.

Posner, S.F., Vaslet, C.A., Jurofcik, M., Lee, A., Seidah, N.G. and Nillni, E.A. (2004) Stepwise posttranslational processing of progrowth hormone-releasing hormone (proGHRH) polypeptide by furin and PC1. *Endocrine*, **23**, 199–213.

Puente, X.S., Sanchez, L.M., Overall, C.M. and Lopez-Otin, C. (2003) Human and mouse proteases: a comparative genomic approach. *Nature Reviews. Genetics*, **4**, 544–558.

Rawson, R.B., Cheng, D., Brown, M.S. and Goldstein, J.L. (1998) Isolation of cholesterol-requiring mutant Chinese hamster ovary cells with defects in cleavage of sterol regulatory element-binding proteins at site 1. *Journal of Biological Chemistry*, **273**, 28261–28269.

Refaie, S., Gagnon, S., Gagnon, H., *et al.* (2012) Disruption of proprotein convertase 1/3 (PC1/3) expression in mice causes innate immune defects and uncontrolled cytokine secretion. *Journal of Biological Chemistry*, **287**, 14703–14717.

Revest, J.M., Dupret, D., Koehl, M., *et al.* (2009) Adult hippocampal neurogenesis is involved in anxiety-related behaviors. *Molecular Psychiatry*, **14**, 959–967.

Roebroek, A.J., Umans, L., Pauli, I.G., *et al.* (1998) Failure of ventral closure and axial rotation in embryos lacking the proprotein convertase Furin. *Development*, **125**, 4863–4876. [*This is the first evidence that furin is an essential gene.*]

Roebroek, A.J., Taylor, N.A., Louagie, E., *et al.* (2004) Limited redundancy of the proprotein convertase furin in mouse liver. *Journal of Biological Chemistry*, **279**, 53442–53450.

Rousselet, E., Benjannet, S., Hamelin, J., Canuel, M. and Seidah, N.G. (2011a) The proprotein convertase PC7: unique zymogen activation and trafficking pathways. *Journal of Biological Chemistry*, **286**, 2728–2738.

Rousselet, E., Benjannet, S., Marcinkiewicz, E., Asselin, M.C., Lazure, C. and Seidah, N.G. (2011b) The proprotein convertase PC7 enhances the activation of the EGF receptor pathway through processing of the EGF precursor. *Journal of Biological Chemistry*, **286**, 9185–9195.

Rousselet, E., Marcinkiewicz, J., Kriz, J., *et al.* (2011c) PCSK9 reduces the protein levels of the LDL receptor in mouse brain during development and after ischemic stroke. *Journal of Lipid Research*, **52**, 1383–1391.

Scerri, T.S., Brandler, W.M., Paracchini, S., *et al.* (2011) PCSK6 is associated with handedness in individuals with dyslexia. *Human Molecular Genetics*, **20**, 608–614.

Schjoldager, K.T. and Clausen, H. (2012) Site-specific protein O-glycosylation modulates proprotein processing – deciphering specific functions of the large polypeptide GalNAc-transferase gene family. *Biochimica et Biophysica Acta*, **1820**, 2079–2094.

Schnegelsberg, B., Seidah, N.G. and Schachner, M. (2014) Processing and sorting of the NCAM-like proteins L1 and CHL1. *Journal of Biological Chemistry* [submitted].

Seidah, N.G. (2013) Proprotein convertase subtilisin kexin 9 (PCSK9) inhibitors in the treatment of hypercholesterolemia and other pathologies. *Current Pharmaceutical Design*, **19**, 3161–3172.

Seidah, N.G. and Prat, A. (2007) The proprotein convertases are potential targets in the treatment of dyslipidemia. *Journal of Molecular Medicine*, **85**, 685–696.

Seidah, N.G. and Prat, A. (2012) The biology and therapeutic targeting of the proprotein convertases. *Nature Reviews. Drug Discovery*, **11**, 367–383.

Seidah, N.G., Marcinkiewicz, M. Benjannet, S., *et al.* (1991) Cloning and primary sequence of a mouse candidate prohormone convertase PC1 homologous to PC2, Furin, and Kex2: distinct chromosomal localization and messenger RNA distribution in brain and pituitary compared to PC2. *Molecular Endocrinology*, **5**, 111–122.

Seidah, N.G., Benjannet, S., Pareek, S., Chretien, M. and Murphy, R.A. (1996a) Cellular processing of the neurotrophin precursors of NT3 and BDNF by the mammalian proprotein convertases. *FEBS Letters*, **379**, 247–250.

Seidah, N.G., Hamelin, J., Mamarbachi, M., *et al.* (1996b) cDNA structure, tissue distribution, and chromosomal localization of rat PC7, a novel mammalian proprotein convertase closest to yeast kexin-like proteinases. *Proceedings of the National Academy of Sciences of the United States of America*, **93**, 3388–3393. [*First identification and characterization of PC7.*]

Seidah, N.G., Mowla, S.J., Hamelin, J., *et al.* (1999) Mammalian subtilisin/kexin isozyme SKI-1: a widely expressed proprotein convertase with a unique cleavage specificity and cellular localization. *Proceedings of the National Academy of Sciences of the United States of America*, **96**, 1321–1326. [*The first cloning and characterization of SKI-1/S1P.*]

Seidah, N.G., Benjannet, S., Wickham, L., *et al.* (2003) The secretory proprotein convertase neural apoptosis-regulated convertase 1 (NARC-1): liver regeneration and neuronal differentiation. *Proceedings of the National Academy of Sciences of the United States of America*, **100**, 928–933. [*Original identification and cloning of PCSK9.*]

Seidah, N.G., Mayer, G., Zaid, A., *et al.* (2008) The activation and physiological functions of the proprotein convertases. *International Journal of Biochemistry & Cell Biology*, **40**, 1111–1125.

Seidah, N.G., Sadr, M.S., Chretien, M. and Mbikay, M. (2013) The multifaceted proprotein convertases: their unique, redundant, complementary and opposite functions. *Journal of Biological Chemistry*, **288**, 21473–21481.

Seidah, N.G., Awan, Z., Chretien, M. and Mbikay, M. (2014) PCSK9: a key modulator of cardiovascular health. *Circulation Research*, **114**, 1022–1036.

Senturker, S., Thomas, J.T., Mateshaytis, J. and Moos, M. Jr. (2012) A homolog of subtilisin-like proprotein convertase 7 is essential to anterior neural development in Xenopus. *PLoS One*, **7**, e39380.

Stein, E.A. (2013) Low-density lipoprotein cholesterol reduction by inhibition of PCSK9. *Current Opinion in Lipidology*, **24**, 510–517.

Stein, E.A. and Raal, F. (2014) Reduction of low-density lipoprotein cholesterol by monoclonal antibody inhibition of PCSK9. *Annual Review of Medicine*, **65**, 417–431.

Steiner, D.F. (1998) The proprotein convertases. *Current Opinion in Chemical Biology*, **2**, 31–39.

Szumska, D., Pieles, G., Essalmani, R., *et al.* (2008) VACTERL/caudal regression/Currarino syndrome-like malformations in mice with mutation in the proprotein convertase Pcsk5. *Genes & Development*, **22**, 1465–1477.

Tassew, N.G., Charish, J., Seidah, N.G. and Monnier, P.P. (2012) SKI-1 and Furin generate multiple RGMa fragments that regulate axonal growth. *Developmental Cell*, **22**, 391–402.

Turpeinen, H., Oksanen, A., Kivinen, V., *et al.* (2013) Proprotein convertase subtilisin/kexin type 7 (PCSK7) is essential for the zebrafish development and bioavailability of transforming growth factor beta1a (TGFbeta1a). *Journal of Biological Chemistry*, **288**, 36610–36623.

Urata, S., Yun, N., Pasquato, A., Paessler, S., Kunz, S. and de la Torre, J.C. (2011) Antiviral activity of a small-molecule inhibitor of arenavirus glycoprotein processing by the cellular site 1 protease. *Journal of Virology*, **85**, 795–803.

Wardman, J.H., Zhang, X., Gagnon, S., *et al.* (2010) Analysis of peptides in prohormone convertase 1/3 null mouse brain using quantitative peptidomics. *Journal of Neurochemistry*, **114**, 215–225.

Warner-Schmidt, J.L. and Duman, R.S. (2006) Hippocampal neurogenesis: opposing effects of stress and antidepressant treatment. *Hippocampus*, **16**, 239–249.

Wetsel, W.C., Rodriguiz, R.M., Guillemot, J., *et al.* (2013) Disruption of the expression of the proprotein convertase PC7 reduces BDNF production and affects learning and memory in mice. *Proceedings of the National Academy of Sciences of the United States of America*, **110**, 17362–17367. [**This is the first evidence that in mice PC7 regulates behavior in hippocampus and amygdala.**]

Wu, Q., Tang, Z.H., Peng, J., *et al.* (2014) The dual behavior of PCSK9 in the regulation of apoptosis is crucial in Alzheimer's disease progression (Review). *Biomedical Reports*, **2**, 167–171.

Yang, J., Goldstein, J.L., Hammer, R.E., Moon, Y.A., Brown, M.S. and Horton, J.D. (2001) Decreased lipid synthesis in livers of mice with disrupted Site-1 protease gene. *Proceedings of the National Academy of Sciences of the United States of America*, **98**, 13607–13612.

Ye, J., Rawson, R.B., Komuro, R., *et al.* (2000) ER stress induces cleavage of membrane-bound ATF6 by the same proteases that process SREBPs. *Molecular Cell*, **6**, 1355–1364.

Zaid, A., Roubtsova, A., Essalmani, R., *et al.* (2008) Proprotein convertase subtilisin/kexin type 9 (PCSK9): hepatocyte-specific low-density lipoprotein receptor degradation and critical role in mouse liver regeneration. *Hepatology*, **48**, 646–654.

Zhang, X., Pan, H., Peng, B., Steiner, D.F., Pintar, J.E. and Fricker, L.D. (2010) Neuropeptidomic analysis establishes a major role for prohormone convertase-2 in neuropeptide biosynthesis. *Journal of Neurochemistry*, **112**, 1168–1179.

Zhu, X., Orci, L., Carroll, R., Norrbom, C., Ravazzola, M. and Steiner, D.F. (2002a) Severe block in processing of proinsulin to insulin accompanied by elevation of des-64,65 proinsulin intermediates in islets of mice lacking prohormone convertase 1/3. *Proceedings of the National Academy of Sciences of the United States of America*, **99**, 10299–10304.

Zhu, X., Zhou, A., Dey, A., *et al.* (2002b) Disruption of PC1/3 expression in mice causes dwarfism and multiple neuroendocrine peptide processing defects. *Proceedings of the National Academy of Sciences of the United States of America*, **99**, 10293–10298. [**First in vivo link of PC1/3 to body growth and neuroendocrine functions.**]

CHAPTER 10

Neuropeptide Receptors

Stephen J. Lolait,[1] James A. Roper,[2] Georgina G.J. Hazell,[1] Yunfei Li,[3] Fiona J. Thomson,[4] and Anne-Marie O'Carroll[1]

[1] School of Clinical Sciences, University of Bristol, Bristol, UK
[2] School of Biochemistry, University of Bristol, Bristol, UK
[3] Department of Orthopaedics, General Hospital of Pingmeishenma Group, Pingdingshan, Henan, China
[4] Institute of Cancer Sciences, University of Glasgow, Glasgow, UK

10.1 Neuropeptides as signaling molecules

Neuropeptides are rich in diversity. Based on a strict definition of neuroepeptides as 'small proteinaceous substances produced and released by neurons through the regulated secretory route and acting on neural substrates' (Burbach, 2011), there are likely to be at least 100 neuropeptide precursor genes in the mammalian genome. These genes are subject to transcriptional regulation and are typically translated in neurons as biologically inactive, large molecules (preprohormones) that have a 20–25 amino acid signal peptide at the NH_2 terminus that is required for precursor entry into the regulated secretory pathway of the endoplasmic reticulum (ER)-Golgi apparatus. The precursor molecules also usually contain pairs of dibasic amino acids which are the targets for enzymatic cleavage by prohormone convertases to yield small polypeptide products (around 3–50 amino acids in length). The enzymatically generated peptides are often post-translationally modified by amidation, acetylation, glycosylation, and phosphorylation to alter the stability and activity of the final bioactive peptide. Once synthesized in the secretory pathway, neuropeptides are usually released in an active, stimulus (for example, elevated intracellular Ca^{2+})-dependent process (see Burbach, 2011). In contrast with the major small molecule, 'fast' neurotransmitters, such as acetylcholine, glutamate, and catecholamines with which they co-exist in many neurons, neuropeptides tend to be released slowly from dendrites, somata, and axon terminals commensurate with their storage and secretion from vesicles in the regulated secretory pathway.

The processing of some neuropeptide precursors is often tissue- or brain region-specific and can lead to one (as in the precursors for the antidiuretic hormone, the nonapeptide **arginine vasopressin [AVP]**, or the 'love' or 'cuddle' hormone oxytocin [OXT]) or multiple copies of the active neuropeptide, or to a number of different peptides with different biological functions. An example of the latter is cleavage of the pro-opiomelanocortin precursor. This undergoes extensive posttranslational processing via a series of enzymatic steps in a tissue-specific manner to generate a number of neuropeptides of varying size. These bioactive peptides include γ- and β-melanocyte stimulating hormone (MSH), the endogenous opioid β-endorphin (plays a vital role in pain control), and adrenocorticotropin hormone (ACTH: released from

Molecular Neuroendocrinology: From Genome to Physiology, First Edition. Edited by David Murphy and Harold Gainer.
© 2016 John Wiley & Sons, Ltd. Published 2016 by John Wiley & Sons, Ltd.
Companion website: www.wiley.com/go/murphy/neuroendocrinology

the anterior pituitary gland, it stimulates cortisol production from the adrenal gland) which can be further processed to α-MSH (active in skin pigmentation, inflammatory responses, adrenal steroidogenesis, adipocyte lipolysis, and appetite regulation). There are at least 130 known mature neuropeptides, including their truncated/posttranslationally modified forms, that activate **G protein-coupled receptors (GPCRs)** in neuroendocrine systems (see the International Union of Basic and Clinical Pharmacology/British Pharmacological Society, Guide to Pharmacology; www.guidetopharmacology.org/ and section 10.2).

10.1.1 Neuropeptides – neurohormones, neuromodulators or neurotransmitters?

Neuropeptides are critical mediators of cell-to-cell communication in neuroendocrine and other systems. Some neuropeptides that are synthesized predominantly in the brain, such as AVP, have central effects on behavior, act as neuroendocrine hormones at the level of the anterior pituitary, and are endocrine hormones in the periphery. With the advent of molecular biological techniques such as the polymerase chain reaction (PCR) and RNA-Seq, it is not surprising to find that many, if not most, neuropeptide genes are expressed in multiple brain regions and also outside the CNS. An issue of ongoing debate is whether neuropeptides act centrally mainly as neurohormones, neuromodulators or neurotransmitters. Neuropeptides can display differential synthesis and release from dendrites, somata, and axon terminals. They often appear to act over long distances from their site(s) of synthesis (the so-called 'neuropeptide-neuropeptide receptor mismatch'), supporting the concept of 'volume transmission' or non-synaptic dendritic release in a neurohormonal mode of action (for example, see Ludwig and Leng, 2006). It should be emphasized, however, that in a number of cases the extent of neuropeptide fiber projections (or indeed neuropeptide receptor expression – see section 10.3), as revealed by molecular biological methods (for example, tract tracing in green-fluorescent protein-neuropeptide/neuropeptide receptor transgenic animals), may be greater than that previously demonstrated by **immunohistochemistry** (IHC) with neuropeptide-specific antibodies alone. A recent example of this is the elegant demonstration of AVP fiber projections from the hypothalamic paraventricular nucleus (PVN) to the **CA2, hippocampus** region of the mouse dorsal hippocampus, previously considered devoid of AVP innervation (Cui *et al.*, 2013) – it is assumed that AVP is released from these CA2-projecting neurons and is biologically active.

The distribution of peptide-containing neuronal fibers and their cognate receptors (see section 10.3) can also overlap, thus supporting a local action of neuropeptides. There is a substantial body of literature showing that neuropeptides are produced in multiple brain regions to modulate their own release (in an autocrine or paracrine fashion), and/or the release of another neuropeptide(s) or classic neurotransmitter (for example, by a presynaptic action), and can target specific synapses. In addition, the release of some neuropeptides (for example, gonadotropin-releasing hormone [GnRH]) is phasic or rhythmic so it is possible that local levels can vary dynamically. Furthermore, plasticity in neuropeptide expression is well documented and may vary depending on gender or species and during development, or be caused by numerous physiological manipulations including stress during early life and adulthood (for example, increasing the expression of AVP in corticotropin-releasing factor [CRF] neurons in the parvocellular region of the hypothalamic PVN with the ensuing synergistic effects of AVP and CRF on pituitary ACTH release) and changes in hydromineral or energy homeostasis which could result in focal neuropeptide release in close proximity to their corresponding receptors (see section 10.2). Therefore, the mechanisms by which neuropeptides can influence neuronal activity are complex and likely involve their action(s) at multiple levels in a functional network (for review, see van den Pol, 2012).

10.1.2   Other peptides act like neuropeptides

The catalog of neuropeptides could be expanded to embrace other 'neuroactive' peptides. These appear to display a number of the distinctive features of classic neuropeptides, including derivation from larger precursors, and secretory and neuronal activity, but have not been traditionally regarded as neuropeptides. Some of these are shown in Box 10.1. Neuroactive peptides include:

- those that are synthesized mainly by non-neuronal cells but have more recently been found to be expressed in neuroendocrine neurons (for example, chemokines)
- peptides produced predominantly outside the brain (for example, in the gut or adipose tissue) but which are neuroactive (for example, adiponectins and various other adipokines)
- peptides such as prosaptide, neuronostatin, BigLEN, and R-spondins which have been proposed as ligands for a number of orphan receptors
- metabolites of mature neuropeptides (for example, GnRH-(1-5))
- endocrine hormones such as insulin and prolactin, and neurotropins such as brain-derived neurotropic factor that are centrally active in brain regions such as the hypothalamus
- a whole host of other peptides, including some that have only recently been identified (for example, augurin, nesfatin-1, neuroendocrine regulatory peptides [NERPS]) for which receptors have not been categorically identified.

Box 10.1   Neuropeptide receptors

Neuropeptide GPCR families

Angiotensin	Kisspeptin	*Parathyroid hormone*
Apelin	Melanin-concentrating hormone	Peptide P518
Bombesin	Melanocortin	Prokineticin
Bradykinin	Motilin	Prolactin-releasing peptide
Calcitonin	Neuromedin U	Relaxin
Cholecystokinin	Neuropeptide FF/neuropeptide AF	Somatostatin
Corticotropin-releasing factor	Neuropeptide S	Tachykinin
Endothelin	Neuropeptide W/neuropeptide B	Thyrotropin-releasing hormone
Galanin	Neuropeptide Y	Urotensin
Ghrelin	Neurotensin	Vasopressin/oxytocin
Glucagon	Opioid	*VIP/PACAP*
Gonadotropin-releasing hormone	Orexin	

Neuroactive peptide GPCRs

Chemokines*
Chemerin
Complement peptide
GPR37/GPR37L1 (proposed ligand for these closely related GPCRs is prosaptide, derived from prosaposin†)
GPR101/173 (both are reported to be activated by the GnRH metabolite, GnRH-(1-5))
GPR107 (putative ligand neuronostatin, derived from somatostatin preprohormone)
GPR146 (proposed ligand proinsulin C)
GPR171 (proposed ligand BigLEN, derived from proSAAS [Ser-Ala-Ala-Ser] gene)
LGR4 (proposed ligands R-spondin-1/3)

(Continued)

Box 10.1 (Continued)

Non-GPCR neuropeptide receptors

Natriuretic peptide
Sortilin (multifunctional single transmembrane domain protein; neurotensin NTS3 receptor)

Examples of other neuroactive peptide (non-GPCR) receptors

Adiponectin
Insulin
Prolactin
Neurotropins (e.g. brain-derived neurotropic factor)

Examples of neuroactive peptides with uncharacterized receptors

Augurin (derived from esophageal cancer-related gene 4 – ECRG4)
Cocaine- and amphetamine-regulated transcript (CART)
Cerebellin-1/2/3/4 (hexadecapeptides derived from the cerebellin 1 precursor protein)
Diazepam-binding inhibitor (DBI)
Granins (e.g. chromogranins, secretogranins)
Neuroendocrine regulatory peptides (NERPs)-1/2/3/4 (and possibly other peptides encoded by the gene for the neurosecretory protein VGF)
Neuron-derived neurotropic factor (NENF)
Nesfatin-1 (derived from nucleobindin2/NEFA)
NPQ/spexin (and possibly other peptides derived from the proNPQ/spexin precursor)
Salusins-α/β (derived from the torsion dystonia-related gene – TORC2A)
Vaspin (visceral adipose tissue-derived serpin [A12])
Visfatin (derived from pre-B-cell colony-enhancing factor 1)

There are 67 mammalian GPCR families (not including the chemosensory taste, vomeronasal, extra-eye opsins, and odorant GPCRs) listed in the IUPHAR/BPS Guide to Pharmacology (www.guidetopharmacology.org/), of which 35 are neuropeptide GPCR subfamilies, comprising approximately 80 individual GPCRs. The majority of neuropeptide GPCRs belong to the Class A (rhodopsin-like) family – Class B GPCRs (secretin receptor family) are in *italics*. All of the neuropeptide GPCRs listed are present or active in rat and/or mouse hypothalamic neuroendocrine neurons. Natriuretic peptide (e.g. atrial natriuretic peptide) receptors are termed catalytic receptors and have a single transmembrane domain and guanylyl cyclase activity. Sortilin is a multifunctional single transmembrane receptor that interacts with many proteins, including neurotensin. Some neuropeptides or neuroactive peptides (e.g. adipokines such as adiponectin, chemerin, vaspin) are predominantly synthesized in peripheral tissues and act in the brain. A list of neuroactive peptides and their precursors can be found in Burbach (2011) or at www.neuropeptides.nl. There is evidence that the action of some neuroactive peptides is receptor (sometimes G protein) mediated. PACAP, pituitary adenylate cyclase-activating peptide; VIP, vasoactive intestinal peptide.

*Some chemokines, peptides classically involved in central and peripheral inflammatory reactions, and their receptors (e.g. the chemokines CXCL12 and CX3CR1 and their receptors CXCR4 and CX3CL1, respectively) are expressed in and modulate the activity of neuroendocrine neurons such as the magnocellular cells of the PVN.

†Updates on possible GPCR deorphanization are given at: www.guidetopharmacology.org/.

We would expect that the list of peptides that are synthesized (note that the demonstration of cellular peptide expression by methods such as IHC or mass spectrometry is not definitive proof of synthesis) by, and activate, neurons will increase substantially with the application of proteome/peptidomic (see Chapters 7 and 8) and transcriptomic/gene mining/bioinformatic techniques. The latter approach has recently led to the identification of previously uncharacterized, peptide/hormone-encoding genes in the introns of other genes and in transcripts that were annotated in the DNA databases as non-protein-coding RNAs; for example, see National Center for Biotechnology Information Gene database entries for the genes encoding betatropin (www.ncbi.nlm.nih.gov/gene/55908) and APELA (or toddler; an apparent apelin receptor ligand in early development) (www.ncbi.nlm.nih.gov/gene/100506013). Neuroactive peptides can even be derived from molecules like hemoglobin (Gelman *et al.*, 2010).

The challenge is to identify the function of existing and novel neuroactive peptides. Established neuropeptides have fundamental, extensive central functions ranging from neural stem cell growth, proliferation and differentiation, and CNS neuroprotection and remodeling, to the regulation of neuroendocrine activity (for example, in the ACTH response to stress; the control of hydromineral homeostasis and appetite), cardiovascular function, thermoregulation, sleep/wake cycles, circadian rhythms, learning and memory, nociception, and social behavior. To subserve these roles, neuropeptides must bind to specific receptors which act as major biological control points of the cell.

10.2   Most neuropeptide receptors are G protein coupled

The vast majority of neuropeptide actions are mediated by GPCRs, otherwise known as 7-transmembrane (TM), serpentine or heptahelical receptors (see Box 10.1 and Figure 10.1), and these will be the focus of this chapter. In a few instances, neuropeptides or neuroactive peptides activate non-GPCRs – examples of these are the single-TM receptors for atrial natriuretic peptides and neurotensin. There are several hundred GPCRs. For example, in the rat genome, there are approximately 240 GPCRs with known extracellular signals such as photons, catecholamines, hormones, amino acids, lipid-derived mediators, proteases, and neuropeptides, and approximately 90 'orphans' for which the endogenous ligand is unknown. These numbers exclude chemosensory GPCRs such as the vomeronasal, taste, and very large odorant receptor families (see www.guidetopharmacology. org/). GPCRs are the targets for over 30% of all currently marketed drugs. They comprise the largest receptor superfamily in the mammalian genome and can be divided into five main family classes in mammals: Class A (rhodopsin-like), Class B (secretin-like), Class C (metabotropic glutamate-like), Adhesion, and Frizzled. There are a number of salient features shared by all GPCRs:

- they have an extracellular NH_2-terminus of variable length followed by 7-TM α-helices clustered in the form of a bundle, linked by three alternating intracellular domains (or loops) and three extracellular domains, and an intracellular COOH-terminus (see Figure 10.1)
- in the brain, they tend to be expressed at very low levels, often less than hundreds of receptors/cell, and are activated by lower ligand concentrations compared to those observed for receptors for major neurotransmitters such as ligand-gated ion channels
- as a result of activating G proteins, there is considerable downstream amplification of a number of sometimes overlapping intracellular signaling cascades (for an overview, see Hazell *et al.*, 2012).

Most neuropeptide GPCRs belong to the Class A family (see Box 10.1).

10.2.1   Structure of neuropeptide GPCRs

Following the cloning of rhodopsin, the prototypical GPCR, and subsequently the β2-adrenoceptor as the first hormone-binding GPCR and two of the acetylcholine muscarinic receptors, the first neuropeptide GPCR, the substance K receptor (now termed the tachykinin NK_2 receptor), was cloned in 1987 (Masu *et al.*, 1987). Before the end of the last millennium,

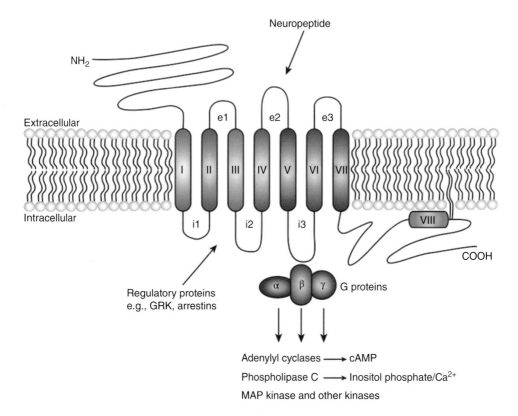

Figure 10.1 Simplified schematic of a prototypical GPCR embedded in the plasma membrane. Not shown is the α-helical coil arrangement of, or interaction between, the transmembrane (TM: numbered I–VII) domains in the basal or agonist-stimulated state. The architecture of GPCRs comprises an extracellular region including the NH_2-terminus and three extracellular (ec) domains, the seven TM helical bundles, and three intracellular (i) domains including the COOH-terminus. The NH_2- and COOH-termini can vary widely in length. The ec regions and especially the ec-TM interfaces and residues in the TM domains (and perhaps the NH_2-terminus) are responsible for ligand binding whereas the i regions and i-TM interfaces are involved in binding signaling molecules such as the Gα proteins, GPCR kinases (GRK), and arrestins. The predicted amphipathic VIII helix in the COOH-terminus present in most GPCRs is anchored to the membrane by palmitoylation and is required for GPCR expression and function. Upon ligand binding, there are conformational changes in the ec region and TM core that lead to structural rearrangements in the i-TM interface to facilitate interaction with intracellular effectors. GPCR activation involves GDP-GTP exchange on the Gα subunit of the G protein heteromer (α, β, γ), leading to downstream changes in intracellular levels of cyclic AMP (cAMP), inositol phosphates and Ca^{2+} ions, and a variety of enzymes such as mitogen-activated protein (MAP) kinase. There is considerable evidence that some GPCRs, including neuropeptide GPCRs, can activate intracellular cascades that do not involve G proteins. For a detailed diagram and description of the various pathways in the GPCR signaling network see: www.sabiosciences.com/pathway.php?sn=GPCR_Pathway. For details of GPCR structure see Katrich *et al.* (2012) and Venkatakrishnan *et al.* (2013).

the cDNAs and/or genes for most of the GPCRs for known neuropeptides had been isolated, principally by homology-based cloning methods such as those utilizing PCR, and aided by computer-based mining of a number of mammalian genomes. This enterprise was facilitated by the common structures of GPCRs, namely conserved 7-TM domains composed largely of hydrophobic amino acids, and high amino acid homology in these and other regions (for example, putative ligand-binding domains), especially between members of a particular GPCR subfamily.

Over the past few years, a number of exciting, high-resolution X-ray crystallography studies elucidating 26 GPCR structures (as of August 2014), that started with the β2-adrenoceptor, have reinforced many of our ideas about the structural features of GPCRs obtained from earlier experiments involving GPCR mutagenesis, domain swapping in chimeric GPCRs and the occasional computational modeling, in addition to revealing new nuances about GPCR ligand binding and activation (Katrich *et al.*, 2012; Venkatakrishnan *et al.*, 2013). The crystal structures have provided a snapshot of GPCRs in the agonist-bound state, lacking the bound G protein, the active state, representing the agonist-GPCR-G protein ternary complex, and the inactive (antagonist- or inverse agonist-bound) state. Of the GPCR crystal structures obtained to date, seven belong to neuropeptide GPCRs mostly in the inactive state: the Class A opioid receptors μ, δ, κ, nociceptin (NOP), neurotensin NT_1, and the Class B CRF_1 and glucagon receptors. The structures for three chemokine GPCRs have also been determined, one of which was solved by nuclear magnetic resonance (NMR) spectroscopy which, unlike X-ray crystallography, can reveal the GPCR structure in a cellular environment, and a number of other neuropeptide GPCRs are in the pipeline (for example, see http://gpcr.scripps.edu/).

We can draw some general conclusions on how and where ligands bind to GPCRs. The ligand-binding pockets are quite diverse, reflecting the different sizes, shapes, and chemical properties of various ligands, and the binding of different ligands (for example, in the case of the β2-adrenoceptor) appears to confer different conformational states (Katrich *et al.*, 2012; Venkatakrishnan *et al.*, 2013). Binding sites for orthosteric ligands, which directly interact with the primary endogenous ligand-binding site, but not allosteric compounds (invariably exogenous ligands that generally enhance or decrease the affinity and/or responsiveness of orthostatic ligands by binding to a different site from the endogenous ligand) appear to be well conserved within a GPCR family. Apart from the neurotensin NT_1 receptor, most of the ligand-contacting amino acid residues are present mainly in the TM helices which create a ligand-binding cradle that extends deep into the cavity formed by the TM bundle. Residues in the extracellular loops can also contribute to the binding pocket. For example, the NT_1 agonist-binding pocket is composed of residues in the NH_2-terminus, three extracellular domains, and six TM helices (TMs 2–7). The binding pocket of the Class B glucagon receptor is also large, with contributions by residues in the three extracellular loops and TM domains 1, 2, 3, 5, 6, and 7. In other neuropeptide GPCRs, such as the opioid receptors, the ligand-binding pocket is formed mainly from residues in TMs 2, 3, 6, and 7 and is exposed due to the β-hairpin loop structure created by the second extracellular loop, allowing ligands easy access to the primary binding pocket. The NH_2-terminus, which is often glycosylated, appears to contribute to the ligand-binding pocket of a number of neuropeptide GPCRs, particularly those which have extended NH_2-termini (for example, glucagon receptor), and may also be vital for GPCR trafficking and membrane expression.

The extracellular domains between GPCR families are highly diverse in secondary structure and disulfide cross-linking patterns – the latter occur between cysteine residues mainly at the top of TM 3 and in the second extracellular loop and are thought to stabilize GPCR structure. The intracellular domains are involved in binding G proteins and other effectors such as GPCR kinases (GRKs) and arrestins, which modulate GPCR activities such as sensitization, desensitization, and internalization. While there are no X-ray structures on neuropeptide GPCRs complexed to G proteins to date, the structure of the agonist-bound β2-adrenoceptor-$Gα_s$βγ complex has been determined. Here the receptor-$Gα_s$ protein interface is formed by the second intracellular loop and cytoplasmic faces of TMs 5 and 6, and there is no direct interaction with Gβγ subunits (Rasmussen *et al.*, 2011). The COOH-terminus is invariably palmitoylated (covalent attachment of fatty acids often to one or more cysteine residues) to facilitate interaction with the plasma membrane, and can often be phosphorylated, which is important in GPCR internalization and intracellular signaling.

10.2.2   Signaling of neuropeptide GPCRs

G protein-coupled receptors are coupled to G proteins that are composed of three subunits: Gα, Gβ, and Gγ. GPCR activation is both modular and dynamic. Upon agonist binding, there is a large-scale conformational shift in the arrangement of TM helices, the most pronounced of which is an outward swing of TM 6 on the intracellular side that is accompanied by movement in TM 5. In addition, there are also shifts in TMs 3 and 7. This conformational overhaul catalyzes the exchange of GDP for GTP on the Gα subunit followed by the functional dissociation of Gα into Gα-GTP and Gβγ subunits, both of which modulate the activity of a plethora of intracellular effectors. Gα's intrinsic GTPase activity hydrolyzes GTP to GDP, leading to reassociation of the Gα-GDP and Gβγ subunits, and signal termination. The Gα proteins are differentiated by their structure and function: the three main Gα proteins involved in neuropeptide signal transdcution are Gα$_s$ which activates adenylyl cyclases to catalyze the production of cAMP; Gα$_{q/11}$ which stimulates phospholipase C activity that raises the levels of intracellular effectors such as inositol triphosphate and Ca^{2+}, and can also activate the mitogen-activated protein kinase/extracellular signal-regulated kinase (MAPK/ERK) pathway (see Figure 10.2); and Gα$_{i/o}$ which activates a variety of phospholipases and phosphodiesterases and often inhibits adenylyl cyclase. In addition, Gβγ can act independently of Gα$_s$ to stimulate adenylyl cyclases, phospholipases, MAP kinases, protein tyrosine kinases, and ion channels (for details, see www.sabiosciences. com/pathway.php?sn=GPCR_Pathway).

The primary signaling pathways activated by most neuropeptide GPCRs are transduced by Gα$_{q/11}$ (see Supplementary Table 4 in Hazell *et al.*, 2012, and www.guidetopharmacology. org/). However, many GPCRs, including neuropeptide GPCRs, are 'promiscuous' and can couple to multiple Gα proteins, depending on factors such as:

- the number of receptors expressed per cell
- the cellular or subcellular distribution of receptor or signaling effectors in the CNS and/or peripheral tissues, for example, within specialized compartments of the plasma membrane such as endosomes or in intracellular organelles such as the nucleus or Golgi
- whether the GPCRs exist as monomers or oligomers (see section 10.4.2)
- the nature of the ligand, for example, different selective agonists, often termed 'biased' ligands, can confer functional selectivity by eliciting a different signaling responses from an individual GPCR.

There is extensive, multi-level regulation to modulate the amplitude, duration, and spatial aspects of GPCR signaling. Signal amplification is prominent whether it is mediated by the activation of overlapping intracellular signaling cascades by a single GPCR or by cross-talk between GPCRs. A classic example in neuroendocrinology is AVP **V1b receptor**-Gα$_{q/11}$-mediated enhancement of CRF$_1$ receptor-Gα$_s$-stimulated cAMP in anterior pituitary corticotropes. Signal amplification can also be mediated by transactivation of other cell surface proteins, including receptor tyrosine kinases such as the epidermal growth factor receptor to converge on MAPK/ERK activation.

A potentially important facet of GPCR signaling that should not be overlooked is that some GPCR ligands can activate G protein-independent, often arrestin-dependent pathways in addition to G protein-dependent pathways (another example of biased signaling). While this attribute of some neuropeptide GPCRs is likely based on the ability of different ligands to stabilize different active GPCR conformations (Venkatakrishnan *et al.*, 2013), its physiological relevance in terms of endogenous, native ligands is not known.

We have only touched here on a few aspects of signaling that are pertinent to neuropeptide GPCRs; it is apparent that GPCRs dynamically interact with an expansive array of proteins that includes activators and regulators of G protein signaling as part of a tightly regulated signaling network.

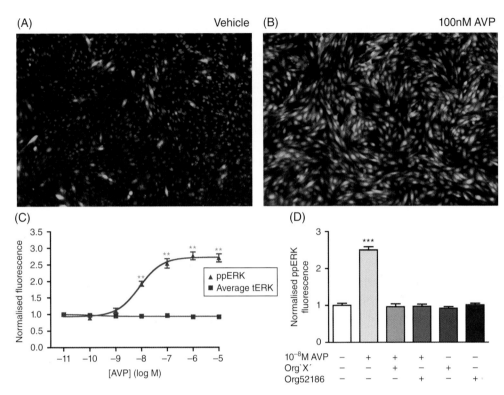

Figure 10.2 AVP-stimulated extracellular signal-regulated kinase (ERK) in cells expressing the AVP V1b receptor. Chinese hamster ovary (CHO) cells expressing the cloned rat V1b receptor were grown in 96-well plates and serum-starved (0.1% v/v fetal calf serum). After 10 minutes stimulation with 10^{-7}M AVP or vehicle, the cells were immunostained for dually phosphorylated (pp)ERK1/2 and total (t) ERK1/2. Cell images were obtained by the automated IN Cell Analyzer 1000 and analysis performed by IN Cell Analyzer Workstation 3.5 software (GE Healthcare) as in Caunt *et al.* (2006). The nuclear and cytoplasmic fluorescence intensities in individual cells were quantified (in arbitrary fluorescence units); data were normalized to vehicle control(s) after subtraction of 'no primary antibody' backgrounds. (A, B) Fused images of vehicle- or AVP-treated cells, respectively, acquired by the IN Cell Analyzer showing intracellular ppERK (*green*); the workstation can automatically demarcate nucleus from cytoplasm according to DAPI (*blue*) nuclear staining. (C) A dose–response curve for 10-minute ppERK stimulation by AVP (EC_{50} ppERK = 8.29×10^{-9}M) – there was no effect of AVP on tERK levels. (D) Preincubating the cells for 30 minutes with the V1b receptor antagonists Org'X' and Og52816 (1 µM) blocks AVP (10^{-7}M for 10 minutes) stimulated ppERK in V1b receptor-CHO cells. The V1b receptor antagonists (up to 1 µM) had no effect on AVP-induced ERK in CHO cells expressing the cloned rat V1a or V2, or OXT receptors (not shown). High-content screening studies of this type on ERK-coupled neuropeptide receptors are amenable to primary neuronal cultures (for example, from fetal or neonatal animals), and ERK IHC can be performed on *ex vivo* slices of different brain regions, or *in vivo* (for example, ERK expression pattern following icv- or intra-brain administration of neuropeptide receptor agonist). Similar approaches can be used for other intracellular signal transduction molecules such as components of the PI3K/Akt pathway. In (C) and (D), the data are mean ± SEM, N = 6 (approx. 1200 cells imaged from three wells of a 96-well plate; experiments repeated at least twice). ** P<0.01; *** P<0.001 vs vehicle-treated control (one-way ANOVA followed by Dunnett's *post hoc* test).

10.3   Neuropeptide receptor expression in the brain

There are approximately 330 non-chemosensory rat GPCRs, of which at least 25% respond to endogenous neuropeptides (www.guidetopharmacology.org/). The neuropeptide GPCR families (groups of GPCRs that respond to the same or similar agonists and usually show a higher degree of homology to other members of the group than to unrelated groups; see Box 10.1) range in size from 1 (for example, apelin, ghrelin, and neuropeptide S) to 4 or 5 (neuropeptide Y, opioid, and somatostatin) members. The cloning of neuropeptide receptor cDNAs/genes has been a significant boost to structure-function studies (for example, establishing pharmacological profiles of endogenous ligands and new drugs in cell lines engineered to express a cloned receptor; mutational and structural studies); it has also been an important aid in the development of most neuropeptide receptor antibodies (generated to protein sequences deduced from the cloned receptor mRNA) and obviously provided the platform for mapping receptor gene expression in the CNS.

10.3.1   Neuropeptide receptor numbers and anatomical localization

What is the total repertoire of neuropeptide and other GPCRs in the CNS? Based on estimates from **receptor autoradiography** (ARG), IHC, *in situ* **hybridization histochemistry** (ISHH), and transcriptomic (DNA arrays) studies, at least 80% of neuropeptide GPCRs are expressed in the neuroendocrine PVN and supraoptic nucleus (SON) (Hazell *et al.*, 2012). As discussed previously (Hazell *et al.*, 2012), this is likely to be an underestimate reflecting the sensitivity and limitations of the assays used. Receptor ARG using radiolabeled receptor ligands and IHC with fluorescent or chromogenically labelled receptor-specific antibodies are both powerful techniques for visualizing ligand-binding sites and immunoreactive receptor protein, respectively, in an anatomical context. IHC usually provides greater sensitivity and resolution but critically, both methods require important controls to ensure specificity, and selective ligands/antibodies to bind to the receptor target. The importance of antibody integrity cannot be overstated since many GPCR antibodies are not considered specific (Hazell *et al.*, 2012). If the appropriate GPCR antibody can be produced or sourced, it can be used in combination with antibodies to other proteins (for example, a neuropeptide itself or neuronal marker) to perform multicolored IHC to define the neuronal cell phenotype. The need for a GPCR antibody can be obviated where reporter- or epitope-tagged GPCR-transgenic mice have been developed and validated.

Two such examples are Cre recombinase-GPCR-reporter lines and the bacterial artificial chromosome (BAC)-enhanced green fluorescent protein (EGFP) transgenic mice developed by the Gene Expression Nervous System Atlas (GENSAT; www.gensat.org) project and elsewhere. In the latter model, EGFP expression is targeted to cells expressing functional neuropeptides and/or their receptors. Other methods that reflect neuropeptide functionality *in vivo* are increasingly being employed, for example functional magnetic resonance imaging (fMRI) and blood oxygenation level-dependent (BOLD) imaging which allows non-invasive, repeated, real-time assessments across multiple brain regions in conscious animals (Febo and Ferris, 2014).

As previously mentioned, most neuropeptide receptors are not abundantly expressed at either the mRNA or protein level in the CNS (Hazell *et al.*, 2012). In addition, the number of neuropeptide receptor mRNAs and proteins can vary dramatically across the brain, within a specific brain region, and even between individual cells (Bartfai *et al.*, 2012; Hazell *et al.*, 2012). ISHH on tissue sections with high specific-activity ^{35}S-labeled DNA or RNA probes (used in multiplex formats and/or with signal amplification methods) can be used to map neuropeptide GPCR gene expression in the brain. This technique can detect as few as

5–10 mRNA copies/cell at microscopic resolution that may take weeks or months of exposure against X-ray film or photographic emulsion; perhaps even single RNA molecules can be detected within a day with fluorescent or chromogenic detection (such as RNAscope; see www.acdbio.com/). ISHH is thus sensitive enough to visualize most of the rarest neuropeptide receptor mRNAs. Like all methods used to detect proteins and mRNAs, specificity is a key issue, and it is critical to establish that a neuropeptide receptor probe only binds to a single receptor isoform and does not cross-react with other genes. This is possible because even where there is high nucleotide identity between closely related neuropeptide GPCR families, subtype specificity can be enhanced by using probes targeting 5′ or 3′ untranslated (UTR) regions. It is also possible to map alternatively spliced receptor isoforms (see section 10.4.1) with the judicious choice of probes, and, like IHC, with which it can be combined, ISHH is amenable to co-localization studies to detect multiple, distinct mRNA transcripts. For example, numerous neuropeptide GPCRs have been detected in AVP-, OXT-, and CRF-expressing neurons in the PVN and SON, or in other neuropeptide-expressing cells (see Hazell *et al.*, 2012, for details) and these neuroanatomical findings can guide further studies on the function of the neuropeptide and its receptor, and possibly the neuropeptide and/or neuro-transmitter it modulates.

Examples of neuropeptide receptor ARG and ISHH are shown in Figures 10.3 and 10.4. AVP V1b receptor binding sites are prominent in one tissue, the anterior pituitary, where the V1b receptor has been pharmacologically characterized and its function established, and are striking in another tissue, the hippocampal CA2 region, where V1b receptor mRNA is also detected by ISHH (see Figure 10.3) and where this receptor appears to be critical for social aggression (Pagani *et al.*, 2014). In fact, the V1b receptor appears to be one of the most selective CA2 markers (for example, see the Allen Brain Atlas, http://brain-map.org, for gene expression mapping in the mouse brain). The CA2 is a region of the brain that hitherto has been underexplored but very recently has been shown to be part of a new trisynaptic pathway (dentate gyrus→CA2→CA1) in the hippocampus (Kohara *et al.*, 2014) and is essential for social memory (Hitti and Siegelbaum, 2014). In addition to employing the usual controls for receptor ARG (for example, binding is blocked by the unlabeled ligand) and ISHH (for example, absence of labeling with the sense RNA probe), we have verified that there is no labeling in the brains and pituitaries of V1b receptor knockout mice, an important negative control (see Figure 10.3). While the regional anatomical distribution of V1b receptor binding sites and mRNA overlaps, this particular example highlights how a site of neuropeptide GPCR synthesis (V1b receptor mRNA expression in CA2 pyramidal cell bodies) can clearly differ from where neuropeptide GPCR protein is expressed (V1b receptor protein in fibers projecting from the pyramidal cells to other hippocampal strata). It also accentuates the differences in sensitivity in methods used to detect neuropeptide (or any protein) gene expression, since V1b receptor mRNA is detected in virtually all brain regions by reverse-transcription PCR (Roper *et al.*, 2011).

Another example of a high correlation between neuropeptide GPCR gene and protein expression from our laboratories is shown in Figure 10.4. In this case, receptor ARG and ISHH for the neuropeptide **apelin** GPCR (**APJ**) reveals a strong overlap between APJ binding sites and APJ gene (Aplnr) expression in the hypothalamus (where apelin modulates AVP and CRF activity – see Hazell *et al.*, 2012) and thalamus. APJ was one of a number of orphan GPCRs, including the opioid NOP, orexin OX_1, prolactin-releasing peptide, and ghrelin receptors, that was deorphanized by reverse pharmacology, which entails expressing orphan receptors in cell lines as baits to screen tissue homogenates for new neuropeptides (for review, see Civelli *et al.*, 2013).

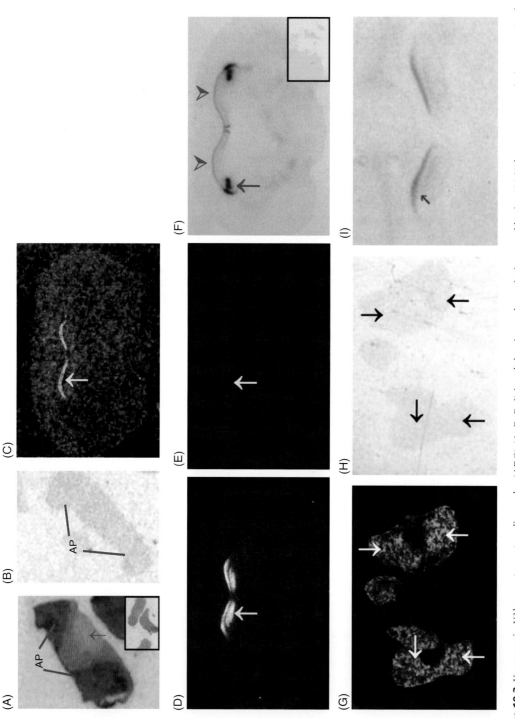

Figure 10.3 Vasopressin V1b receptor autoradiography (ARG) (A, B, D–I) in adult mice and rat pituitary and brain. (A) V1b receptor protein is present in the anterior (AP) but not intermediate and posterior (*red arrow*) lobes of the pituitary from V1b receptor wild-type (WT) mice. (B) Binding is absent from the

10.3.2 Expression profiling of the neuropeptide GPCRome

More accurate estimates of the number of GPCRs genes expressed in a tissue (GPCRome) have been provided by transcriptomic profiling methods, the two most widely used being reverse-transcription PCR (often quantitative real-time PCR; Regard *et al.*, 2008) and **DNA microarrays** (see www.genome.gov/10000533 for a brief overview). Both these methods rely on previous knowledge of an mRNA sequence and have revealed that tissues and brain regions express over 100 GPCRs (Hazell *et al.*, 2012). We have previously noted that micro-array studies, in which cDNA from reverse-transcribed cell or tissue mRNA is hybridized to 'chips' containing thousands of probes to the genes of a large proportion of the genome, are useful in identifying the expression of orphan GPCRs (Hazell *et al.*, 2012). In the past few years, high-throughput sequencing of single cells has begun to more accurately reveal the total number of mRNA transcripts expressed by neurons. These studies have shown that the vast majority of the transcriptome, including many cell surface and intracellular signaling molecules, is expressed at <50 copies/cell, with highly variable expression between neurons (Bartfai *et al.*, 2012). For example, recent RNA-Seq (Illumina) studies on single neurons from the preoptic area of the mouse anterior hypothalamus have detected around 6100 expressed mRNAs (that is, mRNA for protein-coding genes), of which up to 8% encode generic receptors and a further 1% encode ion channels (Bartfai *et al.*, 2012). In these neurons, some receptors expressed at very low levels (<10 copies/cell) were functional as assessed by electrophysiological responses. Further **single cell transcriptomics** of yellow fluorescent protein-tagged 5-HT dorsal raphe neurons (Spaethling *et al.*, 2014) identified

---

pituitary of V1b receptor knockout mice. (C) shows *in situ* hybridization histochemistry (ISHH) of a coronal brain section from a V1b receptor WT mouse with a radiolabeled V1b receptor (antisense) riboprobe – arrow indicates labeling of the pyramidal layer of the far-rostral dorsal hippocampal CA2. (D) V1b receptor protein is present in the oriens, radiatum, and lacunosum moleculare strata of the CA2 of a WT mouse brain in a similar region as in (C) and absent in a V1b receptor KO mouse brain at the same level (E). The images in (C–E) are pseudo-colored (labeling in yellow-red and red-orange for C, D, E, respectively). (F) Intense labeling of the more caudal dorsal CA2 (*arrow*) with weaker labeling in the oriens of the CA1 (*red arrowheads*). All ARG labeling was completely displaced by co-incubation with excess cold ligand or the V1b receptor antagonist Org52186 (see bottom right insets in A and F). In the Sprague Dawley rat, intense V1b receptor protein is present in the anterior pituitary (pseudo-colored red-orange labeling arrowed in G); non-specific binding in the presence of excess Org52186 is shown in H. V1b receptor protein in the dorsal CA2 (*arrowed*) of the rat hippocampus is shown in I. While the neuroendocrine role of the pituitary V1b receptor is well established, it is likely that the CA2 V1b receptor is the anatomical substrate for behaviors such as aggression that are reduced in V1b receptor KO animals (Pagani *et al.*, 2014; Roper *et al.*, 2011). Interestingly, it has been thought that the dorsal hippocampus lacks vasopressinergic innervation, suggesting that AVP may reach the CA2 via 'volume transmission' or non-synaptic diffusion, perhaps from AVP axons and/or PVN/SON neuron dendrites that release large amounts of peptide into the extracellular space. However, recent studies have shown that AVP PVN neurons may project directly to the CA2 (Cui *et al.*, 2013). For V1b receptor ARG 20 μm sections of brains and pituitaries from adult (8–10 weeks) V1b receptor knockout and corresponding WT mice and adult (~200 g) Sprague Dawley rats were fixed in 0.1% (w/v) paraformaldehyde and preincubated in Hepes buffer at room temperature (RT). Sections were then incubated for 2 hours with 0.5–5 nM iodinated V1b receptor antagonist (Org 'X') ± 1000-fold excess unlabeled Org 'X,' Org52816 (Roper *et al.*, 2011) or AVP at RT and washed in Hepes buffer at 4°C. Following a brief dip in ice-cold H_2O, slides were dried in a cold stream of air and then opposed to X-ray film (Hyperfilm MP, Amersham) for 2–10 days. For ISHH, a ^{35}S-UTP-labeled 518bp mouse V1b receptor riboprobe (DNA template corresponding to bp112-629 of GenBank Accession number NM_011924 (extending from the 5'-untranslated region to the initiating methionine) obtained by PCR from a 129 Sv genomic V1b receptor clone (Acc#AF152533S1)) was hybridized to 12 μm sections as described at http://www.wsyacy.com/SNGE/Protocols/ISHH/ISHH.html. Sections were opposed to X-ray film for 5 weeks. Sections hybridized with the corresponding sense probe did not give any labeling above background (not shown). This work was presented in part at the World Congress on Neurohypophysial Hormones in Kitakyushu, Japan (2009), and Bristol, UK (2013).

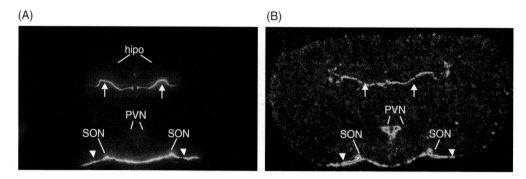

Figure 10.4 Representative receptor ARG (A) and *in situ* hybridization histochemistry (ISHH) (B) photomicrographs of apelin receptor (Aplnr) expression in the brains of adult male Sprague Dawley rats. Apelin binding sites (pseudocolored red-orange in A) correspond well with Aplnr mRNA expression (pseudocolored yellow-red in B) in the PVN (weakly labeled in A), SON which is obscured by the intense labeling of the basal (free) surface of the hypothalamic diencephalon (*arrowheads*), and the dorsal surface of the thalamus (*arrows*). The hippocampus (hipo) is very weakly labeled in A and B. Apelin receptor ARG was performed with [125]I-(pyr1)apelin-13 on 20 μm sections as previously described (Pope *et al.*, 2012). For ISHH, an Aplnr riboprobe (bp369-942 of Genbank Accession number AF184883), extending from transmembranes 3–7 of Aplnr as previously described in O'Carroll *et al.* (2000), was used. The integrity of the riboprobe was verified by DNA sequencing. ISHH on 12 μm sections with ^{35}S-labeled riboprobes was performed as at http://www.wsyacy.com/SNGE/Protocols/ISHH/ISHH.html. Sections hybridized with the corresponding sense probes gave negligible background staining (not shown). Sections were opposed to X-ray film (Hyperfilm MP, Amersham) for 3–5 weeks.

hundreds of GPCR transcripts, including some that were previously unreported on these types of neurons. Functional verification of all the GPCRs and other receptors identified in single neurons awaits further studies, but it is likely that the total neuropeptide receptor repertoire will be greater than 100.

10.3.3 The location of neuropeptide receptors can make a difference

The regulation, desensitization, and function of neuropeptide receptors can differ depending on the cellular context, such as neuronal cell bodies, dendrites or axon terminals. Even in the plasma membrane, the receptors can be located in specialized microdomains (for example, lipid 'rafts' and caveolae) that can compartmentalize receptor signaling. For neuropeptide GPCRs like the opioid receptors, the nature of the agonist dictates whether the receptors translocate from lipid-containing rafts to non-rafts and whether they activate $G\alpha_i$- or arrestin-dependent (G protein-independent) signaling (functional selectivity; for example, see Gonzáles-Maeso and Sealfon, 2012). Although the precise intracellular localization of GPCRs may be somewhat distorted by the questionable specificity of some GPCR antibodies used in IHC, it is clear from functional studies that many neuropeptide GPCRs are not only found clustered on pre- and postsynaptic membranes, or in internalized vesicles (for example, for recycling and/or degradation), but in some instances are also located in organelles such as the nucleus where they may have novel roles such as regulating histone modifications (Re *et al.*, 2010).

Neuropeptide receptors can also operate at multiple levels within a neuroendocrine system, for example, within the hypothalamic-pituitary-adrenal axis or neurohypophysial system at the level of the PVN/SON cell bodies and dendrites, and axon terminals in the median eminence or posterior pituitary, where the receptors may influence neuropeptide secretion. In addition, it is well established that the expression of neuropeptide GPCRs, like neuropeptides themselves, can deviate between cells, and be species and gender dependent, developmentally regulated and altered by a variety of physiological manipulations (for example, adrenalectomy, dehydration; Hazell *et al.*, 2012). A remarkable example of the importance of neuropeptide GPCR localization will be familiar to many. The montane and prairie voles differ both in the central distribution of the AVP V1a receptor (for example, higher in montane than prairie vole septum; higher in prairie than montane vole ventral pallidum) and their behavioral responses to AVP (elevated affiliative behavior in prairie voles). The montane and prairie voles are described as socially 'polygamous' and 'monogamous,' respectively. Young and co-workers showed that the AVP V1a receptor distribution is dependent, at least in part, on the type of DNA microsatellites present in the V1a receptor gene promoter, and that transgenic insertion of the prairie vole (monogamous) V1a receptor gene into the house mouse (polygamous) genome leads to a prairie-like V1a receptor expression and altered social behavior (for example, increased affiliative responses to AVP; Donaldson and Young, 2008). These revelations have provided the impetus for many studies on the role of AVP (and its V1a receptor) in normal and aberrant human behavior.

10.4 Functional diversity of neuropeptide receptors

It is perhaps difficult to imagine that all of the potentially hundreds of neuropeptide receptors that are expressed in a single neuron are functionally active, or have important roles in regulating the activity of neurons or supporting cells like astrocytes. However, functional studies on normal or physiologically or experimentally manipulated animals suggest that most if not all neuropeptide GPCRs have individual roles in regulating neuroendocrine activity. Such studies include measuring the levels of intracellular signaling molecules (for example, Ca^{2+} or ERK) and early genes (for example, c-*fos*) as indices of neuronal activation, neuropeptides and/or their receptors in cells, and neurotransmitters/neuropeptides released from dendrites and/or axon terminals following neuropeptide receptor agonist administration *in vivo* or *ex vivo*. Neuronal excitability which frequently shows electrophysiological heterogeneity (especially in the large magnocellular neuroendocrine neurons of the PVN and SON) has often been investigated, as has behavior and parameters associated with the response to stress (for example, plasma ACTH levels), reproduction, temperature regulation, and water, energy, and cardiovascular homeostasis (Hazell *et al.*, 2012). There are numerous ways in which neuropeptide receptors can potentially be modified at the transcriptional and translational levels to modify their function, and the spatial distribution of these receptors and their associated signaling molecules may be functionally relevant.

10.4.1 Regulation of neuropeptide receptor gene expression

Neuropeptide receptor genes are subject to transcriptional control in the same way as any other gene. For example, transcriptional rates are often regulated by steroids such as glucocorticoids and estrogen, and potential sites for specificity protein 1 (Sp1), activator protein-1 (AP1), and cAMP regulation via promoter/enhancer elements in the 5'-flanking region are not uncommon. The genes for some neuropeptide GPCRs such as the AVP V1b receptor are unique in that they also have prominent DNA microsatellites (dinucleotide CA and CT repeats in the mouse and rat, but not human V1b receptor genes) in the upstream promoter that are involved in transcriptional control. In the case of the mouse V1b receptor, these microsatellites are severely truncated in C57BL mice but not in other mouse strains (Roper *et al.*, 2011). It is not known whether microsatellite length has an impact on V1b receptor expression *in vivo* in a comparable fashion to that observed for variations in microsatellite composition on rodent AVP V1a receptor expression (Donaldson and Young, 2008). It is becoming increasingly clear, however, that DNA methylation and histone modifications contribute to the epigenetic control of gene expression for some neuropeptides (for example, AVP) and neuropeptide receptors (for example, μ opioid receptor; Hazell *et al.*, 2012).

Mutations in neuropeptide receptor genes are relatively rare and can lead to either loss of function or, less frequently, receptor activation (for example, increased agonist sensitivity/ numbers of receptors on cell surface; decreased desensitization) as in the case of kisspeptin receptor gene (KISS1R) mutations and central hypogonadotropic hypogonadism and central precocious puberty, respectively (Vassart and Costagliola, 2011) (see GPCR Natural Variants database, http://nava.liacs.nl/; www.guidetopharmacology.org/). Neuropeptide receptor (and neuropeptide) genes are replete with other common (frequency >1%) genetic variations such as single nucleotide polymorphisms (SNPs; for example, see SNP database, www.ncbi.nlm.nih.gov/projects/SNP/MouseSNP.cgi to search for SNPs between mouse strains) that could also contribute to functional heterogeneity. In addition to potentially influencing disease risk and progression, SNPs in the receptor gene 5'- and 3'-UTR, intronic, and receptor-coding (non-synonymous SNPs) regions can also directly alter receptor expression and pharmacology (http://nava.liacs.nl/). Furthermore, these and other variations can furnish different neuropeptide GPCR configurations. About 50% of GPCRs have introns and display consensus acceptor/donor sites for **alternative splicing** that can alter receptor gene expression patterns and increase neuropeptide receptor functional diversity. For GPCRs, alternative splicing primarily recasts the COOH-terminus but it can also manifest highly truncated neuropeptide GPCRs, examples of which include the ghrelin, growth hormone-releasing hormone, μ and NOP opioid, tachykinin NK$_2$, neuropeptide Y$_1$, neurotensin NTS$_2$, and somatostatin SST$_5$ receptors. In many cases, these splice variants have deleted TM domains and act, at least *in vitro*, as dominant-negative mutants that mediate compromised signaling properties (Wise, 2012).

Two further post-transcriptional mechanisms that can influence neuropeptide receptor gene expression and regulation are non-coding RNAs and RNA editing. Regulatory elements present in the 3'-flanking region of proteins including neuropeptide receptors may play an important role in mRNA stability and present a target for non-coding RNAs such as microRNAs (Hazell *et al.*, 2012). MicroRNAs are short, single-stranded RNAs that can suppress target gene expression by binding to their complementary mRNA sequences, usually in introns or exons of the 3'-UTR, and have been shown to inhibit the expression of neuropeptide GPCRs such as the μ-opioid receptor (Hazell *et al.*, 2012). Some natural antisense transcripts may act as non-coding RNAs to regulate receptor gene expression. RNA editing (for example, adenosine to inosine conversion by the enzyme adenosine deaminase) is another post-transcriptional mechanism that has the potential to change codons, and modulate splice site choice and microRNA targets in neuropeptide receptors. There is at least one example of RNA editing of a neuropeptide GPCR: editing of the endothelin ET$_B$ receptor produces a novel splice variant that results in a COOH-terminally truncated receptor (Tanoue *et al.*, 2002). In a non-peptide GPCR, the serotonin 5HT2c receptor, RNA editing has profound functional implications where edited isoforms code the second intracellular loop differently, leading to a change in the efficiency by which this GPCR couples to G proteins (www.guidetopharmacology.org/). There are substantial data demonstrating that neuropeptide receptor gene expression, which often translates to changes in receptor protein levels, is regulated by or contributes to the plasticity of neuroendocrine systems (Hazell *et al.*, 2012).

10.4.2   Regulation of neuropeptide receptor proteins

Some neuropeptide or orphan GPCRs exhibit constitutive activity that may be signaling pathway dependent and result from receptor overexpression and/or changes in the receptor-encoding DNA or RNA (Hazell *et al.*, 2012). Neuropeptide receptors generally have extensive sites for potential post-translational modification, including glycosylation and phosphorylation that can alter the interactions between receptor-ligand(s) and receptor-intracellular signaling molecules. As mentioned in section 10.2, a key feature of GPCR signaling diversity is that these receptors appear to adopt multiple active states, i.e. there are likely ligand-specific GPCR conformations that can be stabilized by orthosteric ligands and allosteric modulators that differ in their capacity to activate specific signaling pathways (Katrich et al., 2012; Venkatakrishnan et al., 2013; Wootten et al., 2013a).

Many GPCRs, including those that are activated by neuropeptides, appear to exist as dimers or higher-order oligomers, and homo/heterodimerization of two (or occasionally the functional interaction between three) GPCRs has been demonstrated predominantly in in vitro experiments (Milligan, 2013; see the G protein-coupled receptor-oligomerization knowledge base project, http://data.gpcr-okb.org/gpcr-okb/). GPCR oligomerization may be mediated principally by interaction between TM domains and/or intracellular regions (Katrich et al., 2012; Milligan, 2013). It has proved difficult to detect in vivo, although 'heteromer-specific' monoclonal antibodies have been used to detect μ–δ opioid receptor oligomerization in brain sections, and OXT receptor dimerization has been demonstrated using fluorescently labeled OXT agonists in breast tissue where these receptors are highly expressed (Milligan, 2013). We know from the examples of GABA$_B$ and taste receptors (both requiring the interaction between two GPCRs for activity) that stable, functional GPCR oligomerization does occur in vivo, and that heterodimerization can alter the pharmacological profiles (for example, desensitization, G protein coupling, signaling pathways) of one or both members of the heterodimer (Milligan, 2013). With likely over 100 neuropeptide receptors co-expressed in individual neurons, the potential for heterodimerization is enormous, and in in vitro studies oligomerization is prevalent between GPCRs found in neuroendocrine systems (Kamal and Jockers, 2011). Even constitutively active orphan GPCRs may heterodimerize with neuropeptide GPCRs to alter the latter's function – an example of this is the orphan GPCR GPR50 that heterodimerizes with the melatonin MT$_1$ receptor to inhibit its activity (Hazell et al., 2012). In addition, neuropeptide receptors may also heterodimerize with other membrane signaling molecules such as ion channels (Hazell et al., 2012).

For alternatively spliced GPCRs, it has been proposed that truncated mutants act in a dominant-negative manner in a homodimer between mutant and normal GPCR (Wise, 2012). Moreover, it is possible for ligand binding to an orthosteric site on one monomer of a dimer to modulate the binding and/or function of an orthosteric ligand on the other monomer (Wootten et al., 2013a), that ligand activation increases dimer abundance, and/or that dimers act as ligand concentration-dependent switches where the two GPCRs in a heterodimer are activated independently by either low or high agonist concentrations. It should be noted, however, that while allosteric (mostly exogenous) modulators abound for GPCRs such as members of the adenosine, muscarinic, and metabotropic glutamate receptor families, there are relatively fewer of these compounds targeting neuropeptide GPCRs (Melancon et al., 2012). Interestingly, there are instances where the activity of neuropeptide metabolites, that may have altered affinity for the endogenous neuropeptide receptor, can be allosterically modulated (Wootten et al., 2013a). Neuropeptides and their metabolites may also have varying affinities for more than one GPCR. For example, AVP activates the three AVP (V1a, V1b, and V2) and OXT receptors with high affinity (Roper et al., 2011), while two orphan GPCRs (GPR101 and GPR173, both expressed in the hypothalamus; www.guidetopharmacology.org/) are reported to bind the GnRH metabolite GnRH-(1-5) (Cho-Clark et al., 2014).

Ultimately, neuropeptide receptor expression and function rely on a host of factors that influence the activity of the receptor on the cell surface. These include neuropeptide receptor interactions with a variety of signaling molecules such as receptor-modifying proteins, for example, the receptor activity-modifying proteins (RAMPs) that confer pharmacological specificity to adrenomedullin, amylin, and calcitonin gene-related peptide activity on a calcitonin receptor or calcitonin receptor-like receptor GPCR backbone, and also modulate the function of other Class B neuropeptide GPCRs (www.guidetopharmacology.org/) (Wootten *et al.*, 2013b). Neuropeptide receptors also interact with GRKs, the multifunctional arrestins, and other scaffold/scaffold-related proteins that enhance the specificity and/or efficiency of GPCR signaling, including changes in agonist affinity, and receptor turnover as a composite of receptor desensitization to modify signal quality and dampen neuropeptide receptor signaling (for example, following acute, high agonist doses or chronic activation of some receptors), redistribution, and degradation.

10.5 Perspectives

It is clear that neuropeptide receptors mediate diverse physiological and in some cases pathophysiological effects. We have witnessed the discovery of a number of new neuropeptides and neuroactive peptides, and an explosion of data on the number of neuropeptide receptors expressed within a given tissue or cell. For the majority of neuroendocrine systems, we still do not know the total repertoire of the neuropeptide GPCRs, by far the most common type of neuropeptide receptor, in single neurons within and between specific brain regions. A major question is whether neuropeptides and their receptors play a supporting role in orchestrating and/or fine-tuning cellular responsiveness or, given the hundreds of receptors, whether some (or most) are redundant. Some outstanding questions that will be addressed by future studies, and that may require more refined methods or the development of new techniques, include the following.

- Are all neuropeptide receptors in single neurons active? For the moment, the demonstration of functionality relies primarily on single cell electrophysiology. The increasing use of imaging techniques such as 2-photon microscopy will assist in uncovering the spectrum of intracellular signaling dynamics (for example, using Ca^{2+} sensors or genetically encoded fluorescent reporters) at a single cell level.
- Do neuropeptide receptors (particularly GPCRs) oligomerize *in vivo*, and does this alter their function? The development of bivalent or oligomer-biased ligands to detect and activate neuropeptide receptor heteromers would represent a considerable advance in uncovering their function.
- What are the receptor-mediated intracellular signaling cascades that are responsible for the short- and long-term effects of neuropeptides *in vivo*?
- What are the species differences in neuropeptide receptor structure, expression, regulation, and function?
- Are there gender differences in neuropeptide receptor function? Most studies are performed in males to avoid the possible confounding effects of the estrous cycle.
- What are the various hormonal and neurotransmitter inputs onto neuropeptide and/or neuropeptide receptor neurons? Which neuropeptide receptor pathways interact and how do they modulate classic neurotransmitter function? Taking advantage of recent developments in the generation of gene-specific promoter-driven Cre recombinase mouse lines, pharmacogenetics (for example, use of designer receptor exclusively activated by designer drug) and optogenetics will probably further our understanding of the functional plasticity of neuropeptides and their receptors in the circuitry of neuroendocrine pathways.
- What are the relative roles of axonally versus dendritically released neuropeptides in the local regulation of neuropeptide receptor activity?

- Can drugs targeting GPCRs meet unmet clinical need? Although there are increasing numbers of neuropeptide GPCR X-ray structures being elucidated, and the promise of NMR-based receptor structures in a native environment, computational GPCR modeling, docking, and mining of compounds to obtain new GPCR ligands with high affinity and selectivity (including perhaps allosteric compounds), we still need more studies in which neuropeptide receptor circuits have the potential to address an unmet medical need. Neuropeptide GPCRs are implicated in the pathophysiology of conditions such as obesity (for example, ghrelin, melanin-concentrating hormone, melanocortin, orexin, neuropeptide B/neuropeptide W), neurodegenerative disorders such as Alzheimer disease (for example, CRF_1, chemokine CXCR2, δ-opioid, PACAP), anxiety and/or depression (for example, CRFRs, OXT, AVP V1b, and many others), and narcolepsy (for example, orexin OX_2) (Heng *et al.*, 2013), and there are widespread neuropeptide GPCR haplotypes (usually SNPs) associated with many conditions. In terms of pharma, ongoing considerations include how tractable the neuropeptide receptor target is for drug development to treat humans (for example, behavioral disorders), and whether a feasible path towards proving clinical efficacy exists.

It is apparent that a multiple-level system of regulation exists to dynamically modulate neuropeptide receptor signaling. Understanding the functions of these receptors is key to unlocking the complexity of neuroendocrine communication.

Acknowledgments

SJL received funding from the Wellcome Trust (UK), Neuroendocrinology Charitable Trust (UK) and Merck; AMO'C received funding from the British Heart Foundation. SJL and JR collaborated with Merck in studies on AVP V1b receptor antagonists. The authors thank Mark Craighead (Redx Pharma, Liverpool, UK) for his contribution to research collaborations to explore neuropeptide GPCRs as drug targets, and Craig McArdle (University of Bristol) for assistance in imaging cellular ERK.

References

Bartfai, T., Buckley, P.T. and Eberwine, J. (2012) Drug targets: single-cell transcriptomics hastens unbiased discovery. *Trends in Pharmacological Sciences*, **33**, 9–16.

Burbach, J.P.H. (2011) What are neuropeptides? *Methods in Molecular Biology*, **789**, 1–36. [*Provides a comprehensive list of neuropeptides and 'neuroactive' peptides, and the gene structure of their precursors.*]

Caunt, C.J., Finch, A.R., Sedgley, K.R., Oakley, L., Luttrell, L.M. and McArdle, C.A. (2006) Arrestin-mediated ERK activation by gonadotropin-releasing hormone receptors: receptor-specific activation mechanisms and compartmentalization. *Journal of Biological Chemistry*, **281**, 2701–2710.

Cho-Clark, M., Larco, D.O., Semsarzadeh, N.N., Vasta, F., Mani, S.K. and Wu, T.J. (2014) GnRH-(1-5) trans-activates EGFR in Ishikawa human endothetrial cells via an orphan G protein-coupled receptor. *Endocrinology*, **28**, 80–98.

Civelli, O., Reinscheid, R.K., Zhang, Y., Wang, W., Fredriksson, R. and Schiöth, H.B. (2013) G protein-coupled receptor deorphanizations. *Annual Review of Pharmacology and Toxicology*, **53**, 127–146. [*The history of GPCR deorphanization is detailed, highlighting the methods used and difficulties encountered in the discovery of endogenous ligands for orphan GPCRs.*]

Cui, Z., Gerfen, C.R. and Young, W.S. III (2013) Hypothalamic and other connections with dorsal CA2 area of the mouse hippocampus. *Journal of Comparative Neurology*, **521**, 1844–1866.

Donaldson, Z.R. and Young, L.J. (2008) Oxytocin, vasopressin, and the neurogenetics of sociality. *Science*, **322**, 900–904.

Febo, M. and Ferris, C.F. (2014) Oxytocin and vasopressin modulation of the neural correlates of motivation and emotion: results from functional MRI studies in awake rats. *Brain Research*, **1580**, 8–21.

Gelman, J.S., Sironi, J., Castro, L.M., Ferro, E.S. and Fricker, L.D. (2010) Hemopressins and other hemoglobin-derived peptides in mouse brain: comparison between brain, blood, and heart peptidome and regulation in Cpe*lat/fat* mice. *Journal of Neurochemistry*, **113**, 871–880.

Gonzáles-Maeso, J. and Sealfon, S.C. (2012) Functional selectivity in GPCR heterocomplexes. *Mini Reviews in Medicinal Chemistry*, **12**, 851–855.

Hazell, G.G.J., Hindmarch, C.C., Pope, G.R., *et al.* (2012) G protein-coupled receptors in the hypothalamic paraventricular and supraoptic nuclei – serpentine gateways to neuroendocrine homeostasis. *Frontiers in Neuroendocrinology*, **33**, 45–66. [*Provides an overview of some methods used to detect GPCR expression, and the number, regulation, and function of GPCRs found in the paraventricular and supraoptic nuclei of the hypothalamus.*]

Heng, B.C., Aube, D. and Fussenegger, M. (2013) An overview of the diverse roles of G-protein coupled receptors (GPCRs) in the pathophysiology of various human diseases. *Biotechnology Advances*, **31**, 1676–1694.

Hitti, F.L. and Siegelbaum, S.A. (2014) The hippocampal CA2 region is essential for social memory. *Nature*, **508**, 88–92.

Kamal, M. and Jockers, R. (2011) Biological significance of GPCR heterodimerization in the neuro-endocrine system. *Frontiers in Endocrinology*, **2**, 2.

Katrich, V., Cherezov, V. and Stevens, R.C. (2013) Diversity and modularity of G protein-coupled receptor structures. *Trends in Pharmacological Sciences*, **33**, 17–27. [*Summary of the GPCR structures resolved by X-ray crystallography, and comparison of domains involved in ligand binding.*]

Kohara, K., Pignatelli, M., Rivest, A.J., *et al.* (2014) Cell type-specific genetic and optogenetic tools reveal hippocampal CA2 circuits. *Nature*, **17**, 269–279.

Ludwig, M. and Leng, G. (2006) Dendritic peptide release and peptide-dependent behaviours. *Nature Reviews. Neuroscience*, **7**, 126–136. [*Discussion and examples of the differential regulation, mechanisms, and functions of dendritic neuropeptide release.*]

Masu, Y., Nakayama, K., Tamaki, H., Harada, Y., Kuno, M. and Nakanishi, S. (1987) cDNA cloning of bovine substance-K receptor through oocyte expression system. *Nature*, **329**, 836–838. [*Describes the cloning of the substance K GPCR by functional expression, the first neuropeptide GPCR cloned, and one of only a handful of GPCRs that were cloned by this method which does not rely on previous structural information about the receptor.*]

Melancon, B.J., Hopkins, C.R., Wood, M.R., *et al.* (2012) Allosteric modulation of seven transmembrane spanning receptors: theory, practice, and opportunities for central nervous system drug discover. *Journal of Medicinal Chemistry*, **55**, 1445–1464.

Milligan, G. (2013) The prevalence, maintenance, and relevance of G protein-coupled receptor oligomerization. *Molecular Pharmacology*, **84**, 158–169. [*Discussion of GPCR oligomerization including comments on recent studies using different methods to demonstrate such interactions.*]

O'Carroll, A.M., Selby, T.L., Palkovits, M. and Lolait, S.J. (2000) Distribution of mRNA encoding B78/apj, the rat homologue of the human APJ receptor, and its endogenous ligand apelin in brain and peripheral tissues. *Biochimica et Biophysica Acta*, **1492**, 72–80.

Pagani, J.H., Zhao, M., Cui, Z., *et al.* (2014) Role of the vasopressin 1b receptor in rodent aggressive behavior and synaptic plasticity in hippocampal area CA2. *Molecular Psychiatry*, **20**, 490–499.

Pope, G.R., Roberts, E.M., Lolait, S.J. and O'Carroll, A.M. (2012) Central and peripheral apelin receptor distribution in the mouse: species differences with rat. *Peptides*, **33**, 139–148.

Rasmussen, S.G.F., DeVree, B.T., Zou, Y., *et al.* (2011) Crystal structure of the β2 adrenergic receptor-Gs protein complex. *Nature*, **469**, 236–240. [*Key study revealing the first crystal structure of a GPCR bound to the heteromeric G protein complex.*]

Re, M., Pampillo, M., Savard, M., *et al.* (2010) The human gonadotropin releasing hormone type I receptor is a functional intracellular GPCR expressed on the nuclear membrane. *PLoS One*, **5**, e11489.

Regard, J.B., Sato, I.T. and Coughlin, S.R. (2008) Anatomical profiling of G protein-coupled receptor expression. *Cell*, **135**, 561–571. [*Real-time (quantitative) PCR used to give a profile of GPCR gene expression in 41 adult mouse tissues.*]

Roper, J.A., O'Carroll, A.M., Young, W.S. III and Lolait, S.J. (2011) The vasopressin Avpr1b receptor: molecular and pharmacological studies. *Stress*, **14**, 98–115.

Spaethling, J.M., Piel, D., Dueck, H., *et al.* (2014) Serotonergic neuron regulation informed by single-cell transcriptomics. *FASEB Journal*, **28**, 771–780.

Tanoue, A., Koshimizu, T.A., Tsuchiya, M., *et al.* (2002) Two novel transcripts for human ebdothelin B receptor produced by RNA editing/alternative splicing from a single gene. *Journal of Biological Chemistry*, **277**, 33205–33212.

Van den Pol, A.N. (2012) Neuropeptide transmission in the brain. *Neuron*, **76**, 98–105. [*General review on the synthesis, secretion and function of neuropeptides, and the concepts of non-synaptic peptide transmission.*]

Vassart, G. and Costagliola, S. (2011) G protein-coupled receptors: mutations and endocrine diseases. *Nature Reviews. Endocrinology*, **7**, 362–372.

Venkatakrishnan, A.J., Deupi, X., Lebon, G., Tate, C.G., Schertler, G.F. and Babu, M.M. (2013) Molecular signatures of G protein-coupled receptors. *Nature*, **494**, 185–194. [*Review comparing key similarities and differences among diverse GPCRs whose structures have been solved by X-ray crystallography.*]

Wise, H. (2012) The roles played by highly truncated splice variants of G protein-coupled receptors. *Journal of Molecular Signaling*, **7**, 7–13.

Wootten, D., Christopoulos, A. and Sexton, P.M. (2013a) Emerging paradigms in GPCR allostery: implications for drug discovery. *Nature Reviews. Drug Discovery*, **12**, 630–644. [*Discussion of the concepts of allosteric GPCR modulation including bivalent/bitopic compounds, biased signalling, and dimerization and allostery.*]

Wootten, D., Lindmark, H., Kadmiel, M., *et al.* (2013b) Receptor activity modifying proteins (RAMPS) interact with the $AVPAC_2$ receptor and CRF1 receptors and modulate their function. *British Journal of Pharmacology*, **168**, 822–834.

PART C
The Tool Kit

CHAPTER 11

Germline Transgenesis

Jim Pickel

National Institute of Mental Health, National Institutes of Health, Bethesda, Maryland, USA

11.1 Introduction

Humans have been continuously engineering the genomes of other organisms, starting with the earliest domesticated plants and animals. Only the tools have changed. The first tool was guided selection. Bread, cheese or wine cultures were selected for quality and could be transferred from batch to batch. In the field, plants that gave better production could be propagated. Domesticated animals could be selectively mated to produce the best livestock. Now we have a much more detailed view of the genome and the techniques for engineering, but actually, the goals are similar. Our tools are still designed to alter an organism, causing a change of its characteristics. Genome engineering, even using the latest techniques, is a way to change the genotype with the intent of discovering a new phenotype.

Laboratory observations of naturally 'transgenic' bacteria were observed more than half a century ago (Box 11.1). Such organisms could acquire resistance to bacteriophages. This adaptive and selectable process generates a bacterium with a new characteristic: the ability to survive a phage attack. Remarkably, that resistance can be manipulated in the laboratory. By dissecting these systems, the restriction endonucleases were discovered. Systems that perform this same protective function have been most recently developed into tools for manipulating the genome of a range of organisms, both prokaryotes and eukaryotes. Like restriction endonucleases, this new system is only the most recent mechanism of bacterial adaptive immunity to be expropriated for scientific research. The CRISPR/cas9 system was named for its striking gene structure even before its natural function of adaptive protection – or the subsequent application of that function to genome engineering in the laboratory – was known.

Transgenic tools have been instrumental in linking a phenotype to a gene. In the clinic, the phenotype is a genetic disease that may be attributed to a variant allele. By testing the variant in an animal, a link may be established. In the laboratory, the phenotype may be a specific function of a neuron that can be linked to the expression of a neurotransmitter receptor. Only by manipulating the genes of a living organism by producing transgenic animals can we obtain such significant insight into the function of genes on cells, in complex systems, in behavior or in disease. The range of applications stretches from inherited neurological disorders to metabolic diseases to agricultural production of both crops and animals. This technology is being constantly refined, but basic techniques have been available for 25 years. Even with this history of success, in the last several years the applications of innovative new technology have been realized. Genome engineering using sequence-directed nucleases has created an unprecedented opportunity to manipulate genomes of organisms

---

Box 11.1 The development of transgenic tools

The earliest use of artificial transgenes was in the 1970s. Boyer and Cohen transferred kanamycin resistance in a transgene. Rudolf Jaenish used the SV40 virus as a vector to deliver a transgene. In the 1980s Palmiter and Brinster showed that DNA could be injected into single cell embryos.

---

and specific cells more efficiently, precisely, and quickly. This field of endeavor is moving at such a pace that new techniques and applications are being published monthly.

In this chapter, the range of transgenic methods will be outlined, from a practical approach with a focus on the application of the method to specific goals. The standard repertoire of transgenic techniques will be reviewed first. These methods are still often the most straightforward and efficient designs for an experiment and they form a basis for understanding the mechanisms of more complex techniques. Those basic methods, including inserting a gene into the germline using either oocyte injections or embryonic stem cell (ESC)-mediated trangenesis, will be described in section 11.2. Introducing a transgene can be achieved by one of two types of procedures: by random integration or by homologous recombination into a specific locus in the genome. These methods are commonly used in laboratories. In section 11.2, more elaborate multiple-component systems will be outlined that control when a gene will be rendered null or control when a transgene is expressed. Control of transcription of either an endogenous gene or of the transgene is still a challenge, but with combinatorial systems progress is being made. Each method is reviewed below and its application is addressed.

Section 11.3 describes nuclease-mediated gene modification systems that have been progressively developed to change the genes of animals and cells. The capability and efficiency of these systems are explained along with their potential for transcriptional control and therapy, either in cultured cell therapy or, potentially, in patients.

In section 11.5, broader issues of species or strain selection will be addressed. As these genome engineering techniques become more powerful, the design of their application becomes more important. With these new techniques, there is the possibility of manipulating almost any species, so that experimental design must consider the appropriateness of different species for each experiment.

The goals and questions of the investigator will dictate the experimental design, including what species is best. The impressive basic research that described the development of the early mouse embryo made this species particularly amenable to genome engineering. For many experiments, a less complex animal will be a more efficient test-bed for genome engineering. In this section, the major emphasis will be on mammals, and particularly rodents, because of their history in research but mainly because many methods have been optimized for mice and rats. Newer methods do not depend on the unique traits of mice, so an expanded range of species can be used in research.

11.2 A transgene tool kit primer

The repertoire of techniques that can be employed to create a transgenic animal has been dependent on several basic methods. An array of questions can be answered using these tried and true methods, and with each refinement comes the widening of possible applications. These core methods introduce transgenes that are transmitted through the germline from an initial founding animal to its offspring and in turn to their progeny to establish a line. Germline transmission is the most efficient way to produce a colony of transgenic animals that express the transgene at a useful level and in a predictable pattern for experiments in order to yield statistically relevant results. The practical limitations of these methods include the time of generation of the research animal. Worms, flies, and fish can produce lines much more quickly than rodents, for instance.

11.2.1 Inserting transgenes into the genome

The simplest and most efficient method for making a line of transgenic animals is to introduce the gene into an early embryo. The transgene will integrate into the genome of the embryo and be replicated and carried through to the next generation. It will transmit the transgene through the germline to the next generation.

For many animals, this sort of random integration can be reliably and consistently achieved by injecting the transgene directly into fertilized ooctyes. These ooctyes are harvested from animals that have been treated with hormones to stimulate ovarian follicles, inducing ovulation. The transgene in this case is usually introduced by microinjection of a purified DNA transgene fragment into the pronuclei of the fertilized oocyte. In most cases, this results in the integration of the transgene into a single locus of the chromosome, but in multiple copies arranged in a tandem array. Occasionally, the transgene will integrate at multiple sites. A transgenic founder animal in which integration has occurred in multiple loci will give rise to offspring in which the integrants segregate independently and in combinations in the first and following generations. The frequency of transgenics in live births can be expected to be up to 30%, with a lower occurrence with large transgenes, such as bacterial artificial chromosome (BAC) transgenes. These larger transgenes allow the inclusion of more transcriptional control sequences along with the coding sequence. They also allow precise modification of the transgene before it is injected.

A higher frequency of integration, up to 100%, can be achieved by infecting fertilized oocytes or zygotes with integrating viral vectors. These lentiviruses will be incorporated into the germline and so are transmitted to later generations. They are more efficient at producing transgenic animals but, as with DNA injections, they will integrate randomly at multiple locations. Multiple insertions may give a mosaic pattern of expression in the founder. The different inserts will segregate independently. Each offspring may contain a variable number of copies and different insertion sites. For each study, it will be necessary to balance the advantage of increased copy number and the efficiency against the difficulty that comes from multiple insertions. Wrestling with the complications of multiple copy number and the resulting inconsistent expression may make some studies less feasible.

When the virus or DNA fragment integrates randomly, it makes no distinction between benign and disruptive effects on the neighboring endogenous genes. Gene introns, coding exons or controlling sequences may be inactivated or aberrantly activated by the transgene. These limitations of random integration can be circumvented by homologous recombination or targeted genome engineering techniques, discussed in sections 11.2.2 and 11.3.

11.2.1.1 Vector design for oocyte injections

The design of the transgene is crucial to the success of any transgenic project (Figure 11.1). The potential for unintended effects must be kept in mind from the initial design through to the final evaluation of the results. These unintended effects are usually associated with the site or number of insertions in the genome. In the worst case, the phenotype created by a randomly disrupted gene may be more conspicuous and indistinguishable from the effect of the transgene. Yet another consequence of random insertion is that an insertion site may influence the level of transcription. This 'flanking effect' may either increase expression when a transcriptionally active promoter or enhancer is near the transgene or decrease expression when a silencer suppresses transcription. Once these limitations are understood, the design process can focus on arranging the several components that make a transgene construct more likely to succeed.

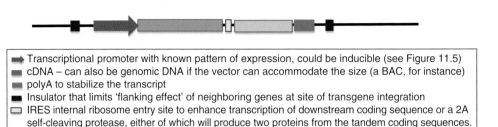

➡ Transcriptional promoter with known pattern of expression, could be inducible (see Figure 11.5)
■ cDNA – can also be genomic DNA if the vector can accommodate the size (a BAC, for instance)
■ polyA to stabilize the transcript
■ Insulator that limits 'flanking effect' of neighboring genes at site of transgene integration
▢ IRES internal ribosome entry site to enhance transcription of downstream coding sequence or a 2A self-cleaving protease, either of which will produce two proteins from the tandem coding sequences.

Figure 11.1 Components of a transgene that could be delivered as DNA injected into the pronucleus of an oocyte. See text for full discussion.

A significant choice is picking the transcriptional promoter that drives the transgene. The ideal promoter is one that is active in the specific place (a subpopulation of cells), at the specific time (when a neuron differentiates or begins expressing an endogenous gene, for instance), and at the correct level. Unfortunately, the ideal promoter that meets all these criteria is rarely available. For many uses, constitutive promoters or ubiquitously active promoters, such as the CAG (CMV enhancer with the chicken β-actin promoter), albumin, PGK (phosphoglycerol kinase) or UbiC (ubiquitin) promoters, can be used. Wide expression at high levels may adequately address the needs of the experiment.

However, in most cases, tissue-specific promoters are required. EOS, a synthetic promoter composed of multiple promoter components including three tandem copies of the Oct3/4 enhancer, is used for early embryo expression. The CAG promoter is another example of a synthetic promoter. In some cases, combining a weak constitutive promoter with a specific enhancer may be useful.

For expression in subpopulations of neural cells, many promoters require a compromise. The GFAP (glial fibrillary acidic protein) and nestin are both intermediate filament proteins whose promoters are expressed in neural stem cells. Unfortunately, they are also expressed in astrocytes (GFAP) or somites (nestin). The camKII α promoter has a long history of use for general cortical expression in the brain, but transgenes driven by this promoter may be expressed in different patterns. Dopamine and serotonin pathway gene promoters have been used to specify expression to cells using these neurotransmitters: dopamine receptor subtypes, serotonin transporter, and tyrosine hydroxylase gene promoters are examples. Other neuronal subtype-specific promoters are useful for specific expression. In any case, the founder animals that result from transgene injections must be screened because each founder's pattern of expression may be different. Finding the line with a pattern of expression that is useful for the experiment is worth the effort.

With luck, a promoter that has been extensively characterized will be available. The original publications that include the transgene and promoter may describe how many founder animals were screened to find the proper expression. This gives some notion of how robust the promoter will be in a new application. If many lines had to be screened to find the correct pattern, the same travail is likely when generating and screening a new line using the same promoters. Especially useful are lines that express effector molecules in defined patterns. Collections of mice and expression patterns are available (e.g. Gensat: www.gensat.org; Allen Brain Atlas: www.brain-map.org). The expression of the CRE recombinase transgene is extensively characterized at the Gensat site. A lox P responder transgene would express the gene of interest in the pattern of the CRE expressing effector line (see section 11.4.1).

Thankfully, there are approaches that limit the likelihood of aberrant expression. For standard transgenics that are inserted randomly into the genome, there are several options. BACs allow the inclusion of a large insertion of up to 200 kb. This much sequence should include almost all the transcriptional control elements of the inserted transgene. In addition, the flanking effect of neighboring genes is muted by the space around the transgene. A transcriptional silencer sequence may be included in the construct to limit the flanking effect of endogenous genes on the transgene. Recombineering is a process of using recombination in bacteria to construct transgenes. For large transgenes, this is a way to make even point mutations in the middle of a 100 kb fragment. The flanking effect should be blunted when the transgene is included in such a large insert.

Components that stabilize the mRNA transcribed from the transgene should also be included in the transgene vector design. Polyadenylation signals should be included to increase the stability of the RNA transcribed from the transgene. Additionally, an artificial, short intron may increase transcript stability and translation of the transgene.

A marker protein may be included as an indicator of the expression from the transgene. A fluorescent protein can be added in tandem with the transgene coding sequence. This is especially useful for imaging live cells *in vitro*, or even *in vivo*. Antibodies that recognize fluorescent proteins can be used if native fluorescence is lost during fixation of tissues. Alternatively, this can take the form of an epitope that is fused to the protein and can be recognized with an available antibody (FLAG, myc or HA are commonly used) for histology or Western blotting. It may be fused to the transprotein or be co-transcribed from an IRES (internal ribosome entry site) or 2A (a self-splicing amino acid sequence). In these cases, the easily detected marker can be used to infer the expression of the transgene.

11.2.1.2 Screening founder animals

Determining the presence of the transgene is usually accomplished by PCR. Primers should be specific for the transgene, amplifying a unique sequence that spans the junction of the promoter and coding sequence or another feature that is not present in the wild-type animal. Before embarking on the production of a transgenic animal, a robust screening procedure should be devised. PCR from serial dilutions of the transgene vector in wild-type genomic DNA will mimic the conditions of low copy number transgenes in the background of genomic DNA.

Screening for the presence of the transgene is only the first step. The pattern, the amount of gene expression, and the presence of the protein produced from the transgene are more relevant indicators of success. It is good to determine benchmark RNA levels by *in situ* hybridization or Northern or dot blotting, but using an antibody that is specific for the protein is more informative. Checking the pattern of a reporter, such as an epitope or a fluorescent protein, is usually easier and a reasonably good indication of expression.

But in the end, to achieve very specific patterns of expression, other approaches may be more feasible alternatives. It is often preferable to insert the transgene into a genomic locus that is transcriptionally neutral, such as the ROSA26 locus or the 'safe haven' AAVS1 integration site. 'Transcriptionally neutral' sites have little flanking effect and allow expression of the transgene that is appropriate for the transcriptional promoter. However, inserting the gene into these sites cannot be achieved by random integration from oocyte injections. This approach requires the use of targeted integration. A better choice may be either homologous insertion of the coding sequence of a transgene into an endogenous locus or the use of inducible promoters. This can be achieved by homologous recombination into ESC or the use of programmable nucleases (discussed later in this chapter).

11.2.2 Embryonic stem cells: targeted and precise transgene integration

Random integration of a transgene is less powerful than either the targeted disruption by deletion (gene knockout) or integration or replacement (knock-in) at a specific genetic locus. The two monumental technical achievements that have been instrumental to genome engineering are, first, the use of homologous recombination to target specific sequences and, second, the culturing of multipotent stem cells. The combination of these two methods (Koller and Smithies, 1989) has until recently been the preferred method of genome manipulation to help determine the role genes play in the complex process of a live animal.

Embryonic stem cells are derived from mouse embryos at the blastocyst stage (around day 3.5 of development). They are captured at a critical window of time. Even as the capacity of stem cells cultured from different stages of embryos has been more finely characterized, the ESC still seem as remarkable as they were when they were first derived decades ago (Box 11.2). At the blastocyst stage, all the cells of the inner cell mass (ICM) are essentially identical in their ability to generate an embryo and then develop into a completely normal animal. They also have a propensity for homologous recombination.

Box 11.2 The development of transgenic tools

Virtuosic techniques and a measure of serendipity were keys to producing the first transgenic mice more than three decades ago. The dissection of the early mouse blastocyst stage embryo into the inner cell mass and the trophoblast cells was a significant step in the understanding of mammalian embryogenesis. The bonus effect was the new mechanisms that could be used to precisely manipulate the mouse genome.

11.2.2.1 ESC cultures

Embryonic stem cell cultures require careful attention to detail. Assessing the state of the cultures is a learned process. Morphology, density, and growth rate are critical measures of the health of the culture. ESCs can be maintained on a 'feeder' layer of fibroblasts. In this scenario, the ESCs form tight junctions with other cells in a colony so that individual cells are difficult to distinguish. These colonies have a smooth refractile border as they form a lens-like shape. They divide every 12–20 hours, so they should be fed every day and passaged after just a few days. The density should be carefully limited and, depending on the size of the colonies, probably not exceed 50% confluence. Cultures without fibroblast feeders can be maintained in a multipotent state when supplements are added to the medium. Serum can be replaced with supplemented medium as well. In these cases, the colony morphology may vary without loss of multipotency. In order to promote ESC establishment and maintenance, inhibitors of pathways that are required for differentiation can be added. Experience is a boon when working with a familiar ESC line using a specific medium and selective drugs. Losing multipotency in the process of making a transgenic line can mean wasted months of work.

Embryonic stem cells should be expanded with as few passages as possible. In healthy, undifferentiated culture, the transgene can be introduced into a large number of cells, usually by electroporation. The process causes a high frequency of cell death. Then the remaining population is selected for those cells that carry the drug resistance marker that is part of the transgene. A negative selection can be applied to enrich the culture for clones that are derived from cells in which the transgene had integrated in the targeted locus.

Finally, the surviving clones can be cryopreserved and duplicated, allowing time to characterize the integration of the transgene to the level of the specific base pair. Once this is confirmed, the individual ESC clones can be thawed and introduced into embryos of the eight-cell to blastocyst stage. At this time in development, the embryo has an ICM that will give rise to the animal, and the trophoblast cells that will differentiate into extraembryonic tissues. ESCs are originally derived from the ICM and so can integrate into the ICM, which contains cells of the same developmental stage. Ideally, ESCs will integrate to such an extent that they will contribute highly to the resulting animal. When generating new lines, male ESCs are usually selected for culturing and targeting. When injected into female embryos, they can convert the resulting animal to a male, which can later mate successfully. But the significant feature of ESCs is that they maintain their multipotency and contribute to the testis. An added benefit of using male ESC lines is that male founders can more rapidly expand the line through serial mating with non-transgenic, wild-type females. The intercalation of transgenic ESCs with the host embryo's ICM will be of varying quality and should be detectible in the resulting chimeric pups. Some will carry a higher concentration of ESC contribution that will be evident in a genetic marker that gives a visible phenotype, such as coat color or fluorescent protein expression.

Because ESCs can be maintained in culture and retain their multipotent state, there is time to evaluate the transgene before an animal is produced (Box 11.3). Therefore, the chance of germline transmission of targeted integration of the transgene is increased. In addition, this method allows for precise genetic engineering, by producing a single base pair change, a large deletion or a conditional change that can be induced by a second component, a recombinase such as CRE or FLP (see section 11.4.1). Balanced against this improvement over random integration are the time and effort required for screening and breeding multiple generations of animals to establish germline transmission.

Box 11.3 Reprogramming

John Gurdon and Shinya Yamanaka shared the 2012 Nobel Prize in Physiology and Medicine for work that bridges the arc of time from the first to the most recent techniques in reprogramming of stem cells. Both showed that a nucleus from a somatic cell could, when influenced by outside factors, direct the development of a stem cell. Early in this remarkable era of developmental biology, Gurdon used microdissection to physically bring together the nucleus and the cytoplasm of a zygote. Yamanaka later brought together a differentiated nucleus and an ensemble of transcriptional factors that could achieve the same task of reprogramming the nuclease. This is a simplification of the remarkable insight of their work and its far-reaching potential. The usefulness of reprogrammed cells, like induced pluripotent stem cells (iPSC), for research and therapy is just lately being fulfilled.

Andra Nagy and colleagues (Tonge *et al*. 2014) have recently shown another sort of stem cell, the F-cell. In these cells, the proteomic analysis has revealed the array of genes that respond to the transcription factors that have been used to reprogram stem cells. This opens the 'black box' that existed between induction and the stem cell phenotype.

More and more, purchasing targeted mouse lines from commercial sources, consortia or other organizations is a viable option. As more lines are produced, published and archived, it is likely that a mouse with the desired targeted allele will be available. Rat transgenics are becoming more available as well. Mouse resources include JAX (www.jax.org), the Knockout Mouse Project (KOMP, www.komp.org) or MMRRC (Mutant Mouse Regional Resource Centers, www.mmrrc.org).

11.2.2.2   Selecting ES cells for transfection

The most important characteristic of an ESC line is that it be germline competent. That is, it can withstand the genetic manipulations, screening and expansion *in vitro*, and it can still contribute to the germ cells of the founder *in vivo* and, months later, be passed on to the next generation. ESC lines with a record of successful germline transmission are a better bet. For convenience, it is best to include a genetic marker in the parental ESC. The marker allows rapid screening of the pups that are produced. The marker is expressed from the parental ESC or from the transgene itself. The traditional method has been to use ESCs with a coat color that contrasts conspicuously from the color encoded from the host blastocyst. For example, an agouti coat color mouse is used to generate ESCs that are then injected into embryos that carry the alleles for a black coat. The skin of the pups will be black, a mix of black and agouti or completely agouti, reflecting an increasing contribution of the ESCs to the skin. Stripes of black and agouti coat will give an approximation of the percent contribution of the ESCs, with a complete agouti coat most desirable. ESCs that express a fluorescent protein can be used for the same purpose. The advantage of the fluorescent marker is that it is visible at birth before coat color is detectable, and that any combination of coat color in ESCs or host blastocysts is possible. Of course, germline transmission is the absolute criterion and it is not perfectly correlated to expression of the marker or skin color alleles in the skin.

11.2.2.3   Targeting vectors for homologous recombination in ES cells

Empty plasmid vectors that contain the necessary components for integration and selection in ESCs are available (Figure 11.2). Usually these have a positive selectable marker (e.g. resistance to neomycin, hygromycin, puromycin or spectinomycin) that is driven by a ubiquitously transcribed promoter (PGK, for instance), and a negative selectable marker (e.g. genes for HSV

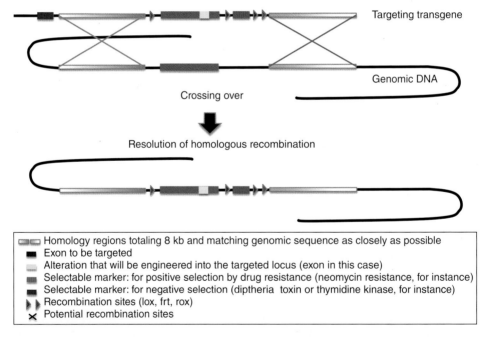

Figure 11.2   General targeting construct and homologous recombination. To target a genomic locus, a transgene can be constructed with homology regions that direct the insertion of the transgene. This targeting is mediated by homologous recombination between matching sequences in the transgene and those in the genome. A positive selectable marker confers resistance to an antibiotic when it is integrated into the ES cell genome and protects the cell from death. A negative selectable marker will kill cells in which it is inserted. This occurs when incorrect 'crossing over' occurs outside the recombination homology sequences.

thymidine kinase or diphtheria toxin A fragment). Convenient endonuclease sites for enzymes that cut infrequently are included to facilitate the insertion of the homology arms that are cloned or amplified from the genomic target. PCR-amplified fragments can be more rapidly inserted using other cloning techniques.

Regardless of the method, the targeting vector should contain between 7 and 20 kb of flanking sequence, with a minimum of 1–1.5 kb on the short arm and the remaining length on the long homology arm. The length of matching sequence in the targeting vector is directly correlated to the efficiency of recombination.

11.3   Programmable nucleases: ZFN, TALEN, CRISPR/cas9 nuclease

The remarkable advantage of homologous recombination in ESCs – the feature that has made it possible to engineer the mouse genome – is that it finds a specific sequence. It searches out and targets a 7 kb needle in the haystack of the almost 3 million kb genome. So there is great excitement at finding other molecular mechanisms that can identify even shorter sequences of as little as 0.02 kb in that same genome. These systems can disrupt the gene that contains that 20 bp target by insertion or deletion, or even replace sequence by homologous recombination from a donor DNA. Initial programmable nucleases (ZFN and TALEN) were based on endogenous DNA binding proteins. Their amino acid sequence could be engineered to scan the genome and bind a specific DNA sequence. But the discovery of the CRISPR gene array and its associated nuclease was significant because the programmable function was not mediated by protein binding to DNA, but nucleic acid binding nucleic acid in a Watson–Crick pairing of hydrogen bonds that is more specific and more robust.

While homologous recombination in ESCs is an incredibly powerful tool, it now seems relatively cumbersome to apply. Using this process (described in section 11.2.2), it can take a year to develop a transgenic mouse line that carries a specific deletion or modification of an endogenous locus. And most research has been limited to mice, since it is difficult to establish naïve ESCs from other species.

Programmable nuclease systems have solved many of these limitations. The adoption and application of these systems have led to a rapid expansion in genome engineering.

These methods represent very compelling improvements over the more traditional methods of genome engineering. Besides the advantages that are gained by homologous recombination over random integration, these methods have utilitarian advantages over the well-developed methods of homologous recombination using ESCs. The most obvious advantage is in the time saved for every step of the process.

- Design: for targeting in ESCs at least 7 kb of genomic sequence must be included in the targeting vector in order to increase the efficiency of targeting a single locus in the genome.
- A selectable marker is required to allow selection of ESC clones that carry the transgene.
- A counter-selectable marker also helps (TK or diphtheria toxin is usually used).
- In many cases, it is beneficial to remove the selectable marker using site-specific recombination (CRE/lox, FRT/flp). This may require backcrossing a transgenic line with the line carrying the targeted allele, a very time-consuming process. None of this is required if guides have been screened for efficiency and are injected directly into mouse oocytes.

On the other hand, nuclease-mediated genome editing requires much less time for design and targeting:

- for CRISPRs, it is much simpler to synthesize the short guides RNAs
- for targeting with programmable nucleases the efficiency is usually high, so simple genotyping by PCR across the target locus is sufficient

- additional screening might be necessary only to confirm modification by homologous recombination
- and similarly, no counter-selection is required.

Crossing in other transgenes to add temporal or spatial specificity of expression might be required. This takes the same amount of time as the similar process with an animal made using homologous recombination in ESCs. However, the frequency of these processes is so high that targeting or adding two transgenes at the same time will be more efficient than backcrossing.

11.3.1   Zinc finger nuclease (ZFN)

The Zn finger nuclease domain structure is one of the most common protein motifs in eukaryotes. They are found in various species, from yeast to humans. These structures are incorporated into trans-acting proteins that bind DNA at regulatory regions to control transcription of adjacent genes. Each domain is built on a 23–27 amino acid backbone that contains two cysteines and two histidines. These residues form a cage that binds the Zn cation to maintain a conformation that can bind double-stranded DNA.

These Zn finger proteins bind DNA by recognizing specific nucleotide sequences. Each domain binds three adjacent nucleotides. The domains are usually arranged in series of binding nucleotides in multiples of three. So a three-domain Zn finger guide would bind nine nucleotides, four domains would bind 12 nucleotides. Adding domains increases the specificity but adds to the complexity of synthesizing the protein. The specificity of the binding is dependent upon several factors besides the specific nucleotides that form the discrete recognition site. Adjacent nucleotides have significant influence on the binding specificity and strength of the recognition domain. The amino acids in the hinge region of the domain can also affect the steric interference at the binding site itself.

To construct a programmable nuclease tool for genome engineering, the Zn finger genes are synthesized to code for a sequence-specific guide protein. It is most efficient to generate a number of Zn finger guides to target several sequences in the proposed target area. The efficiency of these guides to bind the target sequence can be determined in cultured cells. In more innovative methods, the guides can be screened by yeast 2-hybrid or phage array screening. Analysis of the targeting frequency and the resulting modification at the target site can then be evaluated before targeting in a research animal.

A nuclease domain from the truncated Fok I endonuclease is fused to the Zn finger recognition domains. This nuclease only works as a dimer to produce double-strand breaks (DSB). Two Zn finger guides, each with an attached Fok I nuclease domain, are used, one on each side of the intended cut site. Only when the two recognition domains bind and bring the Fok I domains together is the DSB made.

The resolution of the DSB can result in an insertion or deletion of random sequence (an indel) or by the incorporation of a donor sequence by donor homology-directed repair (HDR). This repair can be directed from homologous donor sequence that is delivered as a DNA fragment. In order to drive the insertion of this fragment, it is flanked with arms that are homologous to the insertion site similar to a targeting vector for homologous recombination in ESC. These arms should be at least 100 bp long.

Because the sequence of the protein does not insure the precise binding to a specific nucleotide sequence, the screening of potential zinc finger binding domains is the most critical feature of using this system for genome engineering. To generate a ZFN, there are on-line tools or commercial services that will design a set of guides (www.zincfingers.org; www.jounglab.org/resources.htm). The vectors that code for these guides and the nuclease domain are also available, as are the schemes for constructing the combination of an array of zinc finger domains (www.addgene.org).

Unfortunately, the difficulty of designing multiple guides, building the coding sequence, and screening the resulting components in culture makes this a laborious process for laboratories that are not using the system on a regular basis. Procuring these services and products, if not the animal that results from a successful targeting, is often the cheaper solution.

11.3.2   TALE nuclease

Like the ZFN, TALE nucleases are artificial proteins, each composed of a binding domain fused to a nuclease domain. The nuclease domain is most frequently the truncated Fok I endonuclease domain that is also used with the ZFN. The recognition domain, however, is derived from proteins that bind DNA in *Xanthamon* sp. and others. These proteins have transcriptional control functions and bind to promoter regulatory regions, similar to many ZFN proteins.

The advantage of TALE nucleases over ZFN is in the specificity of binding. In each domain, for 33–34 amino acids there are two adjacent residues called the repeat variable di-residues (RVD) that specify the binding to a single nucleotide in a double-strand DNA target. Three nucleotides have specific RVD: A is bound by NI (asparagine-isoleucine), T by NG (asparagine-glycine), and C by HD (histidine-aspartic acid). The NN (asparagine-asparagine) will bind either a G or A.

By procuring the components, a TALE nuclease can be designed to target a specific sequence. There are a number of kits available to build targeting vectors that will produce the appropriate protein, including the specific recognition guide domain and nuclease domain. These have been deposited in the Addgene collection (www.addgene.org). Each kit uses ever more clever methods to make the synthesis of the gene that codes for the guides efficient and simple. Each domain has a conserved structure with only two amino acids that are different for each nucleotide recognition site. Domains for each of the individual nucleotides can be assembled, but genes that code for two or three domains with the range of di- or trinucleotide sequences are also available, reducing the number of ligation steps.

11.3.3   CRISPR/cas9 nuclease

11.3.3.1   CRISPR/cas9 function

The CRISPR/cas9 system was adapted from bacteria and devised for programmable nuclease-mediated genome engineering (Figure 11.3). It is the most promising method in transgenic technology since homologous recombination was used to reengineer the mouse genome. By 2014, this technique had already been used to generate hundreds of transgenic animals of a diverse range of species. Its initial promise is already leading to practical achievement.

In nature, the CRISPR locus exists in many species of bacteria and archaea. The genes in the *Streptococcus pyogenes* bacteria have been most widely used as a tool for genome engineering. In its natural function, this system protects the bacterium from phage infection, not in a static response (like the function of an endonuclease) but in an adaptive process that copies phage genome sequences to use as targets for the nuclease (Boxes 11.4, 11.5). The genes in the locus have three major functions.

- Adaption: DNA from invading phage is cut into prospacers and inserted into the CRISPR locus as spacers. These spacers are arranged in a striking pattern, giving the CRISPR locus its name: **c**lustered **r**egularly **i**nterspaced **s**hort **p**alindromic **r**epeats.
- Processing: the RNAs that are transcribed from these spacers and from the other cas genes are assembled. The spacer sequences are used as templates for guide RNAs that will bind to the genomic DNA of phage. Associated with these spacers are other coding and non-coding RNAs.
- Interference: the assembled guide RNA, other non-coding RNAs, and the cas9 encoded nuclease cut the genomic DNA or RNA of the invading phage, but spare the host DNA.

This system has been expropriated for laboratory use in order to target virtually any sequence in almost any nucleic acid in a cultured cell or research animal.

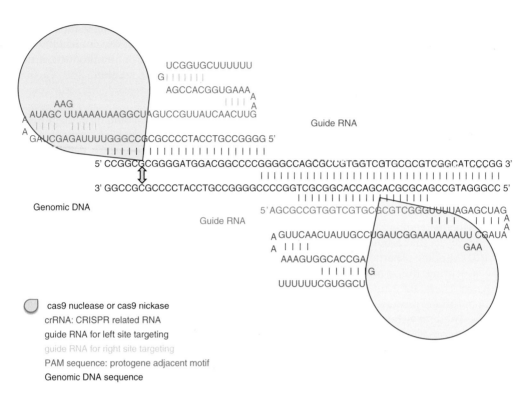

Genomic DNA

cas9 nuclease or cas9 nickase
crRNA: CRISPR related RNA
guide RNA for left site targeting
guide RNA for right site targeting
PAM sequence: protogene adjacent motif
Genomic DNA sequence

Figure 11.3 Nuclease-mediated genome engineering using CRISPR/cas9 nuclease. The CRISPR/ cas9 nuclease system allows specific targeting with as few as 20 bp of matching sequence between the transgene and the target locus. A single guide RNA can direct a cas9 nuclease to the target site where it will make a double-strand break (DSB), here denoted by the double-headed arrow. Alternatively, two guide RNAs can be constructed and a nickase (a nuclease that only cuts one strand of the target) can be used to increase specificity and increase the target size. After targeting, endogenous DNA repair mechanisms are activated to repair the site by inserting or deleting sequences which may result in a disruption of the target site. Alternatively, a targeting transgene may be used to repair the DNA by homologous recombination.

Box 11.4 Adaptive immunity in prokaryotes

Many prokaryotic mechanisms have been adapted for use in molecular biology laboratories. Restriction endonucleases are the most widely exploited example. They were first shown to be a surveillance system to protect bacteria from infection by bacteriophages. These proteins recognized palindromic sequences that occurred in phage, but not in the bacteria's own genome. Phage DNA containing these sequences would be targeted and cut, neutralizing the phage. Similar but more adaptive systems have now been characterized in *Streptococcus thermophiles* and *S. pyogenes* as well as other bacterial and archaeal species. The genes that encode these nucleases and untranslated RNAs were initially discovered as repetitive palindromic sequences that occur in clusters. They were named clustered repetitive interspersed short palindromic repeats or, more memorably, CRISPRs. While restriction endonucleases recognize a specific sequence that is stable, the CRISPRs are a form of *adaptive* immunity because the sequence at the CRISPR locus is a copy of sequences in the invading phage genome. Those bacteria that have generated a phage-specific sequence have adapted, can survive, and in the future pass on the phage-specific targeting sequence to their fortunate daughters.

Box 11.5 CRISPR/cas9: a brief history

1987   Found odd sequence near iap gene
2000   Found in 40% of bacteria, 90% of archaea
2002   CRISPR named
2005   Extrachromosomal sequence origin of direct repeats
2007   Natural role of adaptive immunity
2008   crRNAs used to guide nuclease to DNA and RNA
2008   PAM sequence role
2010   Phage-resistant dairy cultures
2012   Single RNA to serve tracrRNA and crRNA
2013   Genome editing in mammalian cells

11.3.3.2 CRISPR gene locus

Three types of CRISPR cas systems occur across the range of bacteria and archaea. There are two classes of systems, the class 1 using multiple cas proteins in the nuclease complex (these include type I and type III systems). The class 2 uses a single cas nuclease and includes type II and type V. Of these, the type II system has been most commonly used in laboratories for genome engineering because it has a simpler gene array with fewer CRISPR-associated RNAs (crRNAs). Even simpler is the type V Cpf1 that uses a few structural RNAs, a T-rich PAM sequence and creates a DSB with 4–5 nucleotide overhangs (Zentsche *et al.*, 2015). In general terms, the structure of the CRISPR locus is composed of two sections.

- In the first section, the cas genes are transcribed and code for proteins including the nucleases such as cas9. Other genes code for proteins that function in the acquisition of exogenous protospacer sequences which are derived from invading phage. In other species, there are more complex types of CRISPR loci. In these, the function of all the cas genes is not completely clear.

- The second section contains the uniquely arrayed sequences that contain spacers copied from phage genome into the host's CRISPR locus. These sequences do not code for proteins. Instead, their transcripts are used as recognition guides (gRNAs) that hybridize to exogenous phage genomes. This hybridization of recognition sequence to the target triggers the nuclease activity of cas9. Other RNAs make up a trans-activating CRISPR-associated RNA (tracrRNA), which hybridizes with the guide RNA before binding the cas9 nuclease and the target DNA.

CRISPR-associated (cas) nuclease

Target DNA, guide RNA, and nuclease protein form a complex that has been modeled using crystal structure analysis (Nishimura *et al.*, 2014). The nuclease protein component is made of two major functional structures: the DNA recognition lobe and the nuclease lobe. In turn, the nuclease lobe is composed of three nuclease domains that are similar in structure to other nuclease proteins: one HRH and two RuvC domains. Structural analysis of the *S. pyogenes* cas9 suggests that the nuclease is in an inactive conformation – a 'safety' mode – before it is bound to guide RNA or target DNA. Steric blocking of the HNH domain by the RuvC domain prevents nuclease activity. The target binding domain is also blocked by the carboxyl terminus of the protein. This suggests that the protein cannot function without first binding to the RNA hairpin structure and guide RNA, and then being directed to the target site and binding the homologous DNA sequence.

PAM (protospacer adjacent motif) sequence

Target DNA must have a PAM sequence, which varies for different species' CRISPRs in both length and sequence. For *S. pyogenes*, NGG is the preferred PAM sequence, though it will also target the variant NAG at lower frequency. Since this short sequence is very prevalent in genomic DNA, targeting of almost any genome locus is possible. In its native function, the

PAM sequence distinguishes the foreign DNA from the protospacers that code for the guides but have no PAM, thus protecting the host guide templates from being recognized by their own gRNAs and then cut by cas9. The PAM sequence seems to initiate the recognition process, forming a transient pairing with guide structure that allows scanning of the target sequence upstream for matches to the guide RNA. Once the guide sequences match, the nuclease conformation changes and is taken off 'safety mode' and the nuclease cuts the target DNA.

CRISPR guide RNA (gRNA)

The guide RNAs recognize the 20 base pair target sequence and determine the specificity of the cas9 nuclease binding. Mismatches along the length of the guide RNA can be tolerated. Sequences more proximal to the PAM are more critical for specific binding. While the PAM recognition is essential, these mismatches are less critical distal to the PAM. Proximal mismatches are more disruptive to recognition binding. Recent reports hint that mismatches with targets at the distal end of the guide sequence may actually increase activity. Recognition specificity and the resulting targeted nuclease activity can be directed at similar target sequences in the genome. These off-target sites can be predicted based on their sequence similarity. The specificity of binding is much less stringent than targeting or nuclease activity. Conversely, transient binding does not necessarily lead to nuclease activity. Modifications of the system have led to increased specificity and decreased off-target cutting.

11.3.3.3   CRISPRs as genome engineering tools

Many of the limits of zinc finger and TALE nuclease systems have been overcome by the application of CRISPR/cas9 technologies. Because the system uses nucleic acid recognition sequences instead of proteins, it is more specific and simpler to engineer. Much work has been done to explore the limits of the specificity of the gRNA recognition and to develop systems that are as streamlined as possible for convenience and efficiency in laboratory applications (Table 11.1).

The first goal was to increase the specificity of target binding and to limit unintended targeting at similar sequences in the genome. A cas9 'nickase' was engineered, in which one of the nuclease domains (HNH or RuvC) is inactivated by point mutations. This crippled form of cas9 can only make nicks in one strand of the DNA target, compared to the DSB of the wild-type cas9 nuclease. Two target sites offset from each other can each be recognized by homologous gRNAs and cut by the nickase. The sequence specificity is much more stringent when twice as many nucleotides must match. This makes it possible to increase targeting specificity and limit off-target cutting. The result is that unintended targeting may be as infrequent as it is for most other transgenic techniques.

Unintended targeting is not the same as aberrant targeting, since sequences may be similar at many related loci in the genome. The protospacer and PAM sequences can both tolerate

Table 11.1 Comparison of nucleases for genome editing.

	Nuclease	Specificity	Design	Cost
ZfN	Fok I	Lower	Limited	$6000
TALEN	Fok I	High	Flexible	$500 Addgene
				$5000 Complete
CRISPR/cas9	Cas9	High	PAM	$100
CRISPR/nickase	Nickase	Highest	2 X PAM	$100

mismatches and function to guide the nuclease to a target with mismatched sequence. In the unintended target sequence, a single nucleotide mismatch with the PAM or a five nucleotide mismatch with the guide RNA may be sufficient to misdirect the CRISPR complex. In order to restrict targeting to the intended locus, it is necessary to search the genome for sequences that are similar to the target sequence. Several websites allow users to select the species, upload the sequence of their proposed genomic target and quickly find sites with matching sequence (www.genome-engineering.org; www.addgene.org).

Simplifying and reducing the components to a minimum is another goal of genome engineers. The gRNA, PAM, and hairpin RNA that bind the cas9 were fused to allow a single transcript to perform both functions in targeting experiments. cas9 nucleases from many species were screened and several were found to function identically to the *S. pyogenes* nuclease but are less than three-quarters the size. cas9 nucleases with partial function, like the nickases, have been followed by the dcas9 (dead cas9) which has no nuclease activity but can be fused to other proteins to bind a target and control transcription.

Delivery of the CRISPR components has been varied to increase efficiency. Plasmids that carried the genes for cas9, along with the crRNA and tracrRNA, were initially used to target genomic loci. Later, it was determined that direct injection of the RNAs is more efficient and that delivery of the cas9 proteins is more efficient still. Injection of virus that carries either the cas9 or gRNA or both is a useful way to edit the genome in restricted areas. Combination with germline genes that code for another component in specific cells will restrict expression to a target cell population. Mice that express the cas9 nuclease in specific tissues or at specific times in development are available and will become more prevalent and useful.

Engineering the genome of an entire animal so that the modification will be transmitted through the germline is achieved by injecting CRISPR components into fertilized oocytes or early embryos. Injections into either the pronuclei or cytoplasm have been used. By injecting cas9 protein into the nucleus, it should be possible to target at the earliest time, lessening the chance of mosaicism (in which later introduction of the CRISPR will allow some cells to escape targeting). If those cells are more likely to contribute to the germ cell population of the animal, there will be less chance of creating a line of transgenics with a defined modification of the endogenous gene.

Targeting with CRISPR/cas9 will result in a DSB. Using the cas9 nickase will give staggered single-strand cuts. Either can be repaired by the endogenous mechanisms in the cell. The result could be a chewing back of the free end of the DNA or a filling in of more sequence in the gap. These indels (insertions or deletions) are the same as those generated by TALEN or ZFN.

Once animals have been born, they can be screened for these indels by PCR across the target site. A change in size of the PCR product may be enough to signal that the site has been cut and repaired to leave an indel. But usually a single-strand-specific endonuclease, the T7 endonuclease 1, is used to screen PCR products from targeted sites. Genomic DNA from targeted cells is amplified across the targeted region from primers that flank the site. These amplification products are denatured and then reannealed. If the site is targeted, it will leave a single-stranded stretch of DNA, which is the substrate for the T7 endonuclease. When separated by size, the fragments will be smaller than the expected PCR product.

11.3.3.4   Insertion by homologous recombination with CRISPR

Disrupting a gene by targeting with programmable nucleases depends on the imperfect repair of the cut ends of the DNA. The constant surveillance of DNA repair mechanisms results in non-homologous end joining (NHEJ) which results in indels around the cleavage site. Adding sequence from a donor DNA has been a staple of homologous recombination techniques in ESCs, but has also been used with other programmable nuclease-based methods. The efficiency of all these techniques is relatively lower than for targeting itself.

Efficiency of integration from donor plasmids in CRISPRs seems to be much higher than for the other programmable nuclease systems. The frequency is determined, as with other methods of homologous recombination, by the length of homology, with one kilobase sufficient to drive insertion or replacement. This is much shorter than the minimum of 7 kb length that is usually necessary for homologous recombination in ESCs.

11.4   Controlling transgenes with multicomponent systems

11.4.1   Recombinases

Conditional transgenics are those in which a germline transgene is altered by the action of another agent, usually an enzyme. This enzyme is most often a sequence-specific recombinase that catalyzes recombination between short sequences. The restricted spatial and temporal action of the recombinase can be specified by the pattern of expression of another germline transgene in the same animal or a gene that is acutely delivered by a virus.

This system confers another layer of control over the expression or activation or deactivation of the transgene. A conditional knockout (cKO) has two copies of the recombination sequence that flank a critical segment of a targeted gene. When the recombinase is expressed from a cell- or temporally specific promoter, the flanked segment will be looped out and deleted. The arrangement of the recombination sequence will determine the fate of the intervening sequence. Tandem recombination sequences oriented in opposite directions will, after recombination, invert the intervene sequence. Tandem sequences oriented in the same direction will delete the intervening sequence. When the recognition sequences are on different DNA strands, the two strands will be exchanged (Figure 11.4).

The recombinase most often used is the **CRE,** a 38 kDa protein from the P1 bacteriophage. It catalyzes the cutting, strand exchange, and ligation of DNA at the lox recognition sequence. That 34-nucleotide recognition sequence is composed of two palindromic repeats of 13 base pairs each, separated by an 8 bp spacer region. Altered recognition sequences can be used to bias recombination reactions. For instance, Table 11.2 shows five alternative loxP recognition sequences.

The lox5171 and lox2272 will only recombine with identical lox sequences, but not with each other or wild-type loxP. This scheme can be used to create concentric or parallel lox sites that will allow simultaneous but separate deletions, or an inversion and a deletion (depending on the orientation) of two sequences. This is especially useful in the DIO (double-floxed inverted open reading frame) arrangement in which an open reading frame is situated in the inverted orientation and cannot be expressed. It is flanked at each end by loxP and lox2272 sites in a convergent orientation. At one end, the loxP site is internal, at the other end the lox2272 is internal. When lox2272 recombines with its match and loxP with its match, the open reading frame is in the correct orientation for expression. And most importantly, the result is two incompatible lox sites so recombination cannot continue and the gene is expressed.

Lox511 will recombine with a similar lox511 and with wild-type loxP. Lox71 will recombine with lox66 because they have compatible LE/RE (left element/right element) sequence. The resulting recombinant will create one lox 66/77 hybrid and a wild-type loxP that will only recombine with each other at a very low frequency. This is useful for biasing the insertion of a sequence, and almost eliminating the chance of a deletion.

Modification of the CRE recombinase is another approach to controlling gene expression. The estrogen receptor is sequestered in the cytoplasm until it binds estrogen and is translocated into the nucleus, where it activates gene expression. CRE and the estrogen receptor can be fused so that recombination of loxP sites can only occur when estrogen is

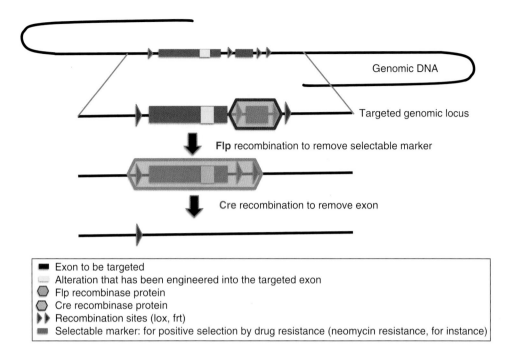

Genomic DNA

Targeted genomic locus

Flp recombination to remove selectable marker

Cre recombination to remove exon

■ Exon to be targeted
▭ Alteration that has been engineered into the targeted exon
⬡ Flp recombinase protein
⬡ Cre recombinase protein
▶▶ Recombination sites (lox, frt)
▬ Selectable marker: for positive selection by drug resistance (neomycin resistance, for instance)

Figure 11.4 Recombinase-mediated control of deletion and transcription. Site-specific recombination by Cre, Flp or Dre enzymes can be used to delete DNA that is flanked by their corresponding target sites: lox, frt or rox. If the target sites are in a tandem array (as in this figure), the intervening sequence will be looped out and deleted. But if the target sites are in a convergent pattern the intervening sequence will be inverted (not shown here).

Table 11.2 Five alternative loxP recognition sequences.

loxP	Sequence	Spacer	Sequence	Specificity
lox 5171	ATAACTTCGTATA	ATGTgTaC	TATACGAAGTTAT	Exclusive
lox 2272	ATAACTTCGTATA	AaGTATcC	TATACGAAGTTAT	Exclusive
lox 511	ATAACTTCGTATA	ATGTATaC	TATACGAAGTTAT	Self & loxP
lox 66	ATAACTTCGTATA	NNNTANNN	TATACGAAcggta	Lox71->stable
lox 71	taccgTTCGTATA	NNNTANNN	TATACGAAGTTAT	Lox66->stable

The use of loxP sites with non-canonical sequences allows specific site-directed recombination between these sites with the specificity shown.

present (CRE-ERT). Tamoxifen, which binds the human estrogen receptor but not the mouse receptor, can be administered in food or drinking water to induce the translocation to the nuclease where the CRE can act on the recognition sequences. A tissue-specific promoter, when combined with the time of drug administration, works to restrict expression as desired. A soluble form of CRE is useful when inducing recombination in cultured cells, as well as mouse embryos that carry loxP recombination sites.

The CRE/lox system is the most developed and has led to the creation of panels of mice that express CRE from different promoters. These have been characterized and archived and are available to use as driver lines to activate floxed responder lines (Gerfen *et al.*, 2013; www.gensat.org).

Table 11.3 Other site-specific recombinases.

Recombinase	Recognition site	Recognition sequence	Source
CRE	Lox	**ATAACTTCGTATA-nnnTAnnn-TATACGAAGTTAT**	P1 phage
FLP	Frt	**GAAGTTCCTATtC-tctagaaa-GtATAGGAACTTC**	*S. cerevisiae*
DRE	Rox	**TAACTTTAAATAAT-tggc-ATTATTTAAAGTTA**	D6 phage

Another recombinase that is used in many transgenics is the flipase (**FLP)** from *Sacccharomyces cerevisiae*. It recognizes the frt site, a 34-nucleotide recognition sequence.

The **DRE** recombinase is less frequently used. It is derived from the D6 phage and functions similarly to FLP and CRE, recognizing and recombining 32-nucleotide rox sites (Table 11.3).

The combinations of different lox recognition variant as well as the other recombinases can be assembled into systems that give more flexibility and control over transcription control, deletion, insertion or inversion in multiple rounds for recombination.

11.4.2   tet expression systems

The tetracycline (tet) transcriptional control systems have been widely used and characterized in cell lines and in transgenic animals. There are two complementary systems. In the first (Tet-Off), transcription is repressed by the drugs tetracycline or doxycycline. The second system (Tet-On) requires doxycycline for transcription (Figure 11.5).

Both systems rely on components of the tetracycline resistance system of *Escherichia coli*. The Tet repressor protein (TetR) is a *trans* activator that binds and regulates genes that carry DNA sequence (tetO) in *cis*, adjacent to the transcriptional promoter. There are three components of the system.

- The transgene of interest with the TRE (tet responsive element) inserted upstream of the gene in order to control transcription.
- A TetR expressed from a tissue-specific promoter, based on the tet repressor, can be either the tTA, the tet-controlled transactivator (for Tet-Off) or the rtTA reverse Tet repressor (for Tet-On).
- Doxycycline to either repress transcription (Tet-Off) or activate transcription (Tet-On).

The endogenous system from the *E. coli* tetracycline resistance genes has been modified to function more efficiently as a transcriptional control system. The TetR is fused to VP16, a transcriptional activator protein from the herpes simplex virus. The rtTR, the transactivator used to induce expression in the Tet-On system, has a four-amino acid change from the native tTA.

Three critical features should be considered in any transcriptional control mechanism in transgenic animals.

- Leakiness, which results from residual binding of the tTA to the tetO site, even in the presence of doxycycline (dox). In the tetO system, it may result from the residual background level of expression that continues even when dox is present. The flanking effect of neighboring promoters on the TRE transgene promoter will also give rise to leakiness.
- The gain, the percent increase in expression when the promoter is induced, may be limited by flanking effects, or it may be that the level of the specific transgene is restricted. Flanking effects might also enhance or suppress gain.
- The response curve depends on the ability of the drug to be diluted out of the system when it is withdrawn from the animal during Tet-Off activation, or to saturate the animal to an extent that it will activate transcription in the Tet-On system.

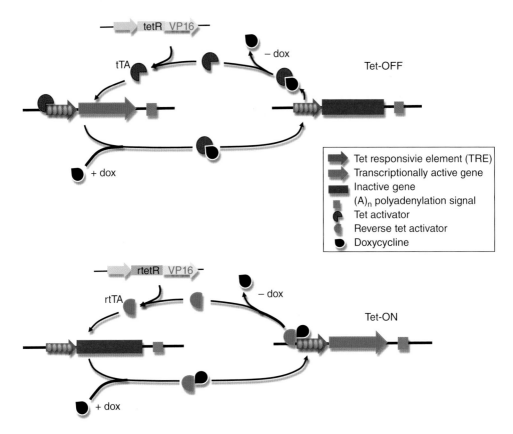

Figure 11.5 Tetracycline-inducible expression by the Tet systems. Transcription of a transgene can be controlled by the Tet-On or Tet-Off systems. Both systems use tetracycline (or doxycycline) to bind a composite protein (TA) that binds both the drug and the tet responsive element (TRE) that is position 5′ upstream of the transgene. In the **Tet-Off** system, the tTA protein binds the TRE to induce expression. When the drug is added, the tTA is inhibited from binding the gene and expression is halted. In the **Tet-On** system, the rtTA modified protein cannot bind the TRE until the drug is added and transcription begins.

In animals, the choice of system will depend on the goal of the experiment. In general, if the gene must be completely silent, then the Tet-Off system might provide the lowest level of leakiness. If a higher gain is required, then the Tet-On system will be a better choice. Flanking effects and the specific responder gene that is expressed will also affect these factors.

To provide the most control, multiple systems can be used together. For instance, to reduce leakiness, the best option may be to use Tet transcriptional control of an inducible CRE, that in turn recombines a floxed STOP and allows transcription of the last gene in the chain of control.

11.4.3   Delivery of effector and responder transgenes

Transcriptional control will also depend on the way that these effector genes and their proteins are brought together with the responder transgene. Using germline transgenic animals with defined effector expression and performance is beneficial. CRE-, Tet-, and cas9-expressing mice are available from repositories, with variable history of success at controlling transcription. Combining these effector mice together with responders with a custom-made transgene is often the simplest way to design an experiment.

Acute delivery of a transgene can also improve temporal and spatial specificity. When one component of the effector or responder genes (either viral or germline transgenic) is also controlled by another system (CRE-ERT, for instance), the control can be quite specific.

11.5 Validity of species and strain choices

Transgenic animals are often seen as substitutes for human subjects, but in this role they are limited in their applicability. The parameters of any experiment, including one with research animals, must be controlled in order to focus the question and critically characterize the results. To increase the likelihood of a useful result, it is especially important to consider the appropriateness of the species (and even the strain of a species) to the experiment. The validity of using research animals for neuropsychiatric disease has been reviewed, and in many cases found problematic (Nestler and Hyman, 2010). The three types of validity (construct validity, face validity, and predictive validity) of animal models for human disease were critiqued. The issue of validity of animal species and strain selection has even reached the popular press with magazine pieces that question the use of the C57BL/6 mouse for biomedical research. Each experiment must meet these criteria, but in transgenic experiments they are especially important. The choice of species is often one that balances ease of use and cost against suitability for the experiment. This is not a trivial concern, since a poor choice can result in wasted efforts, resources, and time.

Fortunately, new genome engineering tools vastly broaden the variety of species available for experiments for which transgenic animals are useful or necessary. And in species previously obtuse to transgenic manipulation, the new tools can be more efficient than they have been in rodents. As the options increase, there are more opportunities to use transgenic animals in research, but the responsibility to carefully consider the validity of those experiments also increases.

11.5.1 The C57BL6 mouse

Transgenic mouse background strains were initially limited to those strains that were easiest to manipulate. Thus the first transgenics (see section 11.2.1) were made in FVB or in hybrid lines (F1[C57BL6 X C3H] or F1[C57BL6 X DBA]). Early ESC lines were initially derived from 129 mice from any of several substrains, including mixed backgrounds. The transgenic techniques were easier to reproduce in these specific strains, so they were favored. These strains were more sensitive to superovulation techniques, producing more oocytes than other strains. And cryopreservation of some lines is more difficult than for others. Strains were methodically categorized for their efficiency at producing transgenics.

Genetically defined animals are critical when a genotype is precisely what is being investigated, so backcrossing mixed lines became a time- and resource-consuming endeavor. Selecting those animals carrying the highest percentage of the desired background was possible using 'speed congenic' analysis. As techniques have improved, it is common to use inbred or completely outbred strains as the background for transgenic mice.

Neuroscientists have generally opted for C57BL6 strains, either the Jackson C57BL/6J or the related substrain C57BL/6N (Simon et al., 2013). Background strains were chosen for the high-throughput projects that have produced large collections of transgenics. The KOMP (Knockout Mouse Project in Europe; www.mousephenotype.org) chose to use C57BL6/N. The Jackson Laboratory and many NIH-funded initiatives use C57BL/6J, though JAX also has a phenotyping project using C57BL/6N mice. These consortia established model transgenic strains and have limited the use of different strains.

As techniques have improved, and strain-specific differences are known, the decision of which strains are most useful and appropriate is much less of a constraint. Also investigators, especially those using behavioral testing, have found experimental limitations in the inbred strains and resorted to outbreeding to produce more relevant results. The use of programmable nucleases for targeting has broadened the strains that can be considered for transgenic projects.

11.5.2   A rodent of another color

When should mice be replaced, or at least supplemented, with other rodent strains or even other species? This decision depends on several factors: historical models, accumulated data, the ease of manipulating the strain, the robustness of the response, the number of animals required, and, of course, the practical considerations – the cost of space and expense of care. Wild mice or mice that were more recently captured and bred in the lab (*M. spretus* or *M. castaneous*, for instance) have been useful for genetic studies because of their divergence from the historical research strains. The suitability of ground squirrels for circadian rhythm experiments requires their use despite the difficulty of working with an animal that is asleep for a third of its life (and, by extension, a third of one's career). Another example of research using an unconventional species is the use of voles to study the genetics of pairing behavior. Two species of vole, the mountain vole and the prairie vole, are polygamous or monogamous, respectively. Now transgenic voles have been produced to further these studies of reproductive behavior (Keebaugh *et al.*, 2012).

Transgenic rats have been made on several background strains. Different fields of study require different strains, whether they are outbred or inbred. Long Evans rats predominate in neuroscience laboratories and are malleable enough for transgenic production. Lewis rats are useful because they are inbred. Rats have proven advantageous over mice in several fields of research. Addictive behavior, for instance, is better modeled in rats than in mice. In some cases, the physical size of brain structures is enough to warrant the use of rats rather than mice. Fewer animals, less time in manipulation, and more robust responses in behavior experiments make rats a more cost-effective choice in many cases. Transgenic rat lines, including those on inbred backgrounds, are becoming more available and are generated by commercial interests. A collection of transgenic and other lines is available through the Rat Resource and Research Center (RRRC; www.rrrc.us).

11.5.3   Non-human primates: marmosets and beyond

Mapping nervous system functions to areas of the brain is a key occupation of non-human primate (NHP) research. Regions that are active in sensory, motor, learning, memory, and specific behaviors can be mapped using imaging and electrophysiological analysis. Neural activity, circuitry, behavior, and genetics can now be analyzed together. The first transgenic non-human primate was produced by Erika Sasaki's group and published in 2009 (Sasaki *et al.*, 2009). Of note was the rapid report of germline transmission, which was made possible because of the short gestation and maturation time of marmoset monkeys. Already, marmosets carrying genes associated with neurological diseases have been produced. Rhesus macaque monkeys have been made transgenic, though a test of germline transmission will require the maturation of these first animals. In addition, the extended gestation of these Old World monkeys makes them less desirable as experimental animals. Cynomolgus monkeys have been made transgenic using the CRISPR/cas9 system (Niu *et al.*, 2014).

Transgenic marmosets are especially useful for research for several practical reasons. They are not endangered, living in a coastal forest habitat of Brazil that is increasing in size. They are smaller, prefer living in family groups, reproduce in research housing, and do not carry human pathogens. For electrophysiology, their lissencephalic brain gives easier access to deeper layers by radially placed electrodes that need not pass tangentially through more peripheral layers. Social behavior is being characterized. Reversal learning, vocal behavior, and visual circuitry are all areas of active study. All of these areas of study would benefit from genetic tools that either report activity or ablate function in specific and reversible ways.

Recently, it has become practical to alter the genome using programmable nucleases, at the same time that transgenic techniques for marmosets and other NHPs have been developed. Already, ZFN and CRISPR transgenics have been made in monkeys. The ethical issues and cost of these experiments militate against any but the most stringent requirements to reduce the number of animals created and refine experiments to most directly and efficiently address significant questions.

11.6   Conclusion: perspectives and opportunities provided by the new toolbox

The history of transgenic techniques has provided powerful, important tools for manipulating the genome of cells and animals to study the genetics of development, of normal function, of disease dysfunction – of life. There have been more than 20 years of research success using these techniques. Just as they appear to be maturing, a new set of techniques has been developed that increases the reach of genetic research. Genome engineering innovations are coming faster than the last month's methods can be applied.

Targeting multiple loci or several exons in a large gene can be achieved simultaneously. Species for which ESCs are not available can be used for gene targeting just as mice have been used in the past. Cultured cells can be targeted as easily as embryos. In short, the potential to use genome engineering methods is still in its earliest stages.

And the technique is accessible to molecular biologists. It can cut the time required to target a gene and expand a colony of mice from almost a year to a matter of weeks. The saving in time and resources is exceptional. The use of transgenic animals can be streamlined and the number of animals required for each experiment can be reduced.

Most importantly, these techniques and their application over the next few years will resolve the limitations of the older techniques and more efficiently explore the role of the genome, coding genes as well as other genomic functions in development, function and disease.

References

Gerfen, C.R., Paletzki, R. and Heintz, N. (2013) GENSAT BAC Cre-recombinase driver lines to study the functional organization of cerebral cortical and basal ganglia circuits. *Neuron*, **80**, 1368–1383.

Keebaugh, A.C., Modi, M.E., Barrett, C.E., Jin, C. and Young, L.J. (2012) Identification of variables contributing to superovulatin efficiency for production of transgenic prairie voles (*Microtus ochrogaster*). *Reproductive Biology and Endocrinology*, **10**, 54.

Koller, B.H. and Smithies, O. (1989) Inactivation of the beta 2-microglobulin locus in mouse embryonic stem cells by homologous recombination. *Proceedings of the National Academy of Sciences USA*, **86**, 8932–8935.

Nestler, E. and Hyman, S. (2010) Animal models of neuropsychiatric disorders. *Nature Neuroscience*, **13**, 1161–1169.

Niu, Y., Shen, B., Cui, Y., *et al.* (2014) Gene of gene-modified cynomolgus monkey via Cas9/RNA-mediated gene targeting in on-cell embryos. *Cell*, **156**, 836–843.

Nishimura, H., Ran, F.A., Hsu, P.D., *et al.* (2014) Crystal structure of cas9 in complex with guide RNA and target DNA. *Cell*, **156**, 935–949.

Sasaki, E., Suemizu, H., Shimada, A., *et al.* (2009) Generation of transgenic non-human primates with germline transmission. *Nature*, **459**, 523–527.

Simon, M.M., Greenway, S,. White, J. and Fuchs, H. (2013) A comparative phenotypic and genomic analysis of C57BL/6J and C57BL/6N mouse strains. *Genome Biology*, **14**, 18–23.

Tonge, P.D., Corso, A.J., Monetti, C., *et al.* (2014) Divergent reprogramming routes lead to alternative stem-cell states. *Nature*, **516**(7530), 198–206.

Zentsche, B., Gottenberg, J.S., Abudayyeh, O.O., *et al.* (2015) Cpf1 is a singel RNA-guided endonuclease of a class 2 CRISPR-cas system. *Cell*, **163**, 1–13.

Further reading

Feil, R., Wagner, J., Metzger, D. and Chambon, P. (1997) Regulation of cre recombinase activity by a mutated estrogen receptor ligand-binding domains. *Biochemical and Biophysical Research Communications*, **237**, 752–757.

Gossen, M. and Bujard, H. (1992) Tight control of gene expression in mammalian cells by tetracycline responsive promoters. *Proceedings of the National Academy of Sciences of the United States of America*, **89**, 5547–5551.

Gossen, M., Freundlieb, S., Bender, G., Muller, G., Hillen, W. and Bujard, H. (1995) Transcriptional activation by tetracycline in mammalian cells. *Science*, **268**, 1766–1769.

Gong, S., Doughty, M., Harbaugh, C.R., *et al.* (2007) Targeting Cre recombinase to specific neuron populations with bacterial artificial chromosome constructs. *Journal of Neuroscience*, **27**, 9817–9823.

Hermann, M., Stillhard, P., Wildner, H., *et al.* (2014) Binary recombinase systems for high-resolution conditional mutagenesis. *Nucleic Acids Research*, **42**, 3894–3907.

Missirlis, P., Smailus, D. and Holt, R. (2006) A high-throughput screen identifying sequence and promiscuity characteristics of the loxP spacer region in Cre-mediated recombination. *BMC Genomics*, **7**, 73.

Nagy, A. (2000) Cre recombinase: the universal reagent for genome tailoring. *Genesis*, **26**, 99–109.

Ran, F.A., Hsu, P.D., Lim, C.Y., *et al.* (2014) Double nicking by RNA-guided CRISPR Cas9 for enhanced genome editing specificity. *Cell*, **154**, 1380–1389.

Sternberg, N. and Hamilton, D. (1981) Bacteriophage P1 site-specific recombination:1.Recombination between loxP sites. *Journal of Molecular Biology*, **150**, 467–486.

Warming, S., Costantino, N., Court, D., Jenkins, N. and Copeland, N. (2005) Simple and highly efficient BAC recombineering using galK selection. *Nucleic Acids Research*, **33**, 109–115.

Yang, X.W., Model, P. and Heintz, N. (1997) Homologous recombination based modification in Escherichia coli and germline transmission in transgenic mice of a bacterial artificial chromosome. *Nature Biotechnology*, **15**, 859–865.

CHAPTER 12

Somatic Transgenesis (Viral Vectors)

Valery Grinevich,[1] H. Sophie Knobloch-Bollmann,[1] Lena C. Roth,[1] Ferdinand Althammer,[1] Andrii Domanskyi,[2] Ilya A. Vinnikov,[2] Marina Eliava,[1] Megan Stanifer,[3] and Steeve Boulant[3]

[1] Schaller Research Group on Neuropeptides, German Cancer Research Center, Heidelberg, Germany
[2] Molecular Biology of the Cell I, German Cancer Research Center, Heidelberg, Germany
[3] Schaller Research Group of Viral Infection Dynamics, German Cancer Research Center, and University of Heidelberg, Heidelberg, Germany

12.1 Introduction

During the last decade, viral vectors have been extensively used for deciphering the anatomy and physiology of higher brain regions, especially the cortex and hippocampus. In contrast, deeper brain areas, such as the **hypothalamus**, have only recently been explored using viral techniques. The hypothalamus is composed of various types of neuropeptidergic neurons, which control diverse endocrine, metabolic, and autonomic functions. Although neuropeptides are expressed in various brain regions, only the hypothalamus carries unique and highly specialized neurons known as neuroendocrine neurons, which produce neurohormones that are directly released into the blood, ventricular system, and local and distal extracellular spaces.

In addition to neurohormones, neuroendocrine neurons can also express classic neurotransmitters such as glutamate or γ-aminobutyric acid (GABA), which can complicate studies of their functions. Furthermore, hypothalamic neuroendocrine neurons can release their neuropeptides from axonal collaterals in various brain regions, affecting various behaviors (Box 12.1). Given this multitude of characteristics and their versatile features affecting homeostasis and behavior, a better understanding of physiology and functional connectomics of neuroendocrine neurons is needed. Viral vectors represent a unique tool, allowing researchers to target specific cells and specific cellular functions. In this chapter, we will first introduce the different viral vectors that can be used to study the structural organization and function of hypothalamic neurons. In the second part, we will illustrate the use of various viral vectors to study the anatomy, physiology, and pathology of neuroendocrine neurons (Figure 12.1).

Box 12.1   Neurohormones versus neuropeptides

Neurohormones are produced by hypothalamic neurons and released in fluid compartments of the body (i.e. cerebrospinal fluid and blood) to act distantly on receptive cells. In contrast, neuropeptides (often neuropeptides of identical chemical nature as neurohormones) are released from axons in neural tissue. The routes of neurohormone/neuropeptide release are shown for oxytocin (OT) in the review of Knobloch and Grinevich (2014).

Molecular Neuroendocrinology: From Genome to Physiology, First Edition. Edited by David Murphy and Harold Gainer.
© 2016 John Wiley & Sons, Ltd. Published 2016 by John Wiley & Sons, Ltd.
Companion website: www.wiley.com/go/murphy/neuroendocrinology

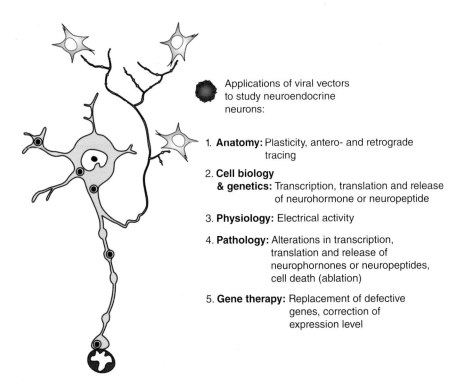

Applications of viral vectors to study neuroendocrine neurons:

1. **Anatomy:** Plasticity, antero- and retrograde tracing

2. **Cell biology & genetics:** Transcription, translation and release of neurohormone or neuropeptide

3. **Physiology:** Electrical activity

4. **Pathology:** Alterations in transcription, translation and release of neurophornones or neuropeptides, cell death (ablation)

5. **Gene therapy:** Replacement of defective genes, correction of expression level

Figure 12.1 Application of viral vectors to study neuroendocrine neurons. Here we depict a generic viral vector and its applications in studying anatomy, cell biology, physiology, and pathological changes in a generic neuroendocrine neuron.

12.2   Overview of viral vectors

12.2.1   Introduction to viruses

Viruses are small infectious agents which range in size from 20 nm to 600 nm. They are termed **obligate intracellular parasites** due to their reliance on host factors to reproduce themselves. Viruses can be classified as either **enveloped** or **non-enveloped/naked**. Enveloped viruses contain a lipid layer, which is taken from the host cell as the virus buds off during the **assembly** and release stage of the viral life cycle. Non-enveloped or naked viruses do not have a lipid envelope and their exterior consists of a protein coat (which can be made of a single or multiple proteins) or **capsid**. The genomes of viruses can be either RNA or DNA but never both. Additionally, the **viral genome**s can be either single stranded or double stranded. Viruses often encode many of their own enzymes to complete their replication process, but they never contain the material required for translation of their proteins (Knoppe and Howly, 2013).

Viruses must complete a series of events to establish infection within a host and produce new progeny (Figure 12.2). Viruses must first bind to a host cell. This binding is mediated by interactions of viral proteins (either in the capsid of non-enveloped viruses or viral proteins that have been inserted into the lipid bilayer of enveloped viruses) and host cell **receptors**. The ability of viruses to attach to cell surfaces can be a very selective process. The type of cells/tissues that a virus can bind to will determine the **tropism** of the virus. Viruses with a broad tropism can bind to many cell and tissue types, such as vesicular stomatitis virus (VSV), which can bind and

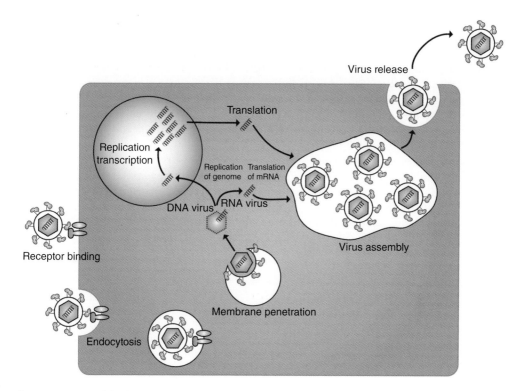

Figure 12.2 Simplified virus lifecycle. Viruses first bind to cells using cellular receptors. They are then endocytosed and release their genome into the cytoplasm after membrane penetration. For RNA viruses, the genomes are transcribed and translated in the cytoplasm. DNA viruses are replicated in the nucleus and viral proteins are translated in the cytoplasm. After new virus genomes and proteins have been produced, new viruses are assembled and released from the cell to start a new infection cycle.

infect all eukaryotic cells. Viruses with a restricted tropism can bind and infect only one or two cell or tissue types, such as hepatitis C virus [HCV], which only infect some hepatic cell lines. After attachment and binding to the cell surface, viruses are endocytosed by a variety of mechanisms. The two most common ways that viruses enter cells are by **clathrin-mediated endocytosis** or macropinocytosis (Yamuuchi and Helenis, 2013). After **endocytosis**, viruses are trafficked into the cell until they reach a compartment where they are able to break out and deliver their genome into the cell. Many viruses will traffic to early or late **endosomes** and either fuse or penetrate the endosomal membrane in a pH-dependent manner. After delivery to the cytoplasm, viral genes of RNA viruses are either directly translated by host ribosomes or transcribed by virally encoded enzymes immediately followed by translation. Viruses which contain DNA genomes will be further transported into the nucleus to use the host cell machinery for transcription (this is true for all DNA viruses except poxviruses). After transcription/translation of viral proteins and replication of the viral genomes, new viruses are assembled. The site of assembly varies depending on the type of virus. Enveloped viruses are commonly assembled at the plasma membrane or the endoplasmic reticulum and bud off from there, releasing hundreds or thousands of new virus particles to infect neighboring cells. Non-enveloped viruses are usually assembled into specialized structures made of viral proteins and are released from the cell either by lysis or through apoptotic/necrotic mechanisms. After release, the new virus particles will bind to new cells and start the process all over again.

12.2.1.1 Viral vectors

As mentioned above, viruses may display very specific tropisms and also affect specific cellular functions. As such, researchers have quickly realized the benefit of using viruses to study cell biology functions. Viral vectors have been created as a mechanism to deliver genetic material to alter cells of both living organisms or in cell culture models. These vectors use the properties of the viruses they were derived from to package and express the gene of interest (proteins or shRNAs); however, they have been altered from the naturally occurring viruses in order to reduce risks and to encode only those elements necessary (Grimm and Kay, 2007).

The use of viruses to deliver genes is termed 'transduction.' This procedure was first used by molecular biologists in the 1970s to create cells that could be maintained in culture for long periods of time. Currently, there are five types of viral vectors used to modify neurons and these are described in more detail in the following sections. Each of these vectors displays advantages and disadvantages, but are under constant modification to produce new vectors with even better characteristics (Table 12.1).

The ideal viral vector:
- targets very specific cell types
- creates little to no immune response in the host
- is maintained over long periods of time
- will not replicate and/or produce new progeny
- has low integration rates (to reduce the risk of cancer) or specific integration sites in safe harbor places.

12.2.2 Retroviruses

Retroviruses are enveloped RNA viruses with a single-stranded genome of 7–11 kb and a virion size of about 80–100 nm (Figure 12.3A). There are two copies of the genome packaged into the viruses. These two copies attach to each other by complementary 5′ regions and form a homodimer. This feature is often referred to as diploid genome. After infection, the RNA genome is reverse transcribed to double-stranded DNA in host cell cytoplasm and subsequently trafficked in the nucleus. Using a virally encoded integrase, the double-stranded DNA is inserted randomly into the host genome. When the viral genome is integrated within the host chromosomes, it is called a **provirus**. Interestingly, recent reports suggest that incorporation hotspots might exist in infected patients. This allows for long-term viral persistence and can lead to viral transmission through the germline.

Table 12.1 Characteristics of families of viruses, which can be used for infection of neurons.

Virus family	Specific virus	Size of insert	Integrated	Purpose
Retroviruses (Gamma)	MoMLV	7–11 kb	Yes, non-specific	Not used in mature neurons due to its need for dividing cells
Retroviruses (Lenti)	HIV-1	9 kb	Yes, non-specific	Exogenous protein expression, shRNA expression, Cre-lox recombination
Rhabdovirus	Rabies	2 kb	No	Monosynaptic retrograde tracer
Parvovirus (Dependo)	AAV	4.2 kb	Low to none	Exogenous protein expression, shRNA expression, Cre-lox recombination
Herpesvirus	Pseudo-rabies	n/a	No	Retrograde tracer

(A) Retrovirus scheme

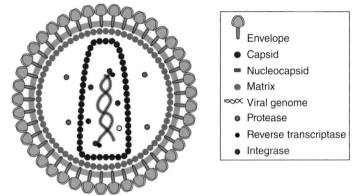

(B) Simple retroviruses genome

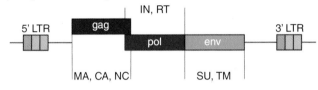

(C) Simple retroviruses vector system

(D) Complex retroviruses genome

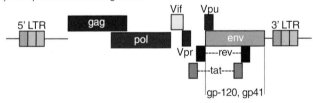

(E) Complex retroviruses vector system

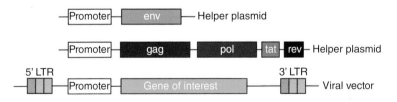

Figure 12.3 Retrovirus genomes and viral vectors. (A) Schematic of a retrovirus virion. Two copies of the viral RNA are found inside the nucleocapsid. The viral enzymes, reverse transcriptase, and integrase are also packaged within the virus. The matrix protein lines the inside of the envelope and the exterior is composed of the env glycoprotein. (B) Schematic of the simple retrovirus genome. CA, capsid; IN, integrase; MA, matrix; NC, nucleocapsid; RT, reverse transcriptase; SU and TM, envelope protein. (C) Schematic of the simple retrovirus vector based on MoMLV. The viral genes are removed and replaced with the gene of interest. The viral plasmids are brought from additional plasmids or packaging cell lines to complete the viral particle. (D) Schematic of a complex retroviral genome (i.e. HIV-1). (E) Scheme used to produce lentivirus vectors. Three plasmids are needed to produce a complete virus. The gene of interest is expressed of its own plasmid. HIV-1 env or an enveloped protein from another virus (i.e. VSV-G) can be used to make the vectors.

There are seven genera of retroviruses, which are classified as either simple or complex. The simple retroviral genome contains three essential genes, termed gag, pol, and env, flanked by long terminal repeats (LTR) (Figure 12.3B). Complex retroviruses such as lentiviruses, described below, contain additional accessory genes that are dependent on the type of retrovirus. The gag gene codes for structural proteins, termed capsid (CA), matrix (MA), and **nucleocapsid** (NC), which form the viral core. The pol gene codes for the viral enzymes, reverse transcriptase (RT) and integrase (IN), and the env gene codes for envelope glycoproteins (SU and TM), which mediate interaction of the virus with host cell surface proteins. All proteins are proteolytically processed from precursor polypeptides. The development of retroviral vectors is based mainly on work with the Moloney murine leukemia virus (MoMLV), which has a simple gag-, pol-, and env-containing genome (Osten *et al.*, 2007). The most recent generations of retroviral vectors include a number of modifications designed to enhance the safety as well as the efficiency of the system (Figure 12.3C). These vectors can be either replication competent or replication defective. Replication-defective viruses are the preferred choice since they will not cause cell lysis or death. These viruses are then produced in a packaging cell line, which expresses the env and gag proteins and just needs to be transfected with the gene of interest to produce the viral vector. The viruses are harvested, purified, and delivered into the neurons of animals by direct infusion (Knoppe and Howly, 2013) (see Figures 12.6 and 12.7).

12.2.2.1 Lentiviral vectors

The most studied lentivirus is the human immunodeficiency virus type 1 (HIV-1), which infects CD4-positive T lymphocytes and macrophages (CD4 antigen acts as the primary surface receptor for HIV-1) and causes chronic immune depletion known as the acquired immune deficiency syndrome (for review see Frankel and Young, 1998). HIV is a complex retrovirus. Its genome is 9.2 kb and contains two additional regulatory genes, termed tat and rev, and four accessory genes, termed vpr, vpu, vif, and nef (Figure 12.3D). The accessory proteins mainly enhance virulence against the host organism. The tat protein enhances transcriptional activity from the promoter region in the 5′ LTR of an integrated provirus; the rev protein enhances transport of HIV unspliced messenger RNA (mRNA) from the nucleus. All of the genes are expressed in one genome in the infectious virus; however, in the lentivirus vector, these genes are split onto multiple different plasmids to provide flexibility and enhance safety. The genes necessary for virion production are expressed from two helper vectors, one containing the gag, pol, tat, and rev genes, and the other expressing the env gene (Figure 12.3E) (Osten *et al.*, 2007).

When developing the lentivirus viral vector system, scientists quickly realized that, when **budding**, HIV particles were able to insert within their viral envelopes heterologous viral protein from distinct viruses. This unique capacity has now been exploited in order to diversify the tropism of lentiviral particles. The tropism of lentiviral vectors is determined by the type of envelope protein used for the viral production; the use of heterologous env proteins is called **pseudotyping**. The most commonly used env protein is the vesicular stomatitis virus glycoprotein (VSV-G), which allows easy concentration of the virus to high titers by ultracentrifugation and gives the virus a broad tropism (VSV-G binds to cells via a yet unknown mechanism but is thought to be through electrostatic interactions) (Albertini *et al.*, 2012). VSV-G-coated lentiviral particles were shown to infect numerous tissues and cell types, including brain (Wiznerowicz and Trono, 2005).

To produce lentivirus, all plasmids are transfected together in 293T cells due to their high rate of transfection (see Figure 12.6). The use of stable packaging cell lines is not common due to the high toxicity of VSV-G, gag, and tat. The drawback of using lentivirus, retroviruses, and rhabdoviruses as a vehicle in the brain is that the size of virions (~90 nm) prevents spread of these viruses via extracellular spaces, and therefore each injection leads to infections of a limited number of neurons. Thus, these viruses may be used for targeting the hypothalamic nucleus in small animals like mice, but may not be suitable for nuclei of bigger animals, like rats or monkeys.

An important difference between lentivirus- and retrovirus-based vectors is that lentiviruses can infect non-dividing cells as they have developed mechanisms to transfer their genome into the nucleus through the nuclear pores. The delivery of retrovirus genome in the nucleus is dependent on cell division that mediates the disassembly of the nuclear envelope, allowing access of the viral DNA/integrase complex to the host chromosomes.

12.2.3   Rabies virus and rabies virus-based viral vectors

Rabies virus (RABV) is an enveloped single-stranded RNA virus of the Lyssavirus genus of the Rhabdoviridae family. RABV is a prototypic neurotropic virus since its life cycle is entirely adapted to survival and retrograde polysynaptic spread in the nervous system. As such, RABV-based vectors might constitute suitable tools to study neurohormone/peptide interneuronal connections in the brain. The RABV genome (Conzelmann *et al.*, 1990) is small, modular, and readily accepts foreign genes without the issues of instability inherent to other RNA viruses (Conzelmann, 1998; Mebatsion *et al.*, 1996). The RABV genome is packaged as a ribonucleoprotein complex in which RNA is tightly bound by the viral nucleoprotein (Figure 12.4A). The RNA genome of the virus encodes five genes whose order is highly conserved. These genes code for nucleoprotein (N), phosphoprotein (P), matrix protein (M), glycoprotein (G), and the viral RNA polymerase (L) (Figure 12.4B). The complete genome sequences range from 11,615 to 11,966 nucleotides (nt) in length.

To decrease cell toxicity, RABV was modified (Wickersham *et al.*, 2007), taking advantage of the capacity of the RABV genome to tolerate additions and deletions. This modification led to the appearance of the deletion mutant pseudotyped rabies virus, often referred as pseudoRABV. The use of this terminology should be revisited as pseudorabies usually refers to a Herpesvirus genus with high neurotropism (see section 12.2.4).

Pseudotyped rabies viruses are derived from WT rabies by deletion of the gene encoding rabies glycoprotein (RG). The RG gene, the most toxic component of the virus, is then replaced with the gene of interest (for example, GFP) (Figure 12.4C). Furthermore, as the viral glycoprotein is now missing (deletion of the RG gene), the virus was pseudotyped with the avian glycoprotein of avian sarcoma and leucosis virus, which does not exist in the mammalian brain. Similar to env in lentiviruses described above, the pseudotyped rabies viruses are produced from two different plasmids, one encoding the genome with the gene of interest and the second coding for the glycoprotein (e.g. avian glycoprotein). To be infected by these pseudotyped rabies particles, mammalian cells must express the avian TVA receptor. This can be achieved either by using viral vector expressing the TVA receptor (lentivirus vector or adeno-associated vector), which are injected in specific locations within the brain, or by specific genetic targeting of a subset of neurons using the Cre-lox system (for details see section 12.3.2 and Figure 12.6). The limitation of this approach is that it only allows for the tracing of monosynaptic circuits since the RG glycoprotein is missing and retrograde transport is now not possible.

An interesting twist to this system was recently established. During the specific genetic targeting of the neuronal subset of interest, the RG glycoprotein can also be expressed together with the receptor TVA. This allows for one round of virus production within the targeted neurons. The newly produced virus particles are now 'pseudotyped' with the WT RG protein and not with the envelope protein of the avian sarcoma and leucosis virus. This

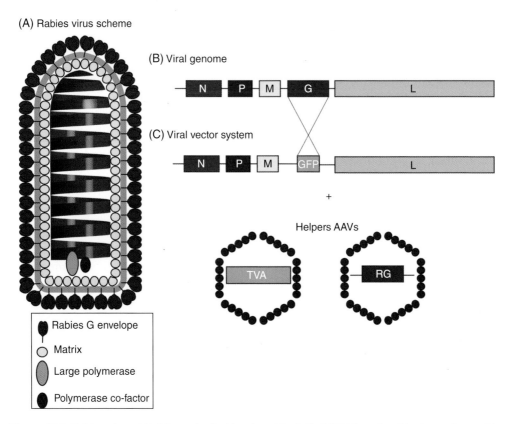

Figure 12.4 Rabies virus. (A) Schematic of rabies virus. The helical RNA is enclosed in the nucleocapsid. The large polymerase (L) and its co-factor (P) bind to the viral RNA. The inside of the lipid envelope is lined with matrix and the capsid is covered with the rabies glycoprotein. (B) Schematic of the rabies virus genome organization. G, glycoprotein; L, large polymerase; M, matrix; N, nucleocapsid; P, polymerase co-factor. (C) Diagram of the rabies virus viral vector. The rabies virus glycoprotein has been replaced with GFP. These particles are then pseudotyped with avian glycoprotein and infected into cells which have been previously infected with AAVs containing TVA and rabies glycoprotein (RG).

allows for one round of retrograde transsynaptic spread from the originally targeted neurons or multisynaptic spread (see also Figure 12.12).

Therefore, to infect mammalian cells and achieve retrograde transport of virus, the cells should be electroporated or infected via viruses with vectors expressing TVA receptor (to allow the virus to enter the cell) and RG (to achieve retrograde monosynaptic spread).

12.2.4 Adeno-associated virus

Adeno-associated viruses (AAVs) belong to the Parvoviridae family, genus Dependovirus. The virion consists of a non-enveloped capsid of about 20 nm in diameter with a 4.7 kb single-stranded DNA genome (Büchen-Osmond, 2001) (Figure 12.5A). AAVs rely on helper viruses, such as adenovirus (Ad) or herpes simplex virus (HSV), for replication of their genome (hence the name adeno-associated virus).

Adeno-associated viruses are highly endemic among humans, primates, and several other species. It is thought that 80–90% of adults are seropositive for AAV serotype 2. The AAV genome consists of two genes, rep and cap, flanked by 145 nt palindromic sequences, called inverted terminal repeats (ITRs). The ITRs contain all *cis*-acting sequences critical

(A) AAV virus scheme

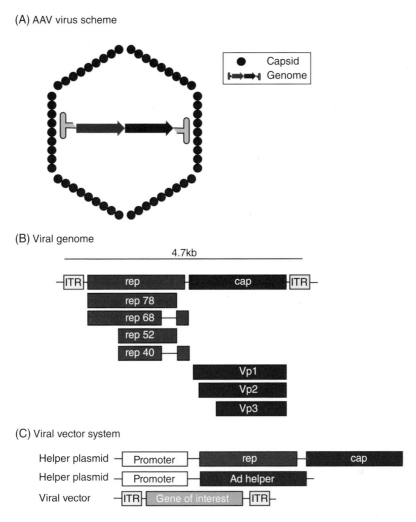

(B) Viral genome

(C) Viral vector system

Figure 12.5 Adeno-associated virus (AAV). (A) Schematic of the AAV capsid. AAV is a non-enveloped virus and its capsid is composed of the Vp1, Vp2, and Vp3 proteins. (B) Diagram of the AAV genome. There are four rep proteins produced (rep 78, 68, 52, 40) and three capsid proteins (Vp1, Vp2, Vp3). (C) Schematic of the AAV viral vector. The rep and cap genes are replaced by the gene of interest. The rep and cap are then complemented from an additional plasmid. Also, a plasmid containing the adenovirus helper functions is co-transfected to allow for proper viral packaging and assembly.

for virus packaging, replication, and integration. The rep gene contains two promoters and each transcript is regulated by internal splicing, resulting in production of four non-structural Rep proteins: Rep78, Rep68, Rep52, and Rep40 (Figure 12.5B). The Reps regulate replication, viral transcription, packaging of the AAV genomes, and site-specific integration. The cap gene encodes three structural proteins, virion protein 1, 2, and 3 (VP1, VP2, VP3), from two alternative transcripts and an alternative translation initiation codon. Upon infection of human cells, in the absence of a helper virus (Ad or HSV), the AAV genome either stays episomal or integrates into a specific region on the human chromosome 19 (19q13.3-qter), a site termed AAVS1 (McCarty *et al.*, 2004). The AAV then remains in a repressed state until reactivated by infection with the helper virus. This allows for replication and packaging of new AAV particles. Additionally, AAV relies on lysis produced by the helper virus to release its newly formed virions.

Recombinant AAV (rAAV) vectors hold great promise in gene therapy, mainly because of their non-pathogenic viral origin. Their main limitation for use in both gene therapy and basic research lies in the small size of the rAAV genome (4.5 kb), and hence small packaging capacity of the vectors. However, the small size (~20 nm) can be an advantage for rAAV as it is easily able to spread within the extracellular space. The rAAV vector design is quite simple. First, the rAAV ITRs contain all *cis*-acting elements necessary for replication and packaging in the presence of a helper virus; this allows the creation of a so-called gutless expression vector where the rep and cap can be completely replaced by the gene of interest as long as it contains the two flanking ITRs. Second, both rAAV rep and cap genes and all necessary adenoviral genes can be expressed by additional plasmids (Grimm *et al.*,1998) (Figure 12.5C). The expression of the helper adenovirus genes from a plasmid thus bypasses the requirement for co-infection with WT adenovirus, and the production of rAAV virions now involves only a simple co-transfection of human embryonic kidney (HEK293) cell line with the gutless expression plasmid (containing the gene of interest) and one or two helper plasmids. In the host cell environment, rAAV vectors stay mostly episomal (in both human and non-human cells). Despite the lack of integration, heterologous expression from these vectors appears to be stable for extended time periods (up to 1 year, according to our observations).

Adeno-associated virus tropism can be controlled at two levels: entry (see later) and cell type-specific promoter. The AAVs exist as 11 different serotypes that can infect different tissue types and different organisms. Interestingly, AAV tropism seems to be mostly determined by the capsid serotype and pseudotyping AAV with different serotype alters the tropism range. To increase the tropism of AAV (make AAV capable of infecting a broader range of cells) or to decrease its tropism (make AAV more specific for a specific cell type or cell line), chimeric AAVs were created and researchers were able to infect cells normally resistant to AAV infection. Recently, DNA shuffling-based approaches were developed to significantly increase the number of AAV variants. In this method, the DNA sequence of the capsid gene of AAV from nine different serotypes was randomly fragmented and reassembled to generate a chimeric capsid library. This library can now be used to screen which chimera is best to specifically infect your favorite cell type/line (Kienle *et al.*, 2012).

12.2.5 Herpesvirus

Herpesviridae is the family of DNA viruses, which infect mammals, birds, fish, and reptiles. There are 130 currently identified herpesviruses. There are five species of herpesviruses which infect humans: HSV-1, HSV-2, varicella zoster, Epstein–Barr, and cytomegalovirus; 90% of adults have been infected with at least one of these. Herpesviruses can exist in both an actively replicating phase and a **latent** phase where small numbers of viruses persist within a host without active production. HSV-1, HSV-2, and varicella zoster have their latent cycles within the neurons of the host (Smith, 2012).

Herpesviruses consist of a linear double-stranded DNA genome that requires relocation to the host cell nucleus for transcription of viral genes. The genome of herpesviruses is very large and encodes for 100–200 genes. The viral genome is encased inside a protein shell, which is surrounded by a tegument, which consists of viral proteins and viral mRNA. The tegument and the viral core are held together by the envelope.

One herpesvirus that has become an important tool in neurobiology is pseudorabies (PRV) which is the causative agent of Aujeszky disease in sheep and mad itch in cattle. PRV infection of livestock is often deadly to the animals. Neuroscientists have used the Bartha strain of PRV as a retrograde and anterograde transneuronal tracer. PRV is transported to the cell body via the axon. After replication and release, PRV is not just lytically released into the extracellular space. Instead, it will infect chains of hierarchically connected neurons via specific transsynaptic passages. PRV is administered by direct injection into the region of interest and has been genetically modified to express either RFP or GFP (Ugolini, 2010).

12.2.6   Viral vector summary

Viral vectors provide a method of delivering peptides, proteins, and shRNA in their vectors to neurons. The type of vector used depends on the application. Lentiviruses provide a vector which accommodates large inserts but integrates into the genome, which can cause unwanted side effects. Additionally, lentiviruses are large and are therefore restricted in their distribution into the brain tissue. AAVs are small and can easily spread within the neuronal space. They stay episomal (i.e. AAVs are not integrated into the genome of the host cell), but the insert is restricted by 5 kb. Although there is not yet any one perfect viral vector, the available constructs provide powerful tools in neurobiology (a review of viral vector characteristics can be found in Table 12.1).

12.2.7   Virus production

The general procedure of virus production, including transfection of permissive cells, harvesting of virus, and purification, is presented in Figure 12.6.

Step-by-step video protocols may be found in the *Journal of Visualized Experiments* (JoVE) (McClue *et al.*, 2011; Wang and McManus, 2009): www.jove.com/video/3348/production-and-titering-of-recombinant-adeno-associated-viral-vectors;   www.jove.com/video/1499/lentivirus-production.

12.2.8   Infusion of viral vectors into the brain

To infect neurons by viral vectors, viral-containing solutions are directly infused into the particular region of the brain using conventional **stereotaxic technique**. Briefly, prior to the infusion of virus, an anesthetized animal is placed into a steretotaxic frame with the head fixed by ear bars (Figure 12.7). Next, the skull is exposed and small holes are drilled

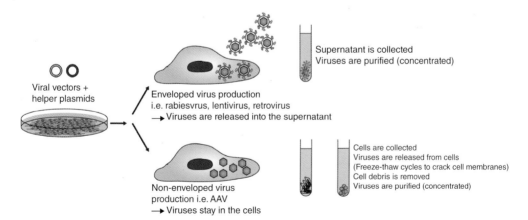

Figure 12.6 General virus production. Viral vectors are transfected into permissive cells along with any additional helper plasmids, which are required for viral production. After allowing for viral proteins to be made and new viruses to be assembled, the newly produced viruses are harvested. Enveloped viruses are directly released into the supernatant, therefore only the supernatant is collected and purified. Non-enveloped viruses require that the cells be harvested, usually by scraping the cells off the plate. The cells and the supernatants are then lyzed by freeze-thaw cycles in the presence of high salt buffers or buffers containing detergent. The cell debris is removed and the viruses can then be centrifuged and purified.

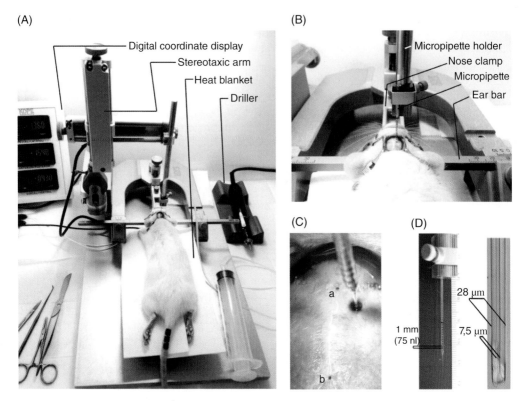

Figure 12.7 Stereotaxic set-up for infusion of viral vectors into the brain of small rodents. Here we demonstrate the instruments used and the main steps of the infusion of viral vectors. (A) After anesthesia, the rat is placed on a heat blanket and its head is fixed with ear bars and an incisor bar. (B, C) The skull is opened and bones cleaned to visualize the skull sutures. Coordinates for injection of specific brain areas are targeted in relation to the crossing points of skull sutures (a: bregma; b: lambda), a hole is drilled at the relevant point and the thin tip of a micropipette is slowly lowered into the injection area. (D) Small volumes of virus are pulled out of the micropipette according to the indicated scale. Panel D is taken from Cetin *et al.* (2006).

above the region of interest. A capillary with an extended tip of small diameter (outer diameter ~10 mm) is filled with viral solution and then slowly lowered into the desired brain region. A defined virus volume is then slowly infused, applying slight positive pressure. The whole procedure of viral infusion can be viewed online here: www.jove.com/video/2140/intracranial-injection-of-adeno-associated-viral-vectors.

12.3 Cell type-specific targeting of neuroendocrine neurons

The ultimate goal of viral vectors is to target a specific cell population expressing a specific neurohormone. There are two strategies that can be applied to achieve this goal.

12.3.1 Direct targeting and minimal promoter selection

Neuroendocrine neurons are preferentially located in the anterior hypothalamus and paraventricular (PVN), periventricular, supraoptic (SON), and arcuate (ArcN) nuclei. These neurons often occur in intermingled assemblies with other types of neuroendocrine neurons, **peptidergic neurons** (lacking the ability to secrete into the circulation), and conventional neurons (i.e. neurons operating only with classic neurotransmitters such as glutamate, glycine or GABA). Direct viral access to distinct neuroendocrine cell types is difficult to accomplish because the regulatory elements and mechanisms for specific expression may differ between similar cell types, are located in different brain regions and can even differ from parvo- to magnocellular variants of the same cell type, located within the same structures (such as the PVN) (Box 12.2).

Box 12.2   Parvocellular versus magnocellular neurons

Magnocellular neurons are typical neuroendocrine neurons, expressing two neurohormones: oxytocin (OXT) and vasopressin (VP). Parvocellular neurons expressing peptides are subdivided into two types: (1) neuroendocrine neurons expressing various releasing factors such as **corticotropin-releasing hormone** (CRH), **gonadotropin-releasing hormone** (GnRH) or thyrotropin-releasing hormones etc., acting via the portal circulation on the anterior pituitary cells; and (2) neuropeptidergic neurons projecting to spinal cord and brainstem (so-called preautonomic neurons) or incorporated in various brain circuits used to control, for example, sleep-wake cycle, pain perception, and hunger.

For some neuropeptides, the promoter design has resulted in successful, cell type-specific intervention. Many viral and genetic studies have involved the use of **oxytocin** (OXT) and **vasopressin** (VP) neuron-specific promoters. Magnocellular hypothalamic neurons in the **supraoptic nucleus** were genetically targeted by employing short promoters of OXT and VP genes first in transgenic mice and later by rAAV (Gainer, 2012). Previous studies had shown that an appropriate expression of transgenes in magnocellular OXT and VP neurons could be achieved with 5′ gene-adjacent sequences of about 600 bp for OXT and 2 kb for VP. Through further testing, these cell type-specific sequences were reduced to 200–300 bp for each gene promoter (Fields *et al.*, 2012; Ponzio *et al.*, 2012).

Another approach to uncover cell type-specific OXT and VP promoter regions was to screen for conserved sequences that surround the gene using the alignment software BLAT from the University of California, Santa Cruz (http://genome.ucsc.edu/cgi-bin/hgBLAT) (Knobloch *et al.*, 2012). Such alignments can reveal functional and specificity-defining sequences of the promoter for these genes. The identified mouse promoter sequences of 2.6 kb length for OXT and 1.9 kb length for VP allowed for cell type-specific expression in hypothalamic neurons (Knobloch *et al.*, 2012) (Figure 12.8). Furthermore, this short promoter preserved responsiveness to physiological stimuli such as lactation and osmotic challenge (Knobloch *et al.*, 2012). Sequence conservations were determined from genome alignments of mouse and eight other mammalian species, and suggested a broad promoter specificity and functionality in rodent species, such as mice and rats, but also in other mammals, including primates and humans.

To our knowledge, among typical neuroendocrine neurons, only OXT and VP cells were directly targeted by rAAV (Atasoy *et al.*, 2012; Fields *et al.*, 2012; Knobloch *et al.*, 2012; Ponzio *et al.*, 2012) and by lentiviruses (Zhang *et al.*, 2011). Using similar strategies, two other types of peptidergic neurons, expressing **orexin** hypocretin (or hypocretin [HCRT]) (Adamantidis *et al.*, 2007) and melanin-concentrating hormone (van den Pol *et al.*, 2004), have been targeted.

12.3.2   Indirect targeting by the use of transgenic animals

Since direct virus-based targeting of neuropeptidergic (and other types of) neurons may be unsuccessful (Nathanson *et al.*, 2009), an alternative technique to achieve the same goal is to design transgenic driver lines. They do not have the space restriction of viral vectors and can allow large promoters with regulatory elements to drive recombinase expression at the best possible specificity. However, the generation of transgenic mice is rather time consuming, expensive, and relatively unpredictable in terms of specificity or level of recombinase expression.

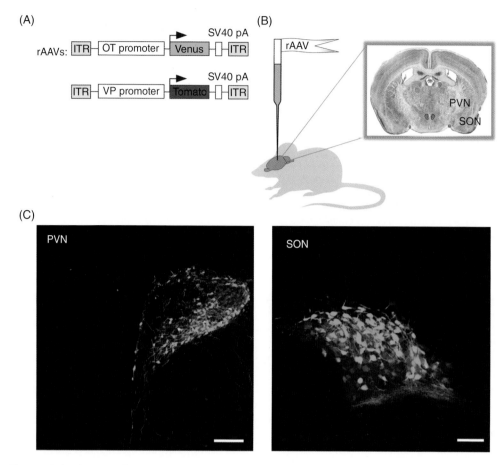

Figure 12.8 Direct viral vector-based targeting of neuroendocrine neurons. (A) The schemas of cell type-specific rAAV vectors, expressing Venus and Tomato (Td) under the control of OXT and VP promoters respectively. (B) Scheme showing injections of these two rAAVs into the PVN and SON. (C) Two weeks post infection, OXT cells were labeled with Venus (*green*), while VP neurons expressed Tomato (*red*). Scale bars = 100 μm (PVN) and 50 μm (SON).

While numerous neuronal-specific Cre drivers have been generated during the last 20 years (see databases of the Allen Institute for Brain Science: http://connectivity.brain-map.org/transgenic/search/basic, and GENSAT: www.gensat.org/cre.jsp), only a few lines properly express Cre in neuroendocrine neurons (see the list of available lines in Knobloch *et al.*, 2014) (Box 12.3).

Box 12.3   Cre recombinase

The recombinase of choice for molecular biology is Cre recombinase. This enzyme was derived from a bacteriophage (a virus which infects bacteria) and is able to perform site-specific recombination events between two DNA sites (called loxP sites). The recombination produced by Cre depends on the location and orientation of the loxP sites. When the loxP sites are in the same orientation, the DNA in between will be removed as a circular piece of DNA. If the loxP sites are in the opposite orientation, the DNA in between will be inverted. This allows researchers to knock-in or knock-out genes of interest.

Despite the limited number of transgenic mice available, a wide range of Cre-responder viruses (i.e. viruses carrying loxP sites) is available. The Cre-responder viruses are exclusive DNA viruses (such as rAAV). RNA viruses are not used because they do not have a DNA intermediate (Cre only works on DNA). Additionally, the genomes of RNA viruses are maintained in the cytoplasm while Cre is nuclear. Initially, these Cre-responder vectors carried the integration of an LSL (loxP-STOP-loxP)-cassette between promoter and gene of interest, which may prevent unintended expression (Kuhlman and Huang, 2008). However, due to size limitations in viral vectors, the shortened STOP cassette (Figure 12.9A) is prone to occasional 'read-through' events, leading to leakiness even in the absence of Cre recombinase (Grinevich, personal observations). A further approach for Cre-dependent targeting of distinct cell types was termed FLEX switch (Atasoy *et al.*, 2008) or DIO (double-floxed inverted open reading frame) (Sohal *et al.*, 2009) (Figure 12.9B). In this approach, the combination of two pairs of heterologous lox sites around the coding region ensures the gene's turn-back in legible, upright orientation only after Cre activity (see such vectors at: www.stanford.edu/group/dlab/optogenetics/ sequence_info.html). An example of targeting neuroendocrine neurons, expressing corticotropin-releasing hormone (CRH) via Cre-dependent rAAV, is demonstrated in Figure 12.9C.

12.4   Application of viral vectors

In this section, we outline the advantages and in many cases even the exclusivity of viral vectors as tools for studying multifunctional properties of neuroendocrine neurons. Furthermore, we will demonstrate the advantages of the viral technique, providing readers with results obtained from the Grinevich team.

12.4.1   Anatomy

12.4.1.1   Cell morphology and plasticity

Viral vectors are an ideal tool for studying the cytology of neuroendocrine neurons and their plastic changes under certain conditions. As a demonstration of the potency of such applications, we will describe here the results of our own experiments on plastic changes of OXT neurons in lactating rats (Box 12.4).

Box 12.4   OXT neurons in lactation

During lactation, OXT neurons change their electric properties such that they start bursting in synchrony with short silent intervals in between. Bursts induce simultaneous release of OXT from all neuroendocrine OXT neurons into the blood, resulting in milk let-down. During the lactation period, OXT neurons undergo drastic morphological changes, which should facilitate their synchronous bursting. Most strikingly, membranes of OXT cell bodies directly and tightly interpose to each other due to the retraction of glial processes. See the palette of morphological changes in OXT neurons during lactation in Theodosis and Poulain (2001).

In the course of our experiment, rAAV expressing Venus (a brighter version of GFP), driven by the cell type-specific OXT promoter, was introduced in the SON of pregnant rats, which were killed either the day before delivery or on the first day of lactation (i.e. on the day of delivery). As can be seen in Figure 12.10, in lactating rats the Venus

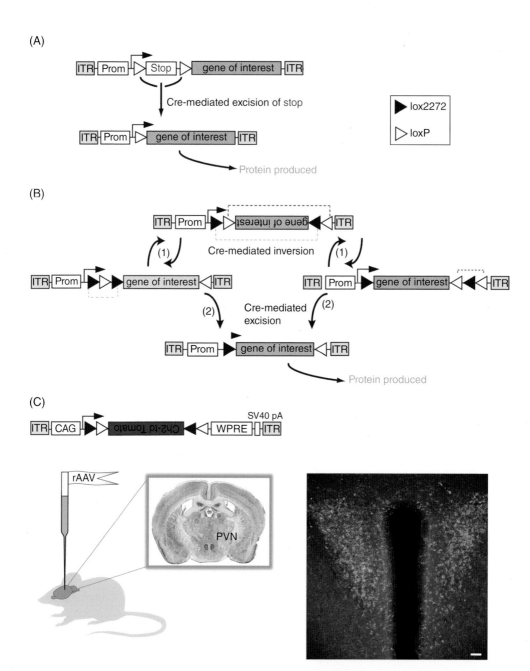

Figure 12.9 Combination of viral vectors with transgenic animals for cell type-specific targeting of hypothalamic neurons. (A) To achieve a Cre-dependent expression of the gene of interest, a STOP cassette is anchored by loxP sites. The STOP cassette is usually inserted between the promoter and the coding sequence (gene of interest). In the presence of Cre, recombination at the loxP sites excises the STOP cassette, thereby activating expression of the gene of interest. (B) Scheme of Cre recombinase FLEX switch system. In this system, the coding sequence (gene of interest) is inverted and flanked by two pairs of heterotypic, antiparallel lox recombination sites. In the presence of Cre, the antiparallel lox sites first undergo an inversion of the coding sequence (1), followed by Cre-mediated deletion of identical parallel lox sites (2), and finally transcription of the gene of interest. Cre can act on either pair of lox sites: loxP (*white triangles*) or lox2272 (*black triangles*). (C) FLEX system used for neuroendocrine neurons. The PVN of CRH-iCre mice was injected with a virus equipped by double-floxed inverted open reading frame of fluorescent marker mCherry under the control of ubiquitous promoter CAG. Four weeks post infection, CRH neurons in the injected site were labeled in red. CRH-positive neurons of the PVN were visualized by green color as can be seen on non-infected CRH neurons residing in the contralateral PVN. Scale bar = 50 μm.

expression 'visualizes' tremendous morphological changes of OXT neurons as cells are closer together and have an increased size, as well as multiple sprouting varicosities and protrusions among dendrites. Without the use of a 'fluorescent' viral vector, these changes are only visible by very laborious and time-consuming immunohistochemical electron microscopy. Even with the availability of very sensitive antibodies, the viral expression of fluorescent proteins is faster and easier. Additionally, the fluorescent markers allow for recording or stimulating with *post hoc* anatomical analysis, which is usually not trivial, with biocytin or neurobiotin filling of cells, widely used in electro-physiological practice.

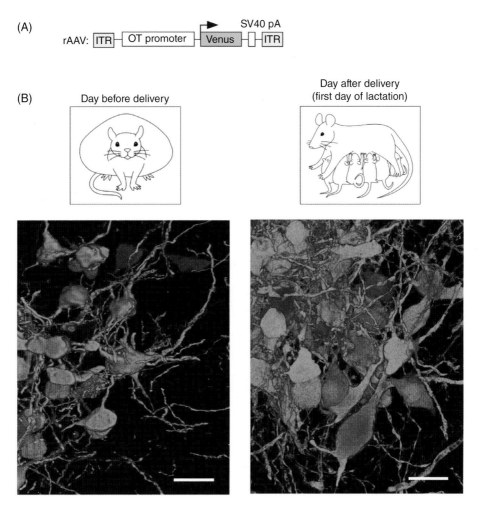

Figure 12.10 Plastic changes in OXT neurons evaluated by viral technique. (A) Schema of cell type-specific rAAV vector, expressing Venus under the control of OXT promoter. (B) The rAAV was injected in the PVN of two rats on day 10 of the pregnancy. In lactating rats, the Venus expression was higher than in rats prior to the delivery of pups (i.e. before onset of milk let-down). Furthermore, in lactating rats, OXT cell somas were bigger, and numerous sprouting processes with varicosities were observed. Scale bars = 50 μm.

12.4.1.2 Anterograde tracing

Viral vectors are excellent tools for tracing anterograde projections of groups of neurons (Harris *et al.*, 2012). Advantages of a viral-based technique are (1) high detectability (in comparison to conventional tracers, such as wheat germ agglutinin, horseradish peroxidase or cholera toxin C); (2) the possibility to follow projections of defined groups of homotypic neurons; and (3) monitoring plastic changes in axonal pathways during, for example, embryonic or early postnatal development. Due to these features, viral vectors allowed the discovery of new monosynaptic brain circuits and their plastic changes (see, for example, Broser *et al.*, 2008; Grinevich *et al.*, 2005). In Figure 12.11, we demonstrate **anterograde tracing** of OXT neurons, infected by rAAV equipped with OXT promoter to drive Venus expression.

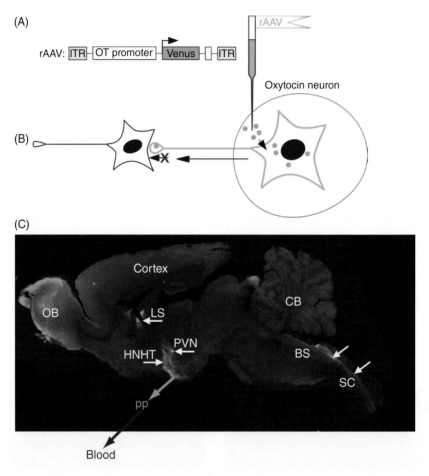

Figure 12.11 Viral vector-based tracing of anterograde projections of OXT neurons. (A) Schema of cell type-specific rAAV vector, expressing Venus under the control of OXT promoter. (B) Injection of rAAV into the PVN resulted in labeling of OXT neuron somas and processes. The labeling is restricted to the infected PVN without further propagation of GFP to postsynaptic cells. (C) Sagittal rat brain section after rAAV injection into the PVN. Prominent fiber distribution is visible within the hypothalamus (area ventral to the PVN), septum, brainstem, and spinal cord. Venus-positive fibers forming bundles are indicated by arrows. BS, brainstem; CB, cerebellum; HNHT, hypothalamo-neurohypophysial tract; OB, olfactory bulb; pp, posterior pituitary; PVN, paraventricular nucleus; S, septum; SC, spinal cord.

The main limitation of rAAV as a retrograde tracer is its lack of transsynaptic transmission, which is essential for labeling of postsynaptic cells. This limitation can be solved by the use of electron microscopy, employing immunohistochemical labeling of the presynaptic terminal and the identification of postsynaptic cells (see, for example, Grinevich *et al.*, 2005; Knobloch *et al.*, 2012). However, this is a technically difficult, time-consuming, and laborious approach. As an alternative, herpesvirus (strain H129) can be used due to its capacity to be transferred via synapses. Furthermore, it may carry in the modified genome 'floxed' STOP cassettes to express a fluorophore in a Cre-dependent fashion (see Figure 12.9A), so initially infected Cre-expressing cells as well as postsynaptic cells become visible (Lo and Anderson, 2011). The main limitation of this method is polysynaptic spread of the herpesvirus, which makes it difficult to dissect first, second, and other orders of connected neurons in the circuit.

12.4.1.3 Retrograde tracing

Box 12.5   First techniques for retrograde tracing

Prior to the application of viral-based vectors, anatomical connections between brain regions and groups of cells were visualized with methods like retrograde degeneration or retrograde spread of substances (such as horseradish peroxidase) within nervous tissue. The method of retrograde degeneration required a lesion of a certain brain region. Due to the damage of axons, distantly located neurons projecting to this region underwent retrograde degeneration. The cell death indicated their monosynaptic projection to the lesioned region. In relation to the neuroendocrine system, the mechanical transection of the pituitary stalk leads to the retrograde degeneration and death of magnocellular OXT and VP neurons. In the 1950s, these observations helped to determine that neurohormones are transported from hypothalamic cells to the posterior pituitary and are released from there into the systemic blood circulation rather than locally synthesized in the posterior pituitary. Indeed, employing this technique at that time deciphered the existence of the hypothalamic-pituitary system. For an excellent history of neuroendocrinology, see the review by Allan Watts (2011).

The first viral-based **retrograde tracing** technique was the implementation of wild-type rabies virus (Box 12.5). As a tracer, the rabies virus has two great limitations: (1) polysynaptic labeling of neurons and (2) high toxicity as neurons and animals die within a week. While the latter obstacle is fatal, the time of infection may roughly correlate with retrograde transmission to first, second, and third orders of neurons. Many publications in the 1990s to early 2000s employed wild-type rabies virus to show the connectivity of non-neuroendocrine (preautonomic) hypothalamic neurons with brainstem and spinal cord.

The next advance in performing retrograde tracing of neural circuits by virus was the employment of pseudorabies vector. Due to its non-toxic nature, this was used in many studies of hypothalamic connections (for example, see Gerendai *et al.*, 2001; La Fleur *et al.*, 2000). This virus was modified to express fluorescent proteins and, hence, can be considered as the first retrograde viral vector. With respect to neuroendocrine neurons, the pseudorabies viral vector (Barta, Ba2001 strain) carrying Cre-dependent (via insertion of short lox-STOP-lox cassette; see Figure 12.9) fluorescent proteins deciphered extensive innervation of neurons expressing GnRH (Yoon *et al.*, 2005). The main limitation of this method is that retrograde spread of this virus is not limited to one synapse, so polysynaptic labeling occurs, which makes it difficult to dissect direct or indirect connections of the primary infected neurons.

A more advanced approach for viral-mediated retrograde tracing is the application of the deletion-mutant pseudotyped rabies virus, which spreads monosynaptically and is less harmful for cells than wild-type rabies virus (Ginger *et al.*, 2013), although it is not completely harmless as it causes instant cell death within 2–3 weeks. The use of this virus helped us to reveal the regions which receive the synapses of hypothalamic OXT neurons (Knobloch *et al.*, 2012) (Figure 12.12).

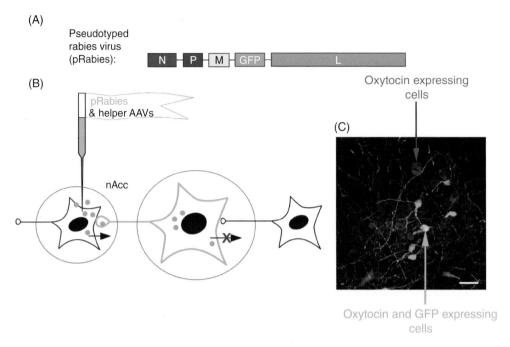

Figure 12.12 Viral vector-based tracing of retrograde projections of OXT neurons. (A) Schema of deletion-mutant pseudotyped rabies virus, PRV, expressing GFP. (B) The injection of PRV followed by (in 1 week) injection of a mixture of two rAAV expressing EnV and rabies glycoprotein (see also Figure 12.4) results in retrograde labeling by GFP of first-order input neurons, transsynaptically connected with initially infected neurons. (C) GFP-expressing neurons that send projections to the nucleus accumbens (nAcc). One cell residing in one of the OXT accessory nuclei (AN) contains both GFP (from retrogradely transmitted PRV) and OXT-immunoreactivity (*red color: indicated by arrow*). The image is from Knobloch *et al*. (2012). Scale bar = 50 μm.

A newly developed viral vector based on the canine adenovirus, which transmits retrogradely and strictly monosynaptically, overcomes the obstacle of toxicity of deletion-mutant pseudorabies virus, and can be used for tracing retrograde connections of peptidergic neurons, as was shown by Wu and colleagues (2012).

In addition to anatomical tracing *per se*, viral vector-expressing proteins can be introduced into cells to identify the nature of their neuronal processes (e.g. labeling axons by fusion of axonal marker Tau with fluorescent markers) as well as to label presynaptic and postsynaptic sites with synapsin or PSD95 proteins, respectively, that are fused with fluorescent markers.

12.4.1.4   Monitoring trafficking and release of neurohormones

Although the well-described mechanism of Ca^{2+}-dependent neurohormone release has been known for some time, no direct monitoring of vesicular trafficking, docking, priming, and exocytosis in hypothalamic neurons has been performed either *in vitro* or *in vivo*. However, due to great progress in the study of vesicular transport and secretion (the 2013 Nobel Prize in Physiology and Medicine was awarded to James E. Rothman, Randy W. Schekman, and Thomas C. Südhof for their great insight into mechanisms of the vesicular transport and secretion), it has become possible to monitor movement of neurohormone-containing **large dense-core vesicles** (LDCV) within neuroendocrine cells. The following technical achievements have been made: (1) generation of functionally neutral cargo (for example, fragment of neuropeptide Y [NPY]), carrying a constant fluorescent marker; and (2) development of pH-sensitive fluorophores (phluorins). Fusion of the cargo marker with pH-dependent fluorophore allows detection of the release event. As the pH increases, the phluorin becomes visible. The interior of the LDCV is acidic and, therefore, phluorin is not visible. However, when fusion of the LDCVs with an external cell membrane occurs after exocytosis, the compartment becomes more basic and the phluorin changes conformation and emits fluorescence in the GFP spectrum. The visualization of processes of trafficking and release of LDCVs was monitored in a culture of hippocampal neurons (van de Bospoort *et al.*, 2012).

There are no doubts that this technique will soon be implemented to help dissect one of the key questions in modern neuroendocrinology: how trafficking of neurohormone-containing LDCVs and their release differ between dendrites (dendritic release), axons contacting blood vessels ('axo-vasal' contacts), and axons in various brain regions (axonal neuropeptide release). See the description of all modes of neuropeptide release in Knobloch and Grinevich (2014).

12.4.2 Manipulation of neuroendocrine neurons

12.4.2.1 Ablation

Box 12.6   Diphtheria toxin to kill neurons

The method of using diphtheria toxin (DT) to kill neurons was invented by Ari Weisman's team (Buch *et al.*, 2005). The method is based on the fact that rodents are not sensitive to DT because they do not express its receptor (iDTR; 'i' stands for humanized, modified version of DTR). Therefore, if this receptor is expressed in cells artificially (via virus or in transgenic mice), injection of its ligand – DT – will induce death. The method is universal so can be used for all types of rodent cells, including neurons.

The common method of **ablation** of neurons is based on interaction of diphtheria toxin (DT) receptor (humanized version is abbreviated as iDTR) with its ligand – DT (Box 12.6).

As an example, expression of iDTR in OXT neurons of the SON with subsequent treatment of rats with DT leads to almost complete loss of OXT neurons shown by *post hoc* staining with antibodies against OXT (Figure 12.13). However, two main obstacles exist to using this technique:

- the ablation of neurons may induce retrograde degeneration of neurons synapsing onto ablated neurons as well as provoking gliosis and altering activity of brain regions innervated by ablated neurons (Wu *et al.*, 2008)
- the death of neurons might affect the physiology of neighboring cells.

These obvious limitations make this method less than ideal for functional studies of neuroendocrine circuits. Therefore, the selective inhibition of the release of a neuropeptide or neurohormone (see later) must be considered as an alternative to the iDTR approach.

12.4.2.2 Alterations of release

In contrast to the manipulation of the release of classic neurotransmitters, such as glutamate and GABA, very little attention has been paid to the mechanisms and manipulation of the release of neurohormones. One agent which can be applied for inhibition of release of LDCVs is tetanus toxin (Tetx) (Box 12.7).

In respect of neuroendocrine cells, it was shown that 'whole Tetx' (composed of heavy and light chains) inhibits calcium-dependent VP release in permeabilized nerve endings (Dayanithi *et al.*, 1991) and suppresses depolarization-evoked release of OXT and VP in a time- and dose-dependent manner in isolated nerve endings (Halpern *et al.*, 1990). These findings opened the possibility to apply Tetx in a viral vector form to overexpress it in neuroendocrine cells and to monitor its effect on transport and release of a neuropeptide.

Box 12.7 Tetanus toxin to inhibit neuropeptide/neurohormone release

Tetanus toxin (Tetx) is an extremely potent neurotoxin synthesized by the anaerobic bacterium *Clostridium tetani*. Tetx has a molecular weight of 150 kDa and is translated from the Tetx gene as a single protein, which is subsequently cleaved into a 100 kDa heavy and a 50 kDa light chain connected via disulfide bonds. While the heavy chain mediates neuroselective binding, internalization, intraneuronal sorting, and translocation of the light chain to the cytosol, the light chain is responsible for the catalytic inhibition of the synaptic transmission. Once it is present in the cytosol, the light chain attacks vesicle-associated membrane proteins (VAMPs), by cleaving synaptobrevin with unique selectivity at single sites. The first demonstration of the mechanism of the intracellular action of the Tetx light chain (TetLC) and its efficiency in blocking synaptic transmission (i.e. synaptic release of synaptic vesicles) was provided in the fruit fly, *Drosophila melanogaster* (Sweeney et al., 1995).

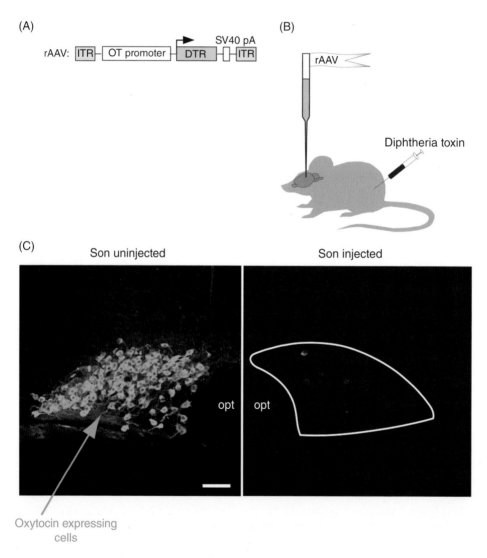

Figure 12.13 Viral vector-based induction of death of OXT neurons. (A) Schema of rAAV vector, expressing iDTR under the control of OXT promoter. (B) Two weeks after infusion of the rAAV in the SON, the rat was subjected to intraperitoneal injection of DT (dose is 50 μg/kg of body weight) each day for five consecutive days. Ten days after the final injection, the rat was killed. (C) In the injected SON, only single OT cells (*green*) were detected, while in the non-injected site numerous OT cells can be seen. Scale bar = 50 μm.

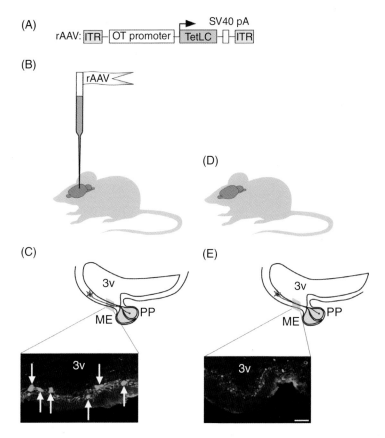

Figure 12.14 Viral vector-based inhibition of OXT release. (A) Schema of rAAV vector, expressing tetanus toxin light chain (TetLC) under the control of OXT promoter. (B) Injection of the rAAV bilaterally into the PVN and SON. (C) Giant swellings, containing OXT-immunoreactivity (*green; indicated by arrows*), were observed in any segment of the tract, especially in the segment at the bottom of the third ventricle (the area called the 'median eminence'). (D) Non-injected control rat. (E) In control rats, there are no swellings found in the median eminence. 3v, third ventricle; ME, median eminence; PP, posterior pituitary. Scale bar = 50 μm.

rAAV-mediated expression of TetLC in OXT neurons led to an accumulation of OXT in the axons of the hypothalamo-neurohypophysial tract, reflecting inhibition of OXT release (Figure 12.14). Importantly, no sign of OXT cell degeneration or death was observed over a period of 6 months (not shown).

12.4.2.3 Alterations of translation

MicroRNA and small hairpin RNA
MicroRNAs are a class of short non-coding RNAs that suppress mRNA translation in all mammalian cell types, including neuroendocrine neurons (Zhang *et al.*, 2011) (Box 12.8).

Viral vectors are the most efficient and most frequently used carriers of microRNAs and shRNAs (Danos, 2008; Grimm *et al.*, 2005). rAAV vectors can be used not only to overexpress, but also to inhibit specific microRNAs. To achieve inhibition, rAAVs are designed to encompass antisense microRNA inhibitory sequences that bind to specific microRNAs (see review by Xie *et al.*, 2012). In order to express microRNAs, shRNAs or microRNA inhibitory

---

Box 12.8 MicroRNAs, small hairpin (sh)RNA, and RNA interference

MicroRNAs are small non-coding RNAs that are transcribed by the RNA Pol II complex in a similar way to mRNAs. MicroRNAs bind roughly half of the protein-coding mRNAs. Each microRNA can bind many mRNA targets, and conversely each transcript usually contains binding sites for several microRNAs. In mammals, precursor microRNA – small hairpin RNA (shRNA) – undergoes cleavage in nucleus (by endonucleases of the Drosha complex) and in the cytoplasm (by Dicer). This leads to the formation of ~22 bp long microRNAs. In combination with the Dicer complex, these small microRNAs recognize the target mRNA transcripts and interfere with their stability and/or translation (the process is termed 'RNA interference').

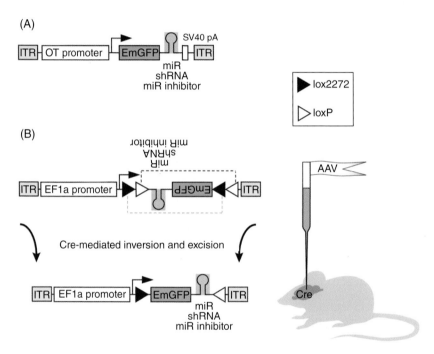

Figure 12.15 Viral vector-based manipulation of microRNA. (A) Direct targeting (shown with OXT promoter) to deliver miRNAs, shRNAs, and miRNAs inhibitors to neuroendocrine neurons. (B) An alternative approach to deliver miRNAs in the inverted orientation into Cre-expressing transgenic mice.

sequences in selected neuroendocrine neurons by viral vectors, direct or indirect strategies can be used, as discussed previously (section 12.3), and can be viewed in Figure 12.15.

12.4.2.4   Alterations of transcription (Box 12.9)

Overexpression

The easiest way to manipulate transcription is via virally mediated overexpression of neurohormones as most neurohormones have a relatively short open reading frame (ORF), which can be suited for both AAV and lentivirus. This approach was used only for CRH-ergic neurons in extrahypothalamic regions (Regev *et al.*, 2011).

Keeping in mind that expression of neurohormones occurs at a very high level (and one LDCV filled with OXT contains ~85,000 molecules of this neuropeptide; Leng and Ludwig,

---

Box 12.9 Transcriptional factors

Transcription factors are classified in five superclasses, 32 classes, and 61 families, comprising hundreds of molecules regulating expression of thousands of peptides and proteins, which are organism and tissue specific (see full list of transcriptional factors at: www.gene-regulation.com/pub/databases/transfac/clSM.html). With respect to the CNS, it was estimated that it contains at least ~1000 proteins. See Supplemental Table S1 for transcription factor classification. Surprisingly, only a small fraction of the CNS proteins shows activity-dependent regulation and most promoters of genes of these proteins contain cAMP-responsive element (CRE) (West *et al.*, 2002). The transcriptional factor which binds CRE is the CRE-binding protein (CREB).

2008), it is technically difficult to overexpress a neurohormone via viral delivery of extra copies of DNA-encoding neuropeptide. However, it can be efficiently done, for example, in animals with deficiency or absence of neurohormonal production such as VP-deficient Brattleboro rats (Box 12.10). Viral-mediated restoration of VP expression was achieved in VP-deficient Brattleboro rats via adenovirus (Geddes *et al.*, 1997), adeno-associated virus (Ideno *et al.*, 2003) or lentivirus (Bienemann *et al.*, 2003). This intervention compensated the symptoms of diabetes insipidus, i.e. normalizing water intake, volume, and osmolarity of urine.

Box 12.10   Brattleboro rats

Due to spontaneous mutation in the VP gene in a rat strain named Brattleboro rats, the VP peptide is not properly packaged into LDCVs and, hence, cannot be released from axonal endings in the posterior lobe of the pituitary to the systemic blood circulation. As a consequence, VP-sensitive water channels – aquaporins (especially aquaporin 2) – do not insert into apical membranes of the epithelial cells of collecting ducts in the kidney. This results in a drastic decrease of the kidney's capacity to concentrate urine, characterized by an increase in water intake and urine volume as well as by a decrease of urine osmolarity. This pathology is named diabetes insipidus.

Manipulation of transcriptional factors

Among transcriptional factors, CREB exerts a multifunctional and universal role in all neurons, including neuroendocrine neurons. For example, one of the members of the CREB family, CREB3L1, was shown to be crucial for induction of VP expression under osmotic challenge (Greenwood *et al.*, 2014). The infection of VP neurons with rAAV carrying CREB3L1 under the control of the VP promoter enhances VP expression in infected neurons (Figure 12.16). However, in implementing such an approach for functional studies (for example, in the modeling of an inappropriate VP secretion), one should be aware of many other proteins which can also be potentially overexpressed in infected neurons. AAV transduction has also been used to selectively target OXT and VP cells in the SON with fluorescence in order to identify them for cell-specific laser dissection to identify selective transcriptional regulators in these cells (Humerick *et al.*, 2013).

12.4.3   Gene deletion in neuroendocrine neurons

In addition to manipulation of translation and transcription of genes, viral vectors can be used to constitutively delete various genes from neuroendocrine neurons. In this case, viral vectors should express Cre recombinase (see section 12.3 and Box 12.3) and should be delivered into transgenic animals carrying a 'floxed' gene of interest. To selectively target certain types of neuroendocrine neurons, viruses equipped with cell type-specific promoters (e.g. OXT promoter) can be used. However, since the number of such 'viral' cell

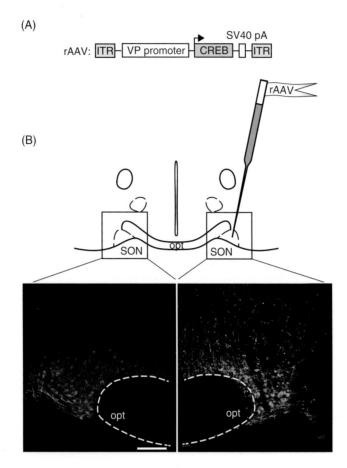

Figure 12.16 Viral vector-based overexpression of transcription factor CREB3L1 in VP neurons.
(A) rAAV vectors, equipped with the CREB3L1 driven by VP promoter. (B) Infusion of the rAAV into one SON resulted in an enormous increase in VP-immunoreactivity to non-injected sites. Scale bar = 200 μm.

type-specific promoters is limited, this approach will require generation of new transgenic lines carrying 'floxed' genes of interest in selected cell types.

12.4.4   Manipulation of electrical activity

12.4.4.1   Optogenetics (Box 12.11)

Box 12.11 Optogenetics

The term 'optogenetics' describes a technology of genetic intervention in biological cells to render them light sensitive and control their activity. Algae and microbes offer blueprints for the required molecular modules, membrane-bound ion pumps and channels, which became a rapidly advancing and indispensable tool in neuroscience (see comprehensive review by Fenno *et al.* (2011) and numerous references therein). Boyden and colleagues (2005) first demonstrated the initiation of light-dependent action potentials in mammalian neuronal tissue via viral delivery of an opsin, channelrhodopsin 2 (ChR2). ChR2 is a 7-transmembrane protein derived from the seaweed *Chlamydomonas reinhardtii*; after stimulation with 473 nm light (i.e. 'blue light'), it rapidly opens a cation-permeable channel, which depolarizes the cell and initiates action potentials under precise temporal control (Boyden *et al.*, 2005).

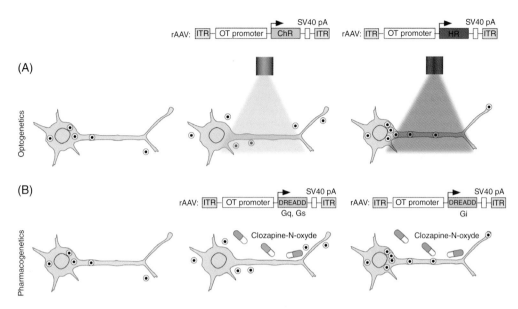

Figure 12.17 General concepts of viral vector-based optogenetics and pharmacogenetics. (A) and pharmacogenetics (B) applied for neuroendocrine neurons. ChR, channelrhodopsin; DREADD, designer receptors exclusively activated by designer drugs; Gi, Gq, Gs, subunits of G-coupled receptors; HR, halorhodopsin.

Viral vectors are most frequently used to introduce **opsins** into neurons, including peptidergic and neuroendocrine neurons, followed by illumination of the opsins *in vitro* and *in vivo* (Figure 12.17A). In the work of Atasoy and colleagues, it was shown that optic stimulation of neurons expressing **agouti-related peptide** in the ArcN or OXT (presumably preautonomic) neurons of the PVN resulted in the inhibition of food intake. Another group showed that the activation of hypocretin neurons located in the lateral hypothalamus induced arousal/awakening effect in mice (Adamantidis *et al.*, 2007). Our group reported on optogenetic manipulation of OXT neurons (Knobloch *et al.*, 2012) and established protocols suitable for neuropeptides (see Knobloch *et al.*, 2014). Other types of neuroendocrine neurons are awaiting application of this technique.

12.4.4.2   Pharmacogenetics

Although the optogenetic approach allows very short (from milliseconds to a few seconds scale) temporal resolution to manipulate neurons with very fast and clear changes in electrophysiological characteristics of neurons and changes in behavior, it requires an invasive step: implantation of an optic fiber (100–200 mm diameter), which may interfere with the normal life of animals. A powerful alternative to optogenetics is pharmacogenetics (particularly DREADD – see Aston-Jones and Diesseroth (2013) and Box 12.12).

Several reports have described the use of viral vector-mediated DREADD in various brain regions, including hypothalamus (see, for example, Silva *et al.*, 2014). With respect to peptidergic and neuroendocrine neurons (Figure 12.17B), viral-mediated activation of hypocretin (orexin) neurons in the lateral hypothalamus promotes wakefulness and decreases the rapid eye movement phase of sleep (Sasaki *et al.*, 2011). Another study using DREADD showed a distinct attribution of pro-opiomelanocortin (POMC)-ergic neurons located in the ArcN and solitary tract nucleus for feeding behavior (Zhang *et al.*, 2013). Furthermore,

---

Box 12.12   Pharmacogenetics

The term 'pharmacogenetics' describes a technique of genetic intervention combined with pharmacology (i.e. external application of chemical drugs). Although pharmacogenetics uses many tools, the best known and most widely used is DREADD: Designer Receptors Exclusively Activated by Designer Drugs (see Farrell and Roth, 2013). The general concept of DREADD is the activation or inhibition of neurons via the expression of specific G coupled receptors (Gs or Gd for activation and Gi for inhibition) followed by injection of their neutral (i.e. having no physiological activity) ligand clozapine n oxide.

---

implementation of DREADD for neuroendocrine VP neurons led to a conclusion that these neurons are involved in the regulation of food intake (Pei *et al.*, 2014).

12.5   Perspectives

Although magnocellular neurons are among the best studied neuroendocrine neurons in the brain and hence have been referred to as Rosetta stones for the fields of neuroendocrinology and neuroscience (Gainer *et al.*, 2002), the use of viral vector-based techniques is now bringing new opportunities to the study of this category of neurons. As indicated in this chapter, viral vectors are very useful tools for the field of neuroendocrinology as they help answer questions about the anatomy, physiology, and pathology of neuroendocrine systems relatively fast and more cheaply than the generation of faithful transgenic mice. The sophisticated viral systems, analogous to tetracycline-dependent systems for tagging experience-activated neurons (Ramirez *et al.*, 2013), should also be applicable to neuroendocrine neurons. Furthermore, viral targeting of neuroendocrine neurons will be further enriched by combinatorial approaches, such as opto- and pharmacogenetics combined with functional magnetic resonance imaging (fMRI) or electrophysiological recording. This will allow *in vivo* monitoring of neuroendocrine neuron activity and the registering of effects of endogenous neuropeptide release on overall brain activity. Finally, due to the fact that viral vectors (especially rAAVs) can be used in various species, including primates (see reviews by Leung and Whittaker (2005) and Asokan *et al.* (2012)), it is expected that in the future they will find new applications in the gene therapy of human neuroendocrine disease.

Acknowledgments

The authors thank Masters students Sadia Oumohand, Lena Heuschmid, and Tim Gruber for their careful reading and important (from their viewpoint as actual students) suggestions on the composition and writing style of the chapter. We thank Ari Weisman, David Murphy, and Mazahir T. Hasan for providing plasmids containing iDTR, CREB, and TTLC, respectively. We are especially grateful to Judith Müller and Heike Böhli for their help in the cloning and production of these viral vectors. The preparation of the manuscript was supported by the Chica and Heinz Schaller Research Foundation (to VG and SB), Olympia Morata Women in Science Fellowship (to MS), German Research Foundation (DFG) grant GR 3619/4-1 (to VG), Royal Society of Edinburgh Award (to VG), and German Academic Exchange Service (DAAD) program for partnership between German universities and the University of Tsukuba, Japan (to VG).

References

Adamantidis, A.R., Zhang, F., Aravanis, A.M., Diesseroth, K., de Lecea, L. (2007) Neural substrates of awakening probed with optogenetic control of hypocretin neurons. *Nature*, **450**, 420–424. [*The authors achieved cell type-specific targeting of hypocretin (orexin) neurons using lentivirus carrying a 3 kb mouse promoter. Using expression of ChR2 driven by hypocretin promoter, the authors showed that activation of this neuronal type induces awakening.*]

Albertini, A.A., Baquero, E., Ferlin, A. and Gaudin, Y. (2012) Molecular and cellular aspects of rhabdovirus entry. *Viruses*, **4**, 117–139.

Asokan, A., Schaffer, D.V. and Samulski, R.J. (2012) The AAV vector toolkit: poised at the clinical crossroad. *Molecular Therapy*, **20**, 699–708.

Aston-Jones, G. and Diesseroth, K. (2013) Recent advances in optogenetics and pharmacogenetics. *Brain Research*, **1511**, 1–5. [*The mini-review 'condensely' compared two approaches to manipulate electrical activity of neurons. In fact, this mini-review is an introduction for a special issue of* **Brain Research**, *which deserves reading!*]

Atasoy, D., Aponte, Y., Su, H.H. and Sternson, S.M. (2008) A FLEX switch targets Channelrhodopsin-2 to multiple cell types for imaging and long-range circuit mapping. *Journal of Neuroscience*, **28**, 7025–7030. [*The authors for the first time proposed an elegant 'FLEX' vector to express via rAAV genes of interest in Cre-expressing neurons of transgenic mice.*]

Atasoy, D., Betley, J.N., Su, H.H. and Sternson, S.M. (2012) Deconstruction of a neural circuit for hunger. *Nature*, **488**, 172–177. [*The authors used a 0.6 kb promoter fragment of OT gene to target OT neurons in mice via rAAV (serotype 1/2).*]

Bienemann, A.S., Martin-Rendon, E., Cosgrave, A.S., *et al.* (2003) Long-term replacement of a mutated non-functional CNS gene: reversal of hypothalamic diabetes insipidus using an EIAV-based lentiviral vector expressing arginine vasopressin. *Molecular Therapy*, **7**, 588–596. [*The authors compensated diabetes insipidus via reintroduction of normal vasopressin gene in the rat hypothalamus, using lentivirus packed with 600 bp vasopressin DNA, driven by the CMV (cytomegalovirus) promoter.*]

Boyden, E.S., Zhang, F., Bamberg, E., Nagel, G. and Deisseroth, K. (2005) Millisecond-timescale, genetically targeted optical control of neural activity. *Nature Neurocience*, **8**, 1263–1268.

Broser, P.J., Grinevich, V., Osten, P., Sakmann, B. and Wallace, D.J. (2008) Critical period plasticity of axonal arbours of layer 2/3 pyramidal neurons in rat somatosensory cortex: layer specific reduction of projections into deprived cortical columns. *Cerebral Cortex*, **18**, 1588–1603.

Buch, T., Heppner, F.L., Tertilt, C., *et al.* (2005) A Cre-inducible diphtheria toxin receptor mediates cell lineage ablation after toxin administration. *Nature Methods*, **2**, 419–426. [*This innovative technical paper provides a tool to ablate cells of immune (T and B lymphocytes) and nervous (oligodendrocytes) systems, using Cre-dependent expression of receptor of diphtheria toxin in respective cell types of transgenic mice. Systemic (intraperitoneal administration) of diphtheria toxin leads to drastic decrease (in some mice complete absence) of number of lymphocytes as well as severe destruction of myelinated structures through the CNS.*]

Büchen-Osmond, C. (2001) The Universal Virus Database of the International Commettee on Taxonomy of Viruses (ICTVdB). www.ncbi.nlm.nih.gov/ICTVdb/index.htm

Cetin, A., Komai, S., Eliava, M., Seeburg, P.H. and Osten, P. (2006) Stereotaxic gene delivery in the rodent brain. *Nature Protocols*, **1**, 3166–3173.

Conzelmann, K.K. (1998) Nonsegmented negative-strand RNA viruses: genetic and manipulation of viral genomes. *Annual Review of Genetics*, **32**, 123–162.

Conzelmann, K.K., Cox, J.H., Schneider, L.C. and Thiel, H.J. (1990) Molecular cloning and complete nucleotide sequence of the attenuated rabies virus SAD B19. *Virology*, **175**, 485–499.

Danos, O. (2008) AAV vectors for RNA-based modulation of gene expression. *Gene Therapy*, **15**, 864–869. [*This focused review summarizes examples of application of shRNA in different neuronal cell types in vitro.*]

Dayanithi, G., Weller, U., Ahnert-Hilger, G., *et al.* (1992) The light chain of tetanus toxin inhibits calcium-dependent vasopressin release from permeabilized nerve endings. *Neuroscience*, **46**, 489–493.

Farrell, M.S. and Roth, B.L. (2013) Pharmacosynthetics: reimagining the pharmacogenetic approach. *Brain Research*, **1511**, 6–20.

Fenno, L., Yizhar, O. and Deisseroth, K. (2011) The development and application of optogenetics. *Annual Review of Neuroscience*, **34**, 389–412.

Fields, R.L., Ponzio, T.A., Kawasaki, M. and Gainer, H. (2012) Cell-type specific oxytocin gene expression from AAV delivered promoter deletion constructs into the rat supraoptic nucleus in vivo. *PLoS One*, **7**, E32085. [*In this study the authors selectively targeted oxytocin neurons of rats by rAAV (serotype 2/6)* *equipped with short oxytocin promoter, driving oxytocin gene with insertion of fluorescent marker in the III exon. The authors showed that the vector containing only 100 kb promoter of oxytocin is sufficient to achieve cell type specificity in oxytocin neurons.*]

Frankel, A.D. and Young, J.A. (1998) HIV-1: fifteen proteins and an RNA. *Annual Review of Biochemistry*, **67**, 1–25.

Gainer, H. (2012) Cell-type specific expression of oxytocin and vasopressin genes: an experimental odyssey. *Journal of Neuroendocrinology*, **24**(4), 528–538.

Gainer, H., Yamashita, M., Fields, R.L., House, S.B. and Rusnak, M. (2002) The magnocellular neuron phenotype: cell-specific gene expression in the hypothalamo-neurohypophysial system. *Progress in Brain Research*, **139**, 1–14.

Geddes, B.J., Harding, T.C., Lightman, S.L. and Uney, J.B. (1997) Long-term therapy in the CNS: reversal of hypothalamic diabetes insipidus in the Brattleboro rat by using an adenovirus expressing arginine-vasopressin. *Nature Medicine*, **3**, 1402–1404.

Gerendai, I., Tóth, I.E., Kocsis, K., *et al.* (2001) Transneuronal labeling of nerve cells in the CNS of female rat from the mammary gland by viral tracing technique. *Neuroscience*, **108**, 103–118.

Ginger, M., Haberl, M., Conzelmann, K.K., Schwarz, M.K. and Frick, A. (2013) Revealing the secrets of neuronal circuits with recombinant rabies virus technology. *Frontiers in Neural Circuits*, **7**, 2.

Greenwood, M., Bordieri, L., Greenwood, M.P., *et al.* (2014) Transcription factor CREB3L1 regulates vasopressin gene expression in the rat hypothalamus. *Journal of Neuroscience*, **34**, 3810–3820. [*This work shows that CREB3L is upregulated under osmotic challenge, binds to VP gene promoter and stimulates its transcriptional activity. The latter was proved by the injection of lentivirus carrying CREB3L1 under the control of CMV promoter into rat SON and PVN with subsequent analysis of VP hnRNA and mRNA.*]

Grimm, D. and Kay, M.A. (2007) RNAi and gene therapy. *Hematology*, 473–481.

Grimm, D., Kern, A., Rittner, K. and Kleinschmidt, J.A. (1998) Novel tool for production and purification of recombinant adeno-associated virus vectors. *Human Gene Therapy*, **9**, 2745–2760.

Grimm, D., Pandey, K. and Kay, M.A. (2005) Adeno-associated vectors for short hairpin RNA expression. *Methods in Enzymology*, **392**, 381–405.

Grinevich, V., Brecht, M. and Osten, P. (2005) Monosynaptic pathway from rat vibrissa motor cortex to facial motor neurons revealed by lentivirus-based axonal tracing. *Journal of Neuroscience*, **25**, 8250–8258. [*Employing lentivirus, equipped with CamKII promoter driving eGFP, the authors discovered a new monosynaptic pathway from the principal neurons of the motor cortex to the motor facial nucleus in the brainstem of rats. This pathway resembles corticospinal projections in primates and human, controlling precise movements of the fingers. In rats, this pathway underlies the precise movements of whiskers as highly sensitive sensory 'organs.'*]

Halpern, J.L., Habig, W.H., Trenvhard, H. and Russell, J.T. (1990) Effect of tetanus toxin on oxytocin and vasopressin release from nerve endings of the heurohypophysis. *Journal of Neurochemistry*, **55**, 2072–2078.

Harris, J.A., Oh, S.W. and Zeng, H. (2012) Adeno-associated viral vectors for anterograde axonal tracing with fluorescent proteins in nontransgenic and Cre driver mice. *Current Protocols in Neuroscience*, 1.20.1–1.20.18.

Humerick, M., Hanson, J., Rodriguez-Canales, J., *et al.* (2013) Analysis of transcription factor mRNAs in identified oxytocin and vasopressin magnocellular neurons isolated by laser capture microdissection. *PLoS One*, **8**, E69407. [*In this study the authors for the first time approached the analysis of transcription factors in two distinct types of neuroendocrine neurons: oxytocin versus vasopressin. Using virally labeled neurons, captured by laser dissector, they performed qRT-PCR analysis of selected transcriptional factors and found that the level of two of them, RORA and c-jun, was substantially higher in OT neurons than VP neurons.*]

Ideno, J., Mizukami, H., Honda, K., *et al.* (2003) Persistent phenotypic correction of central diabetes insipidus using adeno-associated virus expressin arginine-vasopressin in Brattleboro rats. *Molecular Therapy*, **8**, 895–902.

Kienle, E., Senis, E., Boerner, K., *et al.* (2012) Engineering and evolution of synthetic adeno-associated virus (AAV) gene therapy vectors via DNA family shuffling. *Journal of Visualized Experiments*, **62**, pii:3819.

Knobloch, H.S. and Grinevich, V. (2014) Evolution of central oxytocin pathways in vertebrates. *Frontiers in Behavioral Neuroscience*, **8**, 31.

Knobloch, H.S., Charlet, A., Hoffmann, L.C., *et al.* (2012) Evoked axonal oxytocin release in the central amygdala attenuates fear response. *Neuron*, **73**, 553–566. [*In this paper we (1) achieved cell type-specific targeting of OT neurons in rats via rAAV (serotype 1/2), equipped with evolutionarily conserved 2.6 kb sequence of the OT mouse promoter; (2) implemented deletion-mutated pseudotyped rabies virus to trace concomitant projections of OT to the central amygdala and posterior pituitary; (3) expressed ChR2 to elicit blue light-evoked endogenous OT release to modify the amygdala microcircuit* in vitro *and reverse fear behavior (immobility)* in vivo.]

Knobloch, H.S., Charle, A., Stoop, R. and Grinevich, V. (2014) Viral vectors for optogenetics of hypothalamic neuropeptides, in Viral Vectors in Neurobiology and Brain Diseases (ed. R. Brambilla), Humana Press, New York, pp. 311–329.

Knoppe, D. and Howly, P. (2013) Fields Virology, 6th edn. Lippincott Williams and Wilkins, Philadelphia. **[Book series which covers all aspects of the viral lifecycle for all currently known viruses.]**

Kuhlman, S.J. and Huang, Z.J. (2008) High-resolution labeling and functional manipulation of specific neuron types in mouse brain by Cre-activated viral gene expression. *PLoS One*, **3**, e2005. **[In this study, for the first time a lox-STOP-lox cassette of 1.3 kb (designed by Michael Lazarus from Harvard University) was used to target, via rAAV, fast-spiking interneurons in transgenic mice, expressing Cre recombinase under the control of parvalbumin promoter.]**

La Fleur, S.E., Kalsbeek, A., Wortel, J. and Buijs, R.M. (2000) Polysynaptic pathways between hypothalamus, including the suprachiasmatic nucleus, and the liver. *Brain Research*, **871**, 50–56.

Leng, G. and Ludwig, M. (2008) Neurotransmitters and peptides: whispered secrets and public announcements. *Journal of Physiology*, **23**, 5625–5632.

Leung, R.K.M. and Whittaker, P.A. (2005) RNA interference: from gene silencing to gene-specific therapeutics. *Pharmacology & Therapeutics*, **107**, 222–239.

Lo, L. and Anderson, D.J. (2011) A Cre-dependent, anterograde transsynaptic viral tracer for mapping output pathways of genetically marked neurons. *Neuron*, **72**, 938–950.

McCarty, D.M., Young, S.M. and Samulski, R.J. (2004) Integration of adeno-associated virus (AAV) and recombinant AAV vectors. *Annual Review of Genetics*, **38**, 819–845.

McClue, C., Cole, K.L.H., Wulff, P., Klugmann, M. and Murray, A.J. (2011) Production and titering of recombinant adeno-associated viral vectors. *Journal of Visualized Experiments*, **57**, E3348.

Mebatsion, T., Schnell, M.J., Cox, J.H., Finke, S. and Conzelmann, K.K. (1996) Highly stable expression of foreign gene from rabies virus vectors. *Proceedings of the National Academy of Sciences USA*, **93**, 7310–7314.

Nathanson, J.L., Jappelli, R., Scheeff, E.D., *et al.* (2009) Short promoters in viral vectors drive selective expression in mammalian inhibitory neurons, but do not restrict activity to specific inhibitory cell-types. *Frontiers in Neural Circuits*, **3**, 19. **[The article represents a comprehensive study of rAAV- and lentivirus-based cell type-specific targeting of inhibitory and excitatory neurons, using promoters of genes typical to these two types of neurons such as CamKII-expressing (i.e. excitatory, glutamatergic or principal neurons) and calretinin, parvalbumin and calbindin expressing (i.e. inhibitory, GABA-ergic) interneurons. Furthermore, the authors tried to target inhibitory neurons via promoters of neuropeptides, NPY and somatostatin. The promoters used were from either mouse or fugu fish. However, the authors did not achieve selective targeting for inhibitory or excitatory neurons, although partial specificity was observed, for instance in somatostatin neurons.]**

Osten, P., Grinevich, V. and Cetin, A. (2007) Viral vectors: a wide range of choices and high levels of service. *Handbook of Experimental Pharmacology*, **178**, 177–202.

Pei, H., Sutton, A.K., Burnett, K.H., Fuller, P.M. and Olson, D.P. (2014) AVP neurons in the paraventricular nucleus of the hypothalamus regulate feeding. *Molecular Metabolism*, **3**, 209–215. **[In this study, the authors implemented DREADDs for neuroendocrine neurons, for the first time. rAAV (serotype 10), carrying DOI cassettes with sequences of modified muscarinic G-coupled receptors 3Dq and 4Di, was injected in the PVN of VP-Cre mice. Systemic infusion of clozapine-N-oxide activates VP neurons (via 3Dq), resulting in appetite suppression, while inhibiting these neurons (via 4Di) induces appetite, partially reversing melanocortin (powerful suppressor of food intake)-induced anorexia.]**

Ponzio, T.A., Fields, R.L., Rashid, O.M., Salinas, Y.D., Lubelski, D. and Gainer, H. (2012) Cell-type specific expression of the vasopressin gene analyzed by AAV mediated gene delivery of promoter deletion constructs into the rat SON in vivo. *PLoS One*, **7**, E48860. **[In this study the authors selectively targeted vasopressin neurons of rats by rAAV (serotype 2/6) equipped with short vasopressin promoter, exon I fused with a fluorescent marker. They showed that the cell type specificity was maintained by a 288 bp vasopressin promoter. They proposed that the VP promoter sequence between 288 bp and 116 bp restricts expression specifically to VP neurons.]**

Ramirez, S., Liu, X., Lin, P.A., *et al.* (2013) Creating a false memory in the hippocampus. *Science*, **341**, 387–392.

Regev, L., Neufeld-Cohen, A., Tsoory, M., *et al.* (2011) Prolonged and site-specific over-expression if corticotropin-releasing factor reveals differential roles for extended amygdala nuclei in emotional regulation. *Molecular Psychiatry*, **16**, 714–728.

Sasaki, K., Suzuki, M., Mieda, M., Tsujuno, N., Roth, B. and Sakurai, T. (2011) Pharmacogenetic modulation of orexin neurons alters sleep/wakefullness states in mice. *PLoS One*, **6**, E20360.

Silva, B., Matucci, C., Illarionova, A., Grinevich, V., Canteras, N.S. and Gross, C.T. (2013) Independent hypothalamic circuits for social and predator fear. *Nature Neuroscience*, **16**, 1731–1733.

Smith, G. (2012) Herpesvirus transport to the nervous system and back again. *Annual Review of Microbiology*, **66**, 153–176.

Sohal, V.S., Zhang, F., Yizhar, O. and Deisseroth, K. (2009) Parvalbumin neurons and gamma rhythms enhance cortical circuit performance. *Nature*, **459**, 698–702. [*The authors introduced a double-floxed inverted open reading frame (DIO) vector to target via rAAV cells expressing Cre recombinase. The DIO strategy is basically identical to the FLEX strategy proposed by Atasoy and colleagues (2008).*]

Sweeney, S.T., Broadie, K., Keane, J., Niemann. H. and O'Kane, C.J. (1995) Targeted expression of tetanus toxin light chain in Drosophila specifically eliminates synaptic transmission and causes behavioral defects. *Neuron*, **14**, 341–351. [*The authors employed lentivirus to overexpress CRH in the central nucleus of amygdala and the bed nucleus of stria terminalis (BNST). The overexpression of CRH in the central amygdala led to attenuation of stress-induced anxiety-like behavior, while the overexpression of CRH in the BNST provoked depressive-like behavior.*]

Theodosis, D.T. and Poulain, D.A. (2001) Maternity leads to morphological synaptic plasticity in the oxytocin system. *Progress in Brain Research*, **133**, 49–58.

Ugolini, G. (2010) Advances in viral transneuronal tracing. *Journal of Neuroscience Methods*, **194**, 2–20.

Van de Boosport, R., Farina, M., Schmitz, S.K., *et al.* (2012) Munc 13 controls the location and efficiency of dense-core vesicle release in neurons. *Journal of Cell Biology*, **199**, 883–891. [*Using a model of in vitro cultivated hippocampal neurons, expressing cargo (NPY) for labeling large dense-core vesicles (LDCV), the authors showed that LDCVs in cultured neurons usually released at synaptic terminals, while extrasynaptic release occurred only after massive and prolonged stimulation. Since synaptic contacts of neuroendocrine neurons are very modest (see Knobloch and Grinevich, 2014), it is likely that in this type of neuron, most of the release sites are extrasynaptic. However, this assumption requires further elucidation, using similar techniques.*]

Van den Pol, A.N., Acuna-Goycolea, C., Clark, K.R. and Ghosh, P.K. (2004) Physiological properties of hypothalamic MCH neurons identified with selective expression of reporter gene after recombinant virus infection. *Neuron*, **42**, 635–652. [*The authors used 0.7 kb of a mouse promoter of melanin-concentrating hormone, delivered by rAAV (serotype 2), to achieve cell type specificity in the mouse hypothalamus.*]

Wang, X. and McManus, M. (2009) Lentivirus production. *Journal of Visualized Experiments*, **32**, E1499.

Watts, A.G. (2011) Structure and function in the conceptual development of mammalian neuroendocrinology between 1920 and 1965. *Brain Research Review*, **66**, 174–204.

West, A.E., Griffin, E.C. and Greenberg, M.E. (2002) Regulation of transcription factors by neuronal activity. *Nature Reviews Neuroscience*, **3**, 921–930.

Wickersham, I.R., Finke, S., Conzelmann, K.K. and Callaway, E.M. (2007) Retrograde neuronal tracing with a deletion-mutant rabies virus. *Nature Methods*, **4**, 47–49.

Wiznerowicz, M. and Trono, D. (2005) Harnessing HIV for therapy, basic research and biotechnology. *Trends in Biotechnology*, **23**, 42–47.

Wu, Q., Howell, M.P. and Palmiter, R.D. (2008) Ablation of AgRP neurons activates fos in postsynapric target region. *Journal of Neuroscience*, **28**, 9218–9226. [*The 'side effects' of the application of diphtheria toxin receptor approach to kill hypothalamic neurons, expressing agouti-related protein (AgRP), resulted in induction of immediate earlier gene c-fos and gliosis in brain regions, innervated by AgRP neurons.*]

Wu, Q., Clark, M.S. and Palmiter, R.D. (2012) Deciphering a neuronal circuit that mediates appetite. *Nature*, **483**, 594–597.

Xie, J., Ameres, S.L., Friedline, R., *et al.* (2012) Long-term, efficient inhibition of microRNA function in mice using rAAV vectors. *Nature Methods*, **9**, 403–409.

Yamuuchi, Y. and Helenis, A. (2013) Virus entry at a glance. *Journal of Cell Science*, **126**, 1289–1295. [*A review about general viral entry mechanisms. It nicely describes all events from cell attachment to gene expression.*]

Yoon, H., Enquist, L.W. and Dulac, C. (2005) Olfactory inputs to hypothalamic neurons controlling reproduction and fertility. *Cell*, **123**, 669–682. [*The authors, using the technology of bacterial artificial chromosomes, generated a faithful transgenic mouse, expressing Cre recombinase under the control of GnRH promoter. Combining this mouse with Cre-dependent pseudorabies virus Bartha strain, the authors deciphered inputs to GnRH neurons from olfactory epithelium, but not from the vomeronasal organ (pheromone-sensing organ in the nasal cavity). As concluded, GnRH neurons are controlled by olfactory stimuli, but not by pheromones.*]

Zhang, C., Zhou, J., Feng, Q., *et al.* (2013) Acute and long-term suppression of feeding behavior by POMC neurons in the brainstem and hypothalamus, respectively. *Journal of Neuroscience*, **33**, 3624–3632.

Zhang, G., Bai, H., Zhang, H., *et al.* (2011) Neuropeptide exocytosis involving synaptotagmin-4 and oxytocin in hypothalamic programming of body weight and energy balance. *Neuron*, **69**(3), 523–535. [*The authors used a 0.6 kb OT promoter to target – via lentivirus – OT neurons in mice. In addition, they achieved downregulation of OT expression by shRNA (purchased from Sigma) against OT mRNA.*]

CHAPTER 13

Optogenetics Enables Selective Control of Cellular Electrical Activity

Ryuichi Nakajima,[1] Sachiko Tsuda,[2,3] Jinsook Kim,[2,3] and George J. Augustine[1,2,3]

[1] Center for Functional Connectomics, Korea Institute of Science and Technology, Seoul, Korea
[2] Lee Kong Chian School of Medicine, Nanyang Technological University, Singapore
[3] Institute of Molecular and Cell Biology, Singapore

13.1 Introduction: what is optogenetics?

To understand how the brain works, it is essential to define the functions of the many different types of neurons in the brain, as well as to unravel the complex synaptic circuits among neurons that underlie information processing in the brain. Much valuable information about each of these processes has been obtained by using electrodes both for electrical stimulation of neurons and for measurement of neuronal electrical activity. However, such electrophysiological techniques have limitations for examining neuronal circuit function. With extracellular electrode stimulation it is difficult to identify the neurons that are being stimulated, while stimulation via intracellular electrodes is time-consuming and allows only a limited number of neurons to be stimulated at a time. Electrophysiological recording of neuronal activity has high sensitivity and excellent temporal resolution, yet provides limited spatial information and typically examines the activity of only one or at most a few neurons at a time. Because the function of neurons is defined by causally linking neuronal activity to responses such as behavioural output, these limitations make it difficult to identify the function of neuronal populations and the circuits that they form.

Optogenetics provides powerful tools to overcome these problems. It utilizes the advantages of molecular genetics and optical physiology, specifically taking advantage of genetic targeting strategies to express light-sensitive proteins in defined populations of neurons, allowing clear identification of the neurons under investigation. There are two general classes of optogenetic probes that can be distinguished according to their function. While **optogenetic actuators** control neuronal activity, **optogenetic sensors** monitor neuronal activity (Figure 13.1). Many types of rhodopsins that control ion fluxes across membranes via light have been exploited as optogenetic actuators. These include **channelrhodopsin**, which is used to activate neurons, and **halorhodopsin** that serves to silence neurons. Optogenetic sensors include various genetically encoded indicators of ion concentration. Among these are pHluorin, which detects pH changes, and clomeleon and the recently introduced super-clomeleon, which enable measurement of changes in intracellular chloride associated with synaptic inhibition (Grimley *et al.*, 2013; Kuner and Augustine, 2000). A series of calcium indicators, such as cameleon (Miyawaki *et al.*, 1997) and GCaMPs, are now widely used to indirectly measure neuronal electrical activity. Voltage-sensitive

Molecular Neuroendocrinology: From Genome to Physiology, First Edition. Edited by David Murphy and Harold Gainer.
© 2016 John Wiley & Sons, Ltd. Published 2016 by John Wiley & Sons, Ltd.
Companion website: www.wiley.com/go/murphy/neuroendocrinology

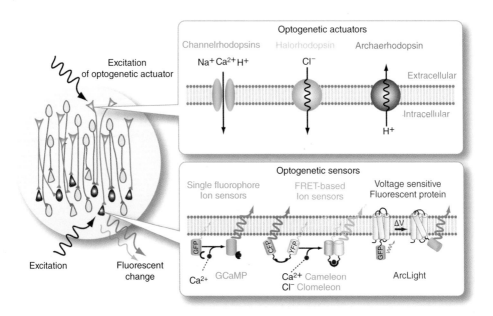

Figure 13.1 Optogenetic tools for control and measurement of electrical activity. Optogenetic actuators are light-sensitive ion channels and pumps that cause ion fluxes across the plasma membrane in response to light. Optogenetic sensors are fluorescent proteins that change their fluorescent properties in response to cellular electrical signals, such as intracellular calcium or chloride ion concentration or changes in membrane potential.

fluorescent proteins enable direct measurement of membrane potential changes occurring during neuronal excitation. Following early pioneering variants such as Mermaid and voltage-sensitive fluorescent proteins (VSFPs), recent probes such as Archaerhodopsins (Arch) (Kralj *et al.*, 2011) and ArcLight (Jin *et al.*, 2012) are finding more widespread use because they have significantly increased sensitivity and signal/noise ratio. In nearly all cases, these optogenetic probes are fluorescent or are tagged with a fluorescent protein to enable convenient identification of the neurons expressing them.

In this chapter, we will focus on optogenetic actuators and their applications. Recent reviews of optogenetic sensors can be found elsewhere (Peterka *et al.*, 2011; Tian *et al.*, 2012). Optogenetic actuators have mostly been applied to neurons thus far, but they are exceedingly valuable for studies of any type of electrically excitable cell. With this in mind, we will close by illustrating the value of optogenetic actuators for studies of the neuroendocrine system by describing some of their applications to study of hypothalamic function.

13.2 Optogenetic actuators allow selective control of cellular activity

Light-sensitive pumps were discovered in the 1970s as a result of their fascinating biophysical properties. Light-driven outward proton pumps (Oesterhelt and Stoeckenius, 1973) and light-driven inward chloride pumps (Lanyi, 1986; Lanyi and Weber, 1980; Matsuno-Yagi and Mukohata, 1977; Mukohata and Kaji, 1981) were found in *Halobacterium salinarum*. Since then, similar pumps have been found in a diversity of archaea, bacteria, fungi, and algae, where they serve photosynthetic and signaling purposes (Boyden, 2011). However, until recently none of these were used as optogenetic actuators for control of cellular electrical activity.

The first report of genetically targeted optical control of neuronal activity was published in 2002 by Gero Miesenböck's group. In this report, rhodopsin and other components of the *Drosophila* phototransduction cascade were expressed in mammalian neurons to photosensitize the neurons and thereby control neural activity via light (Zemelman *et al.*, 2003). In a follow-up study, the P2X2 receptor was transgenically expressed to allow the neurons to be photostimulated by light-uncaged ATP. This permitted the first optogenetically induced behavioral change in *Drosophila* (Lima and Miesenböck, 2005). These first-generation optogenetic actuators were inconvenient because they required expression of multiple components or application of exogenous caged ATP. As a result, application to other systems was not straightforward and they have seldom been used.

In 2005, Karl Deisseroth's group demonstrated that channelrhodopsin-2 (ChR2), a single-component light-activated cation channel from the green algae *Chlamydomonas*, could be expressed in cultured mammalian neurons and used to photostimulate these neurons (Boyden *et al.*, 2005). One month later, Hiromu Yawo's group demonstrated the applicability of channelrhodopsin for photostimulation of hippocampal dentate gyrus neurons in intact mice (Ishizuka *et al.*, 2006). Thus, ChR2 was the first optogenetic actuator of practical utility and its application has launched a technological revolution in physiological analysis of neuronal and neuroendocrine function. Subsequent reengineering of these opsins to optimize their photostimulation properties – such as their kinetics, subcellular localization, and light wavelength sensitivity – has dramatically improved and expanded their utility. These probes are now a standard part of the experimental toolbox for a broad range of neuroscience research areas (Asrican *et al.*, 2013; Tye and Deisseroth, 2012; Yizhar *et al.*, 2011) and we will next summarize the properties of a few of the most widely used optogenetic actuators.

13.2.1   Light-gated cation channels for photostimulation

13.2.1.1   *Chlamydomonas* channelrhodopsin-2

Channelrhodopsin-2 is the most widely used optogenetic actuator. It is a 7-transmembrane cation channel that opens when activated by blue-green light (400–510 nm; Nagel *et al.*, 2003). Expression of ChR2 in neurons allows light to generate large photocurrents that are sufficient to trigger action potentials (Boyden *et al.*, 2005; Zhang *et al.*, 2006) (Figure 13.2, left). Such **photostimulation** is temporally precise, occurring over a time scale of milliseconds, allowing very acute control of neuronal activity. Solution of the crystal structure of ChR2 has aided understanding of the structural underpinnings of its function. ChR2 performance has been improved even further by molecular engineering strategies, and genome or transcriptome sequencing. For example, through these approaches, more rapid gating of ChR2 has been achieved, allowing action potential firing up to 200 Hz (Gunaydin *et al.*, 2010) and increasing its sensitivity to light (Klapoetke *et al.*, 2014).

13.2.1.2   *Volvox* channelrhodopsin

Recent studies demonstrate that a second channelrhodopsin from the algae *Volvox carteri* (VChR1) also can be used to photostimulate neurons in a rapid and precise manner. The advantage of this alternative channelrhodopsin is that it can be excited with yellow light (~590 nm). The red-shifted light spectrum for activation of VChR1 has two advantages. First, because longer wavelength light is scattered less, this means that light can be delivered to deeper areas within brain tissue. Second, its spectrum allows it to be combined with other optogenetic sensor probes that are excited by shorter wavelengths of light, such as ChR2 or optogenetic activity indicator sensors that are activated by blue light (Zhang *et al.*, 2008). However, VChR1 has a couple of limitations. First, its kinetics of deactivation at the end of a light pulse are much slower than ChR2, which makes it more suitable for prolonged

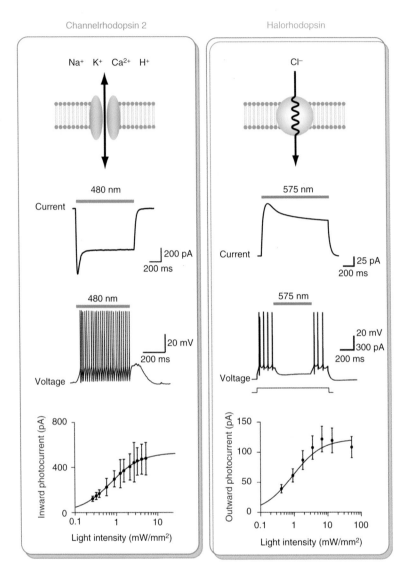

Figure 13.2 Types of optogenetic actuators. (Left) Channelrhodopsin-2 is a light-activated cation channel (*top*). Illumination with blue light (480 nm) causes these channels to open, causing an inward current (second) that can depolarize neurons and cause them to fire action potentials (third). Varying the amount of blue light controls the degree of ion channel opening (*bottom*), thereby controlling the amount of photostimulation. (Right) Halorhodopsin is a light-activated chloride ion pump channel (*top*). Illumination with yellow light (575 nm) causes these pumps to become active, causing an outward current (second) that can hyperpolarize neurons and cause them to stop firing action potentials (third). Varying the amount of yellow light controls the degree of pump activation (*bottom*), thereby controlling the amount of photoinhibition.

photostimulation rather than for brief, repetitive photostimuli. Second, while VChR1 is sensitive to yellow light, it is rather sensitive to blue light as well. This limits its utility as a partner with ChR2 for two-color photostimulation experiments (Zhang *et al.*, 2008).

On the other hand, recent advances in obtaining optogenetic probes have provided variants of red-shifted variants of channelrhodopsins which can be excited with red or far red light, providing better separation from the excitation spectrum of ChR2 (Klapoetke *et al.*, 2014; Lin *et al.*, 2013). By using these channelrhodopsins with different excitation spectra, it is now possible to use optogenetic actuators to control the activity of more than one group of neurons (Inagaki *et al.*, 2014; Klapoetke *et al.*, 2014). Furthermore, the excitation spectra of these novel red-shifted ChR2 do not have an overlap with many of the optogenetic sensors described above. This will enable all-optogenetic approaches in which light is used both to control pre-synaptic neuronal activity and to measure responses of postsynaptic neurons.

13.2.2 Light-gated pumps for photoinhibition

13.2.2.1 Halorhodopsin

Fast optogenetic **photoinhibition** of neuronal activity can be obtained via *Natronomonas* halorhodopsin (NpHR) (Han and Boyden, 2007; Zhang *et al.*, 2007). NpHR is a light-driven chloride pump, inducing an outward chloride current that hyperpolarizes the membrane potential of cells. Such hyperpolarizations photoinhibit action potential firing with great temporal precision (see Figure 13.2, right) (Asrican *et al.*, 2013; Han and Boyden, 2007), causing acute loss of circuit function (Chow *et al.*, 2012; Yizhar *et al.*, 2011). The excitation spectrum for NpHR is in the green/yellow range (~580 nm). Since this is red-shifted compared to the excitation spectrum of ChR2, simultaneous expression of NpHR and ChR2 in neurons allows bidirectional optogenetic modulation of neuronal activity (Han and Boyden, 2007; Mancuso *et al.*, 2011).

Native NpHR exerts relatively modest photoinhibitory effects, with small hyperpolarizing currents of only 40–100 pA elicited by light (Asrican *et al.*, 2013; Han and Boyden, 2007; Zhang *et al.*, 2007). NpHR photocurrent size is mainly limited by intracellular trafficking: NpHR tends to be retained internally within the endoplasmic reticulum (ER) rather than being expressed on the external plasma membrane (Gradinaru *et al.*, 2008; Mancuso *et al.*, 2011; Zhao *et al.*, 2008). To improve membrane trafficking, a second generation 'enhanced' NpHR (eNpHR2.0) was engineered by adding ER export motif onto the C-terminus of NpHR (Gradinaru *et al.*, 2008; Zhao *et al.*, 2008). eNpHR2.0 is more light sensitive and its activation and deactivation kinetics are significantly faster than NpHR, without changing neuronal electrophysiological properties (Asrican *et al.*, 2013). eNpHR2.0 has been successfully applied both *in vivo* and *in vitro* (Asrican *et al.*, 2013; Sohal *et al.*, 2009; Tønnesen *et al.*, 2009). A third-generation NpHR (eNpHR3.0) was created by adding a neurite trafficking sequence from the Kir2.1 potassium channel and has enhanced photoinhibition properties (Gradinaru *et al.*, 2010; Yizhar *et al.*, 2011). eNpHR3.0 expression is sufficiently strong to permit photoinhibition by red/far-red light (~680 nm), enabling deeper penetration of light into brain tissues(Gradinaru *et al.*, 2010). This has allowed definition of the neurons necessary for postsynaptic responses and behaviors in freely moving mice (Tye *et al.*, 2011; Witten *et al.*, 2010). Bidirectional modulation of neuronal activity via optogenetic activators (eg. ChR2) and optogenetic inhibitors (eg. NpHR) will provide optimal insights into neuronal circuit function.

13.2.2.2 Archaerhodopsin

Similar to NpHR, Archaerhodopsin-3 (Arch) also enables photoinhibition of neuronal activity. Arch is a light-driven proton pump originally found in Halobacterium (Mukohata and Kaji, 1981; Mukohata *et al.*, 1988). When activated by 570 nm wavelength light, Arch pumps out protons to yield reliable photoinhibition of neurons (Chow *et al.*, 2010). While both NpHR and Arch are effective for neuronal silencing, their after-effects differ. NpHR, unlike Arch, causes changes in the reversal potential of the $GABA_A$ receptor, due to the accumulation of chloride ions. This can cause inhibitory synapses to become excitatory for a brief period following illumination, thereby increasing the probability of action potential firing in response to synaptic activity (Raimondo *et al.*, 2012). Therefore, for silencing neuronal activity Arch may be a better choice for situations where inhibitory synaptic inputs are present. However, Arch also has its own side effect, transiently increasing intracellular pH due to proton efflux (Chow *et al.*, 2010).

13.3   Using optogenetic actuators to study the function of neurons and circuits

The optogenetic actuators described in the previous section allow optical control of the activity of defined neuronal populations with millisecond time resolution (see Figure 13.2). This enables dissection of the function of neurons and neuronal circuits at organizational levels ranging from single cells *in vitro* to global circuits regulating behavioral and cognitive functions *in vivo*. In this section, we will provide some examples of the types of experiments that are enabled by optogenetic actuators.

13.3.1   *In vivo* optogenetic control
One of the most popular applications of optogenetic actuators is for the control of genetically defined populations of neurons *in vivo*. The brain consists of many different types of neurons, often in close spatial proximity, and these conditions make it difficult to know precisely what is being activated when electrodes are used for stimulation. By genetically targeting expression of optogenetic actuators to only a single cell type, it is possible to selectively turn on or turn off the activity of just this cell type without perturbing neighboring cells. This is a very powerful approach that has been used to study many types of neurons and circuits in living animals (Figure 13.3). For example, ChR2-mediated photostimulation of neurons in the hypothalamus that express orexin/hypocretin facilitates arousal from sleep (Adamantidis *et al.*, 2007) while photoinhibition of these neurons via NpHR has the opposite effect (Tsunematsu *et al.*, 2011). The photostimulation results establish that the activity of these neurons is *sufficient* for arousal, while the photoinhibition results establish that activity of these neurons is *necessary* for arousal. More generally, being able to both increase and decrease neuronal activity via optogenetic actuators can establish cause/effect relationships for the activity of virtually any type of neuron amenable to genetic manipulation. The next section will describe the various approaches that are available for genetic targeting.

In the case of *in vivo* optogenetic control of neuronal activity in deep brain structures, it is possible to record electrical activity in the target region via stereotaxically inserted electrodes, while measuring behavioral changes during optogenetic control of neuronal activity (see Figure 13.3) (Yizhar *et al.*, 2011; Zhang *et al.*, 2010). Investigators have developed 'optrodes,' integrated devices that have both optical fibers and recording electrodes. Recent advances in *in vivo* optogenetic techniques with wireless optrodes enable long-term control of neuronal circuits and various behavioral tests in freely moving animals (Anikeeva *et al.*, 2011; Kim *et al.*, 2013; Sparta *et al.*, 2011). Another *in vivo* monitoring technique uses blood oxygenation level-dependent (BOLD) functional magnetic resonance imaging (fMRI) to detect changes in global neuronal activity in response to optogenetic photostimulation (Kahn *et al.*, 2011; Lee *et al.*, 2010). This optogenetic fMRI (opto-fMRI) technique is attractive because it is non-invasive and can reveal long-range neuronal connections among multiple brain areas (Lim *et al.*, 2013). However, because animals must remain very still within a magnetic coil during fMRI, this imaging approach has limited applicability for behavioral analyses.

13.3.2   Optogenetic probing of connectivity
Another important class of applications of optogenetic-based photostimulation is to establish functional connectivity between neurons in different locations. Neurons expressing ChR2 or other optogenetic actuators can be photostimulated with light spots that are located over the

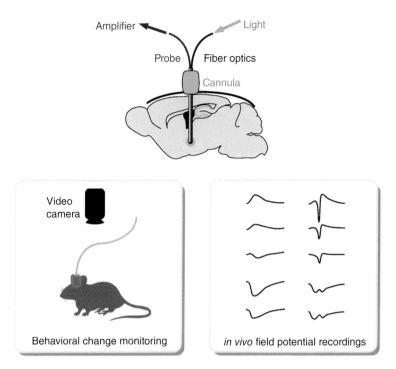

Figure 13.3 Application of optogenetic actuators *in vivo*. For delivery of light into the brain or other tissues, an optical fiber is typically used and is connected to one of the light sources illustrated in Figure 13.14. This light can be used to photostimulate or photoinhibit genetically defined populations of neurons, depending on the type of optogenetic actuator expressed in the brain. The resulting changes in brain function can be monitored in a variety of ways, including changes in behavior (*left*) or neuronal electrical activity (*right*).

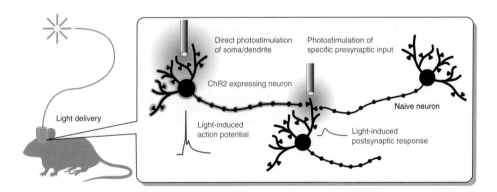

Figure 13.4 Photostimulation strategies permitted by optogenetic actuators such as ChR2. After expressing the actuator in the neurons of interest (*green*), light can be applied onto the cell bodies in the structure where the neurons originate (direct photostimulation) or onto the processes of these neurons at their sites of projection (photostimulation of presynaptic inputs).

cell bodies or their axons (Figure 13.4). The former allows excitation of neurons within a projecting brain structure, while the latter permits activation of their projections within a target structure. As a very recent *in vivo* example of such an approach, expression of ChR2 in the mesencephalic locomotor region (MLR) of the mouse has been used to define the role of this brain structure in locomotion. Direct photostimulation of the MLR increases locomotion, while photostimulation of MLR projections within the basal forebrain enhances responses within the cortex (Lee *et al.*, 2014). These results demonstrate that the MLR regulates both locomotion and cortical state. Similar approaches can be applied to brain slices as well. For example, local photostimulation of neuronal cell bodies has been used to map local circuitry within the cortex (Wang *et al.*, 2007; see Figure 13.5), while photostimulation of projecting axons has been used to selectively activate thalamic projections onto cortical neurons in brain slices (Cruikshank *et al.*, 2010).

13.3.3 Optogenetic mapping of brain circuitry

By scanning small light spots over brain tissue, while recording the activity of postsynaptic neurons electrophysiologically, it is possible to map synaptic connections between pre- and postsynaptic neurons (Figure 13.5). This is a rapid and relatively simple way to gain information about both the function and spatial organization of brain circuitry (Asrican *et al.*, 2013; Kim *et al.*, 2014; Petreanu *et al.*, 2007; Wang *et al.*, 2007). In this approach, ChR2-positive presynaptic neurons are photostimulated by small spots of blue light, while simultaneous whole-cell patch clamp recordings are used to measure the responses of postsynaptic neurons. By correlating the location of photostimulation with the resultant postsynaptic responses, it is possible to map

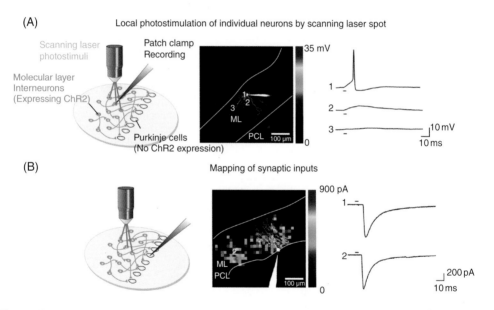

Figure 13.5 Optogenetic circuit mapping. (A) A small spot of laser light can be scanned over a brain slice containing ChR2-expressing neurons, in this case molecular layer interneurons in the cerebellum. (*Left*) Experimental arrangement. (*Right*) Patch clamp recordings reveal that when the light spot is focused on the interneuron cell body (location 1), action potentials are generated. However, moving the light spot away from the cell body causes subthreshold depolarizations (locations 2 and 3). (*Center*) Map of light sensitivity of interneurons. Red pixels indicate locations where action potentials are evoked by light. Under these conditions, action potentials are only evoked near the cell body, due to a larger area of the ChR2-expressing membrane within the laser beam. (B) Preferential photostimulation of molecular layer interneurons by light spots focused near their cell bodies allows mapping of the spatial organization of the circuits formed by these interneurons. (*Left*) Experimental arrangement. Molecular layer interneurons are photostimulated by scanning the laser light spot while recording from a postsynaptic Purkinje cell. (*Right*) Inhibitory postsynaptic currents (IPSCs) evoked when presynaptic molecular layer interneurons innervating the Purkinje cell are photostimulated. (*Center*) Correlating the amplitude of the light-evoked IPSCs (color scale) with the location of the light spot defines the spatial distribution of interneurons innervating this Purkinje cell. Abbreviations: ML, Molecular layer; PCL, Purkinje cell layer. After Kim *et al.* (2014).

the spatial organization of synaptic connectivity. This high-speed optogenetic circuit mapping approach has revealed the spatial organization of local and long-range connections in brain slices from several brain regions, including cortex and cerebellum (Kim *et al.*, 2014; Petreanu *et al.*, 2007; Wang *et al.*, 2007). By measuring motor responses, rather than postsynaptic responses, it has also been used *in vivo* as well (Hira *et al.*, 2009).

13.4   Methods for delivery of optogenetic actuators

There is a variety of molecular genetic approaches that can be used to insert optogenetic actuator genes into cells. In this section, we will introduce the most popular methods for delivery of optogenetic actuators into the mammalian brain. Because the delivery method chosen can affect interpretation of the experimental results, this information will be presented from a user's point of view to allow informed decisions about which delivery strategy is most suitable for your specific experimental goals. We will introduce three major delivery methods: viral vectors, electroporation, and transgenic expression (Figure 13.6, green). The design of the gene vector also determines where a gene is expressed. Thus, expression of optogenetic actuators can be restricted to virtually any cell types of interest (Figure 13.6, red). Indeed, this is one of the main advantages of optogenetic methods over conventional methods for stimulating neurons or other excitable cells: optogenetics allows control over the activity of

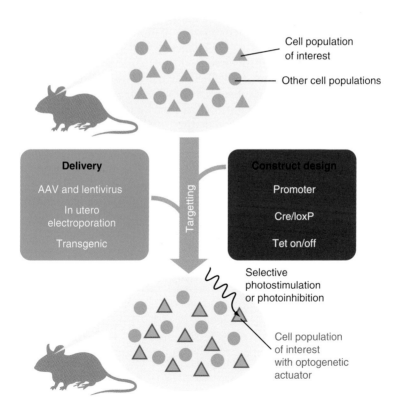

Figure 13.6 Genetic targeting strategies for optogenetic actuators. Different types of delivery methods (*green*) and transgene regulatory strategies (*red*) allow selective expression of optogenetic actuators in genetically defined populations of neurons.

genetically defined populations of cells. Therefore, we will introduce some of the most widely used cell type-specific promoters and the Cre-loxP system for targeted expression of transgenes. Finally, we will discuss the 'tet-on/off' system, which enables temporal control of optogenetic actuator expression.

13.4.1 Viral transduction

Local viral infection of optogenetic actuators is the most popular technique for expression of optogenetic actuators in brain tissue. This strategy is particularly valuable for investigation of functional connectivity in the brain because it allows expression of optogenetic actuators to be restricted to a limited volume, thereby enhancing selectivity when photostimulating brain structures.

13.4.1.1 Adeno-associated virus and lentivirus

Viral vectors based on adeno-associated virus (AAV) and lentivirus can be used to target optogenetic actuator gene expression in a wide range of experimental subjects, ranging from rodents to primates, from *in vivo* to *in vitro* cultures. These viruses can be easily produced in 1–2 weeks within biosafety level 2-certified facilities (McClure *et al.*, 2011; Zhang *et al.*, 2010). Because of this convenience, AAV and lentivirus are frequently utilized when the viral expression is the method of choice. Compared to AAV, lentivirus can carry exogenous transgenes of larger size, but can sometimes be pathogenic. In addition, there are some other advantages and disadvantages of each viral approach, as summarized in Table 13.1.

For *in vivo* injection of virus, an anesthetized animal is placed in a stereotaxic device to allow the virus to be injected into the region of interest via a microsyringe (Figure 13.7). It takes several days or weeks for a sufficient amount of opsin to be expressed. This incubation time period depends on the virus, infection efficiency, and strength of promoter, as well as the kinetics and light sensitivity of the optogenetic actuator. In the case of AAV infection, typically a month is required before optogenetic actuators are expressed on remote axonal fibers at sufficient levels to permit axonal photostimulation (as in Figure 13.4).

Adeno-associated and lentiviral vectors are selective for certain cell types (see Table 13.1). When combined with appropriate promoters, these viruses can also target specific types of neurons. For a given promoter, lentiviral vectors are biased to transduction of excitatory neurons whereas low-titer AAV2 vectors reportedly express more in inhibitory neurons in the mouse somatosensory cortex (Nathanson *et al.*, 2009; Yizhar *et al.*, 2011). Online video material for *in vivo* virus injection is available at: Nakamura *et al.* (2013)

www.jove.com/video/50291/a-method-for-high-fidelity-optogenetic-control-individual-pyramidal.

Table 13.1 Comparison between adeno-associated virus (AAV) and lentivirus expression in neurons.

	AAV	Lentivirus
Pathogenicity	Not reported	Yes
Virus genome	ssDNA	RNA
Virus particle	Very stable	Unstable
Titer	Can be high	Usually low
Integration to host genome	Sometimes	Random
Target cell types	Non-dividing neuron, glia	Dividing & non-dividing neuron
Insert size/genomic size	3 Kb/4.8 Kb	7 Kb/8 Kb
Timing for construction and production	2 weeks	2 weeks
Timing for expression after infection	3 weeks	2 weeks
Expression period	Years	Years

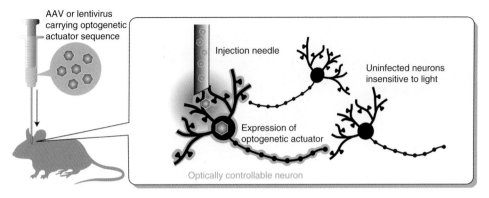

Figure 13.7 Expression of optogenetic actuators via AAV or lentivirus.

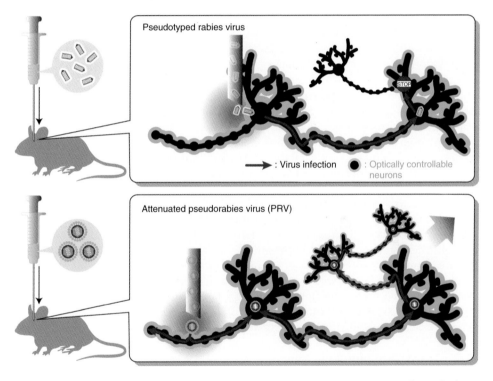

Figure 13.8 Retrograde expression of optogenetic actuators via rabies viruses. (*Top*) Pseudotyped rabies virus causes expression of actuators in targeted neurons and their presynaptic partners (*red arrow*), a property that is very useful for establishing synaptic connectivity between different types of neurons. (*Bottom*) Attenuated pseudorabies virus can cross multiple synapses, in either the retrograde or anterograde direction (*red arrows*). This allows long-distance and polysynaptic pathway tracing.

13.4.1.2   Pseudotyped rabies virus and pseudorabies virus

By using pseudotyped rabies virus containing a glycoprotein deletion and modified envelope proteins, targeted neurons and their immediate presynaptic partners are infected, and then viral spread is terminated (Figure 13.8, top). This is a very useful property for establishing synaptic connectivity between different types of neurons and for studying the function of these

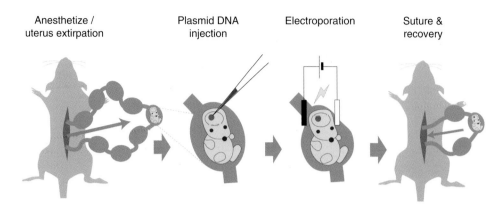

Anesthetize /   Plasmid DNA   Electroporation   Suture &
uterus extirpation   injection   recovery

Figure 13.9 Procedures required for expression of optogenetic actuators via *in utero* electroporation.

synaptic connections. On the other hand, attenuated pseudorabies virus has a high affinity for axon terminals and can cross multiple synapses; depending on the strain, infection can spread in either the retrograde or anterograde direction. This property of attenuated pseudorabies virus allows long-distance and polysynaptic pathway tracing (Figure 13.8, bottom).

13.4.2 *In utero* electroporation

Electroporation is a simple and efficient method for expressing optogenetic actuators in neural tissue with cell type specificity. By appropriate selection of the embryonic stage and injection site used for electroporation, it is possible to achieve quite specific subregional transfection. Another advantage of this method is the sparseness of the transfection. This allows better visualization of individual neurons that have been transfected with fluorescently tagged optogenetic actuators (Druckmann *et al.*, 2014; Matsui *et al.*, 2011).

Figure 13.9 illustrates the *in utero* electroporation technique. For *in utero* electroporation, a pregnant mouse is anesthetized at embryonic days 9.5–15.5 and the uterine horns are carefully removed. With appropriate illumination, the brain of the mouse embryo can be identified. After injecting plasmid DNA into the brain using a sharp glass pipette, electrical pulses are applied across the embryo to transiently electroporate cells within the brain. After electroporation, the uterine horn is restored to its original location and the mice with electroporated brains are born a few days later. Online video material for *in utero* electroporation is available at: Matsui *et al.* (2011)
    www.jove.com/video/3024/mouse-utero-electroporation-controlled-spatiotemporal-gene.

13.4.3 Transgenic expression

Transgenic (tg) expression is a powerful and efficient way to express optogenetic actuators in target cells (Figure 13.10). To make optogenetic actuator tg mice, the gene for the actuator is microinjected into a fertilized egg. The injected gene is randomly integrated into one or more locations within the host genome and suitable tg mouse lines are obtained by selection of founder mice that express the transgene. Once a tg mouse is generated, the mouse line can be maintained over many generations via breeding, yielding a permanent supply of tg mice.

Compared to the other approaches described above, the main advantage of tg mice is that the level of expression of the optogenetic actuator is homogenous among animals. This allows easy comparison of results across experiments done with different individual mice. This contrasts with viral expression and *in utero* electroporation, where there can be considerable variability in expression between individuals. The use of tg mice also does not need preparatory surgery, as is required for virus injection or *in utero* electroporation. Another

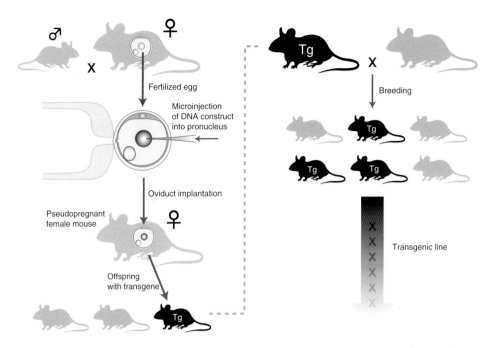

Figure 13.10 Procedures required for expression of optogenetic actuators via generation of transgenic mice.

advantage is that many optogenetic tg mice are available from commercial sources (for example, ChR2-tg from Jackson Laboratories) or other laboratories (Asrican *et al.*, 2013). However, the tg approach has some disadvantages, too. One is that it is difficult to control the expression pattern of the optogenetic actuator, because the randomness of the transgene integration sites make expression susceptible to positional effect variegation. Thus, each line must first be characterized to identify the brain regions and cell types that express the optogenetic actuator. It is also time-consuming and expensive to make transgenic mouse lines and to maintain these lines once established.

13.4.4 Promoters for cell type-specific targeting

By restricting the expression of optogenetic actuators to specific types of cells, it is possible to selectively photostimulate or photoinhibit these neurons and thereby investigate their functions. Indeed, this specificity is one of the main advantages of optogenetic photostimulation. For this purpose, the use of cell type-specific promoters is a valuable and well-established approach for controlling expression of optogenetic actuators (Figure 13.11).

Activity of specific brain cell populations (such as excitatory, inhibitory, glia cells) can be selectively controlled by choosing cell type-specific promoters that drive expression of actuators via viral vectors or transgenesis. A few of the promoters that have been used for viral targeting of optogenetic actuators are listed in Table 13.2. Likewise, a plethora of neuron-specific promoters have been used to drive expression of optogenetic actuators in transgenic mice (Asrican *et al.*, 2013; Kim *et al.*, 2014; Wang *et al.*, 2007; Zhao *et al.*, 2008).

13.4.5 Cre-loxP recombination

Recombination mediated by the Cre-loxP system is another useful approach to control expression of optogenetic actuators in targeted cell types. For example, ChR2 expression can be restricted to hypothalamic Agrp neurons, by crossing a transgenic mouse that expresses

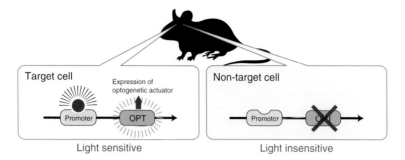

Light sensitive   Light insensitive

Figure 13.11 Cell type-specific promoters to control expression of the optogenetic actuators (OPT). (*Left*) In target cells, a Cell type-specific regulatory signal (*red*) can bind to its cognate promoter (*yellow*), enabling expression of the downstream optogenetic actuator. (*Right*) In non-target cells, the regulatory signal is absent so that the optogenetic actuator is not expressed.

Table 13.2 Characterized viral promoters for optogenetic targeting.

Vector	Promoter (serotype)	Size	Organism	Cell type specificity
Lentivirus				
	EF1α	1.2 Kb	Rat, mouse	Neuron-specific only in lentiviral vectors
	CMV	0.6 Kb	Rat, mouse	Non-specific
	Human synapsin I (hSynI)	0.5 Kb	Rat, mouse	Panneuronal, but a tropism for excitatory cells in lentiviral vectors
	CaMKIIα	1.3 Kb	Rat, mouse, macaque	Excitatory neurons in cortex and hippocampus
	hGFAP	2.2 Kb	Rat, mouse	Astrocytes
	TPH-2	2 Kb	Rat	Raphe serotonergic neurons
Adeno-associated virus				
	CaMKIIα (AAV5)	1.3 Kb	Rat, mouse, macaque	Excitatory CaMKIIα neurons in cortex, amygdala
	hSynI (AAV2)	0.5 Kb	Rat, mouse, macaque	Panneuronal, but a tropism for inhibitory cells at low titers
	hThy1 (AAV5)	5Kb	Macaque	Panneuronal
	fSST (AAV1)	2.6 Kb	Rat, mouse, macaque	Inhibitory neurons (no subtype specificity)
	hGFAP (AAV5, AAV8)	2.2 Kb	Rat	Astrocytes
	MBP (AAV8)	1.35 Kb	Rat	Oligodendrocytes
	SST (AAV2)	2 Kb	Rat	Pre-Bötzinger C somatostatin neurons

Table modified from Yizhar *et al.* (2011).

Cre recombinase (a Cre driver line) under control of the Agrp promoter with another transgenic mouse that expresses ChR2 behind a floxed stop cassette. When these two mice are mated, Cre removes the stop signal and ChR2 will be expressed exclusively in hypothalamic Agrp neurons (Figure 13.12A). A similar approach could be used by injecting into the Cre transgenic mice a virus that expresses floxed ChR2.

The advantages of this approach are relatively high restriction of expression of the optogenetic actuators to the targeted neurons and the availability of many hundreds of different Cre driver lines, as well as transgenic mice containing floxed optogenetic actuator genes. A few

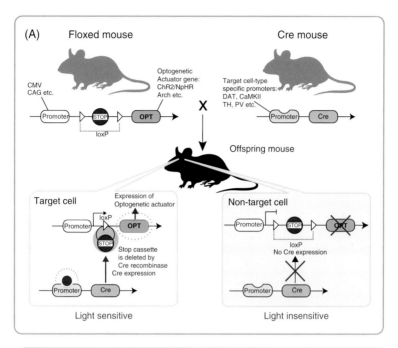

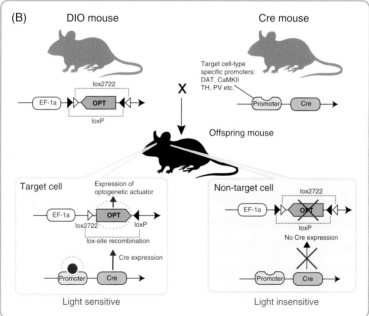

Figure 13.12 Cre-lox systems. (A) Recombination mediated by the Cre-loxP system relies on a transgenic mouse that expresses Cre recombinase under control of a cell type-specific promoter and another transgenic mouse that expresses an optogenetic actuator (*green*) behind a floxed stop cassette (*red*). When these two mice are mated, Cre removes the stop signal and the actuator is expressed in targeted cells. (B) Even more specific expression can be achieved by a double-floxed inverted open reading frame (DIO) to restrict expression of the optogenetic actuator (*green*). Recombination restores the correct orientation to the open reading frame, yielding expression of the optogenetic actuator in target cells.

Table 13.3 Cre driver mouse lines.

Mouse line	Expression in
PV::Cre	Cortical fast-spiking inhibitory interneurons
D1-Cre, D2-Cre	Striatal medium spiny neurons
CaMKIIa-Cre	Excitatory neurons in cortex, hippocampus
Six3-Cre	Mostly cortical layer 4 neurons
ChAT-Cre	Cholinergic neurons
TH-Cre	Dopaminergic neurons, noradrenergic neurons
DAT-Cre	Dopaminergic neurons
ePet-Cre	Serotonergic neurons
Gad2::Cre-ER^{T2}	Cortical inhibitory neurons
Agrp-Cre,	Hypothalamic Agrp neurons
pomc-Cre	Hypothalamic pomc neurons
PKCδ-GluCl-IRES-Cre	Amygdala PKCδ+ neurons

Table modified from Yizhar et al. (2011).

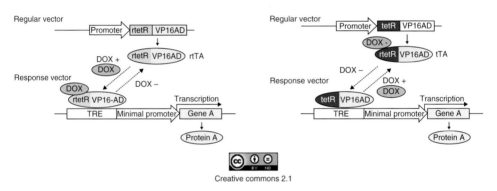

Creative commons 2.1

Figure 13.13 Temporal regulation of optogenetic actuator expression by tet-on (*left*) and tet-off (*right*) strategies. From Takahiro Hirabayashi and Takeshi Yagi: Tet on/off system, Brain science dictionary DOI: 10.14931/bsd.1219 (2012).

of the well-known Cre-driver lines are listed in Table 13.3. One limitation of this approach is that Cre expression can be somewhat 'leaky,' meaning that transgene expression can occur in cells other than the intended targets. This problem can be minimized by using a double-floxed inverted open reading frame (DIO) to restrict expression of the optogenetic actuator (Figure 13.12B). Leakiness is reduced because the floxed version of the actuator gene is inverted and thus unable to lead to consititutive transgene expression, to your target cell population. While there are other caveats to the Cre-lox approach (Smith, 2011; see also jaxmice.jax.org/news/2013/cre_facts.html), currently this is the method of choice for targeting expression of optogenetic actuators to virtually any type of neuron.

13.4.6 Tet-on/off system for temporal control of gene expression

The tet-on/off system (Gossen and Bujard, 1992) is a useful tool for inducible optogenetic actuator gene expression. By feeding tetracycline derivatives (such as doxycycline, DOX) to mice, expression of optogenetic actuators can be reversibly turned on or off *in vivo* (Figure 13.13).

Thus, this method allows temporal control of actuator expression. An example of the advanced application of this system is the selective expression of ChR2 only in hippocampal dentate gyrus (DG) neurons that were active during contextual fear learning. Subsequent selective photostimulation of this subpopulation of DG neurons could reproduce recall of contextual fear memory. For such selective expression of ChR2, the promoter of c-fos (an immediate early gene often used as a marker of recent neuronal activity) was used in combination with the tet-off system (Kubik *et al.*, 2007; Liu *et al.*, 2012). This method is also applicable in other brain regions and could be used to investigate the functions of specific subpopulations of neurons with a known history of prior activity.

13.5   Light delivery strategies for optical control

Once optogenetic probes have been expressed within specific types of neurons in targeted brain areas, the next step is to determine the best approach for delivering light to activate these probes. Two factors enter into this decision: the type of light source (Figure 13.14, top) and the means of delivering light to the tissue expressing the optogenetic actuators (Figure 13.14, bottom). When choosing the proper light source, several factors need to be considered. These include the intensity and frequency of light according to the experimental plan (e.g. *in vivo* versus *in vitro*) and purpose (e.g. measuring behavioral outcome, mapping synaptic connectivity, etc.). For photostimulation of brain slices or brain surface structures, light sources such as arc lamps, light-emitting diodes (LEDs), and lasers are often delivered through a microscope light objective (Yizhar *et al.*, 2011; Zhang *et al.*, 2010). Arc lamps can

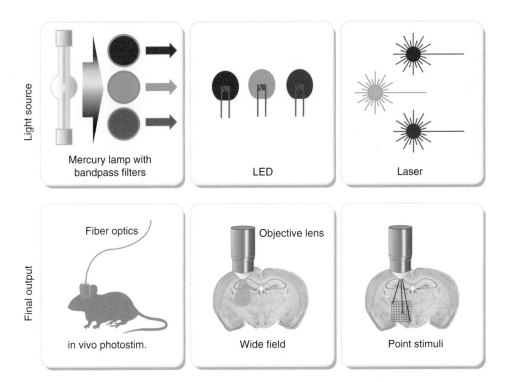

Figure 13.14 Light sources (*top*) and light delivery strategies (*bottom*) for control of neuronal activity via optogenetic actuators.

photostimulate a relatively large area (~0.4 mm^2) (Asrican *et al.*, 2013) and can provide pulse durations as little as 1 ms with pulse frequencies up to 500 Hz under the control of a high-speed shutter (Yizhar *et al.*, 2011). LEDs are a low-cost, high-intensity light source whose intensity and frequency can be well controlled via standard electronic means (Campagnola *et al.*, 2008; Yizhar *et al.*, 2011). Lasers are also optimal for high-speed optogenetic circuit mapping, providing a small spot of light that is scanned throughout the specimen (Kim *et al.*, 2014; Petreanu *et al.*, 2007; Wang *et al.*, 2007).

For *in vivo* studies, lasers and LEDs have been proven to be useful light sources (Carter and de Lecca, 2011; Yizhar *et al.*, 2011; Zhang *et al.*, 2010). In order to deliver light to deep brain regions, a thin optical fiber is stereotaxically inserted into the targeted area while minimizing tissue damage. The success of optogenetic control *in vivo* depends on the level of optogenetic probe expression and on the light intensity delivered to the target region. Heavy light scattering typically occurs in brain tissue and the degree of scattering depends on the light wavelength: longer wavelength light tends to penetrate more deeply due to reduced scattering (Yizhar *et al.*, 2011). The transmission of different wavelengths of light in brain tissue has been thoroughly measured and there is a calculator available to estimate light power density as a function of tissue thickness/depth (Yizhar *et al.*, 2011; www.optogenetic. org/calc). The volume of tissue over which photostimulation occurs can be determined by detecting neuronal activation by electrophysiological recordings and/or expression of activity markers, such as c-fos (Adamantidis *et al.*, 2007; Gradinaru *et al.*, 2009; Tye *et al.*, 2011). Such studies suggest that 5–12 mW/mm^2 of light is sufficient for *in vivo* activation of ChR2 (Boyden *et al.*, 2005; Zhang *et al.*, 2006). While photostimulation of ChR2-expressing neurons through a single fiber can elicit significant functional changes, bilateral illumination is usually required for examining loss-of-function circuit effects *in vivo* for photoinhibition via probes such as eNpHR3.0 (Tye *et al.*, 2011; Witten *et al.*, 2010; Yizhar *et al.*, 2011).

For *in vivo* optogenetic control, great caution is needed regarding light-induced heating effects and cellular toxicity possibly induced by viral infection and/or overexpression of optogenetic actuators (Yizhar *et al.*, 2011). High light intensities over prolonged time periods can cause significant heating effects. To avoid such problems, it is valuable to select optogenetic actuators that are more sensitive to light and to use the minimum light intensity required to control neuronal activity. Such side effects can be defined via control experiments where illumination is applied in the absence of optogenetic actuator expression (Yizhar *et al.*, 2011).

13.6   Study of the neuroendocrine system via optogenetics

In the remainder of this chapter, we will describe some applications of optogenetics to analyze endocrine function, focusing on the hypothalamus. The hypothalamus is part of the ventral diencephalon (Tessmar-Raible *et al.*, 2007) that gathers internal state information and regulates diverse behavioral and autonomic responses, thus maintaining homeostatic balance (Berthoud and Munzberg, 2011; Sternson, 2013). The hypothalamus consists of multiple nuclei and various circuits for regulating metabolism, arousal, and stress responses. It also evokes complex motivated behavior such as food intake, fear processing, aggression, and mating for survival and reproduction (Sternson, 2013).

In order to perform their integrative and regulatory functions, hypothalamic regions have extensive interconnections with other brain regions, including the cerebral cortex, hippocampus, amygdala, striatum, thalamus, and brainstem (Berthoud and Munzberg, 2011). Hypothalamic nuclei contain many types of neurons and axonal pathways, making it difficult to tease out the specific neuronal circuit for functional and behavioral responses. Electrical stimulation, lesion and/or pharmacological approaches have been used for such purposes, but they have clear limitations because of their lack of specificity in controlling neuronal activity. This can be a particularly serious obstacle for mapping precise functional circuits in areas, such as the hypothalamus, that contain many types of neurons that are heavily interconnected with other brain regions. By genetically targeting optogenetic probes within a specific population of neurons, it is possible to manipulate the activity of specific neuron types within a heterogeneous population of neurons and provide more precise information about circuit connectivity. Here we review how optogenetic techniques have been applied for dissecting neuronal circuits in the hypothalamus and what such analyses have revealed regarding hypothalamus-associated behaviors.

13.6.1 Feeding

The main circuits for controlling feeding behavior have been found within the hypothalamic arcuate nucleus (Aponte *et al.*, 2011; Atasoy *et al.*, 2012; Sternson, 2013). Activation of neurons expressing agouti-related protein/neuropeptide Y (AGRP/NYP) enhances feeding, whereas pro-opiomelanocortin (POMC)-expressing neurons suppress food intake (Aponte *et al.*, 2011; Gutierrez *et al.*, 2011; Sternson, 2013). Both of these neurons that induce opposite feeding behaviors are intermingled in the same area of the arcuate nucleus. Thus, optogenetic studies were necessary to dissect functional circuits involving these groups of neurons and establish the causal relationship between the activity of these neurons and behavior (Aponte *et al.*, 2011; Atasoy *et al.*, 2012). For cell type-specific expression of ChR2, the Cre recombinase-dependent viral vector rAAV-FLEX-rev-ChR2:tdTomato was stereotaxically injected into the arcuate nucleus of either *agrp-cre* or *pomc-cre* transgenic mice (Aponte *et al.*, 2011). In this case, the flip-excision (FLEX) switch was added for stable transgene inversion (Atasoy *et al.*, 2008; Schnutgen *et al.*, 2003). After injecting virus, selective expression of ChR2 was confirmed by imaging a fluorescent tag (in this case, tdtomato) and electrophysiological recordings (Atasoy *et al.*, 2008). In brain slices, photocurrents and action potentials were reliably induced by photostimulation using laser illumination or optical fibers, indicating that Cre-dependent expression of ChR2 delivered by rAAV-FLEX-rev allows selective activation of AGRP or POMC neurons (Atasoy *et al.*, 2008). In this case, ChR2 was also strongly expressed in axons, so they could map long-range circuits via stimulating axons of AGRP and POMC and recording postsynaptic response simultaneously in the paraventricular nucleus (PVN) (Atasoy *et al.*, 2008). This method might be applicable to other local and long-range circuits involving AGRP and POMC neurons.

After confirming selective expression of ChR2 *in vitro*, photostimulation of neurons *in vivo* established a correlation between the activation of specific types of cells and behavioral outcomes. Because the arcuate nucleus is located deep within the brain, light was delivered by an optical fiber that was implanted into an area near the arcuate nucleus. Photostimulation of AGRP neurons caused voracious feeding to occur within minutes of photostimulation, even in well-fed AGRP-ChR2 mice. The amount of food intake depended on the number of ChR2-expressing AGRP neurons, as well as the frequency and duration of light stimulation. On the other hand, prolonged photostimulation of ChR2-expressing POMC neurons for up to 24 hours inhibited food intake and reduced body weight (Aponte *et al.*, 2011).

Further optogenetic studies have revealed pre- and postsynaptic connections involving AGRP neurons. Monosynaptic inputs to AGRP neurons were first identified through a Cre-dependent helper virus and a modified rabies virus (Krashes *et al.*, 2014; Wall *et al.*, 2010). Presynaptic excitatory inputs to AGRP neurons were then identified by expressing ChR2 selectively in glutamate neurons under the control of the Vglut2 promotor. This was done by stereotaxically injecting Cre-dependent AAV virus expressing ChR2-mCherry (AAV8-DIO-ChR2-mCherry) into Vglut2-IRES-Cre transgenic mice (Krashes *et al.*, 2014). With this approach, neurons expressing thyrotropin-releasing hormone (TRH) and pituitary adenylate cyclase-activating polypeptide (PACAP) in the hypothalamic paraventricular nucleus were found to provide strong excitatory inputs to AGRP neurons (Krashes *et al.*, 2014). Atasoy *et al.* (2012) showed that AGRP neurons evoke feeding behavior by inhibiting oxytocin neurons within the paraventricular hypothalamus (PVH). In this case, strong ChR2 expression in the axons of AGRP neurons made it possible to verify the long-range AGRP neural circuit function in PVH via photostimulating axons of AGRP neurons in PVN area (Atasoy *et al.*, 2012). Through optogenetic synaptic circuit mapping techniques (Lim *et al.*, 2013), further new connections that regulate feeding behavior may be discovered.

13.6.2 Sleep and metabolism

The sleep-wake cycle is controlled by sleep-promoting areas in the anterior hypothalamus and by arousal systems in the posterior hypothalamus, basal forebrain, and brainstem (Adamantidis and de Lecea, 2008; Adamantidis *et al.*, 2007; Berthoud and Munzberg, 2011). Two different types of neurons in the lateral hypothalamus (LH) play a central role in regulating sleep and metabolism through their opposite regulatory mechanisms. Orexin/hypocretin (hcrt)-expressing neurons are known to facilitate arousal and increase metabolism; impairment of the hcrt system can cause narcolepsy (Adamantidis and de Lecea, 2008). On the other hand, neurons expressing melanin-concentrating hormone in the LH and zona incerta (ZI) promote sleep and dampen metabolism (Adamantidis and de Lecea, 2008). These neuronal systems are heavily connected with other hypothalamic neurons and with brain areas that regulate food intake, energy homeostasis, stress, and reward and motivation (Adamantidis and de Lecea, 2008; Berthoud and Munzberg, 2011). Therefore, optogenetic circuit mapping can be a powerful tool for dissecting the circuits involved in homeostatic regulation.

To investigate the role of hcrt neurons in sleep state transitions, ChR2 has been selectively expressed in hcrt neurons by lentivirus, using the hcrt gene promoter (hcrt::ChR2-mCherry) (Adamantidis *et al.*, 2007). Specific expression of ChR2 was confirmed and reliable photostimulation of hcrt neurons was established via fluorescence immunohistochemistry and electrophysiological recordings in brain slices. Scattering of light emitted from an optical fiber was carefully measured *in vivo* and activation of hcrt neurons was confirmed by measuring c-fos expression. The latency of sleep-to-wake transitions was measured by EEG/EMG recordings while hcrt neurons were photostimulated at different frequencies. A causal relationship was observed between the frequency-dependent activity of hcrt neurons and the degree of arousal, showing that the activity of these neurons is sufficient to cause arousal. Opposite effects were observed when hcrt neurons expressing halorhodopsin were silenced by orange light (Tsunematsu *et al.*, 2011), establishing that activity of these neurons is also necessary for arousal.

13.6.3 Stress

The endocrine stress response is mediated by the hypothalamic-pituitary-adrenal axis (HPA). Neurons in the PVN of the rostral hypothalamus express corticotropin-releasing hormone (CRH) that regulates the synthesis and secretion of adrenocorticotropic hormone (ACTH) in the anterior pituitary (Bruhn *et al.*, 1984; Vale *et al.*, 1981). ACTH triggers the release of glucocorticoids (GCs) from adrenal glands and secreted GCs then regulate various stress-related mechanisms in both the central and peripheral nervous systems.

Recent advances have successfully applied optogenetic techniques to activate CRH neurons in PVN (Wamsteeker Cusulin *et al.*, 2013). An AAV containing a floxed ChR2-EYFP was injected into the PVN of CRH-IRES-Cre mice. Photostimulation of ChR2-expressing CRH neurons in the PVN reliably evoked action potentials in brain slices, indicating that the optogenetic approach can be useful for studying the causal relationship between CRH neuronal activity and cognitive and behavioral changes.

Optogenetic manipulation of the stress axis has also been tested in transgenic zebrafish (de Marco *et al.*, 2013). In this case, a different type of optogenetic probe – a photoactivated adenylyl cyclase (bPAC) – was selectively expressed in ACTH-producing pituitary corticotrope cells under the control of the POMC promoter. Increased cAMP, downstream of CRH receptor activation, was induced by blue light, which enhanced ACTH release from pituitary cells. This ACTH induced the secretion of GCs, yielding behavioral stress responses in the zebrafish.

13.6.4 Mating and aggression

Optogenetic and pharmacogenetic approaches have been valuable in dissection of the neural circuits associated with aggressive behavior (Lin *et al.*, 2011; Yang *et al.*, 2013). Previous c-fos labeling and *in vivo* multielectrode recordings showed that the ventrolateral subdivision of the ventromedial hypothalamic nucleus (VMHvl) is associated with both male-male attack and male-female mating behaviors. About 20% of the neurons that are activated in each type of behavior overlapped, suggesting that circuits for aggression and reproduction are closely correlated with each other. To further probe the circuits for attack and mating in VMHvl, ChR2 was delivered into the VMHvl by co-injecting adeno-associated viral vectors (AAV2) expressing Cre recombinase (AAV2 CMV::CRE) and a Cre-dependent form of ChR2 fused to the enhanced yellow fluorescent protein (AAV2 Cre inducible EF1α::ChR2) (Lin *et al.*, 2011). Immunohistochemistry established that ChR2 was expressed in neurons with cell bodies at the virus injection site and optetrode recordings in anesthetized mice verified that photostimulation elicited action potentials in ChR2-expressing neurons in VMHvl. In the presence of an intruder, photostimulation of VMHvl neurons evoked time-locked attack behavior. Interestingly, this was directed not only toward intruder males but also toward inappropriate targets including castrated males, females, and even an inanimate object. These results indicate that the VMHvl is a part of an attack circuit and that activation of the VMHvl is sufficient to evoke attack behavior.

During male-female encounters and mating, attack-promoting neurons in VMHvl tend to be suppressed. Photostimulation of VMHvl neurons elicited attack behavior towards females before mounting and after ejaculation; in contrast, photostimulation during intromission often failed to induce the attack. This observation suggests that reproductive behaviors actively suppress attack circuitry. Recent studies further revealed that neurons in VMHvl expressing progesterone receptors play an important role in female mating behavior as well as in mating and fighting behaviors in males (Yang *et al.*, 2013).

13.7 Future prospects for optogenetics

Recent advances in optogenetics are revolutionizing neuroscience by enabling direct assessment of the causal relationship between the activity of neurons and neuronal circuits with behavior. The ever-expanding varieties of optogenetic actuators and sensors, as well as transgenic mouse lines expressing these actuators, provide powerful tools for elucidating the function of neuronal circuits (Asrican *et al.*, 2013; Yizhar *et al.*, 2011).

Neuronal circuits consist of multiple pre- and postsynaptic neurons that are closely connected with each other. For a deeper understanding of circuit function, it is essential to detect and analyze the activity of these multiple neurons simultaneously. Optogenetics is now capable of such tasks by combining optogenetic detection of postsynaptic neuron responses with optogenetic photostimulation of presynaptic neurons (Mancuso *et al.*, 2011; Wyart *et al.*, 2009). This is a highly promising way to examine the function of neuronal circuits and should be widely applied in the near future. Improved optics for photostimulation and detection of the activity of neuronal populations with high spatiotemporal resolution will also be important in order to examine neuronal function more precisely.

In regard to neuroendocrinology, an increasing number of optogenetic lines have been developed to describe the roles of neurons and circuits in the hypothalamus. However, the number of such lines is still limited and increasing the number of targeted cell types is essential. Likewise, it will be important to expand such studies to non-neuronal components of the endocrine system, such as the pancreas and adrenal gland. Combining such optogenetic tools with gene knockout systems will help to define the functions of specific cells and circuits, thereby leading to a deeper understanding of both the brain and the endocrine system.

Acknowledgments

Supported by the World Class Institute (WCI) program of the National Research Foundation of Korea (NRF) funded by the Ministry of Education, Science, and Technology (MEST) (WCI 2009-003) and a CRP grant from the National Research Foundation of Singapore.

References

Adamantidis, A. and de Lecea, L. (2008) Sleep and metabolism: shared circuits, new connections. *Trends in Endocrinology and Metabolism*, **19**, 362–370.

Adamantidis, A.R., Zhang, F., Aravanis, A.M., Deisseroth, K. and de Lecea, L. (2007) Neural substrates of awakening probed with optogenetic control of hypocretin neurons. *Nature*, **450**, 420–424. [**This study showed that optogenetic activation of hypocretin-expressing neurons in the lateral hypothalamus enhaces arousal in freely moving mice.**]

Anikeeva, P., Andalman, A.S., Witten, I., *et al.* (2011) Optetrode: a multichannel readout for optogenetic control in freely moving mice. *Nature Neuroscience*, **15**, 163–170. [**Describes the optetrode, a device for pairing optical stimulation with simultaneous multichannel electrophysiological recordings in freely moving mice.**]

Aponte, Y., Atasoy, D. and Sternson, S.M. (2011) AGRP neurons are sufficient to orchestrate feeding behavior rapidly and without training. *Nature Neuroscience*, **14**, 351–355. [**This paper used optogenetic actuators to dissect two discrete neural circuits regulating feeding behavior among intermingled neuron populations in the hypothalamic arcuate nucleus. Selective stimulation of ChR2-expressing AGRP neurons induced voracious feeding, while stimulation of POMC neurons inhibited food intake in mice.**]

Asrican, B., Augustine, G.J., Berglund, K., *et al.* (2013) Next-generation transgenic mice for optogenetic analysis of neural circuits. *Frontiers in Neural Circuits*, **7**, 160. [**This paper characterizes several lines of optogenetic actuator transgenic mouse lines useful for analysis of brain circuit function. These mice express optogenetic probes, such as enhanced halorhodopsin or several different versions of channelrhodopsins, behind various neuron-specific promoters. These mice permit photoinhibition or photostimulation both in vitro and in vivo.**]

Atasoy, D., Aponte, Y., Su, H.H. and Sternson, S.M. (2008) A FLEX switch targets Channelrhodopsin-2 to multiple cell types for imaging and long-range circuit mapping. *Journal of Neuroscience*, **28**, 7025–7030.

Atasoy, D., Betley, J.N., Su, H.H. and Sternson, S.M. (2012) Deconstruction of a neural circuit for hunger. *Nature*, **488**, 172–177.

Berthoud, H.R. and Munzberg, H. (2011) The lateral hypothalamus as integrator of metabolic and environmental needs: from electrical self-stimulation to opto-genetics. *Physiology and Behavior*, **104**, 29–39.

Boyden, E.S. (2011) A history of optogenetics: the development of tools for controlling brain circuits with light. *F1000 Biology Reports*, **3**, 11. [**This review gives an account of the development of various kinds of optogenetic actuators (channelrhodopsin, archaerhodopsin, and halorhodopsins) and their applications.**]

Boyden, E.S., Zhang, F., Bamberg, E., Nagel, G. and Deisseroth, K. (2005) Millisecond-timescale, genetically targeted optical control of neural activity. *Nature Neuroscience*, **8**, 1263–1268. [*A landmark paper in the field: the first application of channelrhodopsin for control of neural activity. It demonstrated that neuronal activity could be controlled by light with millisecond precision.*]

Bruhn, T.O., Sutton, R.E., Rivier, C.L. and Vale, W.W. (1984) Corticotropin-releasing factor regulates proopiomelanocortin messenger ribonucleic acid levels in vivo. *Neuroendocrinology*, **39**, 170–175.

Campagnola, L., Wang, H. and Zylka, M.J. (2008) Fiber-coupled light-emitting diode for localized photostimulation of neurons expressing channelrhodopsin-2. *Journal of Neuroscience Methods*, **169**, 27–33.

Carter, M.E. and de Lecea, L. (2011) Optogenetic investigation of neural circuits in vivo. *Trends in Molecular Medicine*, **17**, 197–206.

Chow, B.Y., Han, X., Dobry, A.S., *et al.* (2010) High-performance genetically targetable optical neural silencing by light-driven proton pumps. *Nature*, **463**, 98–102. [**This is the first application of archaerhodopsin-3 for photoinhibition of neuronal activity, showing the potential of this bacterial rhodopsin for optogenetic control.**]

Chow, B.Y., Han, X. and Boyden, E.S. (2012) Genetically encoded molecular tools for light-driven silencing of targeted neurons. *Progress in Brain Research*, **196**, 49–61.

Cruikshank, S.J., Urabe, H., Nurmikko, A.V. and Connors, B.W. (2010) Pathway-specific feedforward circuits between thalamus and neocortex revealed by selective optical stimulation of axons. *Neuron*, **65**, 230–245.

De Marco, R.J., Groneberg, A.H., Yeh, C.M., Castillo Ramirez, L.A. and Ryu, S. (2013) Optogenetic elevation of endogenous glucocorticoid level in larval zebrafish. *Frontiers in Neural Circuits*, **7**, 82.

Druckmann, S., Feng, L., Lee, B., *et al.* (2014) Structured synaptic connectivity between hippocampal regions. *Neuron*, **81**, 629–640.

Gossen, M. and Bujard, H. (1992) Tight control of gene expression in mammalian cells by tetracycline-responsive promoters. *Proceedings of the National Academy of Sciences of the United States of America*, **89**, 5547–5551.

Gradinaru, V., Thompson, K.R. and Deisseroth, K. (2008) eNpHR: a Natronomonas halorhodopsin enhanced for optogenetic applications. *Brain Cell Biology*, **36**, 129–139.

Gradinaru, V., Mogri, M., Thompson, K.R., Henderson, J.M. and Deisseroth, K. (2009) Optical deconstruction of parkinsonian neural circuitry. *Science*, **324**, 354–359.

Gradinaru, V., Zhang, F., Ramakrishnan, C., *et al.* (2010) Molecular and cellular approaches for diversifying and extending optogenetics. *Cell*, **141**, 154–165.

Grimley, J.S., Li, L., Wang, W., *et al.* (2013) Visualization of synaptic inhibition with an optogenetic sensor developed by cell-free protein engineering automation. *Journal of Neuroscience*, **33**, 16297–16309.

Gunaydin, L.A., Yizhar, O., Berndt, A., *et al.* (2010) Ultrafast optogenetic control. *Nature Neuroscience*, **13**, 387–392.

Gutierrez, R., Lobo, M.K., Zhang, F. and de Lecea, L. (2011) Neural integration of reward, arousal, and feeding: recruitment of VTA, lateral hypothalamus, and ventral striatal neurons. *IUBMB Life*, **63**, 824–830.

Han, X. and Boyden, E.S. (2007) Multiple-color optical activation, silencing, and desynchronization of neural activity, with single-spike temporal resolution. *PLoS One*, **2**, e299.

Hira, R., Honkura, N., Noguchi, J., *et al.* (2009) Transcranial optogenetic stimulation for functional mapping of the motor cortex. *Journal of Neuroscience Methods*, **179**, 258–263.

Inagaki, H.K., Jung, Y., Hoopfer, E.D., *et al.* (2014) Optogenetic control of Drosophila using a red-shifted channelrhodopsin reveals experience-dependent influences on courtship. *Nature Methods*, **11**, 325–332.

Ishizuka, T., Kakuda, M., Araki, R. and Yawo, H. (2006) Kinetic evaluation of photosensitivity in genetically engineered neurons expressing green algae light-gated channels. *Neuroscience Research*, **54**, 85–94.

Jin, L., Han, Z., Platisa, J., *et al.* (2012) Single action potentials and subthreshold electrical events imaged in neurons with a fluorescent protein voltage probe. *Neuron*, **75**, 779–785.

Kahn, I., Desai, M., Knoblich, U., *et al.* (2011) Characterization of the functional MRI response temporal linearity via optical control of neocortical pyramidal neurons. *Journal of Neuroscience*, **31**, 15086–15091.

Kim, J., Lee, S., Tsuda, S., *et al.* (2014) Optogenetic mapping of cerebellar inhibitory circuitry reveals spatially biased coordination of interneurons via electrical synapses. *Cell Reports*, **7**, 1601–1613. [**This paper demonstrates how high-speed optogenetic circuit mapping can reveal the spatial organization of synaptic circuits in brain slices.**]

Kim, T.I., McCall, J.G., Jung, Y.H., *et al.* (2013) Injectable, cellular-scale optoelectronics with applications for wireless optogenetics. *Science*, **340**, 211–216.

Klapoetke, N.C., Murata, Y., Kim, S.S., *et al.* (2014) Independent optical excitation of distinct neural populations. *Nature Methods*, **11**, 338–346.

Kralj, J.M., Douglass, A.D., Hochbaum, D.R., Maclaurin, D. and Cohen, A.E. (2011) Optical recording of action potentials in mammalian neurons using a microbial rhodopsin. *Nature Methods*, **9**, 90–95.

Krashes, M.J., Shah, B.P., Madara, J.C., *et al.* (2014) An excitatory paraventricular nucleus to AgRP neuron circuit that drives hunger. *Nature*, **507**, 238–242.

Kubik, S., Miyashita, T. and Guzowski, J.F. (2007) Using immediate-early genes to map hippocampal subregional functions. *Learning and Memory*, **14**, 758–770.

Kuner, T. and Augustine, G.J. (2000) A genetically encoded ratiometric indicator for chloride: capturing chloride transients in cultured hippocampal neurons. *Neuron*, **27**, 447–459.

Lanyi, J.K. (1986) Halorhodopsin: a light-driven chloride ion pump. *Annual Review of Biophysics and Biophysical Chemistry*, **15**, 11–28.

Lanyi, J.K. and Weber, H.J. (1980) Spectrophotometric identification of the pigment associated with light-driven primary sodium translocation in Halobacterium halobium. *Journal of Biological Chemistry*, **255**, 243–250.

Lee, A.M., Hoy, J.L., Bonci, A., *et al.* (2014) Identification of a brainstem circuit regulating visual cortical state in parallel with locomotion. *Neuron*, **83**, 455–466.

Lee, J.H., Durand, R., Gradinaru, V., *et al.* (2010) Global and local fMRI signals driven by neurons defined optogenetically by type and wiring. *Nature*, **465**, 788–792.

Lim, D.H., Ledue, J., Mohajerani, M.H., Vanni, M.P. and Murphy, T.H. (2013) Optogenetic approaches for functional mouse brain mapping. *Frontiers in Neuroscience*, **7**, 54.

Lima, S.Q. and Miesenböck, G. (2005) Remote control of behavior through genetically targeted photostimulation of neurons. *Cell*, **121**, 141–152.

Lin, D., Boyle, M.P., Dollar, P., *et al.* (2011) Functional identification of an aggression locus in the mouse hypothalamus. *Nature*, **470**, 221–226. [*This study used optogenetic actuators to show that activation of neurons in the ventrolateral subdivision of the ventromedial hypothalamic nucleus is necessary for inducing attack behavior.*]

Lin, J.Y., Knutsen, P.M., Muller, A., Kleinfeld, D. and Tsien, R.Y. (2013) ReaChR: a red-shifted variant of channelrhodopsin enables deep transcranial optogenetic excitation. *Nature Neuroscience*, **16**, 1499–1508.

Liu, X., Ramirez, S., Pang, P.T., *et al.* (2012) Optogenetic stimulation of a hippocampal engram activates fear memory recall. *Nature*, **484**, 381–385. [*By combining the tet-on system with activity-dependent ChR2 expression, it was possible to photostimulate hippocampal dentate gyrus granule cells that had been active during fear conditioning. They showed that such selective photostimulation was sufficient to induce a fear response.*]

Mancuso, J.J., Kim, J., Lee, S., *et al.* (2011) Optogenetic probing of functional brain circuitry. *Experimental Physiology*, **96**, 26–33.

Matsui, A., Yoshida, A.C., Kubota, M., Ogawa, M. and Shimogori, T. (2011) Mouse in utero electroporation: controlled spatiotemporal gene transfection. *Journal of Visualized Experiments*, **54**, e3024.

Matsuno-Yagi, A. and Mukohata, Y. (1977) Two possible roles of bacteriorhodopsin; a comparative study of strains of Halobacterium halobium differing in pigmentation. *Biochemical and Biophysical Research Communications*, **78**, 237–243.

McClure, C., Cole, K.L., Wulff, P., Klugmann, M. and Murray, A.J. (2011) Production and titering of recombinant adeno-associated viral vectors. *Journal of Visualized Experiments*, **57**, e3348.

Miyawaki, A., Llopis, J., Heim, R., *et al.* (1997) Fluorescent indicators for Ca2+ based on green fluorescent proteins and calmodulin. *Nature*, **388**, 882–887.

Mukohata, Y. and Kaji, Y. (1981) Light-induced membrane-potential increase, ATP synthesis, and proton uptake in Halobacterium halobium, R1mR catalyzed by halorhodopsin: effects of N,N'-dicyclohexylcarbodiimide, triphenyltin chloride, and 3,5-di-tert-butyl-4-hydroxybenzylidenemalononitrile (SF6847). *Archives of Biochemistry and Biophysics*, **206**, 72–76.

Mukohata, Y., Sugiyama, Y., Ihara, K. and Yoshida, M. (1988) An Australian halobacterium contains a novel proton pump retinal protein: archaerhodopsin. *Biochemical and Biophysical Research Communications*, **151**, 1339–1345.

Nagel, G., Szellas, T., Huhn, W., *et al.* (2003) Channelrhodopsin-2, a directly light-gated cation-selective membrane channel. *Proceedings of the National Academy of Sciences of the United States of America*, **100**, 13940–13945.

Nakamura, S., Baratta, M.V. and Cooper, D.C. (2013) A method for high fidelity optogenetic control of individual pyramidal neurons in vivo. *Journal of Visualized Experiments*, **79**, e50291.

Nathanson, J.L., Yanagawa, Y., Obata, K. and Callaway, E.M. (2009) Preferential labeling of inhibitory and excitatory cortical neurons by endogenous tropism of adeno-associated virus and lentivirus vectors. *Neuroscience*, **161**, 441–450.

Oesterhelt, D. and Stoeckenius, W. (1973) Functions of a new photoreceptor membrane. *Proceedings of the National Academy of Sciences of the United States of America*, **70**, 2853–2857.

Peterka, D.S., Takahashi, H. and Yuste, R. (2011) Imaging voltage in neurons. *Neuron*, **69**, 9–21.

Petreanu, L., Huber, D., Sobczyk, A. and Svoboda, K. (2007) Channelrhodopsin-2-assisted circuit mapping of long-range callosal projections. *Nature Neuroscience*, **10**, 663–668.

Raimondo, J.V., Kay, L., Ellender, T.J. and Akerman, C.J. (2012) Optogenetic silencing strategies differ in their effects on inhibitory synaptic transmission. *Nature Neuroscience*, **15**, 1102–1104. [***This paper demonstrates that silencing activity with halorhodopsin can increase the probability of synaptically evoked spiking, due to intracellular accumulation of chloride ions. On the other hand, the light-driven proton pump, archaerhodopsin, did not have this effect. This point should be kept in mind when designing optogenetic photoinhibition experiments.***]

Schnutgen, F., Doerflinger, N., Calleja, C., *et al.* (2003) A directional strategy for monitoring Cre-mediated recombination at the cellular level in the mouse. *Nature Biotechnology*, **21**, 562–565.

Smith, L. (2011) Good planning and serendipity: exploiting the Cre/Lox system in the testis. *Reproduction*, **141**, 151–161.

Sohal, V.S., Zhang, F., Yizhar, O. and Deisseroth, K. (2009) Parvalbumin neurons and gamma rhythms enhance cortical circuit performance. *Nature*, **459**, 698–702.

Sparta, D.R., Stamatakis, A.M., Phillips, J.L., *et al.* (2011) Construction of implantable optical fibers for long-term optogenetic manipulation of neural circuits. *Nature Protocols*, **7**, 12–23.

Sternson, S.M. (2013) Hypothalamic survival circuits: blueprints for purposive behaviors. *Neuron*, **77**, 810–824.

Tessmar-Raible, K., Raible, F., Christodoulou, F., *et al.* (2007) Conserved sensory-neurosecretory cell types in annelid and fish forebrain: insights into hypothalamus evolution. *Cell*, **129**, 1389–1400.

Tian, L., Hires, S.A. and Looger, L.L. (2012) Imaging neuronal activity with genetically encoded calcium indicators. *Cold Spring Harbor Protocols*, **2012**, 647–656.

Tønnesen, J., Sorensen, A.T., Deisseroth, K., Lundberg, C. and Kokaia, M. (2009) Optogenetic control of epileptiform activity. *Proceedings of the National Academy of Sciences of the United States of America*, **106**, 12162–12167.

Tsunematsu, T., Kilduff, T.S., Boyden, E.S., *et al.* (2011) Acute optogenetic silencing of orexin/hypocretin neurons induces slow-wave sleep in mice. *Journal of Neuroscience*, **31**, 10529–10539.

Tye, K.M. and Deisseroth, K. (2012) Optogenetic investigation of neural circuits underlying brain disease in animal models. *Nature Reviews Neuroscience*, **13**, 251–266. [***Combining viral infection and local photostimulation properly allows selective control of neuronal circuit components. This review is very helpful for designing your own optogenetic experiments.***]

Tye, K.M., Prakash, R., Kim, S.Y., *et al.* (2011) Amygdala circuitry mediating reversible and bidirectional control of anxiety. *Nature*, **471**, 358–362.

Vale, W., Spiess, J., Rivier, C. and Rivier, J. (1981) Characterization of a 41-residue ovine hypothalamic peptide that stimulates secretion of corticotropin and beta-endorphin. *Science*, **213**, 1394–1397.

Wall, N.R., Wickersham, I.R., Cetin, A., De La Parra, M. and Callaway, E.M. (2010) Monosynaptic circuit tracing in vivo through Cre-dependent targeting and complementation of modified rabies virus. *Proceedings of the National Academy of Sciences of the United States of America*, **107**, 21848–21853.

Wamsteeker Cusulin, J.I., Fuzesi, T., Watts, A.G. and Bains, J.S. (2013) Characterization of corticotropin-releasing hormone neurons in the paraventricular nucleus of the hypothalamus of Crh-IRES-Cre mutant mice. *PLoS One*, **8**, e64943.

Wang, H., Peca, J., Matsuzaki, M., *et al.* (2007) High-speed mapping of synaptic connectivity using photostimulation in Channelrhodopsin-2 transgenic mice. *Proceedings of the National Academy of Sciences of the United States of America*, **104**, 8143–8148.

Witten, I.B., Lin, S.C., Brodsky, M., *et al.* (2010) Cholinergic interneurons control local circuit activity and cocaine conditioning. *Science*, **330**, 1677–1681.

Wyart, C., Del Bene, F., Warp, E., *et al.* (2009) Optogenetic dissection of a behavioural module in the vertebrate spinal cord. *Nature*, **461**, 407–410.

Yang, C.F., Chiang, M.C., Gray, D.C., *et al.* (2013) Sexually dimorphic neurons in the ventromedial hypothalamus govern mating in both sexes and aggression in males. *Cell*, **153**, 896–909.

Yizhar, O., Fenno, L.E., Davidson, T.J., Mogri, M. and Deisseroth, K. (2011) Optogenetics in neural systems. *Neuron*, **71**, 9–34. [***This review contains valuable information about many characterized viral promoters for targeting optogenetic actuators, as well as discussing the kinetic properties of optogenetic actuator families. The authors addressed general principles for*** in vivo ***optogenetic experiments and overview various approaches to investigating neural systems via optogenetics.***]

Zemelman, B.V., Nesnas, N., Lee, G.A. and Miesenbock, G. (2003) Photochemical gating of heterologous ion channels: remote control over genetically designated populations of neurons. *Proceedings of the National Academy of Sciences of the United States of America*, **100**, 1352–1357.

Zhang, F., Wang, L.P., Boyden, E.S. and Deisseroth, K. (2006) Channelrhodopsin-2 and optical control of excitable cells. *Nature Methods*, **3**, 785–792.

Zhang, F., Wang, L.P., Brauner, M., *et al.* (2007) Multimodal fast optical interrogation of neural circuitry. *Nature*, **446**, 633–639. [*This paper documents that activation of halorhodopsin by yellow light can photoinhibit neuronal activity with high temporal precision. It also demonstrated that bidirectional optical neural control of activity could be achieved by co-expression of halorhodopsin and channelrhodopsin.*]

Zhang, F., Prigge, M., Beyriere, F., *et al.* (2008) Red-shifted optogenetic excitation: a tool for fast neural control derived from Volvox carteri. *Nature Neuroscience*, **11**, 631–633.

Zhang, F., Gradinaru, V., Adamantidis, A.R., *et al.* (2010) Optogenetic interrogation of neural circuits: technology for probing mammalian brain structures. *Nature Protocols*, **5**, 439–456. [*Provides a detailed protocol for optimizing expression and targeting of optogenetic tools and for combining optical neural control with compatible readouts both* **in vivo** *and* **ex vivo**. *There is a good comparison of the pros and cons of viral and transgenic targeting methods for optogenetic actuators.*]

Zhao, S., Cunha, C., Zhang, F., *et al.* (2008) Improved expression of halorhodopsin for light-induced silencing of neuronal activity. *Brain Cell Biology*, **36**, 141–154.

CHAPTER 14

Non-Mammalian Models for Neurohypophysial Peptides

Einav Wircer,[1] Shifra Ben-Dor,[2] and Gil Levkowitz[1]

[1] Department of Molecular Cell Biology, Weizmann Institute of Science, Rehovot, Israel
[2] Department of Biological Services, Weizmann Institute of Science, Rehovot, Israel

14.1 Historical overview

It is by now well established that the hypothalamo-neurohypophysial system (HNS) is a major neuroendocrine interface between two discrete biological components: central nervous system (CNS) axons and peripheral blood vessels. This interface allows the hypothalamic neuropeptides oxytocin and vasopressin to traverse to the bloodstream and to control peripheral physiology such as water balance and reproduction. The discovery of the structure and function of the HNS was prompted by a question more than a century old: 'How does the brain exert control over peripheral function?' Over 100 years ago, researchers showed that injections of extracts from the posterior part of the pituitary elicit a variety of physiological effects in mammals (Dale, 1909; Howell, 1898; Oliver and Schäfer, 1895; Ott and Scott, 1910; Schäfer and Vincent, 1899). These effects include changes in blood pressure, antidiuretic activity, contraction of the uterus, and increase in the secretion of milk in lactating females. Experiments in non-mammalian vertebrates have demonstrated conservation in the activities of pituitary extracts derived from mammals. Thus, in 1904, Percy T. Herring demonstrated that extracts from the posterior lobe of the pituitary affect the contraction of the frog heart and blood vessels (Herring, 1904). In 1912, D. Noël Paton and Alexander Watson showed that the pituitary extract causes a reduction in blood pressure when injected into birds (Paton and Watson, 1912). Likewise, Lancelot T. Hogben and Walter Schlapp showed that mammalian posterior lobe extract causes a drop in blood pressure in ducks and fowls and also induces changes in blood pressure of tortoises and frogs (Hogben and Schlapp, 1924). This phenomenon of lowering in blood pressure was later attributed by J. H. Gaddum to the activity of oxytocin in the preparations (Gaddum, 1928).

In parallel with the extensive efforts to isolate the elusive posterior hormone extracts, the field of endocrinology has been occupied with the question of the origin of the posterior lobe hormones. Classic endocrine theory maintained that the posterior pituitary lobe secretory material was a hormone and therefore likely synthesized in endocrine cells within the pituitary itself (Gersh, 1939). An opposing theory was based on Ernst Scharrer's study of a non-mammalian species, the European blind minnow, in which Scharrer described the morphology of nerve-gland neurosecretory cells in the fish hypothalamus (Scharrer, 1928). Inspired by his observation in fish, together with the work of his wife Berta on invertebrate neurons, Scharrer effectively established the concept of neurosecretion, arguing that neurohypophysial

Molecular Neuroendocrinology: From Genome to Physiology, First Edition. Edited by David Murphy and Harold Gainer.
© 2016 John Wiley & Sons, Ltd. Published 2016 by John Wiley & Sons, Ltd.
Companion website: www.wiley.com/go/murphy/neuroendocrinology

secretory material was actually being produced in hypothalamic neurons and secreted into the posterior pituitary lobe – the neurohypophysis (Scharrer and Scharrer, 1937, 1945).

Until the 1920s, there had been a scientific debate between physiologists over whether or not the oxytocic, pressor/depressor, and antidiuretic activities of the posterior pituitary gland extracts are caused by a single or multiple substances. Although adrenaline, a known pressor compound, or histamine were considered as the active chemicals, these theories were later disproved (Kamm *et al.*, 1928, and references therein). Eventually, improvement and standardization of the aforementioned extraction method led to the understanding that at least two active factors underlie posterior lobe activities: one with an oxytocic effect (oxytocin) and the other that causes the pressor and antidiuretic effects (vasopressin) (Gaddum, 1928; Kamm *et al.*, 1928). One of the factors, oxytocin, was the first polypeptide hormone to be chemically synthesized by Vincent du Vigneaud in 1953 for which he was awarded the Nobel Prize in Chemistry in 1955 (du Vigneaud *et al.*, 1953b). At approximately the same time, Roger Acher and Jacqueline Chauvet identified the sequence and structure of vasopressin (Acher and Chauvet, 1953). Surprisingly, despite the major differences in their functions, both hormones turned out to be almost identical in their structure.

14.2 Evolutionary perspective on oxytocin and vasopressin peptides sequence and structure

14.2.1 Ligands

The ligands oxytocin and vasopressin are highly conserved peptide hormones consisting of nine amino acids (nonapeptides). Both are synthesized as part of a large polypeptide precursor that includes a signal peptide of 19 amino acids, the oxytocin/vasopressin nonapeptide, a carrier protein named neurophysin I/II (which may also have a role in their folding and trafficking) and sometimes an additional moiety, copeptin (Figure 14.1; discussed in more detail below).

Oxytocin and vasopressin both have a ring structure formed by a disulfide bridge between the cysteine residues at position 1 and 6, and a tail of three amino acids. Nomenclature is particularly problematic in this family, as every peptide discovered with even a single amino acid difference was given a different name, even if their gene structure and sequence are highly homologous and the processed nonapeptides were shown to have the same function (Figure 14.2). In general, the family in vertebrates is divided into two, the oxytocin-like and

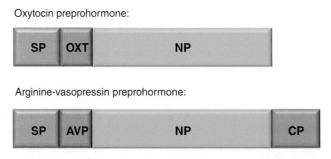

Figure 14.1 Domain structure of neurohypophysial hormones. Schematic representation of the oxytocin (OXT) and vasopressin (AVP) preprohormones. The precursor proteins to the peptide hormones contain a signal peptide (SP), the hormone(s) (OXT or AVP), their carrier protein, neurophysin (NP), and a glycoprotein dubbed copeptin (CP). Before the nonapeptide (i.e. containing nine amino acids) hormones are released from the cell, they become active by means of proteolysis, cyclization via a disulfide bridge between the cysteines at positions 1 and 6, and amidation.

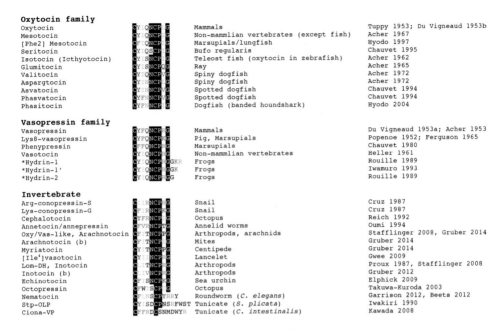

Figure 14.2 Alignment and nomenclature of the neurohypophysial hormone peptide sequences. The nonapeptide sequences are aligned in groups, oxytocin-like, vasopressin-like, and non-vertebrate, according to sequence similarity. The 'backbone' residues, CXXXNCPXG, are highlighted and the other residues colored.

vasopressin-like, while invertebrates have nomenclature related to species, as it is not clear functionally which family they belong to. Oxytocin is the name used for the exact peptide sequence found in mammals, although two non-mammalian species, the elephant shark and ratfish, were found to have an identical oxytocin nonapeptide sequence (Gwee *et al.*, 2009; Michel *et al.*, 1993). Mesotocin is the term for the peptide found in most non-mammalian vertebrates, while isotocin is used for the peptide found in fish, with the exception of zebrafish, where it is called oxytocin (although the sequence is identical to isotocin). The vasopressin family is similarly divided, with the mammalian peptide called vasopressin, the marsupial version phenypressin, and non-mammalian vertebrates (e.g. fish, birds, amphibians, and reptiles) vasotocin.

As shown in Figure 14.2, the only completely invariant positions in the peptide sequence are the cysteine residues at positions 1 and 6, which form a cyclic peptide structure. The next most conserved positions are 5 (asparagine), 7 (proline), and 9 (glycine, which is amidated), which vary only in some invertebrate species. The vertebrate peptides are divided into two families based on position 8, which in the vasopressin family is positively charged, while in the oxytocin family it is not. This change may confer the difference in their biological activity. Position 4 is invariant in the vasopressin family (glutamine), but variable in the oxytocin family. The invertebrate hormones vary more than the vertebrates, but most maintain the core of CXXXNCPXG, though some are longer than nine amino acids.

Oxytocin and vasopressin are highly conserved throughout evolution, and members of the family are found both in vertebrates and invertebrates including annelids, mollusks, arthropods, and chordates. Most invertebrates have a single oxytocin/vasopressin gene whereas most vertebrates have both peptides (Hoyle, 1999). Thus, it is difficult to classify the invertebrate peptides as belonging to a particular mammalian subgroup type (i.e. oxytocin-like or vasopressin-like) based on sequence alone, and functional studies are required to classify them correctly. It is thought that the two genes evolved from a common ancestral gene about 600 million years ago (van Kesteren *et al.*, 1992).The genes have a conserved structure of three exons, with rare exceptions – the zebrafish oxytocin, *C.elegans*, and *S. purpuratus* genes have four, while both the Octopus genes (Kanda *et al.*, 2003) and the *C. commersoni* isotocin genes are single exon genes (Figueroa *et al.*, 1989). The genes are found in close proximity on the same chromosome in all species assessed so far, generally in tandem, except for the teleost lineage (Gwee *et al.*, 2009).

The initial gene duplication event is not the only duplication in the family. Several other duplications and losses have occurred, but to what extent it is species specific as opposed to a more general phenomenon is unclear (Gwee *et al.*, 2008; Hoyle, 1999). The two genes are so closely related in sequence that the question has arisen whether gene conversion has taken place (Ruppert *et al.*, 1984), though current thinking holds that it has not (Gwee *et al.*, 2008, 2009).

Copeptin, the last moiety in the precursor, has a conserved glycosylation site and a leucine-rich region. Oxytocin has lost the copeptin in all but teleosts. In addition, the conserved cleavage site from neurophysin and the glycosylation site are absent in both oxytocin and vasopressin in the teleost lineage (Flores *et al.*, 2007). The cleavage site has been lost in additional non-mammalian lineages, presumably leading to a longer neurophysin peptide, which may function as both neurophysin and copeptin (Flores *et al.*, 2007). Some invertebrates (i.e. insects, snails) have copeptin-like extensions as well.

14.2.2 Receptors

The receptors for oxytocin and vasopressin are 7-transmembrane proteins, and members of the G protein-coupled receptor (GPCR) Class A Rhodopsin-like family. The family has three major branches: the oxytocin family receptors (OXTR) and two vasopressin branches (V1 and V2). The vasopressin branches subdivide further (V1A, V1B, V2A, V2B, V2C) for a total of six classes of receptors in the family (Lagman *et al.*, 2013). OXTR and V1 are more closely related to each other than to V2. No species studied has all six receptors, and different combinations of local duplication and loss occur (Ocampo Daza *et al.*, 2012; Yamaguchi *et al.*, 2012). All vertebrate species have at least one OXTR and one vasopressin receptor (Figure 14.3). Mammals have representatives of four classes: OXTR, V1A, V1B, and V2A. Non-mammalian vertebrates have various patterns of gain and loss: birds have OXTR, V1A, V1B, and V2C, while frogs have V2A and V2C but not V2B. Teleosts have lost V1B, but have duplications in OXTR, V1A, and V2A, and in addition have V2B and V2C. It has recently been proposed that the general duplications of the receptor took place in a larger genomic setting, with several neighboring gene families, many of which are also GPCRs (Lagman *et al.*, 2013). The primordial duplication led to the V1/OXTR and V2 genes in the vertebrate ancestor. After two rounds of genome duplication, in the gnathostome ancestor, the six classes were born, with lineage-specific losses in the process. The original linkage can still be observed, as the V2B and OXTR genes are located on the same chromosomes in various species, as are the V2C and V1A genes. V1B is lost before the split of the teleost lineage, and then more duplication events occur, leading to local duplications of OXTR and V2A. The receptors are remarkably similar at the sequence level to one another, within classes and across classes, as well as within species and across species (Supplemental Figures 14.1 and 14.2). Evidence for the initial duplication can be seen in invertebrates, where several species have more than one receptor (two in snail and nematode worm and three in octopus), while others have one (insects, tunicate, and annelid worms) (Beets *et al.*, 2012; Kanda *et al.*, 2005; Kawada *et al.*, 2004, 2008; Stafflinger *et al.*, 2008; van Kesteren *et al.*, 1996). As

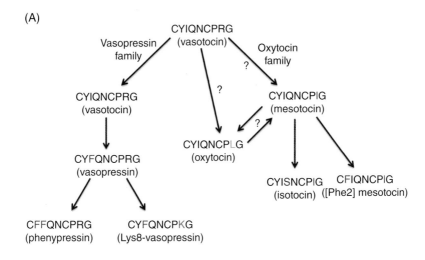

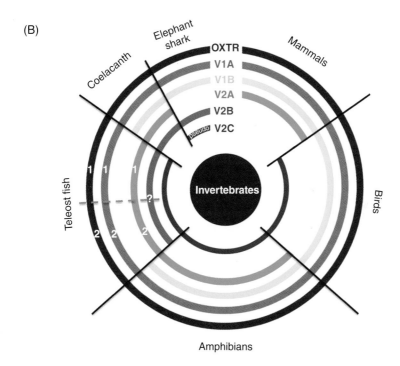

Figure 14.3 Evolutionary development of the hormones and receptors. (A) Neuropeptides. The most likely (least substitution required) tree to describe the peptide hormone sequence development. Note that in the oxytocin family, it cannot be determined on the basis of the DNA sequences what came first in position 8, the switch to leucine or the switch to isoleucine, and if the substitutions happened independently from vasotocin or sequentially from each other (*doubled arrow*). (B) Receptors. Each concentric circle represents one family of receptors. The vertebrate receptors are divided into families, while the invertebrate receptors are not. The dashed line in the teleost fish represents a lineage-specific duplication generating two different paralog genes designated 1 and 2. In V2B, only one species has a duplication so far, and it is not clear if it is species or lineage specific.

with the case of the ligands, it is difficult to classify the invertebrate receptors into oxytocin or vasopressin groups as they differ greatly from the vertebrate sequences. In *Drosophila*, only vestigial remains of the system can be seen – there is one receptor that closely resembles the vasopressin/oxytocin family but there is no ligand, and it responds to a different class of peptide hormones altogether (Cazzamali *et al.*, 2003; Park *et al.*, 2002).

Taking the evolutionary history of the peptide and receptor together, it seems that the receptor duplicated earlier than the peptides did. Most species (currently with the exception only of *C. commersoni*) have more receptors than peptide hormones. The fact that the ligand genes arc always in close proximity also points to a more recent duplication. Vasopressin/vasotocin seems to be the original hormone, with oxytocin/mesotocin/isotocin being the later development. This is supported both by the ligand and receptor evolutionary histories, where species with only one gene generally have a vasopressin family member, and species with multiple receptors having only one receptor preferentially for oxytocin, while having many receptors for vasopressin (see Figure 14.3).

14.2.3   Structure-function

The classification of receptors into oxytocin and vasopressin groups is directly related to their affinities for the various peptides (Busnelli *et al.*, 2013; Chini *et al.*, 2008). While both peptides can bind to all of the receptors, there are differences in ligand specificity. The oxytocin receptor binds to both oxytocin and vasopressin with nearly equal affinity (Postina *et al.*, 1996), but 100-fold more vasopressin is required for downstream activation (Kimura *et al.*, 1994). The vasopressin receptors are more specific to begin with, binding vasopressin with 400-fold higher affinity than oxytocin (Postina *et al.*, 1996).

As the sequences of both the ligands and the receptors are so close, much work has been done to elucidate the structural determinants that determine specificity, both for ligand binding and downstream signaling (summarized in: OXTR (Koehbach *et al.*, 2013); V1A family (Koshimizu *et al.*, 2012); V2 family (Birnbaumer 2001)). Determination of the functional parts of the receptors has been performed with receptor chimeras, disease-causing mutations in V2, and much use of agonists and antagonists.

The most conserved parts of the receptors, across all families, are the transmembrane (TM) domains TM2, TM3, TM6, and TM7. The most variable, in terms of both sequence similarity and length, are the N and C termini, and the third intracellular loop, commonly called ic3.

The exact location of the ligand binding is unknown, as there is no crystal structure of any of the family members. However, various residues have been implicated in ligand binding, both in the extracellular part of the receptor (N-term and loops 1 and 2), and in the extracellular membrane proximal sections of the transmembrane segments, particularly a series of five almost completely conserved glutamine residues in TM2, TM3, TM4, and TM6 (Barberis *et al.*, 1998; Postina *et al.*, 1996). Specificity is achieved by both parts of the ligand, the cyclic portion, and the 3 amino acid 'tail,' or linear portion. The linear portion has been shown to bind to the N-term and first extracellular loop, while the cyclic portion interacts with extracellular loop 2 (Postina *et al.*, 1996), as well as various residues in the membrane, particularly various residues in TM3, TM5, TM6, and TM7 (Koehbach *et al.*, 2013; Rodrigo *et al.*, 2007).

In order to pinpoint putatively important residues in the receptors, we performed an alignment of receptors from many species representing all classes, and have compared them to find differential amino acids (Table 14.1; see Supplemental Figures 14.1 and 14.2). Despite the high similarity of the receptors, they separate easily into groups according to sequence. Some of the residues found have been studied before, while some are novel. This amino acid conservation may be involved in either ligand binding or coupling to specific signal transduction pathways.

Table 14.1 Conserved residues that differentiate neurohypophysial receptor subfamilies.

Ballesteros number*	Protein domain	OXTR	V1A	V1B	V2A	V2B	V2C	V2C2**
1.28	N-term	N	N (D)	D	D (N)	D	D	
1.29	N-term	E	E	E	Not E	E	E	
1.34	TM1	V	I/L	V/A	A/V/W	I/V	I/V	
1.35	TM1	E	E	E	E(Q/N)	E	E	R/K
1.36	TM1	V	I	I/V	L/I	I	I	A
1.38	TM1	V	V	V/I	L/I/V	L	V	
1.43	TM1	L	F	L	L/F	F	F	
1.45–47	TM1	FLAL	VV/AAV				LTAS	XXAT
1.49	TM1	G	G	G (S)	G (S)	L (T)	G (S)	
1.52	TM1	C	S	A/G/I	F/L	G/A/I/S		
1.53	TM1	V	V	V	V	L	L	
1.54	TM1	L	L	L	L	L	I(V)	L
1.56	TM1	A	A				V	
1.57	TM1	L/I	M/L	L/M/I/V	L	L	L	L
1.58	ic1					W	W	
	ic1				Mammals 2 aa addition			
2.39–2.40	ic1	MY/LF/ MF	MH	MH	MH	MR(H)	MY	TQ
2.41	ic1	Y/F	L	L	V/T/L	V	V	L
2.43	TM2	M	I/M	V/I	I/M	V	M	L
2.45	TM2	H	H	H	N/H	H	H	H
2.47	TM2	S	S	A/G	C	C	S	C
2.48	TM2	I	L	L	L/I/V	L	I	L
2.49	TM2	A	A	A	T/S	A	A	A
2.52	TM2	V	V/A	V/G	V/A	V	V	V
2.55	TM2	V/I	F	L	L/F	F	F	F
2.59	TM2	L	L	L	L	C	L	L

(*continued*)

Table 14.1 (Continued)

Ballesteros number*	Protein domain	OXTR	V1A	V1B	V2A	V2B	V2C	V2C2**
2.63	TM2	L/I	C	I/L	A/I/L/V	M/L/I/F	I(L)	
3.26	TM3	L	I/V	A	A/S	L		
3.28	TM3	Y	H	Y	Y	Y	Y	Y
3.32	TM3	V	V	V	I/V/M	V(I)	L	V
3.33	TM3	V	F/L/M	L	V	V/L	L/V	V
3.36	TM3	F	F	F	Y/F	F	F	F
3.38	TM3	T	T/A	T	S	T	T	T
3.42	TM3	M/L	M	M	M	M	M	M
3.47	TM3	S	T	T	T	T	T	T
3.51	ic2	C	Y	Y	H	Y	Y(F)	Y
3.52	ic2	L	I	L/I/V/M	H/Q/R	Q	Q	H
3.56	ic2	Q	H	H	C/R	N(K)	Y	R/K
3.58	ic2	L	L	L	L/M	M	M	M
4.46	TM4	V	I	I	V(I)	V	I	I
4.47	TM4	I/L		G/A		C	C	
4.61	TM4	V/I/L/M	Y	V/I/L/M	L/I/V	V/I	V(I)	L/I
4.62	TM4	Almost no F	Mostly F	F	F	F	F	F
4.66	ec2	L	V/L/M	L(V,M)	R/K/Q	R/Q/K	K	L
Between 4.68-C	ec2	Very variable, both length and sequence						
5.42	TM5	I(M)	I/M	T	I/M/V	T	I	I
5.43	TM5	S/T	T	T	A/T	T	F	T
5.47	TM5	Y	F	F	F	F	F	F
5.48	TM5	I	L/V	V(I/L)	V/I/L	V/I	F	V/I
5.58	TM5	Y	Y	Y	Q	Q	Q	Q
5.59	TM5	G	G	G/S	V(I,F)	V	V	I/M
5.60	TM5	L	F	L	R/L	R	K	R/K

Table 14.1 (Continued)

Ballesteros number*	Protein domain	OXTR	V1A	V1B	V2A	V2B	V2C	V2C2**
	ic3	Very variable, extensively described, both sequence and length, especially expansion in teleosts						
6.30–6.32	ic3	KIR/KIT	KIR/KLR	KIR	MSK V/MAK	RV/IK	MIK	MLK
6.38	TM6	F	F	F	L(M,F)	V	V	F
6.44	TM6	F/Y	Y	Y	Y	Y	Y	Y
6.56	TM6	M	M	M	L	L	L	L
	ec3		3aa addition	3aa addition				
7.30	ec3	R/K	S	S	mix	K/T	T	T
7.39	TM7	M	A	M	M	M	M	M
7.43	TM7	S	S	N/S	S	S	N	S
7.45	TM7	N	N	S	N	N	N	N
7.48	TM7	C	C	C	T(S)	A	A(T)	T
7.51	TM7	W	W	W(C)	W	C	W	W
7.54	TM7	M	M	M	T/A	L	M	L
7.57	C-term	A/T	S	S	S	S	C	S
	C-term	Variable sequence and length				No tail		

*Numbering according to Ballesteros and Weinstein (1995). In this system, the most conserved residue in each transmembrane domain receives the number X.50, where X is the number of the transmembrane domain. Residues before and after are numbered consecutively from 50. This allows comparison of sequences of different lengths, while maintaining the relative position of the particular amino acid.

**V2C2 is the branch of three fish, *D. rerio*, *X. maculatus*, and *G. aculeatus*, which branches separately and prior to the V2B and V2C branches, and is different on the sequence level from both of them.

The receptors are all coupled to G proteins for intracellular signaling. OXTR, the V1 receptors, and V2B/C receptors all signal through $G_{q/11}$ which results in phosphatidylinositol hydrolysis and increase in calcium levels. V2A, in contrast, signals via G_s, and stimulates adenylyl cyclase to produce cAMP. Signaling specificity was delineated by a series of receptor swapping experiments, using mammalian V1A and V2 (Erlenbach and Wess, 1998; Liu and Wess, 1996). They showed that the V2 ic3 is sufficient for cAMP signaling and G_s, and increases the affinity of V1 to match that of V2 (which is slightly higher to begin with). In particular, Q225 and E231 at the junction of TM5 and ic3 are critical (numbering from the human AVPR2). The length of ic3 also affects the efficiency of activation of G_s, with a shorter loop improving the production of cAMP. Part of the cytoplasmic tail, ic4, was also shown to be involved in coupling to G_s. The V1 ic2 is sufficient for calcium signaling, and if both the V1 ic2 and V2 ic3 domains are present in the same protein, it can signal via both pathways.

14.3   Anatomy of neurohypophysial neurons in non-mammalian species

By and large, oxytocin and vasopressin neurons are located in the hypothalamus, from which they send projecting axons to central or peripheral targets. In all vertebrates, a subset of these neurons projects into the posterior pituitary lobe, the neurohypophysis, from which the neuropeptides are secreted into the circulation (Burbach *et al.*, 2001; Lohr and Hammerschmidt, 2011). This connection between the hypothalamus and the pituitary is facilitated by an elongated structure – the infundibular stalk, which contains the descending hypothalamic nerve bundle (Knobloch and Grinevich, 2014; Norris and Carr, 2013).

14.3.1   Fish

In most studied fish species, oxytocin- and vasopressin-like peptides are produced in neurons located bilaterally along the third ventricle in the hypothalamic neurosecretory preoptic area (NPO). In contrast to higher vertebrates in which nonapeptide-producing cell bodies are present in various regions in the brain, fish isotocinergic and vasotocinergic cell bodies are largely confined to the NPO (Goodson *et al.*, 2003; Moore and Lowry, 1998; Saito *et al.*, 2004). Cells of the NPO vary in size. The rostral-ventral preoptic area contains relatively small (parvocellular) neurons and is thought to be the homolog of the supraoptic nucleus (SON) of higher vertebrates. The caudal dorsal region contains large (magnocellular and gigantocellular) neurons and is thought to be the homolog of the paraventricular nucleus (PVN) (Godwin and Thompson, 2012; Herget *et al.*, 2014) (Figure 14.4).

As in mammals, fish oxytocin- and vasopressin- like neurons project to various targets within the brain and to the neurohypophysis (Goodson *et al.*, 2003; Gutnick *et al.*, 2011; Rurak and Perks, 1976; Saito *et al.*, 2004). The neurohypophysis functions as a major storage and source of endocrine release of these neuropeptides. The most primitive pituitary structure is seen in the jawless fishes (agnathans) lampreys and hagfish (Green, 1951; Sower *et al.*, 2009). Only one neurohypophysial hormone, vasotocin, was reported in jawless fish (Rurak and Perks, 1976). Here vasotocin-expressing neurosecretory bipolar cells from the preoptic nucleus project to the neurohypophysis (pars nervosa) (Erhart *et al.*, 1985). In the hagfish, the neurohypophysis has a flat structure laid above the adenohypophysis (Gorbman *et al.*, 1963; Nozaki, 2008). In the lamprey, the neurohypophysis is more closely associated with the pars intermedia of the adenohypophysis, forming a curved neurointermediate lobe (Nozaki, 2008). A putative median eminence is associated with the pars distalis of the lamprey adenohypophysis. No developed median eminence is detected in hagfishes.

A more developed hypothalamic-neurohypophysial system is seen in cartilaginous (chondrichthyes) and bony fishes (osteichthyes). In all of these fish species, excluding teleosts, the neurohypophysis is associated with a well-developed median eminence, which is connected by a portal system to the anterior pituitary (adenohypophysis) (Norris and Carr, 2013; Pogoda and Hammerschmidt, 2007). In teleosts, there is no well-developed portal system and no median eminence. The neurosecretory cells of the hypothalamus directly innervate both the adenohypophysis and the neurohypophysis (Pogoda and Hammerschmidt, 2007). The distribution of fibers and cell sizes may differ between species and within species, between males and females, and according to season (Dewan *et al.*, 2008; Goodson *et al.*, 2003).

(A)

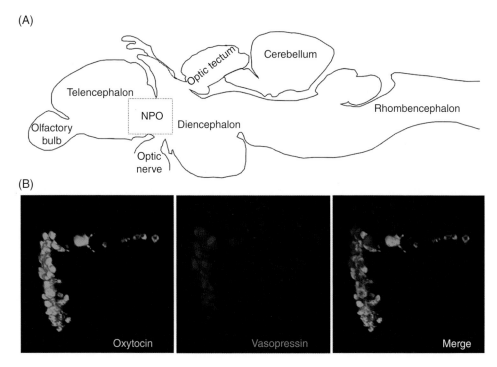

(B)

Oxytocin   Vasopressin   Merge

Figure 14.4 **Expression of oxytocin and vasopressin in adult zebrafish brain.** (A) Schematic representation of a midsagittal section through the zebrafish brain. Rostral to the left. (B) Maximal intensity projection of a confocal microscopy scan through the neurosecretory preoptic (NPO) area showing fluorescent *in situ* hybridization of oxytocin (*green*) and vasopressin (*red*) mRNAs of an adult zebrafish.

14.3.2 Amphibians

The two major neurohypophysial hormones in frogs are mesotocin and vasotocin. The majority of mesotocin and vasotocin cell bodies are localized to the preoptic nucleus. As in fish, magnocellular mesotocin and vasotocin neurons of the NPO project to the neurohypophysis, where these peptides are stored and released to the periphery. However, amphibian vasotocin cell bodies are found in other, extrahypothalamic brain regions such as the nucleus accumbens, olfactory system, bed nucleus of stria terminalis, amygdala, optic tectum, and tegmentum. The level of vasotocin expression and the distribution of vasotocin positive cells vary between sexes and change according to season (Boyd and Moore, 1992; Boyd *et al.*, 1992; Gonzalez and Smeets, 1992; Marler *et al.*, 1999; Moore *et al.*, 2000). For example, there are more vasotocin cells in the BNST, amygdala, and anterior preoptic area in male newts than females (Moore *et al.*, 2000). Extrahypothalamic mesotocin neurons somata are less abundant than vasotocin and can be detected in the medial amygdala in frogs (Gonzalez and Smeets, 1992).

14.3.3 Reptiles

In reptiles, the hypothalamus is organized into distinct nuclei. Instead of having the neurosecretory cells grouped in the NPO, as in fish and amphibians, they are divided into two major nuclei – the SON and the PVN, similar to the anatomical organization of these neurons in mammals. Between these two cell groups, some additional vasotocin and mesotocin cell bodies are scattered. There are also extrahypothalamic mesotocin and/or vasotocin neurons that vary in distribution between reptilian species (Fernandez-Llebrez *et al.*, 1988; Silveira *et al.*, 2002; Thepen *et al.*, 1987). Vasotocin and mesotocin neurons from both the SON and PVN project their axons into the neurohypophysis, forming the neurohypophysial tract. Many additional mesotocin and vasotocin fibers are widely distributed throughout the brain and innervate areas such as the nucleus accumbens, bed nucleus of stria terminalis, septum, organum vasculosum of the lamina terminalis (OVLT), areas in the hindbrain and spinal cord (Fernandez-Llebrez *et al.*, 1988; Silveira *et al.*, 2002; Thepen *et al.*, 1987).

14.3.4 Birds
Similar to reptiles and mammals, avian mesotocin and vasotocin are produced mainly in cells located at the SON and PVN and in small groups of cells distributed in lateral hypothalamic and thalamic areas (Barth *et al.*, 1997; Goossens *et al.*, 1977). Vasotocin and mesotocin are also expressed by cells in the suprachiasmic nucleus of ducks, chickens, and quails (Mikami *et al.*, 1978; Robinzon *et al.*, 1988). The rostral area of the SON contains larger (magnocellular) neurosecretory cells than in the caudal part. Smaller cells are also found in the PVN (Goossens *et al.*, 1977; Viglietti-Panzica and Panzica, 1981). Mesotocin and vasotocin axons from the SON and PVN project to the posterior pituitary (Mikami *et al.*, 1978; Viglietti-Panzica and Panzica, 1981). Notably, only vasotocin nerve endings from the PVN reach the portal capillaries of the external zone of the anterior median eminence (Blahser, 1984; Cornett *et al.*, 2013; Goossens *et al.*, 1977). These cells release arginine vasotocin from fine granules in the anterior median eminence and regulate anterior pituitary functions (Cornett *et al.*, 2013).

14.3.5 Invertebrates
As mentioned above, most invertebrates have a single oxytocin/vasopressin family member, except for octopus which has two genes (cephalotocin and octopressin) (Kanda *et al.*, 2005; Reich, 1992; Takuwa-Kuroda *et al.*, 2003). Neurosecretory brain centers containing vasopressin/oxytocin family members are present in tunicates (Ukena *et al.*, 2008), worms (Beets *et al.*, 2012; Garrison *et al.*, 2012; Tessmar-Raible *et al.*, 2007; Wagenaar *et al.*, 2010), insects (Stafflinger *et al.*, 2008), gastropods (van Kesteren *et al.*, 1995), and cephalopods (Kanda *et al.*, 2005; Takuwa-Kuroda *et al.*, 2003). In some invertebrate species, these peptides are also expressed in peripheral ganglia (Beets *et al.*, 2012; Garrison *et al.*, 2012; van Kesteren *et al.*, 1995). In nematodes, for example, nematocin is expressed in several cell groups including the AFD thermosensory neurons, the DVA mechanosensory neurons, and in male-specific CP motor neurons (Beets *et al.*, 2012; Garrison *et al.*, 2012). The octopus cephalotocin expression is restricted to the brain, while octopressin is expressed in the brain and in certain ganglia (Takuwa-Kuroda *et al.*, 2003). In the gastropods, conopressin-expressing neurons are located in the pedal ganglion and in the cerebral ganglia, in axons that innervate the penis and vas deferens (van Kesteren *et al.*, 1995).

14.4 Function

Historically, vasopressin has been identified as antidiuretic hormone, indicative of its role in the regulation of water balance. Oxytocin, as implied by its name, was known for its role in the contraction of the uterus. As our knowledge progressed, it became clear that these two highly conserved hormones and their family members play an important and much broader role in the facilitation of social interactions and the maintenance of energy and water homeostasis. Thus, these peptides stand at the junction between the organism's wellbeing and its reproductive potential (Figure 14.5).

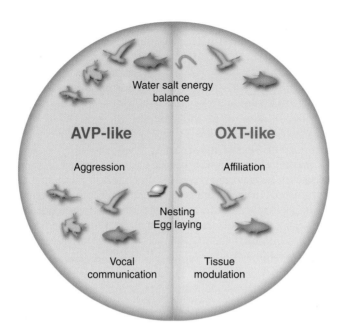

Figure 14.5 Functions of oxytocin (OXT)- and vasopressin (AVP)-like hormones in various models. Schematic summation of processes that are regulated by oxytocin-like (*green*) and vasopressin-like (*blue*) hormones and classes of models in which these functions were demonstrated.

14.4.1 Water and salt balance

Vasopressin-like peptides are involved in the regulation of salts and water balance in various species, although their exact function may vary in specific organisms according to the environmental conditions.

In mammals, vasopressin binds to V2 receptors, thereby activating a signaling cascade which induces the translocation of water channels into the apical plasma membranes of the kidney collecting duct cells. Thus, vasopressin facilitates the reabsorption of water from the kidney collecting duct to the blood (McCormick and Bradshaw, 2006; Nielsen *et al.*, 1995).

The antidiuretic effect of vasopressin or its homologs has been reported in fish (Warne *et al.*, 2005), amphibians (Hillyard and Willumsen, 2011), reptiles (Bradshaw and Bradshaw, 2002), and birds (Goldstein, 2006). Moreover, indications of its osmotic-regulatory function also exist in tunicates, in which there is an elevation in the mRNA expression of oxytocin/vasopressin homolog (called SOP) in response to changes in salinity (Ukena *et al.*, 2008).

Fish inhabit a wide range of environments with different salinities and their hormonal responses to osmotic challenges may vary between species and experimental conditions. While it is not clear whether isotocin has a prominent role in osmoregulation, many studies demonstrate changes in vasotocin expression or release following osmotic or hypovolemic challenges (Haruta *et al.*, 1991; Hyodo *et al.*, 1991; Martos-Sitcha *et al.*, 2013; Pierson *et al.*, 1995; Warne *et al.*, 2005). Vasotocin exerts its action by binding to receptors in the kidney and gills. For example, in the sea water flatfish *Platichthys flesus*, vasotocin is released to the blood and reduces renal water loss following an increase in blood osmolarity (Warne *et al.*, 2005). This is mediated by a decrease in the number of filtering nephrons in the kidney and changes in blood flow and glomerular filtration rate (Warne *et al.*, 2002, 2005).

In amphibians, as in fish, vasotocin is released following hypovolemic stress. It regulates the water permeability by binding to receptors in the kidney glomeruli and to receptors on the skin and bladder (Acher *et al.*, 1997; Boyd and Moore, 1990; Hasegawa *et al.*, 2003; Nouwen and Kuhn, 1985). Interestingly, amphibians also have alternatively processed forms of vasotocin, termed hydrins, which play a role in the rehydration of anurans through skin and bladder (Acher *et al.*, 1997).

Vasotocin blood levels are elevated following hyperosmotic stress in birds (Cornett *et al.*, 2013; Nouwen *et al.*, 1984). It regulates water balance by targeting the renal vasculature, reducing the glomerular filtration rate and inducing the expression of aquaporins in the renal collecting ducts (Goldstein, 2006).

There are some indications for the involvement of oxytocin in the regulation of salt and water homeostasis. Thus, mesotocin was reported to have a diuretic function in amphibians (Galli-Gallardo *et al.*, 1979; Pang and Sawyer, 1978). Oxytocin/isotocin may also be involved in regulation of ion transport through the skin and gills of fish by the regulation of ionocytes formation (Chou *et al.*, 2011). Studies in mammals indicate its function in the regulation of salt appetite (Puryear *et al.*, 2001; Stricker and Verbalis, 1996). In nematodes, the oxytocin/vasopressin-like peptide nematocin facilitates gustatory-associated learning and modulation of salt chemotaxis (Beets *et al.*, 2012).

14.4.2   Energy balance

Mammalian oxytocin neurons are involved in the regulation of energy intake (i.e. appetite) (Atasoy *et al.*, 2012; Sabatier *et al.*, 2013). This is most likely due to a central release of the peptide, rather than its release to the blood. The SON and PVN are innervated by orexinergic neuropeptide-Y cells as well as by pro-opiomelanocortin-positive cells that produce α-MSH, a peptide with anorexigenic effects. Activation of oxytocin cells by α-MSH affects appetite by inducing the dendritic release of oxytocin in the ventromedial nucleus of the hypothalamus (Ludwig and Leng, 2006; Sabatier *et al.*, 2003, 2013). In addition, oxytocin neurons have reciprocal connections with brainstem neurons that are located in the nucleus of the solitary tract (NTS), a brain area known to integrate satiety signals from the gut and hypothalamus (Goodson *et al.*, 2003; Verbalis *et al.*, 1986). The function of oxytocin in the regulation of energy intake exists in non-mammalian vertebrates. In birds, for example, central administration of oxytocin or its homologs has an anorexigenic effect (Jonaidi *et al.*, 2003; Masunari *et al.*, 2013). We showed that the expression of oxytocin/isotocin in the zebrafish is regulated by the intracellular metabolic regulator PGC-1α. PGC-1α forms an *in vivo* complex with the oxytocin promoter in fed but not fasted animals (Blechman *et al.*, 2011). These findings demonstrate that PGC-1α is both necessary and sufficient for the production of oxytocin, implicating PGC-1α in the direct activation of a hypothalamic hormone known to control energy intake.

14.4.3   Sexual and social behavior

Oxytocin vasopressin and their homologs are involved in the regulation of many aspects of intraspecies social behaviors.

14.4.3.1   Vocal communication

Vasotocin and its family members were shown to modulate vocalization in a variety of species, including frogs, fish, and birds. It is involved in the response of song birds to competitive singing (Sewall *et al.*, 2010). In the bullfrog, vasotocin fibers are highly distributed throughout the brain. It innervates areas that are involved in the processing of sensory information and important in sexual behaviors (Boyd and Moore, 1992). Vasotocin also stimulates and modulates the timing, duration, and type of callings in many species of amphibians (Chu *et al.*, 1998; Diakow, 1978; Kime *et al.*, 2007, 2010; Ten Eyck and ul Haq, 2012). Interestingly, there are differences in the distribution of fibers between males and females (Boyd and Moore, 1992; Boyd *et al.*, 1992). These differences may be reflected in their mating behaviors. For example, vasotocin affects advertisement calling in males and calling phonotaxis in female bullfrogs (Boyd, 1994) and in túngara frogs (Kime *et al.*, 2010). In the coqui frog, injection of vasotocin into satellite males induces territorial-related advertisement calling (Ten Eyck, 2005). It also induces aggressive calls in territorial male frogs in the paternal care period (a period in which they do not normally call) (Ten Eyck and ul Haq, 2012).

Courtship and reproduction are highly dependent on vocal communication in the sonic fish, plainfin midshipman. Vasotocin and isotocin differentially regulate neuronal circuits involved in ultrasonic vocalization in neurophysiological tissue preparations derived from territorial (type I) and sneaker (type II) males and from females. Specifically, administration of vasotocin, but not isotocin, inhibits fictive vocalization (that is, it inhibits vocal motor neuron activation) in territorial type I males. On the other hand, isotocin inhibits fictive vocalization in type II sneaker males and in females, with only a limited effect of vasotocin on the calls (Goodson and Bass, 2000a, b).

14.4.3.2 Breeding and egg laying

In mammals, oxytocin is important for proper parturition, and the number of oxytocin receptors on the uterus is increased towards the end of gestation (Fuchs *et al.*, 1982; Soloff *et al.*, 1979). Surprisingly, knockout mice lacking in oxytocin or oxytocin receptor expression are fertile and do not display coarse defects in gestation and parturition, suggesting that other redundant mechanisms promote uterine contraction (Nishimori *et al.*, 1996; Takayanagi *et al.*, 2005).

The involvement of oxytocin/vasopressin family members in reproductive physiology and behaviors is evident in invertebrate organisms, such as worms and mollusks. In leeches, administration of conopressin induces a sequence of stereotypic reproductive behaviors (Wagenaar *et al.*, 2010). Similar phenomena are seen following the administration of nematocin in nematodes (Garrison *et al.*, 2012). The effect of these hormones on reproduction may depend on the season (Fujino *et al.*, 1999). In the gastropods, which are hermaphroditic organisms, the conopressinergic neurons are indicated to be involved in the ejaculation of semen and at the same time inhibition of a female-associated egg-laying function (van Kesteren *et al.*, 1995).

In amphibians, breeding behaviors depend on external and internal signals, and can be promoted by the administration of vasotocin. (Diakow, 1978; Kime *et al.*, 2007, 2010; Marler *et al.*, 1999; Moore *et al.*, 1992; Thompson and Moore, 2003). In the frog *Rana pipiens*, arginine-vasotocin mediates a link between an environmental signal of water availability and reproductive behavior. Thus, injection of vasotocin into the female frog results in an accumulation of water, which causes an increase in the internal pressure and thus decreases the number of release calls. This decline indicates their receptiveness and promotes mating (Diakow, 1978). Injections of vasopressin can also increase courtship clasping and egg deposition in newts, behaviors which also depend on the presence of gonadal sex steroids (Moore *et al.*, 1992; Thompson and Moore, 2003).

Similar to their role in mammals, both oxytocin and vasopressin regulate the contraction of smooth muscles, and are involved in egg laying in both invertebrates (Fujino *et al.*, 1999; Oumi *et al.*, 1996) and non-mammalian vertebrates (Koike *et al.*, 1988; Moore *et al.*, 2000; Saito *et al.*, 1990). In birds and reptiles, vasotocin is much more oxytocic than mesotocin (Atkins *et al.*, 2006; Guillette, 1979; Koike *et al.*, 1988). Vasotocin plasma levels are increased during oviposition and it is important for egg-laying (Fergusson and Bradshaw, 1991; Koike *et al.*, 1988; Nouwen *et al.*, 1984). Mesotocin may have a role in the preparation for egg laying since, in the zebra finch, for example, inhibition of the mesotocin receptor in the periphery reduces nesting behavior (Klatt and Goodson, 2013). In lizards, treatment with vasopressin induces locomotor activity (climbing) in species that lay their eggs on trees (Guillette and Jones, 1982). Yet, at least in birds, vasotocin and mesotocin are not necessary for this process (Opel, 1965). Similarly, administration of vasotocin in fish stimulates the spawning reflex more efficiently than isotocin (Pickford and Strecker, 1977).

14.4.3.3   Aggression and affiliation

As mentioned earlier, the neurohypophysial hormones modulate vocal communication, which is associated with aggressive and territorial behaviors. Many studies have demonstrated the effects of vasotocin and isotocin on fish social, sexual, and aggressive interactions (reviewed in Godwin and Thompson, 2012). In teleost fishes, variation in the levels of expression, number of vasotocin-positive cells, and soma size are associated with specific reproductive and social behaviors (Dewan *et al.*, 2008; Godwin *et al.*, 2000; Miranda *et al.*, 2003; Ramallo *et al.*, 2012). It appears that a more aggressive or territorial behavior is associated with larger vasotocin-positive cells. For example, in a study comparing two related butterflyfish species, it was found that the territorial multiband butterflyfish have larger vasotocin cells in the preoptic region than the shoaling milletseed butterflyfish (Dewan *et al.*, 2008). In African cichlid fish, there is higher expression of vasotocin in the gigantocellular and lower vasotocin expression in parvocellular nuclei in territorial males than non-territorials (Greenwood *et al.*, 2008). In the cichlid fish *Cichlasoma dimerus*, territorial males also have smaller parvocellular preoptic cells than non-territorial males (Ramallo *et al.*, 2012). Similar anatomical differences are found when comparing between two closely related teleost fish with different interspecific social behaviors: the obligate cleaner wrasse and a non-cleaner corallivore wrasse (Mendonca *et al.*, 2013). The cleaner wrasse has fewer gigantocellular vasotocin neurons in the preoptic area than the non-cleaner fish, indicating a function of these cells in interspecies interactions (Mendonca *et al.*, 2013; Soares *et al.*, 2012). Moreover, dominant cichlid fish males have higher vasotocin pituitary levels than subordinate males and subsequently store more urine, which can be used for chemical signaling during male aggression or in courtship (Almeida *et al.*, 2012). Exogenous intraabdominal administration of vasotocin in bluehead wrasse resulted in increased courtship behavior and decreased feeding in non-territorial males (Semsar *et al.*, 2001). Interestingly, experiments done in zebrafish have shown that exogenous administration of vasotocin, vasopressin, oxytocin or isotocin increases shoaling and has anxiolytic and prosocial affects (Braida *et al.*, 2012).

In birds, central administration of mesotocin to female zebra finches increased social preferences of familiar conspecifics and larger groups (Goodson *et al.*, 2009). When comparing several species of estrildid finches, Goodson *et al.* found that there are differences in the binding site distribution of oxytocin receptor antagonists in the lateral septum between territorial and flocking species. Inhibition of the oxytocin receptor in this area caused a reduction in sociality (i.e. the time the bird spent next to the large group), thus indicating mesotocin's role in affiliative behavior (Goodson *et al.*, 2009).

14.4.4   Function in development

Although this chapter focuses on non-mammalian species, for the interest of discussion this section also reviews the function of neurohypophysial hormones in mammals.

Oxytocin has been implicated in the development, regeneration, and remodeling of various tissues, including bone remodeling (Tamma *et al.*, 2009), angiogenesis (Cattaneo *et al.*, 2009; Gutnick *et al.*, 2011; Moll, 1958), neurogenesis (Leuner *et al.*, 2012), and neuronal plasticity (Theodosis, 2002; Zheng *et al.*, 2014). Vasopressin, on the other hand, does not seem to share this proliferative or regenerative potential (Leuner *et al.*, 2012), though it was shown to enhance the formation of osteoclasts and regulate bone mass (Tamma *et al.*, 2013).

Modifying the proximate environment by neuropeptides may serve in the facilitation and modulation of the neuronal activity. In adults, local oxytocin release induces morphological changes upon themselves and their surrounding tissue. These changes affect their interactions with their neighboring glia cells and enable the juxtaposition of somata and dendrites in the SON and PVN, thus leading to a more direct interaction of axons with vessels in the pituitary (reviewed in Theodosis, 2002, Landgraf and Neumann, 2004).

Recently, we have demonstrated that oxytocin functions during embryonic development in the formation of the neurovascular interface in zebrafish (Gutnick and Levkowitz, 2012; Gutnick et al., 2011). Thus, by affecting the morphogenesis of blood vessels, and establishing a tight neurovascular interface between hypothalamic axons and neurohypophysial blood vessels, oxytocin facilitates its own release into the circulation. The development of this neuro-hemal structure in the neurohypophysis depends also on other signaling molecules such as FGFs, as was demonstrated in chick and zebrafish embryos (Liu et al., 2013). A proposed model for the mechanism by which coordinated generation of the neurohypophysis neuro-vascular interface combines our finding with those of Liu et al., and suggests that FGF signaling mediates the guidance of axons and vessels to the ventral diencephalon and then oxytocin signaling regulates the morphogenesis of the hypophysial vascular structure, leading to the formation of a tight neurovascular interface (Liu et al., 2013; Pearson and Placzek, 2012).

14.5 Modes of communication

14.5.1 Axonal release

In vertebrates, oxytocin and vasopressin can be released into the blood circulation by the posterior pituitary, directly into the cerebrospinal fluid (CSF) or locally within the central nervous system. In the mammalian brain, oxytocinergic magnocellular neurons were shown to directly innervate forebrain regions such as the central amygdala and nucleus accumbens where they take part in the regulation of fear responses (Knobloch et al., 2012), sociosexual interactions (Ross et al., 2009), and social reward-associated behavior (Dolen et al., 2013). Interestingly, it has been demonstrated by Knobloch et al. (2012) that a subset of oxytocinergic neurons project into the posterior pituitary and also send collaterals into fore-brain regions. This anatomical feature could presumably underlie the association of complex responses such as in social bonding ('emotions') and mating ('behavioristic output'). Similar collaterals can be identified in fish vasotocin and isotocin neurons (Saito et al., 2004). Even though there is no direct evidence, this may also be the case in a subset of neurons in birds (Mikami et al., 1978) and reptiles (Silveira et al., 2002; Thepen et al., 1987).

14.5.2 Dendritic release

Central release of the neuropeptide in the brain can occur not only from axonal terminals but also from dendrites (Landgraf and Neumann, 2004; Ludwig and Leng, 2006; Moos et al., 1984; Tobin et al., 2012). The secretion of oxytocin and vasopressin from axons or somata/dendrites depends on intracellular Ca^{2+} release (Brown et al., 2013; Ludwig et al., 2002; Tobin et al., 2012). Local dendritic release of oxytocin and vasopressin can modify their own electrical activity (Moos et al., 1984; Morris and Ludwig, 2004). Dendritic release of oxytocin in mammals may underlie the coordinated bursting activity in response to suckling (Ludwig et al., 2002).

There is no direct evidence of dendritic release in lower vertebrates. Nonetheless, anatomical features such as the close proximity and direct membrane appositions of isotocin and vasotocin neurons somata are seen in teleost fish (Cumming et al., 1982; Saito et al., 2004). In addition, at least for isotocin-expressing cells, there is a direct neuronal coupling by gap junctions (Saito *et al.*, 2004). This may indicate similar mechanisms of activity and coordination in these neurons (Saito and Urano, 2001).

In amphibians, many cell bodies of neurosecretory cells of the NPO reside within the ependymal layer of the brain ventricle and have fibers in direct contact with the ventricle (Gonzalez and Smeets, 1992). Similarly, in reptiles and fish, but to a lesser extent, meso-tocin/isotocin and vasotocin neurons of the PVN are located in the subependymal layer and send dendrites which occasionally protrude into the third ventricle (Fernandez-Llebrez *et al.*, 1988; Saito *et al.*, 2004; Silveira *et al.*, 2002). These neuronal protrusions may be the source for secretion of these hormones to the CSF. Release of oxytocin/vasopressin-related peptides into the CSF may function in non-synaptic communication between remote areas in the brain (Veening *et al.*, 2010).

14.5.3 Peripheral release

Peripheral effects of oxytocin or vasopressin on behavior can be seen in many species, including birds nesting (Klatt and Goodson, 2013), egg-laying-related locomotion in lizards (Guillette and Jones, 1982), courtship, shoaling preference, and anxiety reduction in fish (Braida *et al.*, 2012; Semsar *et al.*, 2001), and locomotion and calling in frogs (Boyd, 1994). Studies done mainly in rodents indicate that peripheral administration of oxytocin reduces anxiety (Ayers *et al.*, 2011).

Peripheral release of oxytocin or vasopressin affects behavior by various mechanisms. One mechanism may involve a feedback loop with peripheral-secreted factors such as sex steroids (Thompson and Moore, 2003). Another mechanism is neuronal feedback via spinal pathways (Russell *et al.*, 2003). A third option is that peripheral oxytocin and vasopressin circumvent the blood–brain barrier through a peripheral-to-central feedback mechanism. In the latter case, neurohypophysial hormones may enter and affect the brain via circumventricular organs – areas where there are fenestrated (i.e. permeable) capillaries (such as the median eminence, the OVLT, and area postrema).

Finally, an interesting point regarding the effects of oxytocin and vasopressin in mammals is the debate as to whether nasal administration of these neuropeptides modifies physiological and behavioral responses and how they exert their action. Nasal administration of oxytocin has been shown to have prosocial and anxiolytic effects, particularly in humans (Fischer-Shofty *et al.*, 2010; Kosfeld *et al.*, 2005). Nonetheless, it is not clear if this mode of administration is efficient in other models (Ludwig *et al.*, 2013). There is evidence that they reach relevant areas such as the hippocampus, amygdala, and CSF, yet how they reach these targets within the brain at effective amounts needs to be further investigated (Born *et al.*, 2002; Neumann *et al.*, 2013; Veening and Olivier, 2013).

14.6 Perspectives

Oxytocin and vasopressin are evolutionarily old neuropeptides that are found in species ranging from invertebrates to mammals. Many features in the oxytocinergic-vasopressinergic system, from protein sequence to connectivity and function, are highly conserved throughout evolution. In recent years, we have seen a huge advancement in technological and computational tools in conjunction with new genetic knowledge and tools. Employing a variety of model organisms, we now have the ability to systematically examine oxytocin and vasopressin neurons at the genomic and transcriptomic levels. State-of-the-art genome editing methods are now readily available for targeted and heritable intervention in gene expression in mice (Wang H *et al.*, 2013), fish (Ansai *et al.*, 2013; Bedell *et al.*, 2012; Hwang *et al.*, 2013), frogs (Lei *et al.*, 2012), and various invertebrates (Bannister *et al.*, 2014; Liu *et al.*, 2012; Treen *et al.*, 2014; Wang F *et al.*, 2013). Optogenetics and advanced imaging technics are used for the manipulation and monitoring of neuronal activity (Ahrens and Keller, 2013; Atasoy *et al.*, 2012; Knobloch *et al.*, 2012; Wyart *et al.*, 2009). In this respect, the use of non-mammalian genetic models, such as *C. elegans* and zebrafish, will likely contribute to our understanding of the molecular and cellular processes underlying morphogenesis and function of neurohypophysial neurons. For example, studies in the optically transparent zebrafish present us with the exceptional possibility to analyze the connectivity, molecular composition, and function of neurohypophysial neurons at a single cell resolution (Blechman *et al.*, 2011; Gutnick *et al.*, 2011; Liu *et al.*, 2013). Taken together, the use of non-mammalian model organisms in conjunction with unique genetic tools, fluorescent imaging, and behavioral assays places us in an excellent position to systematically study the embryonic development and function of neurohypophysial circuits. These remarkable capabilities, combined with many of the advantages of relatively simple model organisms, contribute to and will continue to advance our understanding of the way oxytocin and vasopressin shape life.

Acknowledgments

We thank Dr Natalia Borodovsky for performing the fluorescent *in situ* hybridization presented in Figure 14.4 and Dr Savani Anbalagan for comments on the manuscript. The research in the Levkowitz lab is supported by the Simons Foundation Autism Research Initiative (SFARI); Israel Science Foundation; Legacy Heritage Fund for Biomedical Science Partnership; Israel Ministry of Agriculture; Sparr Foundation (in the frame of the Weizmann Research Council); Kirk Center for Childhood Cancer and Immunological Disorders; The Krenter Institute and Estate of Lore Lennon. GL is an incumbent of the Elias Sourasky Professorial Chair.

References

Acher, R. and Chauvet, J. (1953) The structure of bovine vasopressin. *Biochimica et Biophysica Acta*, **12**(3), 487.

Acher, R., Chauvet, J., and Chauvet, M.T. (1972) Phylogeny of the Neurohypophysial Hormones Two New Active Peptides Isolated from a Cartilaginous Fish, Squalus acanthias *Eur. J. Biochem.* **29**, 12–19.

Acher, R., Chauvet, J., Chauvet, M.T. and Crepy, D. (1962). Isolation of a new neurohypophysial hormone, isotocin, present in bony fish. *Biochimica et biophysica acta*, **58**, 624–625.

Acher, R., Chauvet, J., Chauvet, M.T. and Crepy, D. (1965). Phylogeny of neurophyophyseal peptides: isolation of a new hormone, glumitocin (Ser 4-Gln 8-ocytocin) present in a cartilaginous fish, the ray (Raia clavata). *Biochimica et biophysica acta*, **107**(2), 393–396.

Acher, R., Chauvet, J., Chauvet, M.T. and Crepy, D. (1967). Amphibian neurohypophyseal hormones: isolation and characterization of mesotocin and vasotocin from the toad (Bufo bufo). *General and comparative endocrinology*, **8**(2), 337.

Acher, R., Chauvet, J. and Rouille, Y. (1997) Adaptive evolution of water homeostasis regulation in amphibians, vasotocin and hydrins. *Biology of the Cell*, **89**(5–6), 283–291.

Ahrens, M.B. and Keller, P. (2013) Whole-brain functional imaging at cellular resolution using light-sheet microscopy. *Nature Methods*, **10**, 413–420.

Almeida, O., Gozdowska, M., Kulczykowska, E. and Oliveira, R. (2012) Brain levels of arginine-vasotocin and isotocin in dominant and subordinate males of a cichlid fish. *Hormones and Behavior*, **61**(2), 212–217.

Ansai, S., Sakuma, T., Yamamoto, T., *et al.* (2013) Efficient targeted mutagenesis in medaka using custom-designed transcription activator-like effector nucleases. *Genetics*, **193**(3), 739–749.

Atasoy, D., Betley, J., Su, H. and Sternson, S. (2012) Deconstruction of a neural circuit for hunger. *Nature*, **488**(7410), 172–177.

Atkins, N., Jones, S. and Guillette, L. Jr (2006) Timing of parturition in two species of viviparous lizard, influences of beta-adrenergic stimulation and temperature upon uterine responses to arginine vasotocin (AVT). *Journal of Comparative Physiology B*, **176**(8), 783–792.

Ayers, L.W., Missig, G., Schulkin, J. and Rosen, J. (2011) Oxytocin reduces background anxiety in a fear-potentiated startle paradigm, peripheral vs central administration. *Neuropsychopharmacology*, **36**(12), 2488–2497.

Ballesteros, J.A. and Weinstein, H. (1995) Integrated methods for the construction of three-dimensional models and comptuational probing of structure–function relations in G protein-coupled receptors. *Methods in Neuroscience*, **25**, 366–428.

Bannister, S., Antonova, O., Polo, A., *et al.* (2014) TALENs mediate efficient and heritable mutation of endogenous genes in the marine Annelid Platynereis dumerilii. *Genetics*, **197**, 77–89.

Barberis, C., Mouillac, B. and Durroux, T. (1998) Structural bases of vasopressin/oxytocin receptor function. *Journal of Endocrinology*, **156**(2), 223–229.

Barth, S.W., Bathgate, R., Mess, A., Parry, L., Ivell, R. and Grossmann, R. (1997) Mesotocin gene expression in the diencephalon of domestic fowl, cloning and sequencing of the MT cDNA and distribution of MT gene expressing neurons in the chicken hypothalamus. *Journal of Neuroendocrinology*, **9**(10), 777–787.

Bedell, V.M., Wang, Y., Campbell, J., *et al.* (2012) In vivo genome editing using a high-efficiency TALEN system. *Nature*, **491**(7422), 114–118.

Beets, I., Janssen, T., Meelkop, E. *et al.* (2012) Vasopressin/oxytocin-related signaling regulates gustatory associative learning in C. elegans. *Science*, **338**(6106), 543–545. [**Demonstrates the function of nematocin in learning processes in nematodes. In this work the authors identify and characterize an oxytocin/vasopressin-like system in nematodes. Using advanced genetic tools, Beets et al. demonstrate a function of nematocin in sensory processing and modulation of behavior.**]

Birnbaumer, M. (2001) The V2 vasopressin receptor mutations and fluid homeostasis. *Cardiovascular Research*, **51**(3), 409–415.

Blahser, S. (1984) Peptidergic pathways in the avian brain. *Journal of Experimental Zoology*, **232**(3), 397–403.

Blechman, J., Amir-Zilberstein, L., Gutnick, A., Ben-Dor, S. and Levkowitz, G. (2011) The metabolic regulator PGC-1{alpha} directly controls the expression of the hypothalamic neuropeptide oxytocin. *Journal of Neuroscience*, **31**(42), 14835–14840.

Born, J., Lange, T., Kern, W., McGregor, G., Bickel, U. and Fehm, H. (2002) Sniffing neuropeptides, a transnasal approach to the human brain. *Nature Neuroscience*, **5**(6), 514–516.

Boyd, S.K. (1994) Arginine vasotocin facilitation of advertisement calling and call phonotaxis in bullfrogs. *Hormones and Behavior*, **28**(3), 232–240.

Boyd, S.K. and Moore, F. (1990) Autoradiographic localization of putative arginine vasotocin receptors in the kidney of a urodele amphibian. *General and Comparative Endocrinology*, **78**(3), 344–350.

Boyd, S.K. and Moore, F. (1992) Sexually dimorphic concentrations of arginine vasotocin in sensory regions of the amphibian brain. *Brain Research*, **588**(2), 304–306.

Boyd, S.K., Tyler, C. and de Vries, G. (1992) Sexual dimorphism in the vasotocin system of the bullfrog (Rana catesbeiana). *Journal of Comparative Neurology*, **325**(2), 313–325.

Bradshaw, S.D. and Bradshaw, F. (2002) Arginine vasotocin, site and mode of action in the reptilian kidney. *General and Comparative Endocrinology*, **126**(1), 7–13.

Braida, D., Donzelli, A., Martucci, R., *et al.* (2012) Neurohypophyseal hormones manipulation modulate social and anxiety-related behavior in zebrafish. *Psychopharmacology (Berl)*, **220**(2), 319–330.

Brown, C.H., Bains, J., Ludwig, M. and Stern, J. (2013) Physiological regulation of magnocellular neurosecretory cell activity, integration of intrinsic, local and afferent mechanisms. *Journal of Neuroendocrinology*, **25**(8), 678–710.

Burbach, J.P., Luckman, S., Murphy, D. and Gainer, H. (2001) Gene regulation in the magnocellular hypothalamo-neurohypophysial system. *Physiological Reviews*, **81**(3), 1197–1267.

Busnelli, M., Bulgheroni, E., Manning, M., Kleinau, G. and Chini, B. (2013) Selective and potent agonists and antagonists for investigating the role of mouse oxytocin receptors. *Journal of Pharmacology & Experimental Therapeutics*, **346**(2), 318–327.

Cattaneo, M.G., Lucci, G. and Vicentini, L. (2009) Oxytocin stimulates in vitro angiogenesis via a Pyk-2/Src-dependent mechanism. *Experimental Cell Research*, **315**(18), 3210–3219.

Cazzamali, G., Hauser, F., Kobberup, S., Williamson, M. and Grimmelikhuijzen, C. (2003) Molecular identification of a Drosophila G protein-coupled receptor specific for crustacean cardioactive peptide. *Biochemical and Biophysical Research Communications*, **303**(1), 146–152.

Chauvet, M.T., Hurpet, D., Chauvet, J. and Acher, R. (1980). Phenypressin (Phe2-Arg8-vasopressin), a new neurohypophysial peptide found in marsupials.

Chauvet, J., Michel, G., Ouedraogo, Y., Chou, J., Chatt, B.T. and Acher, R. (1995). A new neurohypophysial peptide, seritocin ([Ser5, Ile8]-oxytocin), identified in a dryness-resistant African toad, Bufo regularis. *International journal of peptide and protein research*, **45**(5), 482–487.

Chauvet, J., Rouille, Y., Chauveau, C., Chauvet, M.T. and Acher, R. (1994). Special evolution of neurohypophysial hormones in cartilaginous fishes: asvatocin and phasvatocin, two oxytocin-like peptides isolated from the spotted dogfish (Scyliorhinus caniculus). *Proceedings of the National Academy of Sciences*, **91**(23), 11266–11270.

Chini, B., Manning, M. and Guillon, G. (2008) Affinity and efficacy of selective agonists and antagonists for vasopressin and oxytocin receptors, an easy guide to receptor pharmacology. *Progress in Brain Research*, **170**, 513–517.

Chou, M.Y., Hung, J., Wu, L., Hwang, S. and Hwang, P. (2011) Isotocin controls ion regulation through regulating ionocyte progenitor differentiation and proliferation. *Cellular and Molecular Life Sciences*, **68**(16), 2797–2809.

Chu, J., Marler, C. and Wilczynski, W. (1998) The effects of arginine vasotocin on the calling behavior of male cricket frogs in changing social contexts. *Hormones and Behavior*, **34**(3), 248–261.

Cornett, L.E., Kang, S. and Kuenzel, W. (2013) A possible mechanism contributing to the synergistic action of vasotocin (VT) and corticotropin-releasing hormone (CRH) receptors on corticosterone release in birds. *General and Comparative Endocrinology*, **188**, 46–53.

Cruz, L.J., De Santos, V., Zafaralla, G.C., Ramilo, C.A., Zeikus, R., Gray, W.R. and Olivera, B.M. (1987). Invertebrate vasopressin/oxytocin homologs. Characterization of peptides from Conus geographus and Conus straitus venoms. *Journal of Biological Chemistry*, **262**(33), 15821–15824.

Cumming, R., Reaves, T. Jr and Hayward, J. (1982) Ultrastructural immunocytochemical characterization of isotocin, vasotocin and neurophysin neurons in the magnocellular preoptic nucleus of the goldfish. *Cell and Tissue Research*, **223**(3), 685–694.

Dale, H.H. (1909) The action of extracts of the pituitary body. *Biochemical Journal*, **4**(9), 427. [***Describes biological activities that are induced by posterior lobe extracts.***]

Dewan, A.K., Maruska, K. and Tricas, T. (2008) Arginine vasotocin neuronal phenotypes among congeneric territorial and shoaling reef butterflyfishes, species, sex and reproductive season comparisons. *Journal of Neuroendocrinology*, **20**(12), 1382–1394.

Diakow, C. (1978) Hormonal basis for breeding behavior in female frogs, vasotocin inhibits the release call of Rana pipiens. *Science*, **199**(4336), 1456–1457. [***An example of the association between an environmental signal, the corresponding hormonal response, and the effect on breeding behavior.***]

Dolen, G., Darvishzadeh, A., Huang, K. and Malenka, R. (2013) Social reward requires coordinated activity of nucleus accumbens oxytocin and serotonin. *Nature*, **501**(7466), 179–184.

Du Vigneaud, V., Lawler, H.C. and Popenoe, E.A., (1953a) Enzymic cleavage of glycinamide from vasopressin and a proposed structure for this pressor-antidiuretic hormone of the posterior pituitary. *J. Am. Chem. Soc.* **75**, 4880–4881.

Du Vigneaud, V., Ressler, C., Swan, C., Roberts, C., Katsoyannis, P. and Gordon, S. (1953b) The synthesis of an octapeptide amide with the hormonal activity of oxytocin. *Journal of the American Chemical Society*, **75**(19), 4879–4880.

Elphick, M.R. and Rowe, M.L. (2009). NGFFFamide and echinotocin: structurally unrelated myoactive neuropeptides derived from neurophysin-containing precursors in sea urchins. *Journal of Experimental Biology*, **212**(8), 1067–1077.

Erhart, G., Jirikowski, G., Pilgrim, C. and Adam, H. (1985) Hypothalamo-neurohypophysial system in hagfish, localization of neurophysin-, arg-vasopressin- and oxytocin-like immunoreactivity. *Basic Applied Histochemistry*, **29**(4), 289–296.

Erlenbach, I. and Wess, J. (1998) Molecular basis of V2 vasopressin receptor/Gs coupling selectivity. *Journal of Biological Chemistry*, **273**(41), 26549–26558.

Fergusson, B. and Bradshaw, S. (1991) Plasma arginine vasotocin, progesterone, and luteal development during pregnancy in the viviparous lizard Tiliqua rugosa. *General and Comparative Endocrinology*, **82**(1), 140–151.

Fernandez-Llebrez, P., Perez, J., Nadales, A., *et al.* (1988) Immunocytochemical study of the hypothalamic magnocellular neurosecretory nuclei of the snake Natrix maura and the turtle Mauremys caspica. *Cell and Tissue Research*, **253**(2), 435–445.

Figueroa, J., Morley, S., Heierhorst, J., Krentler, C., Lederis, K. and Richter, D. (1989) Two isotocin genes are present in the white sucker Catostomus commersoni both lacking introns in their protein coding regions. *EMBO Journal*, **8**(10), 2873–2877.

Fischer-Shofty, M., Shamay-Tsoory, S., Harari, H. and Levkovitz, Y. (2010) The effect of intranasal administration of oxytocin on fear recognition. *Neuropsychologia*, **48**(1), 179–184.

Flores, C.M., Munoz, D. Soto, M., Kausel, G., Romero, A. and Figueroa, J. (2007) Copeptin, derived from isotocin precursor, is a probable prolactin releasing factor in carp. *General and Comparative Endocrinology*, **150**(2), 343–354.

Fuchs, A.R., Fuchs, F., Husslein, P., Soloff, M. and Fernstrom, M. (1982) Oxytocin receptors and human parturition, a dual role for oxytocin in the initiation of labor. *Science*, **215**(4538), 1396–1398.

Fujino, Y., Nagahama, T., Oumi, T., *et al.* (1999) Possible functions of oxytocin/vasopressin-superfamily peptides in annelids with special reference to reproduction and osmoregulation. *Journal of Experimental Zoology*, **284**(4), 401–406.

Gaddum, J. (1928) Some properties of the separated active principles of the pituitary (posterior lobe). *Journal of Physiology*, **65**(4), 434–440.

Galli-Gallardo, S.M., Pang, P. and Oguro, C. (1979) Renal responses of the Chilean toad, Calyptocephalella caudiverbera, and the mud puppy, Necturus maculosus, to mesotocin. *General and Comparative Endocrinology*, **37**(1), 134–136.

Garrison, J.L., Macosko, E., Bernstein, S., Pokala, N., Albrecht, D. and Bargmann, C. (2012) Oxytocin/vasopressin-related peptides have an ancient role in reproductive behavior. *Science*, **338**(6106), 540–543. [*Identification of oxytocin/vasopressin-like proteins and receptors in* **C. elegans**. *This work demonstrates a role for oxytocin/vasopressin-like peptide – nematocin – in the activation of motor patterns that lead to mating behavior.*]

Gersh, I. (1939) The structure and function of the parenchymatous glandular cells in the neurohypophysis of the rat. *American Journal of Anatomy*, **64**(3), 407–443.

Godwin, J. and Thompson, R. (2012) Nonapeptides and social behavior in fishes. *Hormones and Behavior*, **61**(3), 230–238. [*An extensive review of the effects of oxytocin- and vasopressin-like peptides on sociosexual behaviors in teleost fish.*]

Godwin, J., Sawby, R., Warner, R., Crews, D. and Grober, M. (2000) Hypothalamic arginine vasotocin mRNA abundance variation across sexes and with sex change in a coral reef fish. *Brain, Behavior and Evolution*, **55**(2), 77–84.

Goldstein, D.L. (2006) Regulation of the avian kidney by arginine vasotocin. *General and Comparative Endocrinology*, **147**(1), 78–84.

Gonzalez, A. and Smeets, W. (1992) Comparative analysis of the vasotocinergic and mesotocinergic cells and fibers in the brain of two amphibians, the anuran Rana ridibunda and the urodele Pleurodeles waltlii. *Journal of Comparative Neurology*, **315**(1), 53–73.

Goodson, J.L. and Bass, A. (2000a) Forebrain peptides modulate sexually polymorphic vocal circuitry. *Nature*, **403**(6771), 769–772. [*An example of the central effect of isotocin and vasotocin on behavior, demonstrating how these peptides modulate auditory and vocal circuits in sonic fish. The nonapeptides differentially activate vocal motor activity according to the fish's social status and sex.*]

Goodson, J.L. and Bass, A. (2000b) Vasotocin innervation and modulation of vocal-acoustic circuitry in the teleost Porichthys notatus. *Journal of Comparative Neurology*, **422**(3), 363–379.

Goodson, J.L., Evans, A. and Bass, A. (2003) Putative isotocin distributions in sonic fish, relation to vasotocin and vocal-acoustic circuitry. *Journal of Comparative Neurology*, **462**(1), 1–14.

Goodson, J.L., Schrock, S., Klatt, J., Kabelik, D. and Kingsbury, M. (2009) Mesotocin and nonapeptide receptors promote estrildid flocking behavior. *Science*, **325**(5942), 862–866. [*Demonstrates the prosocial effects of mesotocin in the zebra finch brain. By using agonists and antagonists, the authors show that mesotocin, but not vasotocin, causes an increase in the time the birds spend with familiar conspecifics and enhances the preference of females towards large rather than small groups. In addition, there are behavior-related differences in the distribution of mesotocin-receptor throughout the brain according to the social strategy in territorial vs flocking species.*]

Goossens, N., Blahser, S., Oksche, A., Vandesande, F. and Dierickx, K. (1977) Immunocytochemical investigation of the hypothalamo-neurohypophysial system in birds. *Cell and Tissue Research*, **184**(1), 1–13.

Gorbman, A., Kobayashi, H. and Uemura, H. (1963) The vascularization of the hypophysial structures of the hagfish. *General and Comparative Endocrinology*, **14**, 505–514.

Green, J.D. (1951) The comparative anatomy of the hypophysis, with special reference to its blood supply and innervation. *American Journal of Anatomy*, **88**(2), 225–311.

Greenwood, A.K., Wark, A., Fernald, R. and Hofmann, H. (2008) Expression of arginine vasotocin in distinct preoptic regions is associated with dominant and subordinate behaviour in an African cichlid fish. *Proceedings of the Royal Society B: Biological Sciences*, **275**(1649), 2393–2402.

Gruber, C.W. (2014). Physiology of invertebrate oxytocin and vasopressin neuropeptides. *Experimental physiology*, **99**(1), 55–61.

Gruber, C.W. and Muttenthaler, M. (2012). Discovery of defense-and neuropeptides in social ants by genome-mining. *PLoS One*, **7**(3), e32559.

Guillette, L.J. Jr (1979) Stimulation of parturition in a viviparous lizard (Sceloporus jarrovi) by arginine vasotocin. *General and Comparative Endocrinology*, **38**(4), 457–460.

Guillette, L.J. Jr and Jones, R. (1982) Further observations on arginine vasotocin-induced oviposition and parturition in lizards. *Journal of Herpetology*, **16**(2), 140–144.

Gutnick, A. and Levkowitz, G. (2012) The neurohypophysis, fishing for new insights. *Journal of Neuroendocrinology*, **24**(6), 973–974.

Gutnick, A., Blechman, J., Kaslin, J., *et al.* (2011) The hypothalamic neuropeptide oxytocin is required for formation of the neurovascular interface of the pituitary. *Developmental Cell*, **21**(4), 642–654. [**Demonstrates a role for oxytocin in the formation of neurovascular interface during zebrafish development. The authors propose a model by which oxytocin plays an important role in the facilitation of its own release into the circulation.**]

Gwee, P.C., Amemiya, C., Brenner, S. and Venkatesh, B. (2008) Sequence and organization of coelacanth neurohypophysial hormone genes, evolutionary history of the vertebrate neurohypophysial hormone gene locus. *BMC Evolutionary Biology*, **8**, 93.

Gwee, P.C., Tay, B., Brenner, S. and Venkatesh, B. (2009) Characterization of the neurohypophysial hormone gene loci in elephant shark and the Japanese lamprey, origin of the vertebrate neurohypophysial hormone genes. *BMC Evolutionary Biology*, **9**, 47.

Haruta, K., Yamashita, T. and Kawashima, S. (1991) Changes in arginine vasotocin content in the pituitary of the Medaka (Oryzias latipes) during osmotic stress. *General and Comparative Endocrinology*, **83**(3), 327–336.

Hasegawa, T., Tanii, H., Suzuki, M. and Tanaka, S. (2003) Regulation of water absorption in the frog skins by two vasotocin-dependent water-channel aquaporins, AQP-h2 and AQP-h3. *Endocrinology*, **144**(9), 4087–4096.

Heller, H. and Pickering, B.T. (1961). Neurohypophysial hormones of non-mammalian vertebrates. *The Journal of physiology*, **155**(1), 98–114.

Herget, U., Wolf, A., Wullimann, M. and Ryu, S. (2014) Molecular neuroanatomy and chemoarchitecture of the neurosecretory preoptic-hypothalamic area in zebrafish larvae. *Journal of Comparative Neurology*, **522**, 1542–1564.

Herring, P.T. (1904) The action of pituitary extracts on the heart and circulation of the frog. *Journal of Physiology*, **31**(6), 429–437. [**Describes biological activities that are induced by posterior lobe extracts.**]

Hillyard, S.D. and Willumsen, N. (2011) Chemosensory function of amphibian skin, integrating epithelial transport, capillary blood flow and behaviour. *Acta Physiologica (Oxf)*, **202**(3), 533–548.

Hogben, L.T. and Schlapp, W. (1924) Studies on the pituitary. iii. the vasomotor activity of pituitary extracts throughout the vertebrate series. *Experimental Physiology*, **14**(3), 229–258.

Howell, W.H. (1898) The physiological effects of extracts of the hypophysis cerebri and infundibular body. *Journal of Experimental Medicine*, **3**(2), 245–258.

Hoyle, C.H. (1999) Neuropeptide families and their receptors, evolutionary perspectives. *Brain Research*, **848**(1–2), 1–25. [**A classic bioinformatics analysis of the evolution of neuropeptides and receptors.**]

Hwang, W.Y., Fu, Y., Reyon, D., *et al.* (2013) Efficient genome editing in zebrafish using a CRISPR-Cas system. *Nature Biotechnology*, **31**(3), 227–229.

Hyodo, S., Kato, Y., Ono, M. and Urano, A. (1991) Cloning and sequence analyses of cDNAs encoding vasotocin and isotocin precursors of chum salmon, Oncorhynchus keta, evolutionary relationships of neurohypophysial hormone precursors. *Journal of Comparative Physiology B*, **160**(6), 601–608.

Hyodo, S., Ishii, S. and Joss, J.M. (1997). Australian lungfish neurohypophysial hormone genes encode vasotocin and [Phe2] mesotocin precursors homologous to tetrapod-type precursors. *Proceedings of the National Academy of Sciences*, **94**(24), 13339–13344.

Hyodo, S., Tsukada, T. and Takei, Y. (2004). Neurohypophysial hormones of dogfish, Triakis scyllium: structures and salinity-dependent secretion. *General and comparative endocrinology*, **138**(2), 97–104.

Iwakiri, M., Sugiyama, A., Ikeda, T., Muneoka, Y. and Kubota, I., (1990). A novel oxytocin-like peptide isolated from the neural complexes of tunicate, Styele plicata. *Zool. Sci.*, **7**, pp. 1035–1041.

Iwamuro, S., Hayashi, H. and Kikuyama, S. (1993). An additional arginine-vasotocin-related peptide, vasotocinyl-Gly-Lys, in Xenopus neurohypophysis. *Biochimica et Biophysica Acta (BBA)-Molecular Cell Research*, **1176**(1), 143–147.

Jonaidi, H., Oloumi, M. and Denbow, D. (2003) Behavioral effects of intracerebroventricular injection of oxytocin in birds. *Physiology & Behavior*, **79**(4–5), 725–729.

Kamm, O., Aldrich, T., Grote, I., Rowe, L. and Bugbee, E. (1928) The active principles of the posterior lobe of the pituitary gland. 1 i. the demonstration of the presence of two active principles. ii. the separation of the two principles and their concentration in the form of potent solid preparations. *Journal of the American Chemical Society*, **50**(2), 573–601. [**First identification of two active principles from posterior pituitary extracts.**]

Kanda, A., Takuwa-Kuroda, K., Iwakoshi-Ukena, E. and Minakata, H. (2003) Single exon structures of the oxytocin/vasopressin superfamily peptides of octopus. *Biochemical and Biophysical Research Communications*, **309**(4), 743–748.

Kanda, A., Satake, H., Kawada, T. and Minakata, H. (2005) Novel evolutionary lineages of the invertebrate oxytocin/vasopressin superfamily peptides and their receptors in the common octopus (Octopus vulgaris). *Biochemical Journal*, **387**(Pt 1), 85–91.

Kawada, T., Kanda, A., Minakata, H., Matsushima, O. and Satake, H. (2004) Identification of a novel receptor for an invertebrate oxytocin/vasopressin superfamily peptide, molecular and functional evolution of the oxytocin/vasopressin superfamily. *Biochemical Journal*, **382**(Pt 1), 231–237.

Kawada, T., Sekiguchi, S., Itoh, Y., Ogasawara, M. and Satake, H. (2008) Characterization of a novel vasopressin/oxytocin superfamily peptide and its receptor from an ascidian, Ciona intestinalis. *Peptides*, **29**(10), 1672–1678.

Kime, N.M., Whitney, T., Davis, E. and Marler, C. (2007) Arginine vasotocin promotes calling behavior and call changes in male tungara frogs. *Brain, Behavior and Evolution*, **69**(4), 254–265.

Kime, N.M., Whitney, T., Ryan, M., Rand, A. and Marler, C. (2010) Treatment with arginine vasotocin alters mating calls and decreases call attractiveness in male tungara frogs. *General and Comparative Endocrinology*, **165**(2), 221–228.

Kimura, T., Makino, Y., Saji, F., *et al.* (1994) Molecular characterization of a cloned human oxytocin receptor. *European Journal of Endocrinology*, **131**(4), 385–390.

Klatt, J.D. and Goodson, J. (2013) Sex-specific activity and function of hypothalamic nonapeptide neurons during nest-building in zebra finches. *Hormones and Behavior*, **64**(5), 818–824.

Knobloch, H.S. and Grinevich, V. (2014) Evolution of oxytocin pathways in the brain of vertebrates. *Frontiers in Behavioral Neuroscience*, **8**, 31.

Knobloch, H.S., Charlet, A., Hoffmann, L., *et al.* (2012) Evoked axonal oxytocin release in the central amygdala attenuates fear response. *Neuron*, **73**(3), 553–566. [***Anatomical and functional mapping of oxytocinergic circuits involved in the attenuation of fear response.***]

Koehbach, J., Stockner, T., Bergmayr, C., Muttenthaler, M. and Gruber, C. (2013) Insights into the molecular evolution of oxytocin receptor ligand binding. *Biochemical Society Transactions*, **41**(1), 197–204.

Koike, T.I., Shimada, K. and Cornett, L. (1988) Plasma levels of immunoreactive mesotocin and vasotocin during oviposition in chickens, relationship to oxytocic action of the peptides in vitro and peptide interaction with myometrial membrane binding sites. *General and Comparative Endocrinology*, **70**(1), 119–126.

Kosfeld, M., Heinrichs, M., Zak, P., Fischbacher, U. and Fehr, E. (2005) Oxytocin increases trust in humans. *Nature*, **435**(7042), 673–676.

Koshimizu, T.A., Nakamura, K., Egashira, N., *et al.* (2012) Vasopressin V1a and V1b receptors, from molecules to physiological systems. *Physiological Reviews*, **92**(4), 1813–1864. [***A comprehensive review of the arginine vasopressin V1A and V1B receptors.***]

Lagman, D., Ocampo Daza, D., Widmark, J., Abalo, X., Sundstrom, G. and Larhammar, D. (2013) The vertebrate ancestral repertoire of visual opsins, transducin alpha subunits and oxytocin/vasopressin receptors was established by duplication of their shared genomic region in the two rounds of early vertebrate genome duplications. *BMC Evolutionary Biology*, **13**, 238. [***Evolutionary history of the oxytocin/vasopressin receptor family in a wider genomic context.***]

Landgraf, R. and Neumann, I. (2004) Vasopressin and oxytocin release within the brain, a dynamic concept of multiple and variable modes of neuropeptide communication. *Frontiers in Neuroendocrinology*, **25**(3–4), 150–176.

Lei, Y., Guo, X., Liu, Y., *et al.* (2012) Efficient targeted gene disruption in Xenopus embryos using engineered transcription activator-like effector nucleases (TALENs). *Proceedings of the National Academy of Sciences USA*, **109**(43), 17484–17489.

Leuner, B., Caponiti, J. and Gould, E. (2012) Oxytocin stimulates adult neurogenesis even under conditions of stress and elevated glucocorticoids. *Hippocampus*, **22**(4), 861–868.

Liu, F., Pogoda, H., Pearson, C., *et al.* (2013) Direct and indirect roles of Fgf3 and Fgf10 in innervation and vascularisation of the vertebrate hypothalamic neurohypophysis. *Development*, **140**(5), 1111–1122.

Liu, J. and Wess, J. (1996) Different single receptor domains determine the distinct G protein coupling profiles of members of the vasopressin receptor family. *Journal of Biological Chemistry*, **271**(15), 8772–8778.

Liu, J., Li, C., Yu, Z., *et al.* (2012) Efficient and specific modifications of the Drosophila genome by means of an easy TALEN strategy. *Journal of Genetics & Genomics*, **39**(5), 209–215.

Lohr, H. and Hammerschmidt, M. (2011) Zebrafish in endocrine systems, recent advances and implications for human disease. *Annual Review of Physiology*, **73**, 183–211.

Ludwig, M. and Leng, G. (2006) Dendritic peptide release and peptide-dependent behaviours. *Nature Reviews Neuroscience*, **7**(2), 126–136.

Ludwig, M., Sabatier, N., Bull, P., *et al.* (2002) Intracellular calcium stores regulate activity-dependent neuropeptide release from dendrites. *Nature*, **418**(6893), 85–89. [***Mechanism of dendritic peptide release within the brain.***]

Ludwig, M., Tobin, V., Callahan, M., *et al.* (2013) Intranasal application of vasopressin fails to elicit changes in brain immediate early gene expression, neural activity and behavioural performance of rats. *Journal of Neuroendocrinology*, **25**(7), 655–667.

Marler, C.A., Boyd, S. and Wilczynski, W. (1999) Forebrain arginine vasotocin correlates of alternative mating strategies in cricket frogs. *Hormones and Behavior*, **36**(1), 53–61.

Martos-Sitcha, J.A., Wunderink, Y., Gozdowska, M., Kulczykowska, E., Mancera, J. and Martinez-Rodriguez, G. (2013) Vasotocinergic and isotocinergic systems in the gilthead sea bream (Sparus aurata), an osmoregulatory story. *Comparative Biochemistry & Physiology A: Molecular & Integrative Physiology*, **166**(4), 571–581.

Masunari, K., Khan, M., Cline, M. and Tachibana, T. (2013) Central administration of mesotocin inhibits feeding behavior in chicks. *Regulatory Peptides*, **187**, 1–5.

McCormick, S.D. and Bradshaw, D. (2006) Hormonal control of salt and water balance in vertebrates. *General and Comparative Endocrinology*, **147**(1), 3–8.

Mendonca, R., Soares, M., Bshary, R. and Oliveira, R. (2013) Arginine vasotocin neuronal phenotype and interspecific cooperative behaviour. *Brain, Behavior and Evolution*, **82**(3), 166–176.

Michel, G., Chauvet, J., Chauvet, M., Clarke, C., Bern, H. and Acher, R. (1993) Chemical identification of the mammalian oxytocin in a holocephalian fish, the ratfish (Hydrolagus colliei). *General and Comparative Endocrinology*, **92**(2), 260–268.

Mikami, S., Tokado, H. and Farner, D. (1978) The hypothalamic neurosecretory systems of the Japanese quail as revealed by retrograde transport of horseradish peroxidase. *Cell and Tissue Research*, **194**(1), 1–15.

Miranda, J.A., Oliveira, R., Carneiro, L. Santos, R. and Grober, M. (2003) Neurochemical correlates of male polymorphism and alternative reproductive tactics in the Azorean rock-pool blenny, Parablennius parvicornis. *General and Comparative Endocrinology*, **132**(2), 183–189.

Moll, J. (1958) The effect of hypophysectomy on the pituitary vascular system of the rat. *Journal of Morphology*, **102**(1), 1–21. [*Reports increase in angiogenesis near neurosecretory material derived from the regenerated axon fibers following pituitary stalk lesions.*]

Moore, F.L. and Lowry, C. (1998) Comparative neuroanatomy of vasotocin and vasopressin in amphibians and other vertebrates. *Comparative Biochemistry & Physiology C: Toxicology, Pharmacology & Endocrinology*, **119**(3), 251–260.

Moore, F.L., Wood, R. and Boyd, S. (1992) Sex steroids and vasotocin interact in a female amphibian (Taricha granulosa) to elicit female-like egg-laying behavior or male-like courtship. *Hormones and Behavior*, **26**(2), 156–166.

Moore, F.L., Richardson, C. and Lowry, C. (2000) Sexual dimorphism in numbers of vasotocin-immunoreactive neurons in brain areas associated with reproductive behaviors in the roughskin newt. *General and Comparative Endocrinology*, **117**(2), 281–298.

Moos, F., Freund-Mercier, M., Guerne, Y., Guerne, J., Stoeckel, M. and Richard, P. (1984) Release of oxytocin and vasopressin by magnocellular nuclei in vitro, specific facilitatory effect of oxytocin on its own release. *Journal of Endocrinology*, **102**(1), 63–72.

Morris, J.F. and Ludwig, M. (2004) Magnocellular dendrites, prototypic receiver/transmitters. *Journal of Neuroendocrinology*, **16**(4), 403–408.

Neumann, I.D., Maloumby, R., Beiderbeck, D., Lukas, M. and Landgraf, R. (2013) Increased brain and plasma oxytocin after nasal and peripheral administration in rats and mice. *Psychoneuroendocrinology*, **38**(10), 1985–1993.

Nielsen, S., Chou, C., Marples, D., *et al.* (1995) Vasopressin increases water permeability of kidney collecting duct by inducing translocation of aquaporin-CD water channels to plasma membrane. *Proceedings of the National Academy of Sciences USA*, **92**(4), 1013–1017.

Nishimori, K., Young, L., Guo, Q., *et al.* (1996) Oxytocin is required for nursing but is not essential for parturition or reproductive behavior. *Proceedings of the National Academy of Sciences USA*, **93**(21), 11699–11704.

Norris, D.O. and Carr, J. (2013) Vertebrate Endocrinology, Academic Press, New York.

Nouwen, E.J. and Kuhn, E. (1985) Volumetric control of arginine vasotocin and mesotocin release in the frog (Rana ridibunda). *Journal of Endocrinology*, **105**(3), 371–377.

Nouwen, E.J., Decuypere, E., Kuhn, E., *et al.* (1984) Effect of dehydration, haemorrhage and oviposition on serum concentrations of vasotocin, mesotocin and prolactin in the chicken. *Journal of Endocrinology*, **102**(3), 345–351.

Nozaki, M. (2008) The hagfish pituitary gland and its putative adenohypophysial hormones. *Zoological Science*, **25**(10), 1028–1036.

Ocampo Daza, D., Lewicka, M. and Larhammar, D. (2012) The oxytocin/vasopressin receptor family has at least five members in the gnathostome lineage, including two distinct V2 subtypes. *General and Comparative Endocrinology*, **175**(1), 135–143.

Oliver, G. and Schäfer, E. (1895) On the physiological action of extracts of pituitary body and certain other glandular organs, preliminary communication. *Journal of Physiology*, **18**(3), 277–279. [*Injected pituitary extracts, termed 'posterior lobe hormones,' caused increased blood pressure.*]

Opel, H. (1965) Oviposition in chickens after removal of the posterior lobe of the pituitary by an improved method. *Endocrinology*, **76**, 673–677.

Ott, I. and Scott, J. (1910) The action of infundibulin upon the mammary secretion. *Experimental Biology and Medicine*, **8**(2), 48–49. [*Pituitary extracts induce milk ejection.*]

Oumi, T., Ukena, K., Matsushima, O., Ikeda, T., Fujita, T., Minakata, H. and Nomoto, K. (1994). Annetocin: an oxytocin-related peptide isolated from the earthworm, Eisenia foetida. *Biochemical and biophysical research communications*, **198**(1), 393–399.

Oumi, T., Ukena, K., Matsushima, O., *et al.* (1996) Annetocin, an annelid oxytocin-related peptide, induces egg-laying behavior in the earthworm, Eisenia foetida. *Journal of Experimental Zoology*, **276**(2), 151–156.

Pang, P.K. and Sawyer, W. (1978) Renal and vascular responses of the bullfrog (Rana catesbeiana) to mesotocin. *American Journal of Physiology*, **235**(2), F151–155.

Park, Y., Kim, Y. and Adams, M. (2002) Identification of G protein-coupled receptors for Drosophila PRXamide peptides, CCAP, corazonin, and AKH supports a theory of ligand–receptor coevolution. *Proceedings of the National Academy of Sciences USA*, **99**(17), 11423–11428.

Paton, D.N. and Watson, A. (1912) The actions of pituitrin, adrenalin and barium on the circulation of the bird. *Journal of Physiology*, **44**(5–6), 413–424.

Pearson, C. and Placzek, M. (2012) Development of the medial hypothalamus, forming a functional hypothalamic-neurohypophyseal interface. *Current Topics in Developmental Biology*, **106**, 49–88.

Pickford, G.E. and Strecker, E. (1977) The spawning reflex response of the killifish, Fundulus heteroclitus, isotocin is relatively inactive in comparison with arginine vasotocin. *General and Comparative Endocrinology*, **32**(2), 132–137.

Pierson, P.M., Guibbolini, M., Mayer-Gostan, N. and Lahlou, B. (1995) ELISA measurements of vasotocin and isotocin in plasma and pituitary of the rainbow trout, effect of salinity. *Peptides*, **16**(5), 859–865.

Pogoda, H.M. and Hammerschmidt, M. (2007) Molecular genetics of pituitary development in zebrafish. *Seminars in Cell & Developmental Biology*, **18**(4), 543–558.

Postina, R., Kojro, E. and Fahrenholz, F. (1996) Separate agonist and peptide antagonist binding sites of the oxytocin receptor defined by their transfer into the V2 vasopressin receptor. *Journal of Biological Chemistry*, **271**(49), 31593–31601.

Popenoe, E.A., Lawler, H.C. and Vigneaud, V.D. (1952). Partial purification and amino acid content of vasopressin from hog posterior pituitary glands. *Journal of the American Chemical Society*, **74**(14), 3713–3713.

Proux, J.P., Miller, C.A., Li, J.P., Carney, R.L., Girardie, A., Delaage, M. and Schooley, D.A. (1987). Identification of an arginine vasopressin-like diuretic hormone from Locustamigratoria. *Biochemical and biophysical research communications*, **149**(1), 180–186.

Puryear, R., Rigatto, K., Amico, J. and Morris, M. (2001) Enhanced salt intake in oxytocin deficient mice. *Experimental Neurology*, **171**(2), 323–328.

Ramallo, M.R., Grober, M., Canepa, M., Morandini, L. and Pandolfi, M. (2012) Arginine-vasotocin expression and participation in reproduction and social behavior in males of the cichlid fish Cichlasoma dimerus. *General and Comparative Endocrinology*, **179**(2), 221–231.

Reich, G. (1992) A new peptide of the oxytocin/vasopressin family isolated from nerves of the cephalopod Octopus vulgaris. *Neuroscience Letters*, **134**(2), 191–194.

Robinzon, B., Koike, T., Neldon, H. and Kinzler, S. (1988) Distribution of immunoreactive mesotocin and vasotocin in the brain and pituitary of chickens. *Peptides*, **9**(4), 829–833.

Rodrigo, J., Pena, A., Murat, B., *et al.* (2007) Mapping the binding site of arginine vasopressin to V1a and V1b vasopressin receptors. *Molecular Endocrinology*, **21**(2), 512–523.

Ross, H.E., Cole, C., Smith, Y., *et al.* (2009) Characterization of the oxytocin system regulating affiliative behavior in female prairie voles. *Neuroscience*, **162**(4), 892–903.

Rouille, Y., Michel, G., Chauvet, M.T., Chauvet, J. and Acher, R. (1989). Hydrins, hydroosmotic neurohypophysial peptides: osmoregulatory adaptation in amphibians through vasotocin precursor processing. *Proceedings of the National Academy of Sciences*, **86**(14), 5272–5275.

Ruppert, S., Scherer, G. and Schutz, G. (1984) Recent gene conversion involving bovine vasopressin and oxytocin precursor genes suggested by nucleotide sequence. *Nature*, **308**(5959), 554–557.

Rurak, D.W. and Perks, A. (1976) The neurohypophysial principles of the western brook lamprey, Lampetra richardsoni, studies inthe adult. *General and Comparative Endocrinology*, **29**(3), 301–312.

Russell, J.A., Leng, G. and Douglas, A. (2003) The magnocellular oxytocin system, the fount of maternity, adaptations in pregnancy. *Frontiers in Neuroendocrinology*, **24**(1), 27–61.

Sabatier, N., Caquineau, C., Dayanithi, G., *et al.* (2003) Alpha-melanocyte-stimulating hormone stimulates oxytocin release from the dendrites of hypothalamic neurons while inhibiting oxytocin release from their terminals in the neurohypophysis. *Journal of Neuroscience*, **23**(32), 10351–10358.

Sabatier, N., Leng, G. and Menzies, J. (2013) Oxytocin, feeding, and satiety. *Frontiers in Endocrinology (Lausanne)*, **4**, 35.

Saito, D. and Urano, A. (2001) Synchronized periodic Ca2+ pulses define neurosecretory activities in magnocellular vasotocin and isotocin neurons. *Journal of Neuroscience*, **21**(21), RC178.

Saito, D., Komatsuda, M. and Urano, A. (2004) Functional organization of preoptic vasotocin and isotocin neurons in the brain of rainbow trout, central and neurohypophysial projections of single neurons. *Neuroscience*, **124**(4), 973–984.

Saito, N., Kinzler, S. and Koike, T. (1990) Arginine vasotocin and mesotocin levels in theca and granulosa layers of the ovary during the oviposition cycle in hens (Gallus domesticus). *General and Comparative Endocrinology*, **79**(1), 54–63. [*Neuroanatomical characterization of isotocin and vasotocin neurons in the rainbow trout brain.*]

Schäfer, E. and Vincent, S. (1899) The physiological effects of extracts of the pituitary body. *Journal of Physiology*, **25**(1), 87–97.

Scharrer, E. (1928) Die Lichtempfindlichkeit Blinder Elritzen.(Untersuchungen Über das Zwischenhirn der Fische I.). *Journal of Comparative Physiology A, Neuroethology, Sensory, Neural, and Behavioral Physiology*, **7**(1), 1–38. [*Reports the unique morphology of neurosecretory cells in the fish hypothalamus. Served as the inspiration for the Scharrers' neurosecretion concept.*]

Scharrer, E. and Scharrer, B. (1937) Uber drüsen-nervenzellen und neurosekretorische organe bei wirbellosen und wirbeltieren. *Biological Reviews*, **12**(2), 185–216.

Scharrer, E. and Scharrer, B. (1945) Neurosecretion. *Physiological Reviews*, **25**(1), 171–181. [*An elegant review of the the concept of neurosecretion based on their earlier work on fish and invertebrate neurosecretory neurons.*]

Semsar, K., Kandel, F. and Godwin, J. (2001) Manipulations of the AVT system shift social status and related courtship and aggressive behavior in the bluehead wrasse. *Hormones and Behavior*, **40**(1), 21–31.

Sewall, K.B., Dankoski, E. and Sockman, K. (2010) Song environment affects singing effort and vasotocin immunoreactivity in the forebrain of male Lincoln's sparrows. *Hormones and Behavior*, **58**(3), 544–553.

Silveira, P.F., Breno, M., Martin del Rio, M. and Mancera, J. (2002) The distribution of vasotocin and mesotocin immunoreactivity in the brain of the snake, Bothrops jararaca. *Journal of Chemical Neuroanatomy*, **24**(1), 15–26.

Soares, M.C., Bshary, R., Mendonca, R., Grutter, A. and Oliveira, R. (2012) Arginine vasotocin regulation of interspecific cooperative behaviour in a cleaner fish. *PLoS One*, **7**(7), e39583.

Soloff, M.S., Alexandrova, M. and Fernstrom, M. (1979) Oxytocin receptors, triggers for parturition and lactation? *Science*, **204**(4399), 1313–1315.

Sower, S.A., Freamat, M. and Kavanaugh, S. (2009) The origins of the vertebrate hypothalamic-pituitary-gonadal (HPG) and hypothalamic-pituitary-thyroid (HPT) endocrine systems, new insights from lampreys. *General and Comparative Endocrinology*, **161**(1), 20–29.

Stafflinger, E., Hansen, K., Hauser, F., *et al.* (2008) Cloning and identification of an oxytocin/vasopressin-like receptor and its ligand from insects. *Proceedings of the National Academy of Sciences USA*, **105**(9), 3262–3267.

Stricker, E.M. and Verbalis, J. (1996) Central inhibition of salt appetite by oxytocin in rats. *Regulatory Peptides*, **66**(1–2), 83–85.

Takayanagi, Y., Yoshida, M., Bielsky, I., *et al.* (2005) Pervasive social deficits, but normal parturition, in oxytocin receptor-deficient mice. *Proceedings of the National Academy of Sciences USA*, **102**(44), 16096–16101.

Takuwa-Kuroda, K., Iwakoshi-Ukena, E., Kanda, A. and Minakata, H. (2003) Octopus, which owns the most advanced brain in invertebrates, has two members of vasopressin/oxytocin superfamily as in vertebrates. *Regulatory Peptides*, **115**(2), 139–149.

Tamma, R., Colaianni, G., Zhu, L., *et al.* (2009) Oxytocin is an anabolic bone hormone. *Proceedings of the National Academy of Sciences USA*, **106**(17), 7149–7154.

Tamma, R., Sun, L., Cuscito, C., *et al.* (2013) Regulation of bone remodeling by vasopressin explains the bone loss in hyponatremia. *Proceedings of the National Academy of Sciences USA*, **110**(46), 18644–18649.

Ten Eyck, G.R. (2005) Arginine vasotocin activates advertisement calling and movement in the territorial Puerto Rican frog, Eleutherodactylus coqui. *Hormones and Behavior*, **47**(2), 223–229.

Ten Eyck, G.R. and ul Haq, A. (2012) Arginine vasotocin activates aggressive calls during paternal care in the Puerto Rican coqui frog, Eleutherodactylus coqui. *Neuroscience Letters*, **525**(2), 152–156.

Tessmar-Raible, K., Raible, F. Christodoulou, F., *et al.* (2007) Conserved sensory-neurosecretory cell types in annelid and fish forebrain, insights into hypothalamus evolution. *Cell*, **129**(7), 1389–1400.

Theodosis, D.T. (2002) Oxytocin-secreting neurons, a physiological model of morphological neuronal and glial plasticity in the adult hypothalamus. *Frontiers in Neuroendocrinology*, **23**(1), 101–135.

Thepen, T., Voorn, P., Stoll, C., Sluiter, A., Pool, C. and Lohman, A.H. (1987) Mesotocin and vasotocin in the brain of the lizard Gekko gecko. An immunocytochemical study. *Cell and Tissue Research*, **250**(3), 649–656.

Thompson, R.R. and Moore, F. (2003) The effects of sex steroids and vasotocin on behavioral responses to visual and olfactory sexual stimuli in ovariectomized female roughskin newts. *Hormones and Behavior*, **44**(4), 311–318.

Tobin, V., Leng, G. and Ludwig, M. (2012) The involvement of actin, calcium channels and exocytosis proteins in somato-dendritic oxytocin and vasopressin release. *Frontiers in Physiology*, **3**, 261.

Treen, N., Yoshida, K., Sakuma, T., *et al.* (2014) Tissue-specific and ubiquitous gene knockouts by TALEN electroporation provide new approaches to investigating gene function in Ciona. *Development*, **141**(2), 481–487.

Tuppy, H. (1953). The amino-acid sequence in oxytocin. *Biochimica et biophysica acta*, **11**, 449–450.

Ukena, K., Iwakoshi-Ukena, E. and Hikosaka, A. (2008) Unique form and osmoregulatory function of a neurohypophysial hormone in a urochordate. *Endocrinology*, **149**(10), 5254–5261.

Van Kesteren, R.E., Smit, A., Dirks, R., *et al.* (1992) Evolution of the vasopressin/oxytocin superfamily, characterization of a cDNA encoding a vasopressin-related precursor, preproconopressin, from the mollusc Lymnaea stagnalis. *Proceedings of the National Academy of Sciences USA*, **89**(10), 4593–4597.

Van Kesteren, R.E., Smit, A., de Lange, R., *et al.* (1995) Structural and functional evolution of the vasopressin/oxytocin superfamily, vasopressin-related conopressin is the only member present in Lymnaea, and is involved in the control of sexual behavior. *Journal of Neuroscience*, **15**(9), 5989–5998.

Van Kesteren, R.E., Tensen, C., Smit, A., *et al.* (1996) Co-evolution of ligand-receptor pairs in the vasopressin/oxytocin superfamily of bioactive peptides. *Journal of Biological Chemistry*, **271**(7), 3619–3626.

Veening, J.G. and Olivier, B. (2013) Intranasal administration of oxytocin, behavioral and clinical effects, a review. *Neuroscience & Biobehavioral Reviews*, **37**(8), 1445–1465.

Veening, J.G., de Jong, T. and Barendregt, H. (2010) Oxytocin-messages via the cerebrospinal fluid, behavioral effects; a review. *Physiology & Behavior*, **101**(2), 193–210.

Verbalis, J.G., McCann, M., McHale, C. and Stricker, E. (1986) Oxytocin secretion in response to cholecystokinin and food, differentiation of nausea from satiety. *Science*, **232**(4756), 1417–1419.

Viglietti-Panzica, C. and Panzica, G. (1981) The hypothalamic magnocellular system in the domestic fowl. A golgi and electron microscopic study. *Cell and Tissue Research*, **215**(1), 113–131.

Wagenaar, D.A., Hamilton, M., Huang, T., Kristan, W. and French, K. (2010) A hormone-activated central pattern generator for courtship. *Current Biology*, **20**(6), 487–495. [**Shows the effect of oxytocin/vasopressin-like peptides on stereotypic mating behaviors in leeches. In addition, the authors demonstrate that administration of the oxytocin/vasopressin-like peptide conopressin onto isolated reproductive-associated ganglia (M4–M6) induces a very slow periodic fictive motor activation.**]

Wang, F., Ma, S., Xu, H., *et al.* (2013) Highefficiency system for construction and evaluation of customized TALENs for silkworm genome editing. *Molecular Genetics & Genomics*, **288**(12), 683–690.

Wang, H., Yang, H., Shivalila, C. *et al.* (2013) One-step generation of mice carrying mutations in multiple genes by CRISPR/Cas-mediated genome engineering. *Cell*, **153**(4), 910–918.

Warne, J.M., Harding, K. and Balment, R. (2002) Neurohypophysial hormones and renal function in fish and mammals. *Comparative Biochemistry & Physiology B: Biochemistry & Molecular Biology*, **132**(1), 231–237.

Warne, J.M., Bond, H., Weybourne, E., Sahajpal, V., Lu, W. and Balment, R. (2005) Altered plasma and pituitary arginine vasotocin and hypothalamic provasotocin expression in flounder (Platichthys flesus) following hypertonic challenge and distribution of vasotocin receptors within the kidney. *General and Comparative Endocrinology*, **144**(3), 240–247.

Wyart, C., del Bene, F., Warp, E., *et al.* (2009) Optogenetic dissection of a behavioural module in the vertebrate spinal cord. *Nature*, **461**(7262), 407–410.

Yamaguchi, Y., Kaiya, H., Konno, N., *et al.* (2012) The fifth neurohypophysial hormone receptor is structurally related to the V2-type receptor but functionally similar to V1-type receptors. *General and Comparative Endocrinology*, **178**(3), 519–528.

Zheng, J.J., Li, S., Zhang, X., *et al.* (2014) Oxytocin mediates early experience-dependent cross-modal plasticity in the sensory cortices. *Nature Neuroscience*, **17**(3), 391–399.

Case Studies – Integration and Translation

CHAPTER 15

Osmoregulation

David Murphy,[1,2] Jose Antunes-Rodrigues,[3] and Harold Gainer[4]

[1] School of Clinical Sciences, University of Bristol, Bristol, UK
[2] Department of Physiology, Faculty of Medicine, University of Malaya, Kuala Lumpur, Malaysia
[3] Department of Physiology, Faculty of Medicine of Ribeirão Preto, University of São Paulo, Ribeirão Preto, SP, Brazil
[4] Molecular Neuroscience Section, Laboratory of Neurochemistry, National Institutes of Health, Bethesda, Maryland, USA

15.1 Body fluid homeostasis

Water accounts for about 50–64% of body weight in humans, depending upon the developmental stage of life. Water and electrolytes are separated into two large fluid compartments: **intracellular fluid** (ICF), which accounts for 30–40% of total water content, and the **extracellular fluid** (ECF), which represents the remaining 18–24% (Figure 15.1). Other constituents of the body are dissolved in these fluid media. Osmolality is expressed as the ratio between the total amounts of solute dissolved in a determined mass of water or, in this case, dissolved in the plasma (see Box 15.1). Sodium (Na^+) is the most common extracellular solute to create an osmotic gradient, which in turn can directly control the movement of water between intracellular and extracellular compartments. Therefore, sodium content, and the content of its accompanying anions (Cl^- and HCO_3^-), determines plasma osmolality and ECF volume. The osmolality of the ICF and the ECF is maintained within narrow limits by vital adaptive mechanisms recruited to maintain homeostasis (Antunes-Rodrigues *et al.*, 2004) (see Figure 15.1).

15.1.1 Animal models of osmotic perturbation

In rats and mice, chronic systemic disturbance of fluid and electrolyte homeostasis is commonly induced by either complete fluid deprivation for up to 3 days (dehydration), which is a hyperosmotic and hypovolemic stimulus (Figure 15.2), or by the obligatory drinking of 2% (w/v) NaCl for up to 7 days (salt loading), which is a hyperosmotic stimulus that does not affect blood volume (see Figure 15.2). Acute osmotic stimulation is usually achieved by the injection into the peritoneum, or into the vasculature, of a single bolus of hypertonic saline. This elicits an immediate and rapid rise in plasma osmolality. To induce hyponatremia and hypoosmolality, rats are infused with the specific arginine vasopressin (AVP) receptor V2 agonist 1-desamino-[8-d-arginine]-AVP (dDAVP) using subcutaneously implanted osmotic minipumps for 7 days, and are simultaneously water loaded by feeding with a dilute preparation liquid formula (Verbalis, 1993) (see Figure 15.2).

15.1.2 Neuroendocrine osmoregulatory mechanisms

The neuroendocrine reflexes regulating osmolality are centered on the hypothalamo-neurohypophysial system (HNS), which consists of the large peptidergic magnocellular neurons (MCNs) of the supraoptic nucleus (SON) and the paraventricular nucleus (PVN),

---

Molecular Neuroendocrinology: From Genome to Physiology, First Edition. Edited by David Murphy and Harold Gainer.
© 2016 John Wiley & Sons, Ltd. Published 2016 by John Wiley & Sons, Ltd.
Companion website: www.wiley.com/go/murphy/neuroendocrinology

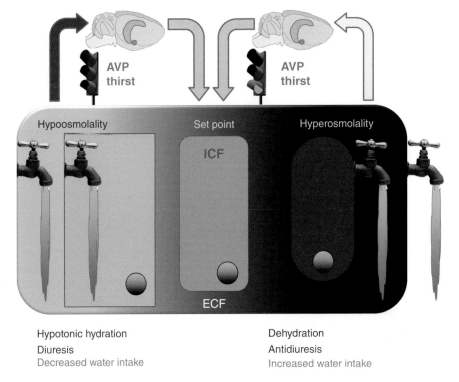

Figure 15.1 Osmotic homeostasis in mammals. The fluids of the body are divided into two compartments: the extracellular fluid (ECF) and the intracellular fluid (ICF). The maintenance of normal osmotic status (~300 mOsm/Kg H_2O) is crucial for the proper functioning of the cells of the body. Under isotonic (set point) conditions, there is no net movement of water either in or out of a cell. In this diagram, the faucets indicate the net flow of water between compartments. Thus, during dehydration, water leaves the organism (ECF) via the faucet leading to the exterior. The resulting increase in ECF hypertonicity in turn leads to an efflux of water from cells (via the faucet from the ICF to the ECF), resulting in an increase in ICF osmolality and cell shrinkage. Under hypoosmotic/hypotonic conditions, the reverse is true; water enters the body from the exterior, lowering tonicity. Thus water enters the cell, reducing osmolality and causing the cell to swell. ECF hyperosmolality and hypertonicity are detected by brain mechanisms that increase thirst and activate AVP secretion (*green traffic light*), resulting in increased water intake and decreased urine output (antidiuresis) respectively. In contrast, hypoosmolality activates brain mechanisms that reduce thirst and AVP secretion (*red traffic light*), thus decreasing water intake and increasing urine output (diuresis).

Box 15.1 Osmotic pressure

The osmotic pressure of a solution is determined by the number of colligative particles in solution, i.e. the ratio of solute particles to the number of solvent molecules in solution, and is independent of the types of molecules involved. The osmotic pressure of a solution is expressed as osmoles of solute per volume of solvent (or per kg of water) in the solution. For example, one mole of NaCl per liter of water will dissociate into two ionic particles (Na^+ and Cl^-) in water, and will therefore be approximately 2 osmol/L (however, such solutions are not ideal and the solute concentrations need to be corrected by empirically derived correction factors known as osmotic coefficients). In contrast, 1 mol per liter of sucrose in water will remain as one osmolyte in solution and therefore will equal around 1 osmol/L (and similarly will require correction by the osmotic coefficient for sucrose). In addition to osmotic pressure, other colligative properties of solutions are melting points, vapor pressure, and freezing points, and these properties are often used as surrogates in commercial 'osmometers' to determine the osmotic pressures of solutions. Solutions are considered isoosmotic if they have the same osmotic pressures, hyperosmotic if the osmotic pressure is greater, and hypoosmotic if it is lower. However, the osmotic effects of solutions on cells is dependent on the solutions' tonicity. Tonicity is not a physiochemical definition but a biological definition dependent on the permeability properties of the cell's membrane. A solution is isotonic if the cell's volume does not change when the cell is immersed in it, hypertonic if a cell shrinks in it, and hypotonic if the cell swells in it. Isoosmotic solutions are not necessarily isotonic, and can be hypotonic if the impermeant solute (e.g. NaCl) is replaced by a permeant solute (KCl) which will enter the cell and cause it to swell. This relationship between osmotic pressure and volume is governed by the classic 'ideal gas law' and when applied to living cells and physiological solutions, it is represented by $P(V-b) = $ constant, where P equals osmotic pressure, V equals the total volume of the cell, and b equals the osmotically inactive volume of the specific cell, the latter also known as the 'osmotic dead space.' The above relationship is referred to as the Boyle–Van't Hoff Law.

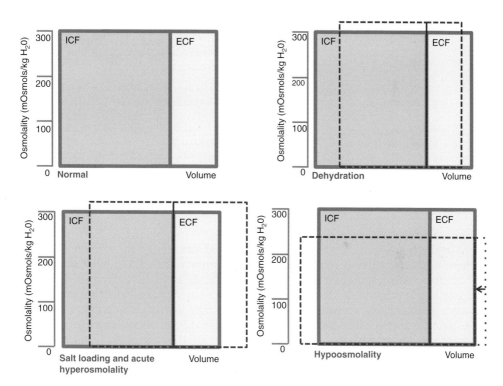

Figure 15.2 Effects of different animal models on ECF and ICF osmolality and volume (Darrow-Yannet plots). In a normal (i.e. at isotonic set point) young adult rat (~12 weeks old) weighing 225 g, total body fluid is around 153 mL, of which 102 mL is in the ICF and 51 mL is in the ECF. Both ECF and ICF have an osmolality of ~300 mOsm/kg H_2O. Chronic fluid deprivation (dehydration) is both hypovolemic (reduced ECF and ICF volume) and hyperosmotic (increased ECF and ICF osmolality). In contrast, both chronic salt loading and acute hyperosmolality resulting from intraperitoneal injection of high salt solution increase the osmolality of the ECF and ICF, but do not change total body fluid volume. The ECF draws water out of the ICF, thus increasing ICF osmolality and decreasing ICF volume while increasing ECF volume. Note that acute hyperosmolality will decrease MCN volume, but that a chronic osmotic stimulus paradoxically increases MCN volume, presumably due to increases in gene expression (see section 15.2.3). Hypoosmolality resulting from dDAVP infusion and water loading decreases the osmolality of both the ECF and ICF. Initially, the volume of both ECF and ICF increases, but eventually, reequilibration returns the ECF volume to normal (*arrow*) and the ICF volume in MCNs to below normal (see section 15.2.3).

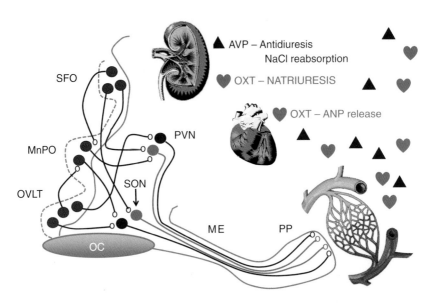

Figure 15.3 The central osmoresponsive circuit controls the secretion of AVP and OXT. AVP and OXT are synthesized in magnocellular neurons in the supraoptic nucleus (SON) and the paraventricular nucleus (PVN), the axons of which course through the internal zone of the median eminence (ME) to terminate on the blood capillaries of the posterior pituitary (PP). Increased plasma osmolality is detected by specialized osmosensitive neurons located in the circumventricular organs, the subfornical organ (SFO), the median preoptic nucleus (MnPO), and the organum vasculosum of the lamina terminalis (OVLT). These neurons project to the magnocellular neurons of the SON and PVN, thereby controlling the release of AVP (*purple triangles*) and OXT (*green hearts*) from the posterior pituitary into the bloodstream. AVP acts at the kidney to promote water and salt reabsorption, whereas OXT induces renal natriuresis and stimulates ANP release from the heart. OC, optic chiasm.

the axons of which course through the internal zone of the median eminence (ME) and terminate on blood capillaries of the posterior pituitary (PP) gland (Brownstein *et al.*, 1980) (Figure 15.3).

The SON is a homogenous collection of MCNs, while the PVN is divided into a lateral magnocellular region and a more medial subdivision of smaller parvocellular neurons. PVN parvocellular neurons mediate the stress response (Smith and Vale, 2006) via projections to the external zone of the median eminence (Vandesande *et al.*, 1977), and, through projections to the brainstem and spinal cord (Sawchenko and Swanson, 1982), are a key premotor component of the central systems that regulate blood pressure and heart rate (Coote, 2005). HNS MCNs produce two major neuroendocrine peptide products, the antidiuretic hormone arginine vasopressin (AVP) and oxytocin (OXT), which are synthesized as separate precursors encoded by highly homologous linked genes (Burbach *et al.*, 2001). These precursors are sorted, via the endoplasmic reticulum and the Golgi apparatus, into **large dense-core vesicles** (LDCVs). The precursors are processed into biologically active peptides during transport of the LDCVs to axon terminals in the PP (Burbach *et al.*, 2001). Mature peptide is stored in the PP until mobilized for secretion into the systemic circulation. Single-cell reverse transcriptase polymerase chain reaction (RT-PCR) enables AVP and OXT transcripts to be detected in the same MCN (Glasgow *et al.*, 1999; Mohr *et al.*, 1988; Xi *et al.*, 1999), and the two neuropeptide RNA levels in most of the MCNs differ by orders of magnitude (Xi *et al.*, 1999). However, a few percent of MCNs in the SON express high, equivalent levels of both peptides (Xi *et al.*, 1999), although it has been reported that the proportion of the latter MCNs increases following dehydration (Telleria-Diaz *et al.*, 2001) and during lactation (Mezey and Kiss, 1991).

An increase in plasma osmolality is a potent stimulus for the release of AVP from the PP. The initial event for osmoregulation is the detection of minimal changes in plasma osmolality by osmoreceptor sensors. Thus, a rise in plasma osmolality is detected by intrinsic MCN mechanisms (see section 15.2.1) and by extrinsic sensory systems found in specialized osmoreceptive neurons in the circumventricular organs (CVOs), such as the subfornical organ (SFO), which provide excitatory inputs to shape the firing activity of MCNs for hormone secretion (see section 15.2.2 and Figure 15.3).

Upon release, AVP travels through the bloodstream to specific receptor targets located in the kidney where it promotes water reabsorption in the kidney collecting duct, and sodium reabsorption in the thick ascending limb (TAL) of the loop of Henle (see Figure 15.3). In the presence of AVP, acting through V2 receptor activation and a subsequent increase in intracellular cAMP, the water aquaporin 2 is mobilized to the apical membrane of the collecting duct cells, increasing water permeability (Fenton *et al.*, 2013). In the TAL, AVP enhances the activity of the $Na^+-K^+-2Cl^-$ co-transporter NKCC2 through phosphorylation and membrane targeting (Markadieu and Delpire, 2013), hence promoting the reabsorption of a considerable proportion of the NaCl filtered by the glomeruli.

In addition to its well-known reproductive roles in parturition, lactation,and associated behaviors (Lee *et al.*, 2009), OXT is also active in fluid balance (see Figure 15.3). OXT secretion from the HNS is stimulated by increased plasma osmolality, promoting increased renal sodium excretion (Huang *et al.*, 1996). OXT has also been described to stimulate atrial natriuretic peptide (ANP) secretion by its direct action in the heart (Haanwinckel *et al.*, 1995).

15.2 Osmosensory mechanisms

15.2.1 Intrinsic osmosensory mechanisms

That MCNs are intrinsically osmosensitive was first shown by Mason (1980) who used intracellular microelectrodes to record from MCNs in the SON. Subsequent studies have confirmed that the firing rate of MCNs is increased or decreased under hypertonic or hypotonic conditions respectively (Figure 15.4).

Extensive studies by Bourque and colleagues showed that the excitation of MCNs in the SON by hypertonicity was due to the activation of non-selective cation channel conductances (Bourque, 1989, 2008) with consequential membrane depolarization and increased action potential firing rate. Interestingly, it was 23 years after the Mason report (Mason, 1980), that Zhang and Bourque (2003) first showed that the MCNs in the SON changed in cell volume inversely and linearly proportional to the osmotic pressures of the extracellular solutions; that is, the MCNs obeyed the Boyle–Van't Hoff Law, with an osmotic dead space of 22.4%. These authors also showed that isolated hippocampal neurons did not change in their steady-state volumes in accordance with the Boyle–Van't Hoff Law when they were exposed to the same osmotic perturbations as the MCNs. Instead, the hippocampal neurons underwent only transient volume changes and rapidly returned to their original volumes that they had in isotonic solutions. The latter transient behavior is typical of most cells, including glia and neurons, when they are exposed to hypotonic and hypertonic conditions and the process is referred to as 'volume regulation' (Fisher *et al.*, 2010; Lang *et al.*, 1998; Pedersen *et al.*, 2011; Toney, 2010). In contrast, the osmosensory MCNs produced stable volume changes in the hypertonic and hypotonic solutions.

Unfortunately, such extensive osmometric-volume experiments have only been done on the MCNs in the SON, and not on other putative osmosensory neurons in the CNS. It would be interesting to know if this resistance to volume regulation in the MCNs is a general property of all osmosensory neurons, and whether this property could be considered a criterion for being identified as an osmosensory neuron.

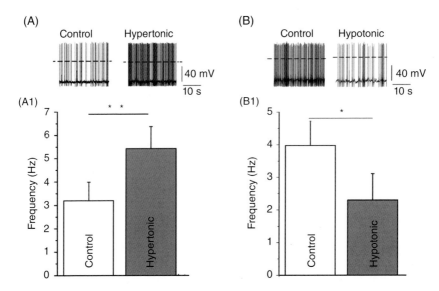

Figure 15.4 Firing rate of magnocellular neurons during osmotic challenges. Representative traces of the firing of MCNs during control conditions, and in the presence of hypertonic (330 mOsm/kg H$_2$O) or hypotonic (275 mOsm/kg H$_2$O) solutions, respectively (A,B). Note that during hypertonic conditions there is a significant increase of firing frequency of neurons (A1) and the opposite is observed in the hypotonic solution (B1). Reproduced with permission of Silva *et al.* 2013.

15.2.2 Extrinsic osmosensory mechanisms

Although MCNs are intrinsically osmosensitive, lesioning studies showed that activation of other regions in the anterior hypothalamus is essential for AVP secretion into blood, the evocation of thirst, and the increased natriuretic responses to hypertonic stimuli (Bourque, 2008; Bourque and Oliet, 1997; Bourque *et al.*, 2002; Johnson and Thunhorst, 1997), and full functional activation of the MCNs in response to osmotic challenges appears to require extrinsic synaptic input from the other osmosensitive brain regions (Bourque *et al.*, 1994; Hollis *et al.*, 2008; Richard and Bourque, 1995). There is now substantial electrophysiological and other data to support the hypothesis that the primary extrinsic osmoreceptive neurons are located in the anteroventral third ventricle (AV3V) region of the lamina terminalis, a forebrain structure that contains the SFO, median preoptic nucleus (MnPO), and the organum vasculosum of the lamina terminalis (OVLT) (see Figure 15.3). These so-called sensory CVOs are unique structures: they are devoid of a blood–brain barrier (BBB), they are rich in small, highly permeable fenestrated capillaries, they contain neuronal cell bodies with compact dendritic trees, and they receive limited neural inputs, which in most cases originate from the same areas that receive efferents (Benarroch, 2011). Not only are the neurons in the lamina terminalis osmosensitive, since they lie outside the BBB, they can integrate this information with endocrine signals borne by circulating hormones, such as angiotensin II (AII; Mimee *et al.*, 2013).

The HNS and the CVOs form an osmoresponsive circuit (Honda *et al.*, 1990; Leng, 1982; Richard and Bourque, 1995) (see Figure 15.3). The HNS receives direct input from the SFO and the OVLT, and indirect input from these via the MnPO (Miselis, 1981; Richard and Bourque, 1995). Connections back to the lamina terminalis from the HNS, perhaps by collateral excitation of interneurons, complete the circuit (Leng, 1982). AII is also a major excitatory neurotransmitter used by SFO neurons that project to the PVN (Li and Ferguson,

1993). In addition, the CVOs provide direct excitatory glutaminergic projections to the HNS (Yang *et al.*, 1994). Paradoxically, both AII (acting through AT_1 receptors) and glutamate (acting through NMDA receptors), upon exciting MCNs, activate an inhibitory feedback system (Ferguson and Latchford, 2000), which diminishes the initial stimulus. It is thought that the AII-mediated, and possibly the glutamate-mediated, increase in inhibitory postsynaptic potentials (IPSPs) in MCNs is mediated by nitric oxide (NO) through activation of soluble guanylyl cyclase (sGC) in GABAergic interneurons (Ferguson and Latchford, 2000).

15.2.3 Molecular mechanisms underlying osmosensory transduction

Pedersen *et al.* (2011) pondered a fundamental question – what is sensed by a cell when it is exposed to osmotic stress? They conclude that while this issue is not yet resolved, there are a number of possible candidate signals in eukaryotic cells. These signals may be changes in intracellular molecular crowding and ionic strength, mechanical (e.g. membrane curvature) changes, chemical changes in the membrane lipid bilayer or extracellular matrix, or changes in the intracellular cytoskeleton organization to which the membrane proteins are connected. As noted earlier, in most cells, including glia and neurons, the osmotic challenges are not presented as steady-state volume changes but instead use volume regulation mechanisms that return the cells to their original volumes in isotonic solutions after the initial volume responses to changed osmotic environments (Fisher *et al.*, 2010; Lang *et al.*, 1998; Pedersen *et al.*, 2011; Toney, 2010; Toney and Stocker, 2010).

In contrast, the only osmosensory neuron that has been studied, the MCN, does not respond to osmotic perturbations by volume regulation, but rather passively maintains the swelling in hypotonic solutions and shrinkage in hypertonic solutions (Zhang and Bourque, 2003) (see Figure 15.1) and transduces these steady-state volume changes into membrane hyperpolarization and depolarization responses, respectively. Zhang and Bourque (2003) also point out that the membrane surfaces of the MCNs, as measured by their membrane capacitances, do not change during the swelling in hypotonic solutions, and conclude that a large membrane reserve is present in the MCNs, probably due to extensive membrane infoldings. Whether this property is shared by other osmosensory neurons is unknown. Other intrinsic properties of osmosensory neurons are discussed later in this chapter.

Electrophysiological studies have shown that MCNs are depolarized when exposed to hypertonic solutions due to activation of a non-selective cationic conductance, and that this is independent of synaptic input (Bourque, 2008; Bourque and Oliet, 1997). It is proposed that this depolarization is directly connected to the decrease in the cell's volume in hypotonic solutions, due to inactivation of stretch inactivated channels (SIC) which are only partially inactivated in resting (isotonic) conditions (Bourque, 2008; Bourque and Oliet, 1997). These channels are opened by the decreased cell volume and possibly by the increased extracellular sodium in the hypertonic solutions, which produces cell depolarization (Bourque, 2008; Bourque and Oliet, 1997; Sharif Naeini *et al.*, 2006). In contrast, swelling of the cells in hypotonic solutions further inactivates the SICs and therefore hyperpolarizes the MCNs, resulting in inhibition of action potential activity. Additional experiments have indicated that it is changes in cell volume and not variations in intracellular solute concentrations that modulate the cation conductances, and concluded that the SICs are mechanosensitive channels (Bourque, 2008; Bourque and Oliet, 1997). This proposal is supported by observations that the trivalent inorganic cation gadolinium ($Gd3^+$), which is known to block many types of cationic mechanical sensor channels, is able to inhibit SIC channel conductances in the MCNs (Bourque, 2008; Bourque and Oliet, 1997). Moreover, it was shown that various neuropeptides, such as AII, cholecystokinin (CCK), and neurotensin, can directly modulate the conductances of the SIC channels in acutely isolated

MCNs (Bourque *et al.*, 2002; Chakfe and Bourque, 2000, 2001). In this regard, it has been suggested that the sensitivity of the SIC channel to osmotic stimulation changes in proportion to the submembranous actin filament density, and that AII acts, in part, by a rapid stimulation of actin polymerization, and that neurotransmitters in general act to reorganize the cytoskeleton in osmosensory neurons and in this manner modulate osmoregulatory gain (Prager-Khoutorsky and Bourque, 2010). Patch-clamp recordings from acutely isolated MCNs have shown that there may be a number of other mechanosensitive potassium channels that could be activated by cell swelling and therefore could contribute to MCN hyperpolarization under these conditions (Han *et al.*, 2003; Honore, 2007).

The above scenario which designates the mechanosensitive SIC channels as the putative mechanism of osmodetection in osmosensory neurons fits very well with observations made during acute osmotic challenges. However, as Verbalis (2007) has pointed out, the MCNs during their responses to chronic changes in tonicity undergo volume changes which are the opposite of those that would be expected from the above model. The MCNs increase substantially in volume in response to chronic hypertonicity (Armstrong *et al.*, 1977) and dramatically decrease in volume in response to chronic hypotonicity (Zhang *et al.*, 2001), while maintaining the same changes in AVP secretion rates to those MCNs experiencing acute osmotic challenges (Burbach *et al.*, 2001). While the changes in MCN volumes in response to the acute osmotic challenges reflects their behaviors as typical 'osmometers,' the reversed volume changes during chronic osmotic stimulations in the MCNs appear to instead reflect their adaptations to greatly increased (in chronic hypertonicity) and decreased (in chronic hypotonicity) commitments to the biosynthesis and secretion of the neurohypophysial peptides and related molecules under these conditions (Burbach *et al.*, 2001). Clearly, the properties and influences of the SICs must be modified during chronic osmotic stimulation, and alternative and still unknown osmoregulatory transduction mechanisms may be being called into play under chronic stimulatory conditions.

Recent studies have implicated the superfamily of transient receptor potential (TRP) channels as possible candidates for being the mechanoreceptive osmosensory receptors in osmosensory neurons (Bourque, 2008). Many subtypes of the TRP channels are blocked by gadolinium ion as are the SIC channels in MCNs and OVLT neurons, and members of the TRP vanilloid (TRPV) family of cation channels are known to evoke non-selective cationic currents with significant permeability to calcium ions, as is found in the MCNs (Liedtke, 2007; Liedtke and Kim, 2005). OVLT neurons express TRPV1 (Ciura and Bourque, 2006) and TRPV4 (Liedtke and Friedman, 2003; Liedtke *et al.*, 2000), and MCNs express TRPV1 (Sharif Naeini *et al.*, 2006) and TRPV2 (Wainwright *et al.*, 2004). Experiments on TRPV1 knockout mice showed defects in osmotically stimulated AVP release and thirst in response to hypertonicity (Bourque, 2008).

Various experiments transfecting TRPV1, TRPV2 or TRPV4 into heterologous cell lines showed that these lines became responsive to hypertonic stimulation (Bourque, 2008). It has recently been found by using gene knockout mice that hypertonicity sensing in the OVLT neurons involves TRPV1 but not TRPV4 (Ciura *et al.*, 2011), and that physical shrinking of the cells is necessary and sufficient to produce excitation in these neurons isolated from adult mice (Ciura *et al.*, 2011). In addition, it has been found that TRPV channels contribute to the osmosensory transduction of thirst, and anticipatory vasopressin release, as well as thermosensation in MCNs in mice (Sharif-Naeini *et al.*, 2008a, b; Sudbury and Bourque, 2013). All of these studies strongly support the proposal that the TRPV family of receptors, alone or in combination, are the molecular candidates for mechanoreceptors in osmosensory neurons. Much more work is still needed to fully clarify the roles of these molecules in the osmoregulatory process *in vivo*.

15.2.4   Sodium sensors in the CNS

For many years, it has been thought that the detection of extracellular sodium was an independent parameter that is integrated with osmoreceptor signals to maintain systemic water and salt homeostasis. Sodium sensors were postulated as being located in the brain (Thrasher and Keil, 1987; Thrasher *et al.*, 1980; Weisinger *et al.*, 1979), particularly in the CVOs (Cox *et al.*, 1987; Denton *et al.*, 1996), and were involved in the control of salt appetite and natriuresis (Noda, 2006, 2007). In the case of the MCNs, the influence of external sodium is believed to act through regulation of the sensitivity (gain) of SIC channels as well as changes in sodium channel gene expression (Voisin and Bourque, 2002; Voisin *et al.*, 1999).

Watanabe *et al.* (2000) discovered a novel sodium channel that was 50% homologous to voltage-gated sodium channels, and that appeared to regulate salt intake behavior. They named this channel Nax and subsequently found that it is expressed in the SFO, OVLT, median eminence, and posterior pituitary in mice (Hiyama *et al.*, 2002; Watanabe *et al.*, 2000, 2006). By performing knockout, rescue, and other experiments in mice, they concluded that this Nax channel is the long sought after brain sodium sensor for the regulation of salt intake (Noda, 2006, 2007). The Nax channel is also known as NaG/SCL11 (in rats), Nav 2.3 (in mice), and Nav2.1 (in humans), and is classified as a subfamily of voltage-sensitive sodium channels (Goldin *et al.*, 2000).

The SFO (not the OVLT) is the primary site of the sodium sensor for the control of salt intake behavior in mice (Noda, 2006, 2007). Interestingly, immunohistochemical studies show that the highest levels of Nax are found in the lamellar processes of astrocytes that often surround GABAergic neuron cell bodies in the SFO. It was proposed that the principal sodium-sensing cells in the SFO are astrocytes and that these glial cells regulate the activity of adjacent neurons by the secretion of lactate (Noda, 2007; Shimizu *et al.*, 2007).

Tremblay *et al.* (2011) described a 'sodium-leak channel,' identified as a Nax channel, that is specifically expressed in the rat MnPO (but not in the SFO) and that acts as a sodium sensor in this species. This differs from the data in the mouse where the SFO is specifically posited as the site of the sodium sensor (Watanabe *et al.*, 2000, 2006), suggesting that there is a species difference in its localization in the brain. In a more recent report from the same laboratory, it was stated that the sodium detection in the rat MnPO requires cooperative action of the Nax and the alpha1 isoform of the Na/K-ATPase (Berret *et al.*, 2013). Unfortunately, it has not yet been possible to study the detailed mechanisms of transduction of the Nax receptor since the generation of functional models by the transfection of Nax cDNAs or genes into cell lines has not yet been successful (Noda, 2007). Clearly, more experimental work will be necessary to clarify the mechanisms of transduction of the sodium sensors and the glial-neuronal interactions that lead to homeostatic sodium regulation *in vivo*.

15.3   Function-related plasticity in the HNS

Dehydration evokes a dramatic functional remodeling of the HNS, a process known as **function-related plasticity** (Hatton, 1997; Theodosis *et al.*, 1998). A plethora of activity-dependent changes in the morphology, electrical properties, and biosynthetic and secretory activity of the HNS have been described (Sharman *et al.*, 2004), all of which might contribute to the facilitation of hormone production and delivery. In terms of changes in biosynthetic activity, studies have focused on the AVP gene, which encodes the major neuroendocrine product regulating osmotic stability. It is well established that, in parallel with AVP release from the posterior pituitary, both dehydration and salt loading engender an increase in the abundance of AVP mRNAs (Burbach *et al.*, 2001). These mRNAs are presumably translated into precursor molecules that are processed into mature biologically active peptides destined to replace depleted pituitary stores. While it is known that the increase in AVP gene expression that accompanies chronic hyperosmotic stress is, at least in part, mediated at the level of transcription (Murphy and Carter, 1990), the mechanisms are not fully understood.

15.3.1   The structure and expression of the AVP gene

The neurohypophysial peptide genes are highly conserved, within and between vertebrate species, at both the structural and sequence level. The AVP and OXT genes both have three exons, and are closely linked, tail to tail, in the genomes of all mammals studied (Figure 15.5). The two genes are transcribed towards each other, from opposite strands of the DNA duplex. In neuroendocrine cells, neuropeptide biosynthesis starts with the production of a prepro-hormone, which is inserted into the lumen of the endoplasmic reticulum with the removal of the signal peptide (SP). The resulting prohormone is folded then routed unidirectionally to the *trans*-Golgi network (TGN), where it is targeted specifically into the granules of the regulated secretory pathway. The structures of the AVP and OXT prepropeptides are shown in Figure 15.5. Following the signal peptide, AVP forms the N-terminal domain of its propeptide. This is followed by the disulfide-rich neurophysin (NP) moiety, a putative intracellular carrier molecule. Neurophysin I (NP-I) is encoded by the OXT gene, whilst neurophysin II (NP-II) relates to the AVP gene. The AVP propeptide contains an additional C-terminal extension consisting of a 39 amino acid glycopeptide (or copeptin, CPP) of unknown function. The AVP and OXT precursors are synthesized in hypothalamic cell bodies. The mature peptides are transported some considerable distance down axons to storage in terminals located in the posterior pituitary, from where they are released into the general circulation in response to physiological cues. Between the AVP and OXT genes is an

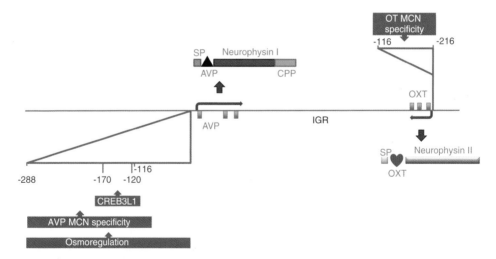

Figure 15.5 Structure, expression, and regulation of the AVP and OXT genes. The highly homologous AVP and OXT genes are closely linked on rat chromosome 3, being separated by an 11 kbp intergenic region (IGR). The AVP and OXT are transcribed towards each other (*red arrows*) from opposite strands of the DNA duplex. Each gene has three exons (*blue boxes*) and two introns. Both genes encode prepropeptides that are processed into mature, biologically active products. Following a signal peptide (SP), the nine amino acid AVP or OXT moieties form the N-terminal domains of their respective propeptides. This is followed by the disulfide-rich neurophysin (NP) moiety, a putative intracellular carrier molecule. Neurophysin I (NP-I) is encoded by the OXT gene, whilst neurophysin II (NP-II) relates to the AVP gene. The AVP propeptide contains an additional C-terminal extension consisting of a 39 amino acid glycopeptide (or copeptin, CPP) of unknown function. Using viral vector transduction strategies, sequences within the AVP gene have been mapped that mediate AVP MCN specificity, AVP gene osmoregulation (see Figure 15.6B), and the binding of the CREB3L1 transcription factor. Similarly, a region within the OXT gene promoter mediates OXT MCN specificity.

intergenic region (the IGR) postulated as being important for the cell-specific expression of both genes (Murphy and Wells, 2003; Young and Gainer, 2003).

15.3.2 Search for the osmoregulatory element in the AVP gene

15.3.2.1 Germline transgenesis
Germline transgenesis in mice and rats has been applied to AVP gene constructs in order to identify genomic elements that are necessary and sufficient to elicit expression in appropriate neurons in the brain and, within these cells, to drive correct physiological regulation. Murphy and colleagues (Davies *et al.*, 2003; Waller *et al.*, 1996; Zeng *et al.*, 1994) generated a series of rat lines carrying transgenes consisting of the AVP structural gene containing chloramphenicol acetyl transferase (CAT) reporter sequences inserted into exon III (Figure 15.6A).

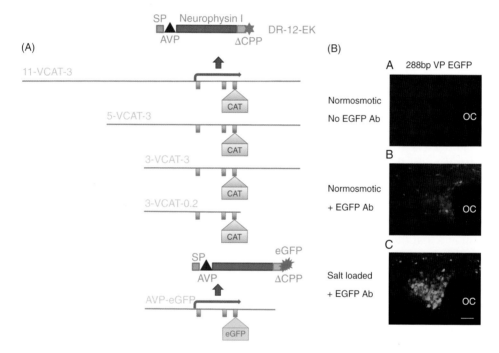

Figure 15.6 Transgenic rat studies on AVP gene osmoregulation. (A) Transgenic rats have been used to study the cell-specific and physiological regulation of the AVP gene. Transgenes consisted of the AVP structural gene containing chloramphenicol acetyl transferase (CAT) reporter sequences inserted into exon III. This transgene is translated into a modified precursor in which the C-terminal CPP sequences are truncated (ΔCPP) and replaced with a novel 16 amino acid epitope (DR-12-EK). This modified, tagged AVP transcription unit (VCAT) was flanked by differing lengths of 5′ and 3′ upstream and downstream contiguous genomic sequences, that contain presumptive regulatory sequences (Zeng *et al.*, 1994). Thus, transgene 11-VCAT-3 consisted of 11 kbp of upstream and 3 kbp of downstream sequences. In transgene 5-VCAT-3, the VCAT gene was flanked by 5 kbp of upstream and 3 kbp of downstream sequences. Transgene 3-VCAT-3 had 3 kbp of upstream and 3 kbp of downstream sequences and, finally, transgene 3-VCAT-0.2 had 3 kbp of upstream, but only 0.2 kbp of downstream sequences. Similarly, Ueta *et al.* (2005) fused ΔCPP with eGFP. In this case, in the AVP-eGFP transgene, this modified precursor fusion is flanked by 3 kbp of upstream and 2 kbp of downstream sequences. (B) AAV transduction was used to localize an osmotic response element in the -288 domain in the AVP gene promoter (Ponzio *et al.*, 2012). In SON transduced with a virus bearing a transgene made up of eGFP coding sequences under the control of 288 bp of AVP proximal promoter sequences, no direct fluorescence is detectable under normosmotic conditions, although an antibody staining is able to detect the low-level presence of the protein. This staining markedly increases with salt loading.

This modified CAT-tagged AVP transcription unit (VCAT) was flanked by differing lengths of 5′ and 3′ upstream and downstream contiguous genomic sequences, that contain presumptive regulatory sequences. Thus, transgene 11-VCAT-3 consisted of 11 kbp of upstream and 3 kbp of downstream sequences. In transgene 5-VCAT-3, the VCAT gene was flanked by 5 kbp of upstream and 3 kbp of downstream sequences. Transgene 3-VCAT-3 had 3 kbp of upstream and 3 kbp of downstream sequences and, finally, transgene 3-VCAT-0.2 had 3 kbp of upstream, but only 0.2 kbp of downstream sequences.

The incorporation of CAT sequences provided a unique nucleic acid reporter of RNA expression, when used as a Northern blot or *in situ* hybridization probe. Further, the insertion of CAT sequences into exon III placed a unique hexadecapeptide (known as DR-12-EK) at the C-terminus of the AVP precursor. Specific antibodies raised against DR-12-EK could be used to distinguish endogenous and transgene-derived AVP gene products using radioimmunoassay or immunocytochemistry. All four transgenes elicited the same expression patterns: low basal expression was seen in AVP MCNs of the SON and PVN, but expression was markedly increased following dehydration, suggesting that the sequences responsible for the physiological regulation of the AVP gene reside within the smallest 3-VCAT-0.2 construct studied. However, the response of all four transgenes to osmotic cues was greatly exaggerated compared to the endogenous AVP gene. Whilst endogenous AVP mRNAs show a doubling in abundance following salt loading or dehydration, 5-VCAT-3 transcripts increased by an order of magnitude up to 20-fold. This suggests that the transgenes lacked silencer or quencher elements that normally attenuate the response of the AVP gene to an osmotic cue. Similar transgenic experiments in mice using an intact CAT gene reporter inserted into exon III in mouse AVP and OXT genes supported the above observations and strategy and showed that the peptides that were expressed were targeted to the secretory pathway in the MCNs (Jeong *et al.*, 2001).

The unique epitope tag (DR-12-EK) on the 5-VCAT-3 peptide product enabled an examination of the effect of a physiological stimulus on transgene RNA translation, and subsequent precursor processing, transport, storage, and release (Waller *et al.*, 1996). Indeed, it was shown that the DR-12-EK tag followed the same processing and transport pathway as the endogenous AVP gene products. Whilst an osmotic stimulus was shown to increase hypothalamic DR-12-EK levels, in parallel with transgene RNA levels, changes in posterior pituitary DR-12-EK levels were more complex. After 5 days of salt loading, DR-12-EK levels fell, as would be expected if its release were coordinate with that of AVP. However, after 10 days of salt loading, posterior pituitary DR-12-EK levels increased, despite the lower level of AVP. This probably reflects the greater response of the transgene to osmotic challenge at the RNA level, increasing the proportion of DR-12-EK-containing translation products transported to the posterior pituitary relative to those derived from the endogenous gene. These observations are further evidence in support of models of neurohypophysial homeostasis that suggest that pituitary AVP peptide levels passively reflect changes in hormone release and synthesis (Fitzsimmons *et al.*, 1994), and that the availability of mRNA is the primary determinant of pituitary AVP content in the basal state (Fitzsimmons *et al.*, 1992).

In a particularly elegant series of experiments, Ueta and co-workers modified the AVP structural gene to incorporate the enhanced form of green fluorescent protein (eGFP) into exon III, again generating a unique fusion protein that has the particular advantage of being readily detectable by fluorescent microscopy (Ueta *et al.*, 2005). Driven by regulatory elements contained within 3 kbp of upstream and 2 kbp of downstream flanking genomic sequences, the AVP-eGFP fusion transgene (see Figure 15.6) is robustly expressed in SON and PVN MCNs. Further, the fusion protein is transported to the posterior pituitary, in parallel with endogenous AVP. The transgene is also subject to exaggerated upregulation by osmotic cues (Fujio *et al.*, 2006). This revolutionary strategy greatly simplifies the process of phenotyping an AVP neuron in a tissue section, or in dissociated cultures, enabling electrophysiological measurements on a specific, identified cell type. As such, thanks to the generosity of the Ueta group, these rats are now routinely used in electrophysiology labs throughout the world. Further, using fiber-optic probes ('optrodes'), it has been possible to directly monitor, in real time, the release of eGFP, as a surrogate for endogenous AVP, from the posterior pituitary of anesthetized rats subjected to an osmotic stimulus (Yao *et al.*, 2011).

The Ueta team have gone on to generate further transgenic rat lines that enable OXT neurons to be visualized by virtue of their expression of an OXT-cyan fluorescent protein (CFP) fusion transgene (Katoh *et al.*, 2010) or an OXT-monomeric red fluorescent protein (mRFP) fusion transgene (Katoh *et al.*, 2011). In double transgenic rats that express both the AVP-eGFP gene and the OXT-mRFP gene, separate populations of green (AVP) and red (OXT) MCNs were clearly observed in the same hypothalamic sections (Katoh *et al.*, 2011). Thus, in the same animal, it is possible to identify AVP and OXT neurons in the hypothalamus by virtue of their different fluorescent characteristics. This is a major boon to scientists studying the neuroanatomy and electrophysiology of these neurons.

Importantly, the Ueta group have also generated transgenic rat lines that express mRFP (Fujihara *et al.*, 2009) or eGFP (Katoh *et al.*, 2014) fusion proteins under the control of regulatory sequences of the c-fos gene. Immediate early genes (IEGs) such as the c-fos gene are transcriptional regulators that have been used routinely as markers of neuronal activity (Kovács, 2008). Similarly, in rats expressing c-fos-eGFP or c-fos-mRFP fusion transgenes, fluorescence can be observed in the nuclei of MCNs in the SON and the PVN after an appropriate stimulus. In double transgenic rats that express both the AVP-eGFP gene and the c-fos-mRFP gene, it was shown that eGFP-positive (AVP) neurons expressed nuclear red fluorescence 90 min after the acute osmotic challenge of an intraperitoneal injection of hypertonic saline (Fujihara *et al.*, 2009). Similarly, in double transgenic rats that express both the OXT-mRFP gene and the c-fos-eGFP gene, it was shown that mRFP-positive (OXT) neurons expressed nuclear green fluorescence 90 min after intraperitoneal administration of CCK-8, which selectively activates OXT-secreting neurons in the hypothalamus (Katoh *et al.*, 2014). Thus, activated neurons in the HNS, defined on the basis of expressing a fluorescent derivative of c-fos, can be identified by their expression of a different fluorescent protein controlled by either AVP or OXT genomic sequences.

Although not directly related to osmoregulation, it should be noted that similar germline transgenesis studies have been done on the expression of the OXT gene in the HNS (reviewed in Murphy and Wells, 2003; Young and Gainer, 2003) and functional green fluorescent protein targeting to oxytocin MCNs by this approach has been achieved (Zhang *et al.*, 2002).

15.3.2.2   Somatic transgenesis

Whilst **somatic transgenesis** using viral gene transfer vectors is routinely used to deliver new genes to the HNS, until recently it has not been possible to successfully use this to elicit specific expression in AVP MCNs. Neither adenoviral nor lentiviral vectors are permissive for the expression of transgenes controlled by AVP gene regulatory sequences (our unpublished results). The recent adoption of adeno-associated virus (AAV) vectors was the breakthrough that will open the door to many new studies in the area of HNS regulation and function. It was found that a 2 kbp AVP promoter functions efficiently within the AAV backbone, and when delivered to the SON or PVN via stereotaxic surgery, can be used to elicit the specific expression of reporter genes, or functional molecules, to AVP MCNs (Knobloch *et al.*, 2012; Ponzio *et al.*, 2012). Further, the 2 kbp AVP promoter segment is able to respond to the physiological cues in this context; the expression of an eGFP reporter is increased by salt loading in the SON.

Promoter deletion constructs were then used to map the sequences within this promoter that are responsible for both the cell type-specific expression of the vasopressin gene and its physiological regulation (Ponzio *et al.*, 2012). Three major enhancer domains were identified, located at -2 kbp to -1.5 kbp, -1.5 bp to -950 bp, and -950 bp to -543 bp. As all three of these constructs showed expression in AVP, but not OXT, MCNs, sequences responsible for this cell type specificity must reside in the proximal promoter region between the start of transcription and position -543. Further mapping showed that this specificity was maintained when eGFP was expressed under the control of a 288 bp proximal promoter segment, but was lost when this was further reduced to -116 bp.

Thus, the promoter region between -288 and -116 appears to contain an element or elements that direct expression to AVP neurons and/or silence transcriptional activity in OXT cells (see Figure 15.5). Further, the promoter region responding to osmotic cues was identified as residing somewhere in the proximal 288 bp region upstream of the start of transcription (see Figures 15.5 and 15.6B). Similar studies using AAV-mediated delivery of eGFP to the SON under the control of the OXT promoter revealed that the key elements that regulate cell type-specific expression are located -216 to -100 bp upstream of the transcription start site (Fields *et al.*, 2012) (see Figure 15.5).

15.3.3   Signaling pathways and transcription factors regulating the osmotic upregulation of AVP gene expression

Hypothesis-driven studies have sought to identify the intracellular signaling systems and transcription factors that mediate the upregulation of AVP gene expression following an osmotic stimulus.

It is known that cAMP levels increase in the rat SON following salt loading, and it has been suggested that this might stimulate AVP gene expression (Carter and Murphy, 1989; Ceding *et al.*, 1990) through the activation of the protein kinase A (PKA) pathway and subsequent phosphorylation of the cAMP response element binding protein (CREB) transcription factor which, in turn, will activate the expression of target genes (Shiromani *et al.*, 1995). **Reporter assays** certainly indicated that CREB positively regulates AVP transcription *in vitro* (Iwasaki *et al.*, 1997). However, in recent *in vivo* studies, injection of a recombinant adeno-associated virus expressing a dominant negative mutant form of CREB into rat SON failed to significantly reduce AVP mRNA expression after acute hyperosmotic stimulation (Lubelski *et al.*, 2012). Thus, clear *in vivo* evidence for a role for CREB in AVP gene transcription is still lacking.

Similarly, the observation that there is an increase in the expression of both Fos and Jun AP1 transcription factor family members in the SON and PVN following both acute and chronic osmotic stimulation has implicated these immediate early genes in AVP gene regulation (Luckman *et al.*, 1996; Yao *et al.*, 2012). The Fos/Jun family of transcription factors is part of the basic leucine zipper (bZIP) superfamily and bind to the AP1cis-acting site on target genes (Shaulian and Karin, 2002). To act as a functional AP1 transcription factor, a Fos family member must heterodimerize with a member of the Jun family. c-Fos protein is associated with actively transcribing chromatin regions in the cell nuclei of SON neurons (Lafarga *et al.*, 1993), and protein extracts of the SON or PVN have increased binding to the consensus AP1 sequence following osmotic stimulation (Borsook *et al.*, 1994; Carrion *et al.*, 1998; Ying *et al.*, 1996). It is also possible that AP1 transcription factors are mediating a cAMP response. The bovine AVP gene is strongly responsive to cAMP agonists (Pardy *et al.*, 1992), and a CRE, which binds to members of the AP1 transcription factor family, is located between -120 to -112 within the proximal promoter. **DNase I footprinting** with SON extracts of salt-loaded and euhydrated rats showed binding to this region, indicating that this element may play a role in the response of the VP gene to changes in water balance (Pardy *et al.*, 1992).

15.3.4   Transcriptomic studies

Hypothesis-driven approaches were unable to identify a principal transcription factor or factors that might mediate either basal or osmotically stimulated AVP expression in SON and PVN MCNs. A fresh strategy was needed, and a number of groups thus adopted a global gene expression profiling approach, with a view to identifying novel regulatory molecules in the HNS. The availability of genome information and the parallel development of transcriptomic technologies provided the means to answer two crucial questions.

- What genes are expressed in the different components of the HNS, namely the PVN, the SON, and the neurointermediate lobe of the pituitary gland?
- What genes are up- or downregulated in these same structures following osmotic stimulation?

Initial studies analyzed the transcriptomic response of the SON to 3 days of dehydration using a small cDNA microarray (the NIA Neurorray) bearing only 1152 human gene features (Ghorbel *et al.*, 2003). Despite the limitations of this approach, interleukin 6 was identified as a novel secretory product of the HNS (Ghorbel *et al.*, 2003). Subsequent studies combined cDNA microarray technology with **suppression subtractive hybridization-polymerase chain reaction** (SSH-PCR) (Ghorbel and Murphy, 2011). SSH-PCR enables the generation of a library of enriched cDNAs corresponding to putative differentially expressed mRNAs. To rapidly screen this library, 1152 of these clones were subjected to custom microarray analysis. This resulted in the identification of 459 putative differentially expressed clones, 56 of which were sequenced (Ghorbel *et al.*, 2006). Four of these clones were then subject to validation using *in situ* hybridization (Ghorbel *et al.*, 2006). Three of these clones are expressed sequence tags (ESTs) that correspond to transcripts that map to the rat genome, but still have no known function (GenBank accession numbers CA865428, CA865430, and CA865432). A fourth clone corresponded to the Ncoa7 (nuclear receptor coactivator 7) gene, which appears to be downregulated in the SON by dehydration (Ghorbel *et al.*, 2006). The Ncoa7 gene product binds to a number of nuclear hormone receptors, including estrogen receptor (ER)α and ERβ, thyroid hormone receptor (TR)β, peroxisome proliferator-activated receptor (PPAR)γ, and retinoic acid receptor (RAR)α, thereby enhancing transcriptional activity (Shao *et al.*, 2002). ERβ is known to be expressed in the SON where levels correlate inversely with dehydration (Sladek and Somponpun, 2008). ERβ inhibits AVP secretion (Sladek and Somponpun, 2004), thus its coordinate downregulation with Ncoa7 following dehydration might contribute to enhanced AVP release.

It was only with the application of the Affymetrix oligonucleotide array platform that a comprehensive catalog of the SON and PVN transcriptomes emerged, as well as descriptions of the extensive changes in the expression of many mRNAs that occur therein as a consequence of dehydration (Hindmarch *et al.*, 2006; Qiu *et al.*, 2007; Yue *et al.*, 2006). Comparable microarray studies were done to evaluate the global changes in mRNAs in the rat SON under conditions of hypoosmolality (Mutsuga *et al.*, 2004, 2005). Although the latter studies employed a different array platform, the results were in excellent agreement with the afore-mentioned studies that used the Affymetrix oligonucleotide array platform. Interestingly, many genes are contraregulated under hyperosmotic or hypoosmotic conditions.

These studies revealed that the expression of a number of novel transcription factors is modulated, either up or down, by an osmotic stimulus. These transcription factors are potentially mediating aspects of the overall transcriptome changes that occur following dehydration, including the upregulation of the AVP mRNA. In this context, it has recently been shown that the transcription factor cAMP responsive element-binding protein-3 like 1(CREB3L1) is a novel regulator of AVP gene transcription in the hypothalamus. Transcriptome analysis revealed that the expression of CREB3L1 transcripts is robustly increased by dehydration (Hindmarch *et al.*, 2006), an observation confirmed by *in situ* hybridization (Qiu *et al.*, 2007) and quantitative PCR (Greenwood *et al.*, 2014). Subsequent studies implicated CREB3L1 in the regulation of AVP gene expression (Greenwood *et al.*, 2014). In the euhydrated animal, CREB3L1 is expressed in glial cells, and at a low level in AVP and OXT neurons, wherein the protein is confined to the endoplasmic reticulum and the Golgi apparatus. Following dehydration, CREB3L1 protein levels increase markedly, but only in AVP neurons. Concomitantly, the membrane resident CREB3L1 precursor protein is cleaved, and the N-terminal transcription factor domain migrates to the nucleus via the cytosol. These data suggested that CREB3L1 might be able to regulate AVP gene expression, and this hypothesis was tested, firstly, *in vitro*, by '**promoter bashing**' and **chromatin immunoprecipitation** (ChIP), where it was shown that a constitutively active CREB3L1 (CREB3L1CA) could robustly activate the expression of AVP promoter reporter constructs through a direct interaction with sequences between -170 and -120 bp upstream of the start of transcription (see Figure 15.5). *In vivo*, ChIP showed that CREB3L1 directly interacts with the AVP promoter, and lentiviral-mediated expression of CREB3L1CA in the SON and PVN increased AVP biosynthesis.

Without the unbiased global approach afforded by transcriptome analysis, CREB3L1 would never have been identified as a novel transcriptional regulator of AVP gene expression. Although related to CREB, there is no evidence that CREB3L1 is directly regulated by cAMP pathways. Thus, it remains to be determined how CREB3L1 is itself regulated; how it responds to external cues and the signaling pathways involved are not yet known.

15.3.5 Closing the loop

It has been hypothesized that the changes in transcription in the SON following dehydration are mediated by transcription factors. This results in a changed transcriptome and, in turn, an altered proteome. Part of the proteome consists of transcription factors that feed back onto the genome to appropriately modulate the constitution of the transcriptome. In order to 'close the loop' (Figure 15.7), Affymetrix Combo **protein-DNA array** analysis was used to identify

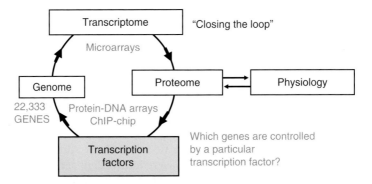

Figure 15.7 'Closing the loop.' The estimated 22,333 protein coding genes are expressed in a spatially and temporally specific fashion. Thus every cell, at a particular moment in time, has a particular defined transcriptome. The transcriptome is, in turn, translated into the proteome, which bidirectionally interacts with the external environment to mediate physiological outcomes. A proportion of the transcriptome is comprised on transcription factors, which interact with the genome to control transcription and hence the composition of the transcriptome. Thus, in order to understand this cycle, it is first necessary to define the transcription factor complement of a cell. This can be achieved using protein-DNA arrays. For a particular novel transcription factor, ChIP-chip can be used to identify specific interactions with the genome. By comparing this information with transcriptome data, it is possible to identify genes that might be regulated at the transcriptional level by that particular *trans*-acting factor.

which transcription factors might be involved in changing patterns of gene expression in the dehydrated SON (Qiu *et al.*, 2014). Nuclear extracts of SON from dehydrated and control male rats were analyzed for binding to the 345 consensus DNA transcription factor binding sequences of the array. Statistical analysis revealed significant changes in binding to 26 consensus elements, and **electrophoretic mobility shift assays (EMSA)** confirmed increased binding to Stat1/Stat3, c-Myc-Max, and Pbx1 sequences following dehydration. Focusing on c-Myc and Max, quantitative PCR was used to confirm previous transcriptomic analysis that had suggested an increase in c-Myc, but not Max, mRNA levels in the SON following dehydration (Qiu *et al.*, 2007), and c-Myc- and Max-like immunoreactivities were identified in SON AVP-expressing cells. Finally, by comparing new data obtained from Roche-NimbleGen ChIP arrays (**ChIP-chip**) with previously published transcriptomic data, putative c-Myc target genes were identified whose expression changes in the SON following dehydration. These include known c-Myc targets such as the Slc7a5 gene, which encodes the L-type amino acid transporter 1 (LAT1), ribosomal protein L24 (Rpl24), histone deactylase 2 (Hdac2), and the Ras-related nuclear GTPase (Ran). Thus, for the first time, a transcriptional circuit in the physiologically activated SON has been described.

15.4 Perspectives

In this chapter, we have described the steady progress in recent years that has resulted in an expansion in our knowledge of two key aspects of the neuroendocrinology of mammalian osmoregulation. Firstly, we are starting to understand, in molecular detail, how the brain senses tiny changes in the osmolality of the body fluids. Secondly, thanks to the application of transcriptomic technologies, we now have a comprehensive catalog of gene expression in key brain osmoregulatory centers, and we know how that transcript population changes following an osmotic challenge. What is still missing is an appreciation of the mechanics of the processes that link osmosensation to transcriptional modulation within the MCN. It is to be anticipated that transcriptomic and, perhaps more importantly, proteomic data sets will be mined to reveal possible candidates. In turn, the hypotheses that emerge from 'omic studies will need to be tested *in vivo*. In this regard, viral gene transfer vectors that enable targeted gene expression in specific neuronal cell types will be an invaluable tool.

Acknowledgments

JAR wishes to thank Drs Lucila L.K. Elias, Silvia G. Ruginsk, and André de Souza Mecawi for their important suggestions during the elaboration of this text.

References

Antunes-Rodrigues, J., de Castro, M., Elias, L.L., Valença, M.M. and McCann, S.M. (2004) Neuroendocrine control of body fluid metabolism. *Physiological Reviews*, **84**, 169–208.

Armstrong, W.E., Gregory, W.A. and Hatton, G.I. (1977) Nucleolar proliferation and cell size changes in rat supraoptic neurons following osmotic and volemic challenges. *Brain Research Bulletin*, **2**, 7–14.

Benarroch, E.E. (2011) Circumventricular organs: receptive and homeostatic functions and clinical implications. *Neurology*, **77**, 1198–204.

Berret, E., Nehme, B., Henry, M., Toth, K., Drolet, G. and Mouginot, D. (2013) Regulation of central Na+ detection requires the cooperative action of the NaX channel and alpha1 Isoform of Na+/K+–ATPase in the Na+–sensor neuronal population. *Journal of Neuroscience*, **33**, 3067–3078.

Borsook, D., Konradi, C., Falkowski, O., Comb, M. and Hyman, S.E. (1994) Molecular mechanisms of stress-induced proenkephalin gene regulation: CREB interacts with the proenkephalin gene in the mouse hypothalamus and is phosphorylated in response to hyperosmolar stress. *Molecular Endocrinology*, **8**, 240–248.

Bourque, C.W. (1989) Ionic basis for the intrinsic activation of rat supraoptic neurones by hyperosmotic stimuli. *Journal of Physiology*, **417**, 263–277.

Bourque, C.W. (2008) Central mechanisms of osmosensation and systemic osmoregulation. *Nature Reviews Neuroscience*, **9**, 519–531. **[Excellent review of osmoregulation and related mechanisms in the central nervous system. Hypertonicity activates of non-selective cation channel conductances involved in MCN activation, as well as the changes in sodium channel gene expression.]**

Bourque, C.W. and Oliet, S.H. (1997) Osmoreceptors in the central nervous system. *Annual Review of Physiology*, **59**, 601–619.

Bourque, C.W., Oliet, S.H. and Richard, D. (1994) Osmoreceptors, osmoreception, and osmoregulation. *Frontiers in Neuroendocrinology*, **15**, 231–274.

Bourque, C.W., Voisin, D.L. and Chakfe, Y. (2002) Stretch-inactivated cation channels: cellular targets for modulation of osmosensitivity in supraoptic neurons. *Progress in Brain Research*, **139**, 85–94.

Brownstein, M.J., Russell, J.T. and Gainer, H. (1980) Synthesis, transport, and release of posterior pituitary hormones. *Science*, **207**, 373–378. **[The classic description of the mechanisms by which AVP and OXT are made and released by HNS neurons. Pulse labeling with radioactive amino acid was used to show that the two hormones and their respective neurophysin carrier proteins are synthesized as parts of separate precursor proteins. These precursors are then processed into smaller, biologically active molecules while they are being transported along the axon towards the posterior pituitary.]**

Burbach, J.P., Luckman, S.M., Murphy, D. and Gainer, H. (2001) Gene regulation in the magnocellular hypothalamo-neurohypophysial system. *Physiological Reviews*, **81**, 1197–1267.

Carrion, A.M., Mellstrom, B., Luckman, S.M. and Naranjo, J.R. (1998) Stimulus-specific hierarchy of enhancer elements within the rat prodynorphin promoter. *Journal of Neurochemistry*, **70**, 914–921.

Carter, D.A. and Murphy, D. (1989) Cyclic nucleotide dynamics in the rat hypothalamus during osmotic stimulation: in vivo and in vitro studies. *Brain Research*, **487**, 350–356.

Ceding, P., Schilling, K. and Schmale, H. (1990) Vasopressin expression in cultured neurons is stimulated by cyclic AMP. *Journal of Neuroendocrinology*, **2**, 859–865.

Chakfe, Y. and Bourque, C.W. (2000) Excitatory peptides and osmotic pressure modulate mechanosensitive cation channels in concert. *Nature Neuroscience*, **3**, 572–579.

Chakfe, Y. and Bourque, C.W. (2001) Peptidergic excitation of supraoptic nucleus neurons: involvement of stretch-inactivated cation channels. *Experimental Neurology*, **171**, 210–218.

Ciura, S. and Bourque, C.W. (2006) Transient receptor potential vanilloid 1 is required for intrinsic osmoreception in organum vasculosum lamina terminalis neurons and for normal thirst responses to systemic hyperosmolality. *Journal of Neuroscience*, **26**, 9069–9075.

Ciura, S., Liedtke, W. and Bourque, C.W. (2011) Hypertonicity sensing in organum vasculosum lamina terminalis neurons: a mechanical process involving TRPV1 but not TRPV4. *Journal of Neuroscience*, **31**, 14669–14676.

Coote, J.H. (2005) A role for the paraventricular nucleus of the hypothalamus in the autonomic control of heart and kidney. *Experimental Physiology*, **90**, 169–173.

Cox, P.S., Denton, D.A., Mouw, D.R. and Tarjan, E. (1987) Natriuresis induced by localized perfusion within the third cerebral ventricle of sheep. *American Journal of Physiology*, **252**, R1–6.

Davies, J., Waller, S., Zeng, Q., Wells, S. and Murphy, D. (2003) Further delineation of the sequences required for the expression and physiological regulation of the vasopressin gene in transgenic rat hypothalamic magnocellular neurons. *Journal of Neuroendocrinology*, **15**, 42–50.

Denton, D.A., McKinley, M.J. and Weisinger, R.S. (1996) Hypothalamic integration of body fluid regulation. *Proceedings of the National Academy of Sciences USA*, **93**, 7397–7404.

Fenton, R.A., Pedersen, C.N. and Moeller, H.B. (2013) New insights into regulated aquaporin-2 function. *Current Opinion in Nephrology & Hypertension*, **22**, 551–558.

Ferguson, A.V. and Latchford, K.J. (2000) Local circuitry regulates the excitability of rat neurohypophysial neurones. *Experimental Physiology*, **85 Spec No**, 153S–161S.

Fields, R.L., Ponzio, T.A., Kawasaki, M. and Gainer, H. (2012) Cell-type specific oxytocin gene expression from AAV delivered promoter deletion constructs into the rat supraoptic nucleus in vivo. *PLoS One*, **7**, e32085.

Fisher, S.K., Heacock, A.M., Keep, R.F. and Foster, D.J. (2010) Receptor regulation of osmolyte homeostasis in neural cells. *Journal of Physiology*, **588**, 3355–3364.

Fitzsimmons, M.D., Roberts, M.M., Sherman, T.G. and Robinson, A.G. (1992) Models of neurohypophyseal homeostasis. *American Journal of Physiology*, **262**, R1121–1130.

Fitzsimmons, M.D., Roberts, M.M. and Robinson, A.G. (1994) Control of posterior pituitary vasopressin content: implications for the regulation of the vasopressin gene. *Endocrinology*, **134**, 1874–1878.

Fujihara, H., Ueta, Y., Suzuki, H., *et al.* (2009) Robust up-regulation of nuclear red fluorescent-tagged fos marks neuronal activation in green fluorescent vasopressin neurons after osmotic stimulation in a double-transgenic rat. *Endocrinology*, **150**, 5633–5638.

Fujio, T., Fujihara, H., Shibata, M., *et al.* (2006) Exaggerated response of arginine vasopressin-enhanced green fluorescent protein fusion gene to salt loading without disturbance of body fluid homeostasis in rats. *Journal of Neuroendocrinology*, **18**, 776–85.

Ghorbel, M.T. and Murphy, D. (2011) Suppression subtractive hybridization. *Methods in Molecular Biology*, **789**, 237–259.

Ghorbel, M.T., Sharman, G., Leroux, M., *et al.* (2003) Microarray analysis reveals interleukin-6 as a novel secretory product of the hypothalamo-neurohypophyseal system. *Journal of Biological Chemistry*, **278**, 19280–19285.

Ghorbel, M.T., Sharman, G., Hindmarch, C., Becker, K.G., Barrett, T. and Murphy, D. (2006) Microarray screening of suppression subtractive hybridization-PCR cDNA libraries identifies novel RNAs regulated by dehydration in the rat supraoptic nucleus. *Physiological Genomics*, **24**, 163–172.

Glasgow, E., Kusano, K., Chin, H., Mezey, E., Young, W.S. III and Gainer, H. (1999) Single cell reverse transcription-polymerase chain reaction analysis of rat supraoptic magnocellular neurons: neuropeptide phenotypes and high voltage-gated calcium channel subtypes. *Endocrinology*, **140**, 5391–5401.

Goldin, A.L., Barchi, R.L., Caldwell, J.H., *et al.* (2000) Nomenclature of voltage-gated sodium channels. *Neuron*, **28**, 365–368.

Greenwood, M., Bordieri, L., Greenwood, M.P., *et al.* (2014) Transcription factor CREB3L1 regulates vasopressin gene expression in the rat hypothalamus. *Journal of Neuroscience*, **34**, 3810–3820. [***Transcriptome analysis identified transcription factor CREB3L1 as being upregulated in the SON and PVN by dehydration. This study demonstrated that CREBL1 can regulate the AVP gene through a direct interaction with sequences in the proximal promoter.***]

Haanwinckel, M.A., Elias, L.K., Favaretto, A.L.V., *et al.* (1995) Oxytocin mediates atrial natriuretic peptide release and natriuresis after volume expansion in the rat. *Proceedings of the National Academy of Sciences USA*, **92**, 7902–7906.

Han, J., Gnatenco, C., Sladek, C.D. and Kim, D. (2003) Background and tandem-pore potassium channels in magnocellular neurosecretory cells of the rat supraoptic nucleus. *Journal of Physiology*, **546**, 625–639.

Hatton, G.I. (1997) Function-related plasticity in hypothalamus. *Annual Review of Neuroscience*, **20**, 375–379.

Hindmarch, C., Yao, S., Beighton, G., Paton, J. and Murphy, D. (2006) A comprehensive description of the transcriptome of the hypothalamoneurohypophyseal system in euhydrated and dehydrated rats. *Proceedings of the National Academy of Sciences USA*, **103**, 1609–1614. [***It does exactly what it says on the tin! Affymetrix microarray analysis of the SON, PVN, and neurointermediate lobe of the pituitary in euhydrated and dehydrated rats identified many novel differentially expressed genes.***]

Hiyama, T.Y., Watanabe, E., Ono, K., *et al.* (2002) Na(x) channel involved in CNS sodium-level sensing. *Nature Neuroscience*, **5**, 511–512.

Hollis, J.H., McKinley, M.J., D'Souza, M., Kampe, J. and Oldfield, B.J,. (2008) The trajectory of sensory pathways from the lamina terminalis to the insular and cingulate cortex: a neuroanatomical framework for the generation of thirst. *American Journal of Physiology; Regulatory, Integrative and Comparative Physiology*, **294**, R1390–1401.

Honda, K., Negoro, H., Dyball, R.E., Higuchi, T. and Takano, S. (1990) The osmoreceptor complex in the rat: evidence for interactions between the supraoptic and other diencephalic nuclei. *Journal of Physiology*, **431**, 225–241.

Honore, E. (2007) The neuronal background K2P channels: focus on TREK1. *Nature Reviews Neuroscience*, **8**, 251–261.

Huang, W., Lee, S.L., Arnason, S.S. and Sjöquist, M. (1996) Dehydration natriuresis in male rats is mediated by oxytocin. *American Journal of Physiology*, **270**, R427–433.

Iwasaki, Y., Oiso, Y., Saito, H. and Majzoub, J.A. (1997) Positive and negative regulation of the rat vasopressin gene promoter. *Endocrinology*, **138**, 5266–5274.

Jeong, S.W., Castel, M., Zhang, B.J., *et al.* (2001) Cell-specific expression and subcellular localization of CAT-reporter protein expressed from oxytocin & vasopressin gene promoter-driven constructs in transgenic mice. *Experimental Neurology*, **171**, 255–271.

Johnson, A.K. and Thunhorst, R.L. (1997) The neuroendocrinology of thirst and salt appetite: visceral sensory signals and mechanisms of central integration. *Frontiers in Neuroendocrinology*, **18**, 292–353.

Katoh, A., Fujihara, H., Ohbuchi, T., *et al.* (2010) Specific expression of an oxytocin-enhanced cyan fluorescent protein fusion transgene in the rat hypothalamus and posterior pituitary. *Journal of Endocrinology*, **204**, 275–285.

Katoh, A., Fujihara, H., Ohbuchi, T., *et al.* (2011) Highly visible expression of an oxytocin-monomeric red fluorescent protein 1 fusion gene in the hypothalamus and posterior pituitary of transgenic rats. *Endocrinology*, **152**, 2768–2774.

Katoh, A., Shoguchi, K., Matsuoka, H., *et al.* (2014) Fluorescent visualisation of the hypothalamic oxytocin neurones activated by cholecystokinin-8 in rats expressing c-fos-enhanced green fluorescent protein and oxytocin-monomeric red fluorescent protein 1 fusion transgenes. *Journal of Neuroendocrinology*, **26**, 341–347.

Knobloch, H.S., Charlet, A., Hoffmann, L.C., *et al.* (2012) Evoked axonal oxytocin release in the central amygdala attenuates fear response. *Neuron*, **73**, 553–566.

Kovács, K.J. (2008) Measurement of immediate-early gene activation – c-fos and beyond. *Journal of Neuroendocrinology*, **20**, 665–672.

Lafarga, M., Martinez-Guijarro, F.J., Berciano, M.T., *et al.* (1993) Nuclear Fos domains in transcriptionally activated supraoptic nucleus neurons. *Neuroscience*, **57**, 353–364.

Lang, F., Busch, G.L., Ritter, M., *et al.* (1998) Functional significance of cell volume regulatory mechanisms. *Physiological Reviews*, **78**, 247–306.

Lee, H.J., Macbeth, A.H., Pagani, J.H. and Young, W.S. III (2009) Oxytocin: the great facilitator of life. *Progress in Neurobiology*, **88**, 127–151.

Leng, G (1982) Lateral hypothalamic neurons: osmosensitivity and influence on activiating magnocellular neurosecretory neurons. *Journal of Physiology*, **326**, 35–48.

Li, Z. and Ferguson, A.V. (1993) Subfornical organ efferents to the paraventricular nucleus utilise angiotensin as a neurotransmitter. *American Journal of Physiology*, **265**, R302–309.

Liedtke, W. (2007) Role of TRPV ion channels in sensory transduction of osmotic stimuli in mammals. *Experimental Physiology*, **92**, 507–512.

Liedtke, W. and Friedman, J.M. (2003) Abnormal osmotic regulation in trpv4-/- mice. *Proceedings of the National Academy of Sciences USA*, **100**, 13698–13703.

Liedtke, W. and Kim, C. (2005) Functionality of the TRPV subfamily of TRP ion channels: add mechano-TRP and osmo-TRP to the lexicon! *Cellular & Molecular Life Sciences*, **62**, 2985–3001.

Liedtke, W., Choe, Y., Marti-Renom, M.A., *et al.* (2000) Vanilloid receptor-related osmotically activated channel (VR-OAC), a candidate vertebrate osmoreceptor. *Cell*, **103**, 525–535.

Lubelski, D., Ponzio, T.A. and Gainer, H. (2012) Effects of A-CREB, a dominant negative inhibitor of CREB, on the expression of c-fos and other immediate early genes in the rat SON during hyperosmotic stimulation in vivo. *Brain Research*, **6**,18–28.

Luckman, S.M., Dye, S. and Cox, H.J. (1996) Induction of members of the Fos/Jun family of immediate-early genes in identified hypothalamic neurons: in vivo evidence for differential regulation. *Neuroscience*, **73**, 473–485.

Markadieu, N. and Delpire, E. (2013) Physiology and pathophysiology of SLC12A1/2 transporters. *Pflugers Archiv*, **466**, 91–105.

Mason, W.T. (1980) Supraoptic neurones of rat hypothalamus are osmosensitive. *Nature*, **287**, 154–157.

Mezey, E. and Kiss, J.Z. (1991) Coexpression of vasopressin and oxytocin in hypothalamic supraoptic neurons of lactating rats. *Endocrinology*, **129**, 1814–1820.

Mimee, A., Smith, P.M. and Ferguson, A.V. (2013) Circumventricular organs: targets for integration of circulating fluid and energy balance signals? *Physiology & Behavior*, **10**,96–102.

Miselis, R.R. (1981) The efferent projections of the subfornical organ of the rat: a circumventricular organ within a neural network subserving water balance. *Brain Research*, **230**, 1–23.

Mohr, E., Bahnsen, U., Kiessling, C. and Richter, D. (1988) Expression of the vasopressin and oxytocin genes in rats occurs in mutually exclusive sets of hypothalamic neurons. *FEBS Letters*, **242**, 144–148.

Murphy, D. and Carter, D. (1990) Vasopressin gene expression in the rodent hypothalamus: transcriptional and posttranscriptional responses to physiological stimulation. *Molecular Endocrinology*, **4**, 1051–1059.

Murphy, D. and Wells, S. (2003) In vivo gene transfer studies on the regulation and function of the vasopressin and oxytocin genes. *Journal of Neuroendocrinology*, **15**, 109–125.

Mutsuga, N., Shahar, T., Verbalis, J.G., *et al.* (2004) Selective gene expression in magnocellular neurons in rat supraoptic nucleus. *Journal of Neuroscience*, **24**, 7174–7185.

Mutsuga, N., Shahar, T., Verbalis, J.G., *et al.* (2005) Regulation of gene expression in magnocellular neurons in rat supraoptic nucleus during sustained hypoosmolality. *Endocrinology*, **146**, 1254–1267.

Noda M (2006) The subfornical organ, a specialized sodium channel, and the sensing of sodium levels in the brain. *Neuroscientist*, **12**, 80–91.

Noda M (2007) Hydromineral neuroendocrinology: mechanism of sensing sodium levels in the mammalian brain. *Experimental Physiology*, **92**, 513–522.

Pardy, K., Adan, R.A., Carter, D.A., Seah, V., Burbach, J.P. and Murphy, D. (1992) The identification of a cis-acting element involved in cyclic 3′,5′-adenosine monophosphate regulation of bovine vasopressin gene expression. *Journal of Biological Chemistry*, **267**, 21746–21752.

Pedersen, S.F., Kapus, A. and Hoffmann, E.K. (2011) Osmosensory mechanisms in cellular and systemic volume regulation. *Journal of the American Society of Nephrology*, **22**, 1587–1597.

Ponzio, T.A., Fields, R.L., Rashid, O.M., Salinas, Y.D., Lubelski, D. and Gainer, H. (2012) Cell-type specific expression of the vasopressin gene analyzed by AAV mediated gene delivery of promoter deletion constructs into the rat SON in vivo. *PLoS One*, **7**, e48860. [***The AVP promoter is functional in an AAV backbone. This strategy enabled the mapping of promoter elements that mediate AVP MCN specificity and osmoregulation* in vivo.**]

Prager-Khoutorsky, M. and Bourque, C.W. (2010) Osmosensation in vasopressin neurons: changing actin density to optimize function. *Trends in Neuroscience*, **33**, 76–83.

Qiu, J., Yao, S., Hindmarch, C., Antunes, V., Paton, J. and Murphy, D. (2007) Transcription factor expression in the hypothalamo-neurohypophyseal system of the dehydrated rat; up-regulation of Gonadotrophin inducible transcription factor 1 mRNA is mediated by cAMP-dependent protein kinase A. *Journal of Neuroscience*, **27**, 2196–2203.

Qiu, J., Kleineidam, A., Gouraud, S., *et al.* (2014) The use of protein-DNA, chromatin immunoprecipitation, and transcriptome arrays to describe transcriptional circuits in the dehydrated male rat hypothalamus. *Endocrinology*, **155**, 4380–4390.

Richard, D. and Bourque, C.W. (1995) Synaptic control of rat supraoptic neurones during osmotic stimulation of the organum vasculosum lamina terminalis in vitro. *Journal of Physiology*, **489**, 567–577. [***In rat hypothalamic explants, inhibitory and excitatory pathways originating from OVLT neurons are tonically active under norm-osmotic conditions. However, osmotically evoked changes in MNC firing are selectively mediated through changes in the intensity of the excitatory component of inputs from the OVLT.***]

Sawchenko, P.E. and Swanson, L.W. (1982) Immunohistochemical identification of neurons in the paraventricular nucleus of the hypothalamus that project to the medulla or to the spinal cord in the rat. *Journal of Comparative Neurology*, **205**, 260–272.

Shao, W., Halachmi, S. and Brown, M. (2002) ERAP140, a conserved tissue-specific nuclear receptor coactivator. *Molecular & Cellular Biology*, **22**, 3358–3372.

Sharif Naeini, R., Witty, M.F., Seguela, P. and Bourque, C.W. (2006) An N-terminal variant of Trpv1 channel is required for osmosensory transduction. *Nature Neuroscience*, **9**, 93–98.

Sharif Naeini, R., Ciura, S. and Bourque, C.W. (2008a) TRPV1 gene required for thermosensory transduction and anticipatory secretion from vasopressin neurons during hyperthermia. *Neuron*, **58**, 179–185.

Sharif Naeini, R., Ciura, S., Zhang, Z. and Bourque, C.W. (2008b) Contribution of TRPV channels to osmosensory transduction, thirst, and vasopressin release. *Kidney International*, **73**, 811–815.

Sharman, G., Ghorbel, M., Leroux, M., *et al.* (2004) Deciphering the mechanisms of homeostatic plasticity in the hypothalamo-neurohypophyseal system – genomic and gene transfer strategies. *Progress in Biophysics & Molecular Biology*, **84**, 151–182.

Shaulian, E. and Karin, M. (2002) AP-1 as a regulator of cell life and death. *Nature Cell Biology*, **4**, E131–136.

Shimizu, H., Watanabe, E., Hiyama, T.Y., *et al.* (2007) Glial Nax channels control lactate signaling to neurons for brain [Na+] sensing. *Neuron*, **54**, 59–72.

Shiromani, P.J., Magner, M., Winston, S. and Charness, M.E. (1995) Time course of phosphorylated CREB and Fos-like immunoreactivity in the hypothalamic supraoptic nucleus after salt loading. *Brain Research, Molecular Brain Research*, **29**, 163–171.

Silva, M.P., Ventura, R.R. and Varanda, W.A. (2013) Hypertonicity increases NO production to modulate the firing rate of magnocellular neurons of the supraoptic nucleus of rats. *Neuroscience*, **250**, 70–79.

Sladek, C.D. and Somponpun, S.J. (2004) Oestrogen receptor beta: role in neurohypophyseal neurones. *Journal of Neuroendocrinology*, **16**, 365–371.

Sladek, C.D. and Somponpun, S.J. (2008) Estrogen receptors: their roles in regulation of vasopressin release for maintenance of fluid and electrolyte homeostasis. *Frontiers in Neuroendocrinology*, **29**, 114–127.

Smith, S.M. and Vale, W.W. (2006) The role of the hypothalamic-pituitary-adrenal axis in neuroendocrine responses to stress. *Dialogues in Clinical Neurosciences*, **8**, 383–395.

Sudbury, J.R. and Bourque, C.W. (2013) Dynamic and permissive roles of TRPV1 and TRPV4 channels for thermosensation in mouse supraoptic magnocellular neurosecretory neurons. *Journal of Neuroscience*, **33**, 17160–17165.

Telleria-Diaz, A., Grinevich, V.V. and Jirikowski, G.F. (2001) Colocalization of vasopressin and oxytocin in hypothalamic magnocellular neurons in water-deprived rats. *Neuropeptides*, **35**, 162–167.

Theodosis, D.T., El Majdoubi, M., Pierre, K. and Poulain, D.A. (1998) Factors governing activity-dependent structural plasticity of the hypothalamoneurohypophysial system. *Cellular & Molecular Neurobiology*, **18**, 285–298.

Thrasher, T.N. and Keil, L.C. (1987) Regulation of drinking and vasopressin secretion: role of organum vasculosum laminae terminalis. *American Journal of Physiology*, **253**, R108–120.

Thrasher, T.N., Brown, C.J., Keil, L.C. and Ramsay, D.J. (1980) Thirst and vasopressin release in the dog: an osmoreceptor or sodium receptor mechanism? *American Journal of Physiology*, **238**, R333–339.

Toney, G.M. (2010) Regulation of neuronal cell volume: from activation to inhibition to degeneration. *Journal of Physiology*, **588**, 3347–3348.

Toney, G.M. and Stocker, S.D. (2010) Hyperosmotic activation of CNS sympathetic drive: implications for cardiovascular disease. *Journal of Physiology*, **588**, 3375–3384.

Tremblay, C., Berret, E., Henry, M., Nehme, B., Nadeau, L. and Mouginot, D. (2011) Neuronal sodium leak channel is responsible for the detection of sodium in the rat median preoptic nucleus. *Journal of Neurophysiology*, **105**, 650–660.

Ueta, Y., Fujihara, H., Serino, R., *et al.* (2005) Transgenic expression of enhanced green fluorescent protein enables direct visualization for physiological studies of vasopressin neurons and isolated nerve terminals of the rat. *Endocrinology*, **146**, 406–413. [*Description of transgenic rats in which eGFP expression is directed to AVP neurons enabling their facile fluorescent detection. This has revolutionized electrophysiological studies on AVP neurons.*]

Vandesande, F., Dierickx, K. and De Mey, J. (1977) The origin of the vasopressinergic and oxytocinergic fibres of the external region of the median eminence of the rat hypophysis. *Cell & Tissue Research*, **180**, 443–452.

Verbalis, J.G. (1993) Hyponatremia induced by vasopressin or desmopressin in female and male rats. *Journal of the American Society of Nephrology*, **3**, 1600–1606.

Verbalis, J.G. (2007) How does the brain sense osmolality? *Journal of the American Society of Nephrology*, **18**, 3056–3059.

Voisin, D.L. and Bourque, C.W. (2002) Integration of sodium and osmosensory signals in vasopressin neurons. *Trends in Neuroscience*, **25**, 199–205.

Voisin, D.L., Chakfe, Y. and Bourque, C.W. (1999) Coincident detection of CSF Na+ and osmotic pressure in osmoregulatory neurons of the supraoptic nucleus. *Neuron*, **24**, 453–460.

Wainwright, A., Rutter, A.R., Seabrook, G.R., Reilly, K. and Oliver, K.R. (2004) Discrete expression of TRPV2 within the hypothalamo-neurohypophysial system: implications for regulatory activity within the hypothalamic-pituitary-adrenal axis. *Journal of Comparative Neurology*, **474**, 24–42.

Waller, S., Fairhall, K.M., Xu, J., Robinson, I. and Murphy, D. (1996) Neurohypophyseal and fluid homeostasis in transgenic rats expressing a tagged rat vasopressin prepropeptide in vasopressinergic magnocellular neurons. *Endocrinology*, **137**, 5068–5077.

Watanabe, E., Fujikawa, A., Matsunaga, H., *et al.* (2000) Nav2/NaG channel is involved in control of salt-intake behavior in the CNS. *Journal of Neuroscience*, **20**, 7743–7751.

Watanabe, E., Hiyama, T.Y., Shimizu, H., *et al.* (2006) Sodium-level-sensitive sodium channel Na(x) is expressed in glial laminate processes in the sensory circumventricular organs. *American Journal of Physiology*, **290**, R568–576.

Weisinger, R.S., Considine, P., Denton, D.A. and McKinley, M.J. (1979) Rapid effect of change in cerebrospinal fluid sodium concentration on salt appetite. *Nature*, **280**, 490–491. [*First suggestion that sodium sensor might be located in the brain.*]

Xi, D., Kusano, K. and Gainer, H. (1999) Quantitative analysis of oxytocin and vasopressin mRNAs in single magnocellular neurons isolated from supraoptic nucleus of rat hypothalamus. *Endocrinology*, **140**, 4677–4682.

Yang, C.R., Senatorov, V.V. and Renaud, L.P. (1994) Organum vasculosum lamina terminalis-evoked postsynaptic responses in rat superoptic neurones in vitro. *Journal of Physiology*, **477**, 59–74.

Yao, S.T., Antunes, V.R., Bradley, P.M., *et al.* (2011) Temporal profile of arginine vasopressin release from the neurohypophysis in response to hypertonic saline and hypotension measured using a fluorescent fusion protein. *Journal of Neuroscience Methods*, **201**, 191–195.

Yao, S.T., Gouraud, S.S., Qiu, J., Cunningham, J.T., Paton, J.F.R. and Murphy, D. (2012) Selective up-regulation of JunD transcript and protein expression in vasopressinergic supraoptic nucleus (SON) neurones in water deprived rats. *Journal of Neuroendocrinology*, **24**, 1542–1552.

Ying, Z., Reisman, D. and Buggy, J. (1996) AP-1 DNA binding activity induced by hyperosmolality in the rat hypothalamic supraoptic and paraventricular nuclei. *Brain Research*, **39**, 109–116.

Young, W.S. III and Gainer, H. (2003) Transgenesis and the study of expression, cellular targeting, and function of oxytocin, vasopressin, and their receptor genes. *Neuroendocrinology*, **78**, 185–203.

Yue, C., Mutsuga, N., Verbalis, J. and Gainer, H. (2006) Microarray analysis of gene expression in the supraoptic nucleus of normoosmotic and hypoosmotic rats. *Cellular & Molecular Neurobiology*, **26**, 959–978.

Zeng, Q., Carter, D.A. and Murphy, D. (1994) Cell specific expression of a vasopressin transgene in rats. *Journal of Neuroendocrinology*, **6**, 469–477. [**The first description of transgenic rats that express an AVP transgene with appropriate cell-specific and physiological regulation.**]

Zhang, B., Glasgow, E., Murase, T., Verbalis, J.G. and Gainer, H. (2001) Chronic hypoosmolality induces a selective decrease in magnocellular neurone soma and nuclear size in the rat hypothalamic supraoptic nucleus. *Journal of Neuroendocrinology*, **13**, 29–36.

Zhang, B.J., Kusano, K., Zerfas, P., Iacangelo, A., Young, W.S. III and Gainer, H. (2002) Targeting of green fluorescent protein to secretory granules in oxytocin magnocellular neurons and its secretion from neurohypophysial nerve terminals in transgenic mice. *Endocrinology*, **143**, 1036–1046.

Zhang, Z. and Bourque, C.W. (2003) Osmometry in osmosensory neurons. *Nature Neuroscience*, **6**, 1021–1022.

CHAPTER 16

Food Intake, Circuitry, and Energy Metabolism

Giles S.H. Yeo
MRC Metabolic Diseases Unit, University of Cambridge Metabolic Research Laboratories, Wellcome Trust-MRC Institute of Metabolic Science, Addenbrooke's Hospital, Cambridge, UK

16.1   Obesity is a problem ... who can we blame?

Since time immemorial, the control of food intake and body weight has been thought to be simply an issue of self-control and will power. So as obesity has become an increasing public health problem, reaching epidemic proportions in most developed and emerging economies, society in turn blames those overweight and obese for a lack of moral fortitude. 'The obese only have themselves to blame, all they have to do is to eat less ...', so the argument goes.

From a thermodynamic standpoint, this view is of course quite accurate. Body weight is clearly a balance between energy intake and energy expenditure. Thus, the only way to gain weight is to eat more than you burn, and the only way to lose weight is to eat less than you burn. There, in one succinct statement, is the cause and cure of the obesity epidemic. However, this sage piece of advice, that your grandmother could have given you, is clearly not working. A far more complex and interesting question to ask is WHY some people eat more than others. Few would dispute that the obesity epidemic has been driven by lifestyle and environmental changes. However, individuals respond differently to these 'obesigenic' environmental changes and this variation in response has a strong genetic element underlying physiological variations. Indeed, studies of BMI (Body Mass Index; weight in kg/height in m^2, a correlate of body fat mass) correlations of monozygotic, dizygotic, biological, and adopted siblings reveal heritability of fat mass to be between 40% and 70% (Stunkard *et al.*, 1986, 1990). Consequently, genetic approaches offer an effective tool for characterizing the molecular and physiological mechanisms of food intake and body weight control, and allow us to understand how these may become defective in the obese state.

16.2   Genetics as a tool

The genetic determinants of body weight in the general population are believed largely to be polygenic and, in spite of the recent spate of genome-wide association studies (GWAS), which have begun to reveal some of the genetic architecture underlying common obesity, remain poorly understood. However, the study of extreme phenotypes in both mice and humans has identified a number of genes that, when mutated, cause severe obesity. Although relatively rare, these monogenic disorders indicate a fundamental failure of the mechanisms of energy homeostasis and, together with genetically modified mouse models,

Molecular Neuroendocrinology: From Genome to Physiology, First Edition. Edited by David Murphy and Harold Gainer.
© 2016 John Wiley & Sons, Ltd. Published 2016 by John Wiley & Sons, Ltd.
Companion website: www.wiley.com/go/murphy/neuroendocrinology

have illuminated the critical role that these molecules play in the physiological control of food intake and body weight. Much of this chapter will be devoted to discussing the lessons learnt from both naturally occurring and genetically modified mouse models of obesity, as well as from human monogenic syndromes of obesity.

16.3   Body weight is homeostatically controlled

The control of energy balance involves a homeostatic feedback control circuitry, whereby peripheral signals communicate energy availability to the central nervous system (CNS), and hypothalamic circuits drive appropriate feeding and fuel partitioning responses. It was first proposed in the 1950s that circulating signals generated in proportion to body fat stores influenced food intake and energy expenditure in a coordinated manner to regulate body weight (Kennedy, 1953). However, it was not until the cloning of leptin in 1994 (Zhang *et al.*, 1994) that the molecular basis for this homeostatic control was identified. Subsequently, insights from human and mouse genetics have characterized specific pathways in the hypothalamus, the brainstem, and higher brain regions that play key roles in the control of food intake (Yeo and Heisler, 2012). Peripheral homeostatic regulators of energy balance can be broadly divided into adipocyte-derived hormones, such as leptin, responsible for signaling long-term energy stores, and gut-derived hormones, which regulate short-term control of food intake (Figure 16.1). These long- and short-term peripheral nutritional

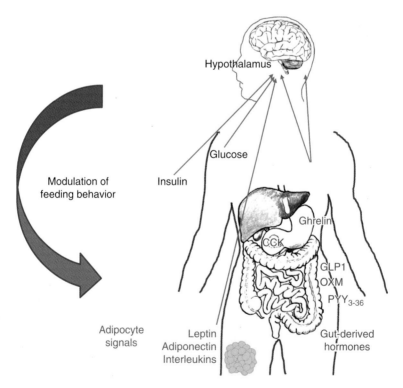

Figure 16.1 Brain sensing of gut- and adipocyte-derived hormones. Gut- and adipocyte-derived hormones, reflecting short- and long-term nutritional status, respectively, circulate in the periphery and signal to specific receptors in the brain. CCK, cholecystokinin; GLP-1, glucagon-like peptide 1; OXM, oxyntomodulin; PYY3–36, peptide YY residues 3–36.

signals are sensed by the brain, in particular by the hypothalamus and the nucleus of the solitary tract in the brainstem, where they are integrated and acted upon to appropriately regulate food intake and energy expenditure.

16.3.1 Peripheral signals (south of the neck)

16.3.1.1 Leptin

The study of two naturally occurring obese mouse models, obese (*ob/ob*) and diabetic (*db/db*), which were shown to harbor mutations in leptin and the leptin receptor respectively, provided the first molecular handle on this homeostatic system controlling food intake and body weight (Friedman and Halaas, 1998). Leptin is secreted from adipose tissue, with its circulating levels directly related to the amount of fat in the body, thereby signaling to leptin receptors in the brain the status of 'long-term' energy stores and, in essence, how long we would survive without food. Consequently, both humans and mice with inherited loss-of-function mutations of the genes encoding either leptin (Montague *et al.*, 1997; Zhang *et al.*, 1994) or its receptor (Chen *et al.*, 1996; Clement *et al.*, 1998) display a pathogenic drive to eat, or 'hyperphagia,' leading to severe early-onset obesity, as well as a broad range of neuroendocrine and some immunological abnormalities.

Leptin deficiency is unique amongst monogenic obesity syndromes identified thus far, in that it is amenable to treatment with recombinant leptin. Thus, leptin administration to ob/ob animals and humans genetically lacking leptin dramatically reversed their hyperphagia, obesity, and all neuroendocrine and immunological phenotypes (Farooqi *et al.*, 1999; Pelleymounter *et al.*, 1995). However, despite understandable and intense initial interest in the therapeutic properties of recombinant leptin, administering the hormone to obese individuals did not induce weight loss as predicted (Heymsfield *et al.*, 1999), which is consistent with observations that circulating leptin is positively correlated with fat mass (Considine *et al.*, 1996).

16.3.1.2 Leptin resistance

Leptin resistance, a term not without controversy, is used to describe the apparent paradox of leptin's action as an anorectic hormone and its elevated levels in the majority of obese individuals. In fact, these observations would suggest that the primary function of the leptin signal is not to prevent excessive weight gain, which makes sense as evolutionary pressures would have selected for us to favor fat storage over fat consumption. Thus, the major physiological role of leptin is not as a 'satiety signal' to prevent obesity in times of energy excess, but as a 'starvation signal' to maintain adequate fat stores for survival during times of energy deficit. This was elegantly demonstrated by Ahima *et al.* (1996) who showed that the full range of neuroendocrine changes seen in nutritionally depleted mice could be rescued in part by leptin administration.

There are a number of proposed molecular mechanisms to explain the phenomenon of leptin resistance. These encompass a spectrum of molecular and functional disorders, which can broadly be classified into impaired leptin transport across the blood–brain barrier (BBB), and impairment of leptin receptor function and signaling. Both of these are discussed later in this chapter. Importantly, compelling evidence suggests that leptin itself may play an important role in the development of resistance to its own effects, so-called 'leptin induced leptin resistance.' Chronically raised leptin levels which characterize obesity decrease the transport of leptin into the CNS and impair the signaling properties of leptin receptors. The resulting resistance to leptin confers increased susceptibility to diet-induced obesity, which in turn raises leptin levels further and worsens existing leptin resistance, leading to a vicious cycle of weight gain. Therefore, in addition to being a major cause of obesity, leptin resistance is also an important consequence (Myers *et al.*, 2008; Scarpace and Zhang, 2009).

Conceptually, because it has evolved as a trigger of the starvation response, we are designed to respond to low levels of leptin, which occurs when our fat stores are depleted, and not when it is circulating at normal or elevated levels.

16.3.2 The role of gut peptides

The gastrointestinal tract is the body's largest endocrine organ and is an important source of regulatory peptide hormones which regulate physiological functions, including gut motility and the release of digestive secretions. Gut hormones are important short-term signals of nutritional state and communicate presence or absence of food in the gastrointestinal tract to the hypothalamus and brainstem, the key areas in the CNS responsible for regulating appetite and satiety (Chaudhri *et al.*, 2008; Murphy and Bloom, 2006).

16.3.2.1 Cholecystokinin

Cholecystokinin (CCK) is the classic gut peptide long established to act as a postprandial satiety signal (Chaudhri *et al.*, 2008). It is released by enteroendocrine cells in the duodenum and jejunum of the small intestine into the circulation, responding in particular to fatty acids. CCK acts at receptors on peripheral vagal afferent terminals, which transmit signal to appetite centers, such as the nucleus of the solitary tract (NTS).

16.3.2.2 Glucagon-like peptide-1, oxyntomodulin and PYY$_{3-36}$

Glucagon-like peptide-1 (GLP-1) is a peptide product of the proglucagon gene, released from the L cells of the small intestine in response to food ingestion (Drucker, 2006). GLP-1 is a so-called 'incretin hormone,' a potent inducer of glucose-dependent insulin release. The GLP-1 analog, exenatide, is thus used in the treatment of type 2 diabetes mellitus (Drucker, 2006). Interestingly, central or peripheral administration of GLP-1 to rodents reduces food intake, and exenatide also reduces body weight in obese type 2 diabetic humans. The effects on body weight may be as a result of induction of satiety via inhibition of gastric emptying, but there is also evidence that GLP-1 reduces food intake via actions in the brainstem and the hypothalamus (Chaudhri *et al.*, 2008).

The gut hormones peptide YY (PYY) and oxyntomodulin (OXM) are produced from the same cells in the intestine. OXM is released postprandially together with PYY from gastrointestinal L cells. OXM, like GLP-1, is a cleavage product of the preproglucagon gene. Chronic peripheral administration of OXM reduces weight gain in rats (Dakin *et al.*, 2004) and intravenous infusion of OXM reduces food intake acutely in man (Cohen *et al.*, 2003). Chronic subcutaneous injection of OXM in overweight and obese humans results in a sustained reduction in energy intake (Wynne *et al.*, 2005, 2006).

Peripheral administration of PYY$_{3-36}$, the circulating form of PYY, to rodents or humans causes a marked inhibition of food intake (Batterham *et al.*, 2002). Significantly, both lean and obese individuals are fully responsive to this administration of exogenous PYY$_{3-36}$ (Batterham *et al.*, 2003). This anorectic effect was also long-lasting, since PYY$_{3-36}$ significantly reduced overall 24-hour caloric intake in both obese and lean individuals (Batterham *et al.*, 2003). Therefore, unlike leptin, obesity does not appear to be associated with resistance to PYY$_{3-36}$. The phenotype of the *Pyy* null mouse indicates that PYY may play a physiological role in eating behavior, in particular mediating the satiating effects of dietary proteins (Batterham *et al.*, 2006). In humans, PYY$_{3-36}$ levels are elevated in many disease states that are characterized by weight loss. Overweight subjects have been reported to have a relative deficiency of postprandial PYY$_{3-36}$ release associated with reduced satiety (le Roux *et al.*, 2006b) and bariatric surgery results in an exaggerated postprandial PYY$_{3-36}$ surge, potentially explaining the effectiveness of such surgery in maintaining weight loss (Le Roux *et al.*, 2006a).

16.3.2.3 Ghrelin

Ghrelin was discovered as an endogenous ligand for the growth hormone secretagogue receptor (GHS-R1A) (Kojima *et al.*, 1999). Ghrelin requires posttranslational acylation with octanoate to confer bioactivity. It is octanoylated by ghrelin octanoyl acyl transferase (GOAT) (Yang *et al.*, 2008), produced and secreted by cells within the oxyntic glands of the stomach (Kojima *et al.*, 1999). Ghrelin is unique amongst gastrointestinal hormones in that it stimulates appetite. Peripheral or central administration of ghrelin powerfully stimulates food intake in rodents, and intravenous or subcutaneous administration stimulates appetite in humans (Tschop *et al.*, 2000). Ghrelin secretion increases with fasting, and is increased by weight loss and insulin-induced hypoglycemia, suggesting it may act as a physiological signal of hunger (Williams and Cummings, 2005).

16.4 The brain (north of the neck)

16.4.1 The hypothalamus as a sensor

The hypothalamus receives and integrates circulating peripheral nutritional cues, including leptin and most of the gut peptides (Murphy and Bloom, 2006; Yeo and Heisler, 2012).

16.4.1.1 Leptin receptor

Mechanistically, the regulation of acute energy balance by leptin is contingent upon its binding to leptin receptors (LepR) within the CNS. Indeed, neuronal specific restoration of LepR in *db/db* mice, which have naturally mutated LepR, entirely rescues the obesity and diabetes phenotype (de Luca *et al.*, 2005). Although the range of clinical phenotypes of congenital leptin and leptin receptor deficiency are similar, leptin receptor deficiency in humans results in a less severe phenotype (Farooqi *et al.*, 2007b). It is also believed to be far more prevalent than leptin deficiency, and may account for up to 3% of all cases of extreme early-onset obesity (Farooqi *et al.*, 2007b).

The leptin receptor is a single membrane-spanning protein which shows structural similarity to the class 1 cytokine receptor family (Lee *et al.*, 1996; Tartaglia *et al.*, 1995). Several different alternatively spliced isoforms of the LepR exist (Lee *et al.*, 1996), each with a characteristic intracellular domain. Depending upon the length of the intracellular domain, the isoforms are classified as either short or long. The short isoforms (LepRa, LepRb, LepRc, LepRd, LepRe, and LepRf) have limited signaling capacity whilst the long isoform LepRb is believed to be the primary signaling form of the receptor (Bjorbaek *et al.*, 1997; Fei *et al.*, 1997; Lee *et al.*, 1996). LepRb is highly expressed in the hypothalamus where it mediates the effects of leptin on energy homeostasis (Elmquist *et al.*, 1998; Fei *et al.*, 1997).

16.4.1.2 Leptin receptor signaling

Activation of the LepRb initiates a cascade of signal transduction pathways, deficits of which may play an important role in the etiology of leptin resistance. Thought to be of particular importance is the JAK/STAT (Janus kinases/Signal transducers and activators of transcription) pathway (Bjorbaek *et al.*, 1997; Ghilardi and Skoda, 1997). The LepRb has no intrinsic tyrosine kinase activity and therefore must recruit cytoplasmic kinases, especially JAK2, which in turn phosphorylate numerous tyrosine residues of the intracellular domain (Ghilardi and Skoda, 1997). Although the precise mechanisms of JAK2 activation and signaling remain to be elucidated, a growing body of evidence supports a model in which leptin binding triggers LepRb aggregation into oligomers such that their constitutively bound JAK2 molecules are brought within close contact of each other, hence enabling them to autophosphorylate (Peelman *et al.*, 2006; White *et al.*, 1997; Zabeau *et al.*, 2004). It appears that there are three conserved tyrosine residues in the intracellular domain which are phosphorylated and contribute to signaling: Y985, Y1138, and Y1077 (Banks *et al.*, 2000; Tartaglia, 1997; White *et al.*, 1997). The phosphorylated domains provide highly specific binding sites for src homology 2 (SH2) domain-containing proteins such as STATs which are activated and translocate to the nucleus where they behave as transcription factors. STAT3 is known to be crucial to energy homeostasis, and after binding to the LepR, becomes the substrate of receptor-associated JAKs and subsequently dissociates from the receptor before forming active dimers (Banks *et al.*, 2000). Mice with a targeted mutation in a key intracellular tyrosine residue (Y1138) of the LepRb which is essential for STAT3 activation (s/s) are markedly hyperphagic and obese (Bates *et al.*, 2003).

The hypothalamic arcuate nucleus (ARC) is a major site for leptin signaling and leptin resistance (Chen *et al.*, 1996; Mercer *et al.*, 1996; Thornton *et al.*, 1997). Within the arcuate, two distinct classes of neurons exist; anorexigenic neurons expressing pro-opiomelanocortin (POMC) and cocaine and amphetamine related transcript (CART), which decrease food intake, and orexigenic neurons expressing neuropeptide Y (NPY)) and (agouti-related protein (AgRP), which increase food intake (Yeo and Heisler, 2012). Leptin receptors are highly expressed on the membranes of both types of neurons, allowing leptin to reciprocally regulate these two populations. For example, using Cre recombinase technology, Xu *et al.* (2007) elegantly demonstrated that STAT3 is necessary for inducing *POMC* transcription. Despite this, however, mice lacking STAT3 in POMC-expressing neurons are responsive to the anorexigenic effects of leptin and display no increased sensitivity to a high-fat diet, highlighting that the STAT3/POMC pathway forms only part of the energy homeostatic response to leptin (Xu *et al.*, 2007). We discuss the melanocortin pathway in detail later in this chapter.

16.4.1.3   Tanycytes and the blood–brain barrier

In order to access the brain, circulating molecules have to first navigate their way past the blood–brain barrier (BBB). The BBB is formed by capillary endothelial cells, which are connected by tight junctions and serve to regulate access of circulating factors into the brain parenchyma. There are, however, discrete regions of the brain known as the circumventricular organs (CVO), in which the brain endothelium is fenestrated. This includes the median eminence (ME) of the basal hypothalamus, situated adjacent to the arcuate nucleus (ARC) and ventral to the third ventricle. The ME also contains a population of highly polarized glial cells called tanycytes, which line the third ventricle and send projections all the way down into the capillary network lying deeper within the ME. It has been reported that there is regional variation in the distribution of tight junction proteins in this area and that the ME may act as a privileged site for peripheral molecules to specifically target the ARC (Langlet *et al.*, 2013b; Mullier *et al.*, 2010). Data now emerging indicate that there is a previously unappreciated degree of plasticity within the ME, particularly at the interface between circulation and brain. Using fluorescently labeled ligand coupled with *in vivo* imaging, Schaeffer *et al.* (2013) demonstrated that circulating ghrelin rapidly binds neurons in the vicinity of fenestrated capillaries, with the amount of ghrelin bound to NPY neurons directly influenced by nutritional state. Thus, fasting significantly increases the number of NPY fluorescent-ghrelin labeled neurons, with this labeling falling away with refeeding. In addition, fasting increases both the tight junction barrier of tanyctes lining the third ventricle and fenestration of the microvasculature at the interface of the ME and ARC, with VEGF-A produced in tanycytes appearing to be critical in modulating these barrier properties (Langlet *et al.*, 2013a).

Intriguingly, Balland *et al.* (2014) show a sequence of events that has to occur for leptin to enter the hypothalamus. Peripherally administered leptin first activates its receptor in ME tanycytes, before activating medial-basal hypothalamic (MBH) neurons, a process requiring tanycytic ERK signaling and the passage of leptin through the cerebrospinal fluid. In mice lacking the signal-transducing LepRb isoform or with diet-induced obesity, leptin taken up by tanycytes accumulates in the median eminence and fails to reach the MBH. Thus, circulating leptin enters the brain through the hypothalamic ME and tanycytes, which capture leptin from the bloodstream, acting as a checkpoint along this route (Balland *et al.*, 2014). These findings suggest that tanycyte-mediated leptin transport between the periphery and the cerebrospinal fluid (CSF) is necessary for its downstream effects.

16.4.1.4   The melanocortin pathway

Once past the BBB tanycytes, a large part of leptin's effects on controlling food intake is mediated by the hypothalamic melanocortin pathway (Cone, 2005). As mentioned earlier, within the ARC, leptin acts directly on anorexigenic POMC/CART neurons and orexigenic NPY/AgRP neurons. Both subsets of neurons are reciprocally regulated by leptin and make numerous connections with other hypothalamic nuclei, such as the lateral hypothalamus (LH), the paraventricular nucleus (PVN), the ventromedial nucleus (VMN), and the dorsomedial nucleus (DMN) (Figure 16.2) (Cone, 2005; Yeo and Heisler, 2012).

16.4.1.5   POMC

The first direct evidence for the involvement of melanocortin peptides in the control of mammalian food intake came from data showing that homozygous loss-of-function mutations in *POMC* produced severe obesity in both mice and humans. POMC is posttranslationally cleaved in a tissue-specific manner to produce an array of smaller biologically active products including β-endorphin, β-lipotrophin, and the melanocortin peptides adrenocorticotropic hormone (ACTH) and α-, β-, and γ-melanocyte stimulating hormone (MSH) (Bertagna, 1994; Castro and Morrison, 1997). The type of peptides produced depends upon the specificity of the endoproteases expressed (Bloomquist *et al.*, 1991; Zhou *et al.*, 1993). For example, pituitary corticotropes express prohormone convertase 1 (PC1) and cleave POMC to form ACTH, β-lipotrophin, N-terminal peptide, and joining peptide (Castro and Morrison, 1997). These are then further cleaved by PC2, which is expressed in the feeding centers of the arcuate nucleus, yielding the melanocortins α-, β-, and γ-MSH (Figure 16.3) (Castro and Morrison, 1997). The actions of melanocortins are mediated by a family of five G protein-coupled receptors known as the melanocortin receptors (MCRs) (Cone, 2006; Mountjoy *et al.*, 1992). The MC1R is expressed on melanocytes where it controls skin and hair pigmentation (Mountjoy *et al.*, 1992; Valverde *et al.*, 1995), whilst the MC2R regulates adrenal steroidogenesis (Mountjoy *et al.*, 1992). The MC5R appears to be important in sebum production (Chen *et al.*, 1997). The MC3 and MC4 receptors are largely expressed in the brain and central nervous system, and compelling genetic evidence over the past few years has shown that the MC4R in particular plays a central role in the control of body weight (Yeo and Heisler, 2012).

In humans, congenital POMC deficiency presents with several key clinical features, including a failure of adrenal steroidogenesis consistent with the loss of ACTH action at the MC2R, pale skin and red hair owing to the loss of α-MSH action at the MC1R and obesity due to the loss of MSH action at the MC4R (Krude *et al.*, 1998). The affected children responded well to physiological replacement with glucocorticoids but all went on to develop severe early-onset obesity in association with hyperphagia. The *Pomc* −/− mice recapitulate all of the phenotypes seen in human POMC deficiency (Challis *et al.*, 2004; Yaswen *et al.*, 1999). *Pomc* +/− mice, although of equivalent weight to wild-type mice on normal chow, are hyperresponsive and become more obese than wild-type mice when exposed to a high-fat diet (Challis *et al.*, 2004).

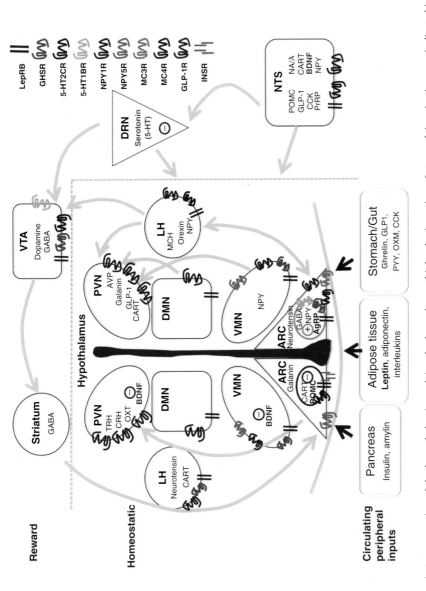

Figure 16.2 Schematic representation of the hypothalamic nuclei and other relevant higher brain regions. Internuclei projections are indicated in light blue arrows. Peptides (−) inhibiting or (+) stimulating food intake. Leptin signaling is indicated in red, ghrelin in green, and serotonin (5HT) in purple. 5HT-1B, 5HT-2C, serotonin receptors; AGRP, agouti-related protein; ARC, arcuate nucleus; AVP, arginine vasopressin; BDNF, brain derived neurotrophic factor; CART, cocaine- and amphetamine-regulated transcript; CRF, corticotrophin-releasing factor; DMN, dorsomedial nucleus; GABA, gamma-aminobutyric acid; GHSR, ghrelin receptor; GLP-1, glucagon-like peptide 1; GLP-1R, glucagon-like peptide 1 receptor; INSR, insulin receptor; LepRB, leptin receptor; LH, lateral hypothalamus; MC3R, MC4R, melanocortin receptors; MCH, melanin-concentrating hormone; NAc, nucleus accumbens; NPY, neuropeptide Y; NPY1R, NPY5R, neuropeptide Y receptors; OXT, oxytocin; POMC, pro-opiomelanocortin; PVN, paraventricular nucleus; TRH, thyrotropin-releasing hormone; VMN, ventromedial nucleus; VTA, ventral tegmental area.

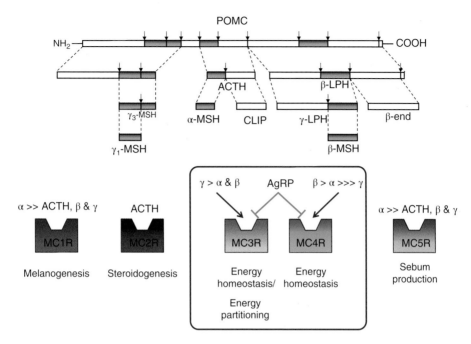

Figure 16.3 Schematic diagram of pro-opiomelanocortin (POMC) and its receptors. POMC is posttranslationally processed to produce a number of biologically active peptides including adrenocorticotropin (ACTH), as well as α-, β-, and γ-melanocyte-stimulating hormone (MSH). The dibasic cleavage sites are indicated by black arrows. The actions of the melanocortin peptides are mediated via 5 receptors (MC1-5Rs). Aside from MC2R, which only binds ACTH, the other receptors are promiscuous for the different peptides. The relative binding efficiency of the different melanocortin peptides for each receptor is indicated. MSHs signal to the MC3R and MC4R to mediate energy homeostasis. Additionally, the endogenous antagonist/inverse agonist agouti-related peptide (AgRP) competes with the MSHs to bind to both receptors.

16.4.1.6 POMC processing: PC1 and carboxypeptidase E

As discussed above, key prohormone convertases are required for the correct tissue-specific processing of POMC into its constituent peptides, and loss-of-function mutations in these enzymes can also cause obesity. Heterozygous inactivating mutations of *PC1* have been identified in human cases of early-onset obesity (Jackson *et al.*, 1997, 2003). The disorder presents with severe obesity, low plasma insulin levels, and elevated insulin levels (reflecting the loss of PC1-dependent cleaving of proinsulin to form insulin), hypogonadotropic hypogonadism and hypocortisolemia with elevated POMC levels. Somewhat paradoxically in light of the obesity, affected individuals also suffer an absorptive dysfunction of the small bowel which may reflect an important role for PC1 in the processing of intestinal propeptides (Jackson *et al.*, 2003). In contrast, mice lacking either PC1 or PC2 are stunted but of normal weight, suggesting that they may not rely upon these enzymes for the maintenance of normal energy homeostasis (Zhu *et al.*, 2002). However a very similar failure of POMC processing occurs in fat/fat mice, which are deficient in carboxypeptidase E (Fricker *et al.*, 1996), an enzyme which, like the prohormone convertases, is also required for peptide processing.

16.4.1.7 MC4R

The body of genetic and functional evidence detailing the role of the MC4R in the control of food intake and body weight is substantial. The MC4R has a widespread distribution within the brain and is expressed at sites including the cortex, striatum, brainstem, hippocampus, and hypothalamus (Kishi *et al.*, 2003; Liu *et al.*, 2003; Mountjoy *et al.*, 1994), reflecting its role in a range of neuroendocrine and autonomic processes (Cone, 2006). Homozygous *Mc4r* knockout mice are hyperphagic, hypometabolic, and obese and also display accelerated linear growth, with hyperinsulinemia and hyperglycemia (Huszar *et al.*, 1997). Crucially, heterozygous null *Mc4r* animals show a phenotype intermediate to that of their homozygous and wild-type litter mates (Huszar *et al.*, 1997).

In 1998, we and others reported that heterozygous mutations in *MC4R* in humans were associated with dominantly inherited obesity (Vaisse *et al.*, 1998; Yeo *et al.*, 1998). Since then, a large number of pathogenic mutations in *MC4R* have been reported in obese humans from various ethnic groups, accounting for up to 6% of all severe cases of early-onset obesity, making this the most common monogenic obesity disorder in humans described to date (Farooqi *et al.*, 2003). Recent studies have reported that in UK and European populations, 1–2.5% of people with a BMI greater than 30 kg/m^2 harbor pathogenic mutations in *MC4R*. This provides an amazing indication of the true population prevalence of this disorder and places it amongst the most common of genetic diseases, with a higher prevalence than more familiar diseases such as cystic fibrosis.

In all cases, as seen in the mice, individuals carrying homozygous mutations had a more severe phenotype than the heterozygous carriers. This gradation of phenotype included not only increased height, hyperinsulinemia, increased fat and fat-free mass, obesity and an increase in bone mineral density, but also included an increase in the amount of food consumed during an *ad libitum* test meal (Farooqi *et al.*, 2003). It has also been demonstrated that the severity of receptor dysfunction seen in *in vitro* assays can actually predict the amount of food ingested by the subject harboring that particular mutation at this same *ad libitum* test meal (Farooqi *et al.*, 2003). Thus this observed genotype-phenotype correlation emphasizes that the melanocortin regulatory system is exquisitely sensitive to quantitative variation in MC4R expression and function.

In mice, Mc4r deficiency is also associated with altered basal metabolism. Measurements of metabolic rate in young non-obese *Mc4r-/-* animals demonstrated a 20% reduction in oxygen consumption and a reduction in locomotor activity as compared to weight-matched WT mice (Ste Marie *et al.*, 2000). These data suggest that MC4R deficiency leads to defective regulation of energy expenditure, with *Mc4r*-null mice being more metabolically efficient than WT mice. Additionally, experiments where *Mc4r-/-* mice were exposed to a high-fat diet also strongly suggested that this receptor is necessary to mediate the appropriate changes in homeostatic mechanisms in response to changes in the caloric content of diet (Butler *et al.*, 2001). On a high-fat diet, WT mice transiently increased their caloric intake, followed by a decrease in food intake by mass, thus retaining an overall isocaloric intake. In contrast, *Mc4r-/-* mice exposed to a high-fat diet, like the POMC-deficient mice discussed earlier (Challis *et al.*, 2004), developed obesity at an accelerated rate due to sustained hyperphagia, a lack of increase in motor activity, and a reduction of diet-induced thermogenesis (Butler *et al.*, 2001).

Interestingly, Balthasar and colleagues have elegantly demonstrated an anatomical divergence in the role of the MC4R in energy homeostasis (Balthasar *et al.*, 2005). By discretely 'rescuing' MC4R expression in the PVN and amygdala of *Mc4r*-null mice utilizing Cre-lox technology, 60% of the obesity phenotype was rescued, while the altered energy expenditure phenotype was unaffected. Thus MC4R in the PVN and amygdala appears to regulate energy intake, but MC4Rs elsewhere, in as yet undefined regions, control energy expenditure (Balthasar *et al.*, 2005). Another interesting role for MCR4, determined by studying patients with pathogenic mutations in the receptor, is to influence heart rate and blood pressure in humans, likely mediated through changes in sympathetic neural activity (Greenfield *et al.*, 2009).

16.4.1.8 Agouti and AgRP

Initial evidence implicating the melanocortin pathway in energy homeostasis came from studies of *Agouti* mice, which have a complex phenotype characterized by obesity, hyperphagia, hyperinsulinemia and yellow coat coloration (Bultman *et al.*, 1992; Lu *et al.*, 1994; Michaud *et al.*, 1994; Miller *et al.*, 1993). Under physiological conditions in mice, agouti is produced in hair follicles, during a small window of time, where it regulates pigmentation by antagonizing the action of α-MSH at the MC1R. The binding of agouti induces a switch in pigment production from black eumelanin to yellow phaeomelanin, resulting in the characteristic mouse brown or black coat coloration with a small subapical yellow band. The genetic basis for the Agouti mouse phenotype is a rearrangement of the agouti gene promoter sequence, resulting in ectopic and constitutive expression of the agouti peptide, thus causing the coat color phenotype. In addition to antagonizing the MC1R, agouti also antagonized MC4R, resulting in the accompanying obese phenotype seen in the Agouti mouse (Bultman *et al.*, 1992; Lu *et al.*, 1994; Michaud *et al.*, 1994; Miller *et al.*, 1993).

Following on the revelations from studying the *Agouti* mice, one question that emerged was why a peptide produced largely in the skin was able to antagonize a brain receptor. The explanation came with the identification of the agouti-related protein or AgRP, a hypothalamic neuropeptide found in both rodents and humans, which shares high sequence homology with Agouti (Ollmann *et al.*, 1997). AgRP was found to be an endogenous melanocortin receptor antagonist that primarily antagonizes the MC3R and MC4R, but not the MC1R. Thus, transgenic overexpression of human AgRP ubiquitously in mice recapitulates the phenotype of the Agouti mouse, except for the coat coloration (Ollmann *et al.*, 1997). AgRP mRNA is expressed exclusively in the ARC and, as mentioned earlier, it appears that AgRP is co-expressed in neurons containing the potent orexigenic peptide NPY (Chen *et al.*, 1999). AgRP mRNA displays dynamic changes in levels of expression, increasing tenfold in fasted animals, and is returned to normal levels after leptin treatment (Chen *et al.*, 1999; Shutter *et al.*, 1997).

However, when *AgRP* was removed from mice by gene-targeted deletion, it was a surprise to many in the field that there was no overt metabolic or body weight phenotype (Qian *et al.*, 2002), raising questions about AgRP's physiological importance. Was the ability of AgRP to induce food intake simply a pharmacological artefact? Or is it possible that the removal of AgRP in early development promoted compensatory mechanisms, such is the importance in maintaining an intact drive to eat for survival, that overrides the loss of AgRP?

2005, as it turns out, was a bad year to be an AgRP neuron, as four labs independently took different approaches to genetically ablate these neurons. Two groups expressed the receptor for diphtheria toxin in AgRP neurons, permitting temporally controlled ablation of these neurons after administration of diphtheria toxin (Gropp *et al.*, 2005; Luquet *et al.*, 2005). A third group deleted the mitochondrial transcription factor A (Tfam) in AgRP neurons (Xu *et al.*, 2005) whilst a fourth transgenically expressed the neurotoxic CAG expanded repeat form of ataxin-3 in AgRP neurons (Bewick *et al.*, 2005), with the latter two approaches both leading to progressive postnatal cell death. Although the details in experimental design differed and the specific phenotypes of these mice varied in their severity (Flier, 2006), all four studies reported that loss of AgRP neurons produced a robust phenotype of hypophagia and leanness, and concluded that AgRP neurons are more essential to feeding and weight gain than is AgRP itself.

These observations were then further refined when it was shown that the anorexia following ablation of Agrp neurons is suppressed by chronic delivery of the GABAA receptor agonist bretazenil to the parabrachial nucleus (PBN) (Wu *et al.*, 2009). Thus, in addition to expression of Agrp, GABAergic tone in the PBN, presumably projecting from Agrp neurons in the ARC, plays a role in feeding behaviors. Intriguingly, chronic delivery of bretazenil during loss of AgRP neurons provides time to establish compensatory mechanisms that eventually restore feeding (Wu *et al.*, 2009). In contrast, while adaptation to loss of signaling by AgRP neurons can occur, adaption to loss of POMC neurons does not occur. Recent optogenetic and pharmacogenetic studies reveal that discrete real-time activation of Agrp neurons stimulates voracious feeding and weight gain in the adult mice (Aponte *et al.*, 2011; Krashes *et al.*, 2011). This optogenetic-evoked feeding appears to be independent of the melanocortin receptors, indicating that Agrp neurons are able to directly engage feeding circuits, possibly via projection of GABA-ergic tone onto PBN neurons.

16.5   Neuronal development and plasticity

Obesity is a state of system dysregulation conceivably sustained through alterations in synaptic connectivity between key energy balance neurons. This notion gained general acceptance following the elegant study by Pinto *et al.* (2004) which demonstrated that leptin induces changes in the balance of excitatory and inhibitory synapses onto both POMC- and NPY-expressing neurons. In this study, a whole cell patch clamp technique was used to demonstrate that obese leptin-deficient ob/ob mice have increased inhibitory postsynaptic currents onto POMC-expressing neurons and increased excitatory postsynaptic currents onto AgRP/NPY-expressing neurons compared to their wild-type littermates. These differences correlated with morphological differences in the ratios of inhibitory and excitatory synapses projecting to the two groups of neurons. Remarkably, leptin treatment in ob/ob mice induced a rapid normalization of the balance of synaptic inputs to these two groups of neurons within just 6 hours (Pinto *et al.*, 2004). Although this rapid rewiring of synaptic inputs occurs over a much shorter time course than the observed change in feeding behavior, its importance to normal energy homeostasis is highlighted by the fact that ghrelin has a similar effect (Abizaid *et al.*, 2006).

In addition to its role in synaptic plasticity, leptin has also been implicated as a neurotropin and may be responsible for the development of specific circuitry both within the ARC and between the ARC and other hypothalamic centers involved in energy homeostasis (Bouret *et al.*, 2004). Perturbed ARC-hypothalamic connectivity can be reversed in *ob/ob* mice by administering leptin perinatally, but not in adulthood. These data support the idea that a perinatal leptin surge acts as a developmental signal to promote ARC connectivity and the formation of pathways that control energy homeostasis (Bouret *et al.*, 2004).

16.5.1   BDNF and TrkB

BDNF *via* TrkB regulates the development, survival, and differentiation of neurons (Tapia-Arancibia *et al.*, 2004). Given that leptin modulates BDNF expression, it is conceivable that the neurotropic role of leptin is mediated via BDNF/TrkB signaling. In the developing brain, BDNF mRNA is expressed in the hippocampus, hypothalamus, septum, cortex, and brainstem (Tapia-Arancibia *et al.*, 2004). In the mature CNS, BDNF is most abundant in the hippocampus and hypothalamus (Tapia-Arancibia *et al.*, 2004). Unlike other neurotropins, BDNF is secreted in an activity-dependent manner that allows for highly controlled release (Bramham and Messaoudi, 2005). BDNF is an important regulator of synaptic transmission and plasticity at adult synapses (Bramham and Messaoudi, 2005), and homozygous deletion of either BDNF or its receptor TrkB is embryonically lethal (Kernie *et al.*, 2000). Therefore, BDNF clearly plays a critical role in brain development and neuron connectivity and it is plausible that this role is relevant to BDNF's effects on energy balance.

Much evidence has emerged implicating BDNF and TrkB in regulation of mammalian eating behavior and energy balance (Kernie *et al.*, 2000; Xu *et al.*, 2003). Mice either lacking one copy of *Bdnf* or with a tissue-specific conditional deletion of *Bdnf* in the postnatal brain (Rios *et al.*, 2001) develop hyperphagia and obesity. Similarly, mice with a hypomorphic mutation in *Trkb*, resulting in 25% normal levels of expression, also develop obesity (Xu *et al.*, 2003). Xu *et al.* have demonstrated that the expression of BDNF in the hypothalamic ventromedial nucleus, an area known to be involved in the control of food intake, is regulated by nutritional status.

In 2004, we reported a *de novo Trkb* mutation in an 8-year-old boy with a complex developmental syndrome including hyperphagia and severe obesity (Yeo *et al.*, 2004). The mutation (Tyr722Cys) replaces one of three highly conserved tyrosine residues within the catalytic kinase domain with a cysteine, resulting in an impairment of receptor phosphorylation and signaling via phospholipase Cγ (PLCγ), Akt, and MAPK. Furthermore, this mutation also affected the ability of TrkB to promote neurite outgrowth in response to BDNF (Gray *et al.*, 2007). Phenotypic features of this disorder include severe hyperphagia, obesity, and impaired learning and memory (Yeo *et al.*, 2004). We also characterized a *de novo* inversion of a region of chromosome 11p encompassing the BDNF locus, that resulted in a functional loss of one copy of the BDNF gene in an 8-year-old girl with obesity and developmental delay (Gray *et al.*, 2006). Clearly, with only one child each with disruption of TrkB and BDNF, it is not possible to make any conclusive comparisons between the two genetic disorders. That said, there is certainly a congruence in the clinical phenotype between the two patients, given the unusual and distinctive combination of impaired cognitive function, hyperactivity, and severe obesity (Gray *et al.*, 2006), all of which recapitulates the phenotypes seen in the BDNF- and TrkB-deficient mice.

16.6   Hedonic control of food intake

The system our brain uses to control food intake consists of two parts: a fuel sensor, which I have discussed in detail earlier in this chapter, that in effect calculates how many calories we have burnt during the day, and therefore how many calories we would need to consume to meet this need. However, given the importance that eating enough has for keeping us alive, our brain has also evolved mechanisms to make sure that it also feels 'good' or rewarding to eat. This is easily illustrated by the all too familiar restaurant paradigm of overindulging in dessert, despite already feeling full from the previous courses. As we evolved over tens of thousands of years to stay alive through multiple famines, any increase in motivation, however small, to continue to search for food was an evolutionary advantage. In addition, the rewarding feedback from food that tasted good, and which was then presumably not poisonous, was useful in guiding the development and entrainment of our eating behavior.

Thus, energy balance signals act in concert with neural systems involved in drive and motivation, and in addition to homeostatic factors, food intake is also influenced by cognition, emotion, and reward. The pleasure of eating is modulated by satiety, and food deprivation increases the reward value of food. Hunger increases the incentive value of food and functional magnetic resonance imaging (fMRI) demonstrates enhanced appetitive cue response in hunger (Dagher, 2012). Complex behaviors such as motivation, learning from errors, and performance monitoring also influence feeding behavior, and can overrule basic homeostatic mechanisms. The dopaminergic mesolimbic reward circuitry and the brain's autonomic networks are critical regulators of these behaviors and are altered following the development of obesity.

Eating for pleasure in the absence of caloric need is referred to as 'hedonic eating.' Human fMRI studies have shown that the homeostatic signals ghrelin, insulin, leptin, and PYY act on components of the appetitive brain network beyond the hypothalamus and the brainstem, including the insula, amygdala, striatum, and orbitofrontal cortex (OFC) (Dagher, 2012). For example, it has been shown that ghrelin enhances the neural response to food pictures in all components of the appetitive network (Malik *et al.*, 2008). In fact, purely hedonic stimuli (such as food pictures) have the ability to trigger ghrelin release from the stomach in humans. Exposure to food cues also increases the desire for food as measured by prospectively chosen portion size, suggesting that ghrelin is part of a gut–brain feed-forward loop matching hunger to food availability and energy balance.

The opposite effect (i.e. a dampened response to cues) is seen after administration of the anorexigenic hormones PYY, GLP-1, insulin, and leptin. Leptin, for example, markedly affects neural responses to visual food stimuli. Using fMRI in patients with congenital leptin deficiency, Farooqi and colleagues demonstrated that leptin administration increases the ability to discriminate between the rewarding properties of food and modulates neuronal activity in the ventral striatum (Farooqi *et al.*, 2007a). These data suggest that leptin acts on neural circuits governing food intake to diminish perception of food reward while enhancing the response to satiety signals generated during food consumption. Using BOLD fMRI, it has been shown that food cueprelated reward system activation was attenuated in the physiological fed state compared to when fasted overnight (de Silva *et al.*, 2012). In the same fasted subjects, gut hormone administration resulted in a similar pattern and degree of signal attenuation in the reward system.

Thus, our current understanding of the regulation of eating behavior suggests that although food intake and food choice are regulated by homeostatic mechanisms, these mechanisms are in turn modulated by the brain's reward system.

16.7   The natural response

There is still a strongly held belief in many quarters that we are in full 'executive control' of our own eating behavior; that the environment is responsible for our shape and size, and that our genes, our 'nature,' has minimal, if any, effect. However, it is crucial to remember that the drive to consume food is one of the most primitive of instincts to promote survival. It has been shaped by many millions of years of evolution and has provided living creatures with powerful and redundant mechanisms to adapt and respond to times of nutrient scarcity. Thus, I would argue that to be overweight in our current environment is indeed the natural (highly evolved even) response. The main issue is that the current environment, in which energy-dense foods and stimulatory food cues are ubiquitous, coupled with concurrent changes in lifestyle, is in dissonance with the millennia of austere surroundings to which we have adapted, and has consequently pushed obesity to become the serious problem it is today. I am fully aware that without this 'obesogenic' environment, most of us would not be overweight or obese but to deny the central role that our genes have played in our response to this environment is unhelpful as we strive to tackle one of the greatest public health challenges of the 21st century.

References

Abizaid, A., Liu, Z.W., Andrews, Z.B., *et al.* (2006) Ghrelin modulates the activity and synaptic input organization of midbrain dopamine neurons while promoting appetite. *Journal of clinical Investigation*, **116**, 3229–3239.

Ahima, R.S., Prabakaran, D., Mantzoros, C., *et al.* (1996) Role of leptin in the neuroendocrine response to fasting. *Nature*, **382**, 250–252. [*A seminal piece of work which establishes leptin NOT as a hormone to* *make you stop eating, but a hormone that functions to let your brain know that you have no fat, and to turn on the starvation response.*]

Aponte, Y., Atasoy, D. and Sternson, S.M. (2011) AGRP neurons are sufficient to orchestrate feeding behavior rapidly and without training. *Nature Neuroscience*, **14**, 351–355.

Balland, E., Dam, J., Langlet, F., *et al.* (2014) Hypothalamic tanycytes are an ERK-gated conduit for leptin into the brain. *Cell Metabolism*, **19**, 293–301.

Balthasar, N., Dalgaard, L.T., Lee, C.E., *et al.* (2005) Divergence of melanocortin pathways in the control of food intake and energy expenditure. *Cell*, **123**, 493–505.

Banks, A.S., Davis, S.M., Bates, S.H. and Myers, M.G. Jr (2000) Activation of downstream signals by the long form of the leptin receptor. *Journal of Biological Chemistry*, **275**, 14563–14572.

Bates, S.H., Stearns, W.H., Dundon, T.A., *et al.* (2003) STAT3 signalling is required for leptin regulation of energy balance but not reproduction. *Nature*, **421**, 856–859.

Batterham, R.L., Cowley, M.A., Small, C.J., *et al.* (2002) Gut hormone PYY(3-36) physiologically inhibits food intake. *Nature*, **418**, 650–654. [*First description of a gut hormone that, when infused into humans (and mice) in physiological concentrations, reduces food intake.*]

Batterham, R.L., Cohen, M.A., Ellis, S.M., *et al.* (2003) Inhibition of food intake in obese subjects by peptide YY3–36. *New England Journal of Medicine*, **349**, 941–948.

Batterham, R.L., Heffron, H., Kapoor, S., *et al.* (2006) Critical role for peptide YY in protein-mediated satiation and body-weight regulation. *Cell Metabolism*, **4**, 223–233.

Bertagna, X. (1994) Proopiomelanocortin-derived peptides. *Endocrinology and Metabolism Clinics of North America*, **23**, 467–485.

Bewick, G.A., Gardiner, J.V., Dhillo, W.S., *et al.* (2005) Post-embryonic ablation of AgRP neurons in mice leads to a lean, hypophagic phenotype. *FASEB Journal*, **19**, 1680–1682.

Bjorbaek, C., Uotani, S., da Silva, B. and Flier, J.S. (1997) Divergent signaling capacities of the long and short isoforms of the leptin receptor. *Journal of Biological Chemistry*, **272**, 32686–32695.

Bloomquist, B.T., Eipper, B.A. and Mains, R.E. (1991) Prohormone-converting enzymes: regulation and evaluation of function using antisense RNA. *Molecular Endocrinology*, **5**, 2014–2024.

Bouret, S.G., Draper, S.J. and Simerly, R.B. (2004) Trophic action of leptin on hypothalamic neurons that regulate feeding. *Science*, **304**, 108–110. [*A seminal paper describing the critical role of leptin in the appropriate development of the hypothalamic feeding circuitry.*]

Bramham, C.R. and Messaoudi, E. (2005) BDNF function in adult synaptic plasticity: the synaptic consolidation hypothesis. *Progress in Neurobiology*, **76**, 99–125.

Bultman, S.J., Michaud, E.J. and Woychik, R.P. (1992) Molecular characterization of the mouse agouti locus. *Cell*, **71**, 1195–1204.

Butler, A.A., Marks, D.L., Fan, W., Kuhn, C.M., Bartolome, M. and Cone, R.D. (2001) Melanocortin-4 receptor is required for acute homeostatic responses to increased dietary fat. *Nature Neuroscience*, **4**, 605–611.

Castro, M.G. and Morrison, E. (1997) Post-translational processing of proopiomelanocortin in the pituitary and in the brain. *Critical Reviews in Neurobiology*, **11**, 35–57.

Challis, B.G., Coll, A.P., Yeo, G.S., *et al.* (2004) Mice lacking pro-opiomelanocortin are sensitive to high-fat feeding but respond normally to the acute anorectic effects of peptide-YY(3-36) *Proceedings of the National Academy of Sciences of the USA*, **101**, 4695–4700.

Chaudhri, O.B., Salem, V., Murphy, K.G. and Bloom, S.R. (2008) Gastrointestinal satiety signals. *Annual Review of Physiology*, **70**, 239–255.

Chen, H., Charlat, O., Tartaglia, L.A., *et al.* (1996) Evidence that the diabetes gene encodes the leptin receptor: identification of a mutation in the leptin receptor gene in db/db mice. *Cell*, **84**, 491–495.

Chen, P., Li, C., Haskell-Luevano, C., Cone, R.D. and Smith, M.S. (1999) Altered expression of agouti-related protein and its colocalization with neuropeptide Y in the arcuate nucleus of the hypothalamus during lactation. *Endocrinology*, **140**, 2645–2650.

Chen, W., Kelly, M.A., Opitz-Araya, X., Thomas, R.E., Low, M.J. and Cone, R.D. (1997) Exocrine gland dysfunction in MC5-R-deficient mice: evidence for coordinated regulation of exocrine gland function by melanocortin peptides. *Cell*, **91**, 789–798.

Clement, K., Vaisse, C., Lahlou, N., *et al.* (1998) A mutation in the human leptin receptor gene causes obesity and pituitary dysfunction. *Nature*, **392**, 398–401. [*The description of humans with homozygous mutations in the gene encoding the leptin receptor resulting in severe early-onset obesity.*]

Cohen, M.A., Ellis, S.M., Le Roux, C.W., *et al.* (2003) Oxyntomodulin suppresses appetite and reduces food intake in humans. *Journal of Clinical Endocrinology and Metabolism*, **88**, 4696–4701.

Cone, R.D. (2005) Anatomy and regulation of the central melanocortin system. *Nature Neuroscience*, **8**, 571–578.

Cone, R.D. (2006) Studies on the physiological functions of the melanocortin system. *Endocrine Reviews*, **27**, 736–749.

Considine, R.V., Sinha, M.K., Heiman, M.L., *et al.* (1996) Serum immunoreactive-leptin concentrations in normal-weight and obese humans. *New England Journal of Medicine*, **334**, 292–295.

Dagher, A. (2012) Functional brain imaging of appetite. *Trends in Endocrinology and Metabolism*, **23**, 250–260.

Dakin, C.L., Small, C.J., Batterham, R.L., *et al.* (2004) Peripheral oxyntomodulin reduces food intake and body weight gain in rats. *Endocrinology*, **145**, 2687–2695.

De Luca, C., Kowalski, T.J., Zhang, Y., *et al.* (2005) Complete rescue of obesity, diabetes, and infertility in db/db mice by neuron-specific LEPR-B transgenes. *Journal of Clinical Investigation*, **115**, 3484–3493.

De Silva, A., Salem, V., Matthews, P.M. and Dhillo, W.S. (2012) The use of functional MRI to study appetite control in the CNS. *Experimental Diabetes Research*, **2012**, 764017.

Drucker, D.J. (2006) The biology of incretin hormones. *Cell Metabolism*, **3**, 153–165.

Elmquist, J.K., Bjorbaek, C., Ahima, R.S., Flier, J.S. and Saper, C.B. (1998) Distributions of leptin receptor mRNA isoforms in the rat brain. *Journal of Comparative Neurology*, **395**, 535–547.

Farooqi, I.S., Jebb, S.A., Langmack, G., *et al.* (1999) Effects of recombinant leptin therapy in a child with congenital leptin deficiency. *New England Journal of Medicine*, **341**, 879–884. **[The first demonstration that leptin replacement therapy works in humans with leptin deficiency.]**

Farooqi, I.S., Keogh, J.M., Yeo, G.S., Lank, E.J., Cheetham, T. and O'Rahilly, S. (2003) Clinical spectrum of obesity and mutations in the melanocortin 4 receptor gene. *New England Journal of Medicine*, **348**, 1085–1095.

Farooqi, I.S., Bullmore, E., Keogh, J., Gillard, J., O'Rahilly, S. and Fletcher, P.C. (2007a) Leptin regulates striatal regions and human eating behavior. *Science*, **317**, 1355. **[The first evidence of leptin playing a role in the 'hedonic' aspects of eating behavior. Leptin-deficient humans are unable to discriminate foods of different nutritional and reward value. However, upon leptin replacement, these patients are then able to make a 'value judgment' on food.]**

Farooqi, I.S., Wangensteen, T., Collins, S., *et al.* (2007b) Clinical and molecular genetic spectrum of congenital deficiency of the leptin receptor. *New England Journal of Medicine*, **356**, 237–247.

Fei, H., Okano, H.J., Li, C., *et al.* (1997) Anatomic localization of alternatively spliced leptin receptors (Ob-R) in mouse brain and other tissues. *Proceedings of the National Academy of Sciences of the USA*, **94**, 7001–7005.

Flier, J.S. (2006) AgRP in energy balance: will the real AgRP please stand up? *Cell Metabolism*, **3**, 83–85.

Fricker, L.D., Berman, Y.L., Leiter, E.H. and Devi, L.A. (1996) Carboxypeptidase E activity is deficient in mice with the fat mutation. Effect on peptide processing. *Journal of Biological Chemistry*, **271**, 30619–30624.

Friedman, J.M. and Halaas, J.L. (1998) Leptin and the regulation of body weight in mammals. *Nature*, **395**, 763–770.

Ghilardi, N. and Skoda, R.C. (1997) The leptin receptor activates janus kinase 2 and signals for proliferation in a factor-dependent cell line. *Molecular Endocrinology*, **11**, 393–399.

Gray, J., Yeo, G.S., Cox, J.J., *et al.* (2006) Hyperphagia, severe obesity, impaired cognitive function, and hyperactivity associated with functional loss of one copy of the brain-derived neurotrophic factor (BDNF) gene. *Diabetes*, **55**, 3366–3371.

Gray, J., Yeo, G., Hung, C., *et al.* (2007) Functional characterization of human NTRK2 mutations identified in patients with severe early-onset obesity. *International Journal of Obesity*, **31**, 359–364.

Greenfield, J.R., Miller, J.W., Keogh, J.M., *et al.* (2009) Modulation of blood pressure by central melanocortinergic pathways. *New England Journal of Medicine*, **360**, 44–52.

Gropp, E., Shanabrough, M., Borok, E., *et al.* (2005) Agouti-related peptide-expressing neurons are mandatory for feeding. *Nature Neuroscience*, **8**, 1289–1291.

Heymsfield, S.B., Greenberg, A.S., Fujioka, K., *et al.* (1999) Recombinant leptin for weight loss in obese and lean adults: a randomized, controlled, dose-escalation trial. *Journal of the American Medical Association*, **282**, 1568–1575.

Huszar, D., Lynch, C.A., Fairchild-Huntress, V., *et al.* (1997) Targeted disruption of the melanocortin-4 receptor results in obesity in mice. *Cell*, **88**, 131–141. **[*Report of Mc4r deficiency in mice and the fact tht it results in a dominantly inherited obesity. All the other previous models have been recessive disorders. Heterozygous mutant mice display an intermediate phenotype compared to homozygous mice.*]**

Jackson, R.S., Creemers, J.W., Ohagi, S., *et al.* (1997) Obesity and impaired prohormone processing associated with mutations in the human prohormone convertase 1 gene. *Nature Genetics*, **16**, 303–306.

Jackson, R.S., Creemers, J.W., Farooqi, I.S., *et al.* (2003) Small-intestinal dysfunction accompanies the complex endocrinopathy of human proprotein convertase 1 deficiency. *Journal of Clinical Investigation*, **112**, 1550–1560.

Kennedy, G.C. (1953) The role of depot fat in the hypothalamic control of food intake in the rat. *Proceedings of the Royal Society of London Series B*, **140**, 578–596.

Kernie, S.G., Liebl, D.J. and Parada, L.F. (2000) BDNF regulates eating behavior and locomotor activity in mice. *EMBO Journal*, **19**, 1290–1300.

Kishi, T., Aschkenasi, C.J., Lee, C.E., Mountjoy, K.G., Saper, C.B. and Elmquist, J.K. (2003) Expression of melanocortin 4 receptor mRNA in the central nervous system of the rat. *Journal of Comparative Neurology*, **457**, 213–235.

Kojima, M., Hosoda, H., Date, Y., Nakazato, M., Matsuo, H. and Kangawa, K. (1999) Ghrelin is a growth-hormone-releasing acylated peptide from stomach. *Nature*, **402**, 656–660.

Krashes, M.J., Koda, S., Ye, C., *et al.* (2011) Rapid, reversible activation of AgRP neurons drives feeding behavior in mice. *Journal of Clinical Investigation*, **121**, 1424–1428.

Krude, H., Biebermann, H., Luck, W., Horn, R., Brabant, G. and Gruters, A. (1998) Severe early-onset obesity, adrenal insufficiency and red hair pigmentation caused by POMC mutations in humans. *Nature Genetics*, **19**, 155–157. [*The direct evidence for pro-opiomelanocortin, the precursor for the melanocortin peptides including α- and β-MSH, having a critical role to play in the control of food intake and body weight, with mutations causing severe obesity, red hair, and isolated ACTH deficiency in humans.*]

Langlet, F., Levin, B.E., Luquet, S., *et al.* (2013a) Tanycytic VEGF-A boosts blood-hypothalamus barrier plasticity and access of metabolic signals to the arcuate nucleus in response to fasting. *Cell Metabolism*, **17**, 607–617.

Langlet, F., Mullier, A., Bouret, S.G., Prevot, V. and Dehouck, B. (2013b) Tanycyte-like cells form a blood-cerebrospinal fluid barrier in the circumventricular organs of the mouse brain. *Journal of Comparative Neurology*, **521**, 3389–3405.

Le Roux, C.W., Aylwin, S.J., Batterham, R.L., *et al.* (2006a) Gut hormone profiles following bariatric surgery favor an anorectic state, facilitate weight loss, and improve metabolic parameters. *Annals of Surgery*, **243**, 108–114.

Le Roux, C.W., Batterham, R.L., Aylwin, S.J., *et al.* (2006b) Attenuated peptide YY release in obese subjects is associated with reduced satiety. *Endocrinology*, **147**, 3–8.

Lee, G.H., Proenca, R., Montez, J.M., *et al.* (1996) Abnormal splicing of the leptin receptor in diabetic mice. *Nature*, **379**, 632–635.

Liu, H., Kishi, T., Roseberry, A.G., *et al.* (2003) Transgenic mice expressing green fluorescent protein under the control of the melanocortin-4 receptor promoter. *Journal of Neuroscience*, **23**, 7143–7154.

Lu, D., Willard, D., Patel, I.R., *et al.* (1994) Agouti protein is an antagonist of the melanocyte-stimulating-hormone receptor. *Nature*, **371**, 799–802.

Luquet, S., Perez, F.A., Hnasko, T.S. and Palmiter, R.D. (2005) NPY/AgRP neurons are essential for feeding in adult mice but can be ablated in neonates. *Science*, **310**, 683–685.

Malik, S., McGlone, F., Bedrossian, D. and Dagher, A. (2008) Ghrelin modulates brain activity in areas that control appetitive behavior. *Cell Metabolism*, **7**, 400–409.

Mercer, J.G., Hoggard, N., Williams, L.M., Lawrence, C.B., Hannah, L.T. and Trayhurn, P. (1996) Localization of leptin receptor mRNA and the long form splice variant (Ob–Rb) in mouse hypothalamus and adjacent brain regions by in situ hybridization. *FEBS Letters*, **387**, 113–116.

Michaud, E.J., Bultman, S.J., Klebig, M.L., *et al.* (1994) A molecular model for the genetic and phenotypic characteristics of the mouse lethal yellow (Ay) mutation. *Proceedings of the National Academy of Sciences of the USA*, **91**, 2562–2566.

Miller, M.W., Duhl, D.M., Vrieling, H., *et al.* (1993) Cloning of the mouse agouti gene predicts a secreted protein ubiquitously expressed in mice carrying the lethal yellow mutation. *Genes & Development*, **7**, 454–467.

Montague, C.T., Farooqi, I.S., Whitehead, J.P., *et al.* (1997) Congenital leptin deficiency is associated with severe early-onset obesity in humans. *Nature*, **387**, 903–908. [*The first report of a human monogenic obesity syndrome, when two cousins were found to have homozygous loss-of-function mutations in their leptin gene, and showing the relevance of leptin signaling in the control of human food intake and body weight.*]

Mountjoy, K.G., Robbins, L.S., Mortrud, M.T. and Cone, R.D. (1992) The cloning of a family of genes that encode the melanocortin receptors. *Science*, **257**, 1248–1251.

Mountjoy, K.G., Mortrud, M.T., Low, M.J., Simerly, R.B. and Cone, R.D. (1994) Localization of the melanocortin-4 receptor (MC4-R) in neuroendocrine and autonomic control circuits in the brain. *Molecular Endocrinology*, **8**, 1298–1308.

Mullier, A., Bouret, S.G., Prevot, V. and Dehouck, B. (2010) Differential distribution of tight junction proteins suggests a role for tanycytes in blood-hypothalamus barrier regulation in the adult mouse brain. *Journal of Comparative Neurology*, **518**, 943–962.

Murphy, K.G. and Bloom, S.R. (2006) Gut hormones and the regulation of energy homeostasis. *Nature*, **444**, 854–859.

Myers, M.G., Cowley, M.A. and Munzberg, H. (2008) Mechanisms of leptin action and leptin resistance. *Annual Review of Physiology*, **70**, 537–556.

Ollmann, M.M., Wilson, B.D., Yang, Y.K., *et al.* (1997) Antagonism of central melanocortin receptors in vitro and in vivo by agouti-related protein. *Science*, **278**, 135–138. [*Report of the cloning of AgRP and a description of how it plays an important role in the control of food intake by competing with α-MSH to bind to the MC4R, where it acts as an antagonist, resulting in an increase in food intake.*]

Peelman, F., Iserentant, H., De Smet, A.S., Vandekerckhove, J., Zabeau, L. and Tavernier, J. (2006) Mapping of binding site III in the leptin receptor and modeling of a hexameric leptin.leptin receptor complex. *Journal of Biological Chemistry*, **281**, 15496–15504.

Pelleymounter, M.A., Cullen, M.J., Baker, M.B., *et al.* (1995) Effects of the obese gene product on body weight regulation in ob/ob mice. *Science*, **269**, 540–543.

Pinto, S., Roseberry, A.G., Liu, H., *et al.* (2004) Rapid rewiring of arcuate nucleus feeding circuits by leptin. *Science*, **304**, 110–115. [*A seminal paper first describing the neurotropic effects of leptin, and how rapid rewiring of synaptic projections represents an important component of signaling circuitry within the brain.*]

Qian, S., Chen, H., Weingarth, D., *et al.* (2002) Neither agouti-related protein nor neuropeptide Y is critically required for the regulation of energy homeostasis in mice. *Molecular and Cellular Biology*, **22**, 5027–5035.

Rios, M., Fan, G., Fekete, C., *et al.* (2001) Conditional deletion of brain-derived neurotrophic factor in the postnatal brain leads to obesity and hyperactivity. *Molecular Endocrinology*, **15**, 1748–1757.

Scarpace, P.J. and Zhang, Y. (2009) Leptin resistance: a predisposing factor for diet-induced obesity. *American Journal of Physiology: Regulatory, Integrative and Comparative Physiology*, **296**, R493–500.

Schaeffer, M., Langlet, F., Lafont, C., *et al.* (2013) Rapid sensing of circulating ghrelin by hypothalamic appetite-modifying neurons. *Proceedings of the National Academy of Sciences of the USA*, **110**, 1512–1517.

Shutter, J.R., Graham, M., Kinsey, A.C., Scully, S., Luthy, R. and Stark, K.L. (1997) Hypothalamic expression of ART, a novel gene related to agouti, is up-regulated in obese and diabetic mutant mice. *Genes & Development*, **11**, 593–602.

Ste Marie, L., Miura, G.I., Marsh, D.J., Yagaloff, K. and Palmiter, R.D. (2000) A metabolic defect promotes obesity in mice lacking melanocortin-4 receptors. *Proceedings of the National Academy of Sciences of the USA*, **97**, 12339–12344.

Stunkard, A.J., Foch, T.T. and Hrubec, Z. (1986) A twin study of human obesity. *Journal of the American Medical Association*, **256**, 51–54.

Stunkard, A.J., Harris, J.R., Pedersen, N.L. and McClearn, G.E. (1990) The body-mass index of twins who have been reared apart. *New England Journal of Medicine*, **322**, 1483–1487. [*These two papers by Stunkard et al. are seminal publications, setting up the now widely accepted notion that genetics plays a critical role in the determination of body shape and size. Stunkard first studied fraternal versus identical twins, then followed this up with a study on twins raised apart, thereby showing that your genes play a larger role in determining your body weight than the environment you are raised in.*]

Tapia-Arancibia, L., Rage, F., Givalois, L. and Arancibia, S. (2004) Physiology of BDNF: focus on hypothalamic function. *Frontiers in Neuroendocrinology*, **25**, 77–107.

Tartaglia, L.A. (1997) The leptin receptor. *Journal of Biological Chemistry*, **272**, 6093–6096.

Tartaglia, L.A., Dembski, M., Weng, X., *et al.* (1995) Identification and expression cloning of a leptin receptor, OB-R. *Cell*, **83**, 1263–1271.

Thornton, J.E., Cheung, C.C., Clifton, D.K. and Steiner, R.A. (1997) Regulation of hypothalamic proopiomelanocortin mRNA by leptin in ob/ob mice. *Endocrinology*, **138**, 5063–5066.

Tschop, M., Smiley, D.L. and Heiman, M.L. (2000) Ghrelin induces adiposity in rodents. *Nature*, **407**, 908–913. [*The very first description of ghrelin's role as a hormone that powerfully stimulates food intake and increasing body weight. It is still the only known circulating orexigenic hormone.*]

Vaisse, C., Clement, K., Guy-Grand, B. and Froguel, P. (1998) A frameshift mutation in human MC4R is associated with a dominant form of obesity. *Nature Genetics*, **20**, 113–114. [*One of two papers published at the same time reporting that mutations in the MC4R lead to dominantly inherited obesity. Mutations in MC4R still represent the most common monogenic cause of human obesity to date.*]

Valverde, P., Healy, E., Jackson, I., Rees, J.L. and Thody, A.J. (1995) Variants of the melanocyte–stimulating hormone receptor gene are associated with red hair and fair skin in humans. *Nature Genetics*, **11**, 328–330.

White, D.W., Kuropatwinski, K.K., Devos, R., Baumann, H. and Tartaglia, L.A. (1997) Leptin receptor (OB-R) signaling. Cytoplasmic domain mutational analysis and evidence for receptor homo–oligomerization. *Journal of Biological Chemistry*, **272**, 4065–4071.

Williams, D.L. and Cummings, D.E. (2005) Regulation of ghrelin in physiologic and pathophysiologic states. *Journal of Nutrition*, **135**, 1320–1325.

Wu, Q., Boyle, M.P. and Palmiter, R.D. (2009) Loss of GABAergic signaling by AgRP neurons to the parabrachial nucleus leads to starvation. *Cell*, **137**, 1225–1234.

Wynne, K., Park, A.J., Small, C.J., *et al.* (2005) Subcutaneous oxyntomodulin reduces body weight in overweight and obese subjects: a double-blind, randomized, controlled trial. *Diabetes*, **54**, 2390–2395.

Wynne, K., Park, A.J., Small, C.J., *et al.* (2006) Oxyntomodulin increases energy expenditure in addition to decreasing energy intake in overweight and obese humans: a randomised controlled trial. *International Journal of Obesity*, **30**, 1729–1736.

Xu, A.W., Kaelin, C.B., Morton, G.J., *et al.* (2005) Effects of hypothalamic neurodegeneration on energy balance. *PLoS Biology*, **3**, e415.

Xu, A.W., Ste Marie, L., Kaelin, C.B. and Barsh, G.S. (2007) Inactivation of signal transducer and activator of transcription 3 in proopiomelanocortin (Pomc) neurons causes decreased pomc expression, mild obesity, and defects in compensatory refeeding. *Endocrinology*, **148**, 72–80.

Xu, B., Goulding, E.H., Zang, K., *et al.* (2003) Brain-derived neurotrophic factor regulates energy balance downstream of melanocortin-4 receptor. *Nature Neuroscience*, **6,** 736–742.

Yang, J., Brown, M.S., Liang, G., Grishin, N.V. and Goldstein, J.L. (2008) Identification of the acyltransferase that octanoylates ghrelin, an appetite-stimulating peptide hormone. *Cell*, **132**, 387–396.

Yaswen, L., Diehl, N., Brennan, M.B. and Hochgeschwender, U. (1999) Obesity in the mouse model of pro-opiomelanocortin deficiency responds to peripheral melanocortin. *Nature Medicine*, **5**, 1066–1070.

Yeo, G.S. and Heisler, L.K. (2012) Unraveling the brain regulation of appetite: lessons from genetics. *Nature Neuroscience*, **15**, 1343–1349.

Yeo, G.S., Farooqi, I.S., Aminian, S., Halsall, D.J., Stanhope, R.G. and O'Rahilly, S. (1998) A frameshift mutation in MC4R associated with dominantly inherited human obesity. *Nature Genetics*, **20**, 111–112. [*One of two papers published at the same time first reporting that mutations in the MC4R lead to dominantly inherited human obesity. Mutations in MC4R still represent the most common monogenic cause of human obesity to date.*]

Yeo, G.S., Connie Hung, C.C., Rochford, J., *et al.* (2004) A de novo mutation affecting human TrkB associated with severe obesity and developmental delay. *Nature Neuroscience*, **7**, 1187–1189. [*The first paper implicating the neurotropin receptor TrkB in the control of human food intake and body weight, with mutations in the gene leading to dominantly inherited obesity and developmental delay.*]

Zabeau, L., Defeau, D., van der Heyden, J., Iserentant, H., Vandekerckhove, J. and Tavernier, J. (2004) Functional analysis of leptin receptor activation using a Janus kinase/signal transducer and activator of transcription complementation assay. *Molecular Endocrinology*, **18**, 150–161.

Zhang, Y., Proenca, R., Maffei, M., Barone, M., Leopold, L. and Friedman, J.M. (1994) Positional cloning of the mouse obese gene and its human homologue. *Nature*, **372**, 425–432. [*The paper describing the cloning of leptin, that mutations in the leptin gene caused the severe obesity seen in ob/ob mice, and that recombinant leptin treatment reversed the obesity.*]

Zhou, A., Bloomquist, B.T. and Mains, R.E. (1993) The prohormone convertases PC1 and PC2 mediate distinct endoproteolytic cleavages in a strict temporal order during proopiomelanocortin biosynthetic processing. *Journal of Biological Chemistry*, **268**, 1763–1769.

Zhu, X., Zhou, A., Dey, A., *et al.* (2002) Disruption of PC1/3 expression in mice causes dwarfism and multiple neuroendocrine peptide processing defects. *Proceedings of the National Academy of Sciences of the USA*, **99**, 10293–10298.

CHAPTER 17

Stress Adaptation and the Hypothalamic-Pituitary-Adrenal Axis

Greti Aguilera

Section on Endocrine Physiology, National Institute of Child Health and Human Development, National Institutes of Health, Bethesda, Maryland, USA

17.1 Stress and stress response

Living organisms are continually challenged by external or internal threats to homeostasis, defined as constancy of the internal environment (McEwen, 2007), and their survival and evolutionary success depend on their ability to adjust to these challenges or stressors. The neuroendocrine system plays a key role in mediating coordinated behavioral, autonomic (sympathoadrenal), and endocrine responses (mainly hypothalamic-pituitary-adrenal axis) necessary to maintain homeostasis. Initiation of the stress response requires sensory mechanisms such as baroreceptors or glucose sensors for biogenic or systemic stressors, or the senses (e.g. hearing, vision) for psychosocial stressors. The neural circuitry involved in processing and integrating this sensory information depends on the nature of the stress (Ulrich Lai and Herman, 2009). Physical and metabolic stressors requiring immediate response, such as loss of blood volume or immune challenge, pain and hypoglycemia, involve adrenergic and noradrenergic monosynaptic ascending pathways from the brainstem and spinal cord, with direct projections to the hypothalamic paraventricular nucleus (PVN). In contrast, psychogenic stressors utilize complex polysynaptic pathways, involving limbic structures, such as the prefrontal cortex, amygdala, bed nucleus of the stria terminalis and hippocampus to activate endocrine, autonomic, and behavioral responses. Behavioral responses including arousal, increased awareness and improved cognition are necessary for avoidance and defense reactions, while autonomic and metabolic responses ensure the increased energy supply and cardiovascular preparedness required for coping with stress. In contrast to the transient and beneficial responses to acute stress, chronic stress requires additional mechanisms that can change the set point of the systems mediating the adaptive processes. Depending on the intensity and duration of the stress, psychosocial environment, genetic background, and previous life experiences, stress can lead to psychological, cardiovascular, and metabolic morbidity (Myers *et al.*, 2014).

Thus uncovering the basic mechanisms underlying normal stress adaptation is essential for understanding the pathophysiology and finding new approaches for the diagnosis and treatment of stress-related disorders.

Molecular Neuroendocrinology: From Genome to Physiology, First Edition. Edited by David Murphy and Harold Gainer.
© 2016 John Wiley & Sons, Ltd. Published 2016 by John Wiley & Sons, Ltd.
Companion website: www.wiley.com/go/murphy/neuroendocrinology

17.1.1 The HPA axis is a key component in stress adaptation

The major neuroendocrine response to stress is activation of the hypothalamic-pituitary-adrenal (HPA) axis, leading ultimately to increases in circulating glucocorticoids: the 11-hydroxysteroid cortisol in humans, and the 11-ketosteroid corticosterone in most rodents. Release of corticotropin-releasing hormone (CRH) and vasopressin (VP) from parvocellular neurons of the hypothalamic paraventricular nucleus into the pituitary portal circulation stimulates adrenocorticotropic hormone (ACTH) secretion from corticotropes in the anterior pituitary gland, which in turn stimulates glucocorticoid synthesis and secretion from the adrenal cortex (Figure 17.1). Glucocorticoids increase energy supply, by stimulating glycolysis and gluconeogenesis, and by providing substrates for gluconeogenesis through lipolysis and proteolysis. In addition, glucocorticoids regulate immune function, memory acquisition, and consolidation, and the synthesis and action of a number of neurotransmitters and hormones involved in the stress response (Witchel and DeFranco, 2006).

Glucocorticoids and other components of the HPA axis, CRH, and VP released from hypothalamic and extrahypothalamic sources play a critical role in controlling autonomic and behavioral responses to stress and in inhibiting non-essential functions including feeding and reproductive behaviors (Beurel and Nemeroff, 2014). The main site of production of CRH is the hypothalamic paraventricular nucleus but the peptide is also expressed in limbic areas of the brain, such as the amygdala, bed nucleus of the stria terminalis, locus coeruleus and cortical

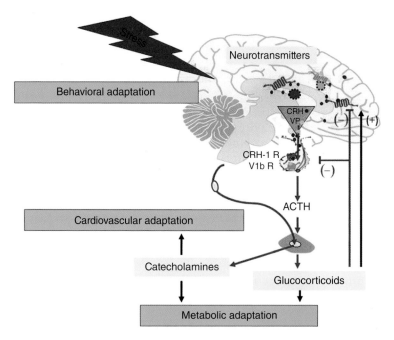

Figure 17.1 Schematic representation of the interactions between activation of the hypothalamic-pituitary-adrenal (HPA) axis and major adaptive responses to stress. Stress stimulates the release of corticotropin-releasing hormone (CRH) and vasopressin (VP) from the paraventricular nucleus of the hypothalamus (PVN) into the pituitary portal circulation. Binding of CRH to CRHR1 and VP to V1b receptors in the pituitary corticotrope stimulates ACTH secretion which stimulates adrenal glucocorticoid secretion. Glucocorticoids are essential for metabolic adaptation, exert feedback inhibition on the HPA axis (*red*), and modulate brain function by positive or negative regulation of a number of genes. In addition to stimulating the HPA axis, CRH released within the brain binds to CRHR1 in the brain and contributes to behavioral and autonomic adaptation to stress.

areas (Swanson *et al.*, 1983). This extrahypothalamic CRH expression and release within the brain plays a critical role in mediating behavioral and autonomic responses to stress, including arousal, decrease in feeding and sexual behaviors and increased sympathoadrenal activity. In addition, glucocorticoids and central CRH have prominent roles in the integration of the HPA axis with the function of other neuroendocrine systems, such as the hypothalamic-pituitary-gonadal, hypothalamic-pituitary-thyroid, and hypothalamic-growth hormone axes, during stress. Considering the broad actions of glucocorticoids as well as CRH and VP, it is clear that HPA axis dysregulation with either deficient responses or failure to limit activation has profound consequences for the wellbeing of the organism (Uchoa *et al.*, 2014) (see Figure 17.1).

17.1.2   Basal and stress-stimulated HPA axis activity

In basal conditions, HPA axis activity follows circadian (or daily) rhythms with increased glucocorticoid levels during the wake phase: daytime in humans and nighttime in nocturnal animals. Overlaying circadian rhythms, pituitary ACTH release and adrenal glucocorticoid secretion are episodic, displaying hourly or ultradian pulses (Spiga *et al.*, 2014). Microdialysis studies in rats have shown that corticosterone pulses in blood parallel pulses in free corticosterone, and that peaks in plasma precede highly synchronized peaks of free corticosterone in subcutaneous tissue and in the hippocampus (Qian *et al.*, 2012).

Acute stress increases HPA axis activity with marked but transient increases in circulating ACTH and glucocorticoids. In contrast to the episodic activation during acute stress, chronic or repeated stress elicits variable HPA axis responses according to the intensity and frequency of the stimulus and the ability of the individual to control the stress. Studies in experimental animals have shown two major patterns of response, the first with habituation or decline of ACTH and glucocorticoid secretion after several hours of stress exposure. Interestingly, in spite of habituation to a persistent stressor, there is an increase in HPA axis responsiveness to a heterotypical (or novel) stressor, indicating that the HPA axis is capable of responding to a new challenge. The second pattern involves no habituation and persistently elevated circulating levels of ACTH and glucocorticoids (Aguilera, 1994). Depending on the intensity and context, prolonged exposure to this type of stressor or failure to habituate can lead to either sustained increases in plasma glucocorticoids with loss of rhythmicity or exhaustion of the response and adrenal failure.

17.2   Molecular mechanisms of glucocorticoid action

17.2.1   Glucocorticoid receptors

The biological effects of glucocorticoids are mediated by evolutionary conserved intracellular receptors belonging to the nuclear receptor family. These include the type 1 glucocorticoid receptor or mineralocorticoid receptor (MR), and the type 2 receptor or glucocorticoid receptor (GR) (Joels and de Kloet, 1994). The genes of MRs and GRs originated from a common ancestor, the corticoid receptor, before the evolutionary emergence of the potent mineralocorticoid aldosterone (Baker *et al.*, 2013). There are two GR splice variants. GR-α is the major subtype responsible for glucocorticoid actions. The GR-β isoform differs on its carboxy terminus and lacks ligand-binding activity. GR-β resides constitutively in the nucleus and can act as a dominant negative (Oakley and Cidlowski, 2013).

Both GR and MR bind glucocorticoids and mineralocorticoid and interact with the same DNA sequence, the palindromic sequence AGAACAnnnTGTTCT, called glucocorticoid response element (GRE), and specificity depends exclusively on the access of the receptors to the steroid. Glucocorticoids circulating concentrations are much higher than mineralocorticoids and since the MR has higher affinity than GR for glucocorticoids, MRs are largely occupied at low circulating glucocorticoid concentrations (Figure 17.2).

(A)

Receptor	kd (nM)		DNA response element
	Cort	Aldo	
GR or NR3C1	~3	0.8	AGAACANNNTGTTCT
MR or NR3C2	~0.5	0.8	AGAACANNNTGTTCT

(B)

Steroid	Blood levels (nM)		Protein bound (%)
	Basal	Stress	
Corticosterone	10–10	100–1000	95
Aldosterone	0.2–0.4	0.8–1.5	65

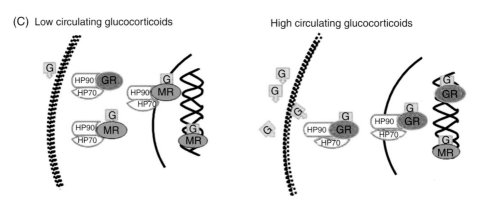

(C) Low circulating glucocorticoids   High circulating glucocorticoids

Figure 17.2 (A) Properties of glucocorticoid (GR) and mineralocorticoid (MR) receptors. Both GR and MR bind both ligands, glucocorticoids and mineralocorticoids, and are recruited by the same response element in the chromatin. (B) Glucocorticoids circulate at concentrations about 1000-fold greater than mineralocorticoids and a large proportion is bound to corticosteroid-binding globulin (CBG) or albumin. (C) In the absence of ligand, MR and GR in the cytoplasm are associated with a protein complex including heat shock proteins (HSP) 90 and 70. Low basal circulating levels of glucocorticoids bind and activate only the high-affinity MR, while higher levels in the circadian peak of stress range bind and activate GR.

The main site of expression of the type 1 receptors, MR, is the distal tubule of kidney glomeruli where they regulate aldosterone-mediated sodium reabsorption (Funder, 2005). In the kidney and other mineralocorticoid target tissues, MRs are protected from circulating glucocorticoids by 11-β-dehydrogenase type 2 (11β-HSD2), which inactivates glucocorticoids by converting them to 11-ketosteroids (Benediktsson *et al.*, 1995) (Figure 17.3). In contrast to mineralocorticoid target tissues, MRs expressed in limbic areas of the brain, such as hippocampus, amygdala, and frontal cortex responds to glucocorticoid since these sites express 11β-HSD type 1, which regenerates active glucocorticoids from inactive 11-ketosteroids (Chapman *et al.*, 2013). Mineralocorticoid receptors of the hippocampus play a key role mediating negative glucocorticoid feedback of the HPA axis in basal conditions at the nadir of the circadian rhythm when hippocampal MRs are significantly occupied by glucocorticoids (Berardelli *et al.*, 2013). Since limbic sites do not express type 11β-HSD2, activation of MR in these areas solely depends on glucocorticoids. The nucleus of the solitary tract (NTS), an area involved in cardiovascular regulation and sodium appetite, is the only brain region co-expressing 11β-HSD2 and MRs (Geerling and Loewy, 2009), and therefore capable of responding to mineralocorticoids.

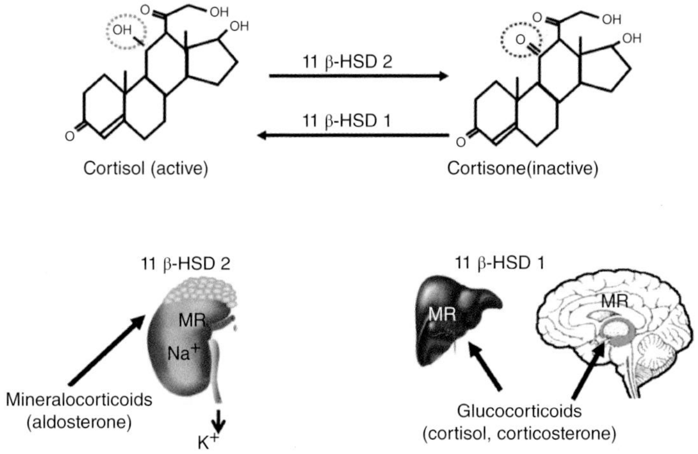

Figure 17.3 The specificity of MR for mineralocorticoids depends on the expression of 11β-hydroxysteroid dehydrogenase type 2 (11β-HSD2) which inactivates glucocorticoids and allows access of mineralococorticoids to the MR in mineralocorticoid target tissues like the kidney. In other tissues, including liver and brain, the presence of the reductase 11β-HSD1 increases tissue glucocorticoid exposure by converting the inactive metabolite to the active glucocorticoid.

Glucocorticoid receptors are widely expressed in the peripheral and central nervous system, including the hypothalamus and hippocampus. Under normal conditions, brain GRs are essentially ligand free at the nadir of the circadian rhythm but occupied by peak circadian or stress-induced levels of glucocorticoids (GCs). Activation of central GRs during stress regulates neuropeptide and neurotransmitter synthesis, memory acquisition, and consolidation, by promoting hippocampal neuron plasticity (Herbert *et al.*, 2006) and negative feedback regulation of HPA responsiveness (Uchoa *et al.*, 2014). Similarly, transcriptional activity of GR in the periphery on metabolic enzymes, circadian clock, and other proteins depends on ultradian pulses of glucocorticoids and tissue expression of 11β-HSD. Upregulation of 11β-HSD1 in the liver during chronic stress and iatrogenic glucocorticoid excess has been implicated in the pathogenesis of the metabolic syndrome (Chapman *et al.*, 2013; Morgan *et al.*, 2014).

An additional factor determining glucocorticoid activity is the levels of corticosteroid-binding globulin (CBG). This is a blood protein produced in the liver with high affinity for glucocorticoids and most circulating glucocorticoids circulate bound to CBG and a small proportion to albumin. Free glucocorticoid concentration is about 5% in humans and 35% in rats and mice. Levels of CBG undergo changes in a number of conditions, and the affinity of glucocorticoids for the protein decreases with elevations of temperature and pH within the physiological range. In addition, the location of CBG in neurons and pituitary gland suggests that CBG has a role in transporting glucocorticoids into the cell. Thus, alterations in steroid/protein dissociation rate or CBG levels can influence free circulating glucocorticoid levels and the access of the target cell (Henley and Lightman, 2011).

17.2.2   Genomic actions of glucocorticoids

Glucocorticoid and MRs are ligand-activated transcription factors. In basal conditions, they are located in the cytoplasm, forming complexes with chaperone proteins such as heat shock proteins (HSP) 90 and 70, the 23 kDa protein p23, and co-chaperones including the immunophilin FK506 binding protein 5 and the non-receptor kinase Src (Figure 17.4A). Glucocorticoids freely cross the cell membrane and upon binding, the ligand-receptor

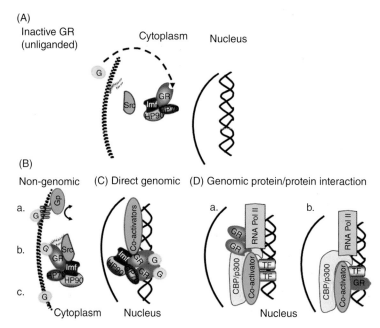

Figure 17.4 Genomic and non-genomic actions of glucocorticoids. In basal conditions, GRs remain in the cytoplasm as part of a protein complex containing heat shock proteins (HSP) 70 and 90, immunophilins (Imf), the not receptor kinase Src and other proteins (A). Upon ligand binding, the GR is activated and the complex translocates to the nucleus, and regulates gene transcription by binding to a glucocorticoid response element (GRE) in the gene promoter (C), or through protein-protein interaction, by forming part of the protein complex (Da), or by binding to the promoter as a dimer with other transcription factors (Db). Proposed mechanisms for rapid non-genomic actions include association of the liganded GR with a membrane protein (Ba), binding of glucocorticoids to a putative G protein-coupled receptor (B-b), and non-specific interaction with the plasma membrane (Bc).

complex translocates to the nucleus. Against the conventional belief, HSP90, other chaperone and co-chaperone proteins are part of the mechanism of receptor activation and translocation, and subsequent GR transactivation effects (Harrell *et al.*, 2004). Once in the nucleus, GRs form dimers and bind to the promoter region of glucocorticoid-dependent genes. This induces recruitment of co-activators or co-repressors, leading to transcriptional activation or repression (Figure 17.4C). In some genes, activated GR does not bind directly to chromatin but regulates gene transcription by protein-protein interaction with a transcriptional complex (Figure 17.4Da), or by forming dimers with other transcription factors (Figure 17.4Db). Several repressor effects of glucocorticoids, including negative feedback for the HPA axis at the pituitary level, operate through these protein-protein interactions (Jenks, 2009).

Although both GRs and MRs bind to the same response element, their mode of interaction with the gene promoters is different, with GRs displaying a highly dynamic interaction with the chromatin and MRs showing prolonged association. Imaging-based studies using genome integrated GR promoter arrays have shown that GR cycles between chromatin bound and unbound within seconds (McNally *et al.*, 2000). This process requires the scaffolding proteins HSP90 and p23 for GR association to the chromatin, while the proteasome is essential for GR dissociation (Stavreva *et al.*, 2004).

The dynamic nature of GR interaction with chromatin is critical for transcriptional activity of glucocorticoid-dependent genes and requires episodic patterns of glucocorticoid secretion. *In vitro* and *in vivo* studies show that ultradian hormone stimulation induces cyclic GR-mediated transcriptional regulation and that receptor occupancy of regulatory elements parallels the ligand pulses (Stavreva *et al.*, 2009). Glucocorticoid receptor activation and reactivation in response to glucocorticoid secretion episodes couple transcriptional activity to physiological fluctuations in hormone levels. This emphasizes the importance of maintaining physiological patterns of glucocorticoid secretion with hourly ultradian pulses in basal conditions and transient increases during stress. Increased circulating levels of glucocorticoids during stress prolong the residence time of GR in the chromatin, increasing transcriptional output necessary for stress adaptation (Voss and Hager, 2014).

During acute stress or chronic stress paradigms with habituation of HPA axis responses, the glucocorticoid surge is transient and transcriptional activity returns to basal levels. However, prolonged elevation of circulating glucocorticoids or loss in ultradian and circadian rhythmicity can lead to severe metabolic alterations due to either prolonged activation or desensitization of glucocorticoid-dependent transcription.

17.2.3 Non-genomic actions of glucocorticoids

Glucocorticoids exert actions that are too rapid to be mediated through the classic genomic pathway. These actions, referred to as the non-classic/non-genomic pathway, occur within seconds to minutes and are insensitive to inhibitors of transcription and translation (Lowenberg *et al.*, 2007; Uchoa *et al.*, 2014). Several studies over the past decade have provided clear evidence for the involvement of rapid non-genomic actions of glucocorticoids on negative feedback in CRH neurons and in regulating neuronal activity in the hippocampus (Sarabdjitsingh *et al.*, 2012).

Potential mechanisms underlying non-genomic effects of glucocorticoids include the following.

- Binding of glucocorticoids to a yet unidentified guanyl nucleotide binding protein (G protein)-coupled membrane glucocorticoid receptor (Figure 17.4Ba). This possibility is supported by studies showing rapid effects of glucocorticoids, insensitive to GR antagonists, such as RU486, sensitive to G protein inhibitors, and reproducible with albumin-linked glucocorticoids (Di *et al.*, 2003).
- Binding of glucocorticoids to the classic GR leading either to activation of chaperone molecules such as Src and/or direct association of GR with the membrane and activation of membrane-associated signaling transduction pathways (Samarasinghe *et al.*, 2011) (Figure 17.4Bb).
- Lastly, non-specific physicochemical interactions of the GC with the cell membrane have been proposed (Figure 17.4Bc) (Buttgereit *et al.*, 1998). This type of interaction occurs with high concentrations of glucocorticoids *in vitro*, but it is unlikely to happen in physiological conditions *in vivo*, even during therapeutic use of glucocorticoids.

17.3 Regulation of HPA axis activity during stress

Maintenance of circadian and ultradian rhythmicity and adequate HPA axis responses to stress requires a delicate balance between stimulatory and inhibitory signals mediating interactions between the hypothalamus, pituitary, and adrenal glands. This involves the participation of numerous molecular events, initiated by activation of receptors for regulatory neurotransmitter and neuropeptides triggering stimulation of signaling transduction pathways, and consecutive changes in membrane potential, hormone secretion, gene transcription, mRNA translation, and posttranslational processing of the many proteins involved in HPA axis regulation.

17.3.1   The hypothalamic paraventricular nucleus
and regulation of CRH and VP

The hypothalamic paraventricular nucleus (PVN) is a major relay nucleus for integrating stress information and initiating HPA axis and autonomic responses to stress. The PVN comprises three main functional regions controlling stress responses (Swanson and Sawchenko, 1980). First, parvocellular neurons in the dorsomedial and anterior divisions of the PVN release peptides to the pituitary portal circulation from axonal terminals in the external zone of the median eminence. These neurons express CRH and are a major target of the neural pathways activated during stress. In all species studied, including rat, mouse and human, a proportion of CRH neurons also express VP, which is co-released with CRH and contributes to HPA axis regulation. The second group comprises magnocellular vasopressinergic and oxytocinergic neurons projecting axons to the posterior pituitary (neurohypophysis), and responsible for releasing VP to the peripheral circulation in response to osmotic stress. Magnocellular vaso-pressinergic axons in the median eminence can release VP to the pituitary portal circulation and probably contribute to the increases in ACTH secretion during acute osmotic stimulation (Holmes *et al.*, 1986). The third group includes neurons located in the lateral parvocellular, dorsal parvocellular, and ventromedial parvocellular divisions, expressing CRH and other neuropeptides. These neurons have afferent and efferent projections from/to nuclei in the brainstem and play an essential role controlling autonomic responses to stress (Swanson and Sawchenko, 1980).

Acute stress induces rapid and transient activation of parvocellular CRH neurons leading to rapid release of CRH and VP into the pituitary portal circulation (Plotsky, 1991). This is associated with increases in early expression genes and CRH transcription as shown by rapid (within minutes) and transient increases in primary transcript or heteronuclear RNA (hnRNA), followed by elevation of steady-state mRNA levels and translation to the precursor peptides, pro-CRH, and arginine vasopressin-neurophysin II, respectively (Aguilera and Liu, 2012). Consistent with the primary role of CRH regulating ACTH secretion, changes in CRH expression during chronic stress parallel the pattern of ACTH secretion, with habituation or sustained activation depending on the stress paradigm. Acute stress also induces a delayed and prolonged increase in VP transcription in CRH neurons, which persist during chronic stress, with sustained increases in mRNA in perikarya and immunoreactive peptide in the external zone of the median eminence (Aguilera *et al.*, 2008). The mechanism of this differential regulation of CRH and VP expression in the same neuron is still unclear.

17.3.2   Molecular mechanisms regulating hypothalamic CRH synthesis

Maintenance of releasable pools of CRH in nerve endings of the median eminence following stress-induced CRH release requires *de novo* synthesis of peptide. In resting conditions, there are relatively high levels of mature CRH mRNA ready to initiate translation. Therefore, the rapid transcriptional activation observed in CRH neurons following stress is necessary for restoring mRNA utilized for translation rather than producing peptide for the immediate secretory response.

The CRH gene has two exons and one short intron located in the 5' untranslated region (UTR). The sequence of the CRH gene including the proximal 5' flanking region is highly conserved between species, and it is almost identical for human and rat. It contains two TATA binding protein sites and a number of responsive elements, including a cyclic AMP responsive element (CRE) at position -247, a C/EBP site, an atypical GRE/AP1 element located at -249 to -248, half estrogen responsive elements (ERE), and a neuron-restrictive silencing element (RE-1/NRSE) element located in the intron (Aguilera and Liu, 2012). Some of these sites, including the CRE, GRE/AP1 and REST, are functional in reporter gene assays in heterologous cell lines but, with the exception of the -247 CRE, their importance in physiological conditions remains unclear. The CRE has been the most studied *in vitro* and *in vivo* and it is clearly essential for positive and negative regulation of CRH transcription. In addition, the proximal promoter contains a number of potential sites for the POU homeodomain protein Brn-2, which is required for the development of CRH neurons (Ingraham *et al.*, 1990).

17.3.2.1 Activation of CRH transcription requires cyclic AMP-dependent signaling

Activation of the CRH neuron during stress is associated with induction of a number of immediate early genes, such as c-fos, Fra-2, zif-268/Egr-1, the Nr4a family factors, Nur-77 and Nor-1 and NGF1B all potentially involved in transcriptional regulation. Stress also induces phosphorylation and nuclear translocation of CREB and the MAP kinase ERK in CRH neurons (Khan *et al.*, 2011). However, with the exception of CREB, the exact interaction of these factors with response elements in the CRH promoter remains unknown.

Activation of CRH transcription depends on cAMP/protein kinase A (PKA) signaling, leading to phosphorylation of cyclic AMP response element binding protein (pCREB) and its recruitment by the CRE at position -247 of the CRH promoter (Liu *et al.*, 2008; Nikodemova *et al.*, 2003; Wolfl *et al.*, 1999). A number of signaling transduction pathways lead to CREB phosphorylation but CRH transcription requires cyclic AMP production and PKA activation, indicating that CREB is required but not sufficient to activate transcription (Figure 17.5). Since the major neurotransmitters mediating activation of the CRH neuron, norepinephrine and glutamate, do not stimulate cyclic AMP, neuropeptides released in the PVN during stress should be considered as potential sources of cyclic AMP. These include:

- CRH released by dendritic projections and acting upon stress-inducible type 1 CRH receptors in the CRH neuron (Luo *et al.*, 1994)
- pituitary adenylate cyclase activating polypeptide (PACAP). The peptide stimulates CRH transcription in hypothalamic neuron primary cultures, and PACAP knockout mice display attenuated HPA axis responses to stress with blunted CRH mRNA responses in the PVN (Stroth *et al.*, 2011)
- glucagon-like peptide 1 (GLP-1), released in the PVN by non-catecholaminergic neurons from the nucleus of the solitary tract. GLP-1 stimulates CRH expression and its blockade inhibits HPA axis activity (Tauchi *et al.*, 2008).

The recently uncovered missing link explaining the cyclic AMP dependence of activation of CRH transcription is the CREB co-activator, transducer of regulated CREB activity (TORC), also known as CREB-regulated transcription co-activator (CRTC) (Aguilera and Liu. 2012) (Figure 17.6). There are three TORC subtypes, TORC 1, 2, and 3, encoded by different genes. In basal conditions, TORC remains inactive in the cytoplasm, phosphorylated and bound to the scaffolding protein 14-3-3. TORC phosphorylation is mediated by members of the AMP-activated protein kinase (AMPK) family of Ser/Thr protein kinases, including salt-inducible kinase (SIK) (Takemori and Okamoto, 2008). Protein kinase A inactivates these kinases and prevents TORC phosphorylation. Dephosphorylation releases TORC from 14-3-3, allowing its translocation to the nucleus and binding to the dimerization domain of CREB, which is necessary for CREB-mediated transcription in a number of genes (Conkright *et al.*, 2003). In addition to cyclic AMP-dependent inhibition of TORC kinases, the calcium/calmodulin-dependent phosphatase calcineurin dephosphorylates and therefore facilitates TORC activation. TORC 1 is the most abundant and widely distributed in the brain, but all three TORC subtypes are present in the PVN (Watts *et al.*, 2011).

There is evidence that TORC, especially TORC 2, is essential for activation of CRH transcription (Aguilera and Liu, 2012). First, studies in rats show specific TORC 2 immunostaining in 100% of CRH neurons in the PVN. Immunostaining is predominantly cytoplasmic in basal conditions but undergoes nuclear translocation by 30 min restraint stress. Second,

(A)

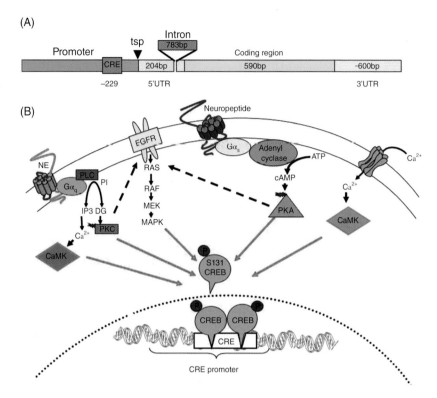

(B)

Figure 17.5 Signaling pathways leading to CREB phosphorylation in CRH neurons. The CRH gene has two exons (*shown in yellow*) and one intron (*shown in blue*) located in the 5′ untranslated region (5′ UTR). The 5′ flanking region is shown in green and emphasizes the cyclic AMP response element (CRE) at -229, which is essential for transcriptional activation and inhibition (A). Activation of G protein-coupled receptors, receptor and voltage-activated calcium channels leads to increases in intracellular calcium and other second messengers such as inositol 3 phosphate (IP_3), diacylglycerol (DG), and cyclic AMP (cAMP), and consequent activation of calcium/calmodulin-dependent protein kinase (CaMK), protein kinase C (PKC), protein kinase A (PKA), and transactivation of the mitogen-activated protein kinase (MAPK). All of these protein kinases can activate cyclic AMP responsive element binding protein (CREB), by phosphorylation. However, CREB phosphorylation alone is not sufficient to initiate CRH transcription (B).

overexpression of TORC 2 or TORC 3 in the hypothalamic cell line 4B potentiates the effect of the cyclic AMP stimulator forskolin on CRH promoter activity. Third, knockdown of endogenous expression of TORC isoform using silencing RNA (siRNA) inhibits the stimulatory effect of forskolin on CRH transcription in a reporter gene assay in 4B cells, or CRH hnRNA production in primary cultures of hypothalamic neurons. Combined knockdown of TORC 2 and TORC 3 completely blocked cyclic AMP-stimulated CRH promoter activity without affecting CREB phosphorylation. Fourth, cyclic AMP induction of CRH transcription is associated with rapid TORC translocation to the nucleus (evident by immunohistochistry or Western blots). Moreover, the inability of CREB phosphorylation by phorbolesters to induce CRH transcription is associated with cytoplasmic sequestration and hyperphosphorylation of TORC. Fifth, co-immunoprecipitation experiments in 4B cells demonstrate that forskolin but not phorbolesters induce association of TORC 2 protein with CREB. Sixth, chromatin immunoprecipitation studies reveal increased association of TORC 2 and CREB proteins with the CRH promoter after treatment of 4B cells with forskolin but not with PMA. Similarly, immunoprecipitation of hypothalamic chromatin of rats subjected to restraint stress revealed

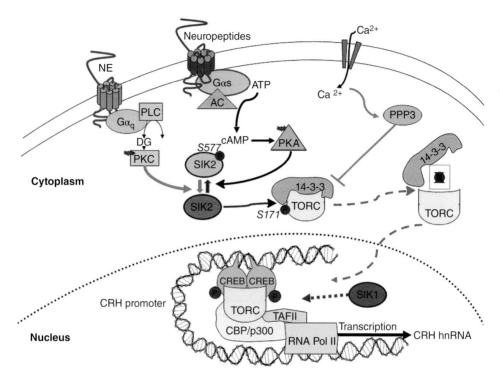

Figure 17.6 The CREB co-activator TORC is required for CREB-dependent stimulation of CRH transcription. In basal conditions, TORC is in the cytoplasm, in an inactive state phosphorylated at Ser 171 and bound to the scaffolding protein 14-3-3. Activation of salt-induced kinase 2 (SIK2) by PKC results in TORC phosphorylation and sequestration in the cytoplasm. Stimulation of AMP/PKA-dependent pathways by neuropeptide (or forskolin *in vitro*) inactivates SIK2 through phosphorylation at Ser 577, blocking TORC phosphorylation. In addition, increase in intracellular calcium stimulates the phosphatase calcineurin (PPP3), which dephosphorylates TORC. Upon dephosphorylation, TORC dissociates from 14-3-3 and translocates to the nucleus (*indicated by the green dotted lines*) where it interacts with CREB at the cyclic AMP response element in the CRH promoter and facilitates the recruitment of the transcription initiation complex, including CREB binding protein (CBP)/p300, TATA binding protein (TBP), TBP associated factors (TAFII), and RNA polymerase II (Pol II). In addition, activation of SIK1 in the nucleus may rephosphorylate and inactivate TORC, thus contributing to the declining phase of transcription. (The latter requires further experimental confirmation.)

recruitment of TORC 2 and phospho-CREB by the CRH promoter at 30 minutes, but 3 hours after stress only phospho-CREB remained associated with the CRH promoter (Aguilera and Liu, 2012).

The mechanism of activation/inactivation of TORC in CRH neurons involves SIK (Liu *et al.*, 2012). There are two major isoforms of SIK, SIK1 and SIK2, and both are present in the dorsomedial PVN, corresponding to the location of CRH neurons. Restraint stress causes dramatic induction of SIK1 mRNA and lesser induction of SIK2 in this region. Overexpression of either SIK1 or SIK2 in the hypothalamic cell line 4B reduces nuclear TORC 2 levels and inhibits forskolin-stimulated CRH transcription. Conversely, the SIK inhibitor staurosporine increases nuclear TORC 2 content and stimulates CRH transcription. Specific shRNA knockdown of endogenous SIK2 induces nuclear translocation of TORC 2 and CRH transcription, in spite of a compensatory increase of SIK1. Suppression of SIK1 has no effect on TORC translocation or CRH transcription. The SIK overexpression experiments indicate that both SIK1 and SIK2 can inhibit TORC translocation and CRH transcription, but the lack of effect of SIK1 blockade and the fact that increases in SIK1 due to SIK2 knockdown cannot compensate for the lack of SIK2 suggest that SIK2 is responsible for maintaining phosphorylated TORC in the cytoplasm. The temporal pattern of induction of SIK1, reaching a maximum at the time when TORC content in the nucleus and CRH transcription decline, suggests that SIK1 induction and activation limit transcriptional responses of the CRH gene. The overall evidence indicates that the CREB co-activator TORC is essential for activation of CRH transcription, and that regulation of the SIK/TORC system by stress-activated signaling transduction pathways acts as a sensitive switch mechanism for rapid activation and inactivation of CRH transcription (see Figure 17.5).

17.3.2.2 Glucocorticoid feedback inhibits CRH secretion and expression

Excessive production of CRH can lead to disease not only by overstimulating HPA axis activity but also through direct effects in the brain. Glucocorticoid feedback plays a critical role in preventing HPA axis hyperactivity and the hypothalamic CRH neuron is an important target. The steroid inhibits CRH secretion and expression through a number of mechanisms. First, the rapid inhibition of CRH secretion depends on non-genomic actions, mediated partly by retrograde inhibition of glutamaergic transmission by endocannabinoids in the PVN (Tasker and Herman, 2011). *In vitro* and *in vivo* evidence indicates that glucocorticoids and stress induce rapid synthesis of anandamide and 2-arachidonoylglycerol in CRH neurons of the PVN, and act as retrograde messengers by binding to presynaptic CB1 receptors. Pharmacological or genetic disruption of CB1 function increases CRH expression in the PVN and plasma levels of ACTH and corticosterone. Some reports suggest that this non-genomic effect of glucocorticoids depends on a G protein-coupled receptor (Di *et al.*, 2003). However, the lack of rapid effects of dexamethasone in hypothalamic slices of mice bearing targeted deletion of GR in the PVN strongly suggests that the classic GR mediates non-genomic effects (Hamm *et al.*, 2010). The immunohistochemical demonstration of membrane localization of GR also supports the involvement of the classic GR (Johnson *et al.*, 2005; Komatsuzaki *et al.*, 2005; Liposits and Bohn, 1993; Samarasinghe *et al.*, 2011).

A second mechanism involves glucocorticoid-induced activation of GR in limbic areas, such as the hippocampus and the frontal cortex, resulting in stimulation of inhibitory neural pathways to the CRH neuron (Uchoa *et al.*, 2014). For example, glucocorticoid implants in the medial frontal cortex inhibit HPA axis responses to psychogenic stressors. Moreover, mice with selective GR deletion in the cortex, hippocampus, and amygdala have elevated basal and prolonged corticosterone responses to psychogenic stress, as well as reduced HPA axis inhibition following exogenous glucocorticoid administration (Arnett *et al.*, 2011).

Thirdly, there is evidence that glucocorticoids modulate stimulatory catecholaminergic and peptidergic afferents pathways to the PVN. Microdialysis studies in rats show that peripheral administration of glucocorticoids inhibits stress-induced norepinephrine release in the PVN (Pacak *et al.*, 1995). Adrenalectomy increases and glucocorticoid administration decreases the content of α-adrenergic receptors in the PVN. Another target of glucocorticoid feedback is the excitatory peptide GLP-1, produced by non-catecholaminergic neurons of the nucleus of the solitary tract and released in the PVN during stress. Stress-induced glucocorticoid surge or glucocorticoid injection in rats causes rapid decreases of the GLP-1 precursor protein, preproglucagon mRNA, suggesting an increase in translation, and increases in hnRNA indicating transcriptional activation (Tauchi *et al.*, 2008).

Lastly, glucocorticoids inhibit CRH transcription but in physiological conditions, most evidence points to indirect mechanisms (Aguilera and Liu, 2012). First, suppression of the glucocorticoid surge by adrenalectomy does not affect the declining phase of stress-induced CRH hnRNA, indicating that factors other than glucocorticoids are responsible for the rapid decline of CRH transcription during persistent stress. Second, injection of high doses of corticosterone at the time of stress exposure does not prevent the increases in CRH hnRNA induced by stress (Shepard *et al.*, 2005). Third, glucocorticoids do not directly affect cAMP-induced CREB phosphorylation and nuclear accumulation of the co-activator TORC *in vitro* (Evans *et al.*, 2013). Fourth, the CRH promoter lacks a consensus GRE, but a conserved sequence upstream of the essential -247 CRE is capable of binding GR in gel shift assays and mediates glucocorticoid-dependent repression in reporter gene assays. However, GR immunoprecipitation in hypothalamic chromatin from rats injected with corticosterone shows a lack of GR recruitment by the CRH promoter, in spite of marked recruitment by the glucocorticoid-dependent period 1 gene (Evans *et al.*, 2013). Although GR interaction at distal promoter sites or with other transcriptional proteins may play a role in the minor inhibition of CRH transcription observed *in vitro*, the evidence suggest that the repressor effects of glucocorticoids on CRH expression *in vivo* are predominantly indirect, through modulation of pathways regulating CRH neuron function.

17.3.2.3 Intracellular feedback mechanisms limiting CRH transcription

In addition to long-loop (endocrine) and short-loop (paracrine or autocrine) feedback, most intracellular mechanisms are autoregulated at the level of signaling cascades leading to gene induction or posttranslational modifications of existing proteins. There is evidence for involvement of at least two intracellular control mechanisms regulating CRH transcription.

First, the rapid induction of CREB phosphorylation, TORC translocation, and CRH hnRNA in CRH neurons of the PVN by stress is followed by delayed increases in expression of the transcriptional repressor, inducible cyclic AMP early repressor (ICER) (Shepard *et al.*, 2005). ICER, a product of activation of the second promoter of the CREM gene, represses cAMP-induced gene transcription in several neuroendocrine tissues, by inhibiting the effect of phospho-CREB at the CRE (Lamas *et al.*, 1996). Gel shift assays and chromatin immunoprecipitation experiments reveal ICER recruitment by the CRH promoter paralleling decreases in Pol II binding. Thus, delayed induction of ICER and binding to the CRE in the CRH promoter probably contribute to the declining phase of stress-induced CRH transcription.

Second, another potential intracellular mechanism limiting the activation of CRH transcription during stress is regulation of the activity of the CREB co-activator, TORC by SIK1 and 2. Activation of TORC by PKA-induced inactivation of SIK2 allows TORC translocation to the nucleus and initiation of CRH transcription during stress. Stress also causes rapid induction and activation of SIK1 which would mediate phosphorylation and export of TORC from the nucleus and termination of CRH transcription (Liu *et al.*, 2012).

17.4 Pituitary targets in HPA axis regulation

Hypothalamic CRH and VP released into the pituitary portal circulation gain access to receptors in the anterior pituitary corticotrope, where they stimulate ACTH secretion (Figure 17.7). ACTH is a 39 amino acid peptide derived from the precursor protein pro-opiomelanocortin (POMC). The main sites of expression of POMC are corticotrope cells in the anterior lobe of the pituitary, melanotropes in the intermediate lobe, and neurons of the arcuate nucleus of the hypothalamus. POMC undergoes diverse posttranslational processing according to the type of endopeptidase expressed in the cell. The anterior pituitary corticotrope expresses prohormone convertase 1 (PC1), which cleaves POMC into ACTH, β-lipotropin, and β-endorphin. In melanotrope and arcuate neurons, prohormone convertase 2 (PC2) cleaves the molecule into α-MSH, corticotropin-like intermediate lobe peptide, and acetyl-β-endorphin (Bicknell, 2008). Prohormone convertase expression increases during stimulation by CRH or stress, suggesting that increased POMC cleavage by PC1 contributes to ACTH release during stress.

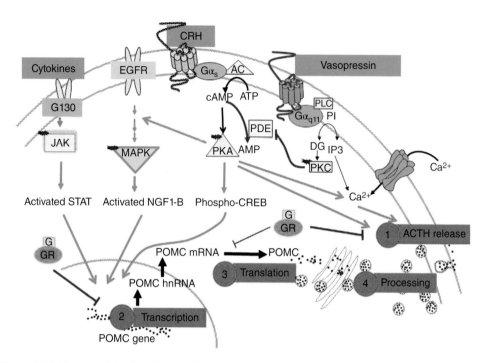

Figure 17.7 Diagram of the signaling transduction pathways regulating ACTH secretion and POMC transcription in the anterior pituitary corticotrope. CRH and VP released into the pituitary portal circulation bind to their specific plasma membrane receptors, type 1 CRH receptor (CRHR1) and type 1b vasopressin receptor (V1bR), in the pituitary corticotrope. Activation of CRHR1, coupled to the stimulatory guanyl nucleotide binding protein (Gs) and adenylate cyclase (AC), activates the cyclic AMP (cAMP)/protein kinase A (PKA) pathway which stimulates ACTH secretion, calcium influx, MAP kinase (MAPK) activity, and CREB phosphorylation. MAPK activates NGFI-B, a major transcription factor involved in activation of POMC transcription. Activation of the V1bR, coupled to the guanyl binding protein Gq/11 and phospholipase C (PLC), cleaves phosphatidylinositol (PI) into inositol-3-phosphate (which releases calcium [Ca++] from intracellular stores) and diacylglycerol, which activates protein kinase C (PKC). PKC potentiates the effect of CRH on cyclic AMP production by inhibiting phosphodiesterase (PDE). Cytokines contribute to activation of POMC transcription through the JAK/STAT pathway. Calcium-activated calmodulin-dependent protein kinase (CaCmK) (not shown), PKC, and PKA contribute to ACTH secretion from secretory granules. Glucocorticoids inhibit POMC transcription and ACTH secretion.

17.4.1   CRH and VP mediate rapid pituitary ACTH responses to stress

Stress stimulation of ACTH secretion results from CRH- and VP-regulated exocytosis of releasable pools in secretory granules of the pituitary corticotrope. The exact mechanism of this regulated exocytosis is not fully understood but it involves cyclic AMP and calcium signaling, resulting in actin reorganization within the cell. CRH is a potent stimulant of ACTH secretion by the pituitary corticotrope *in vitro* and *in vivo*, through activation of cyclic AMP/ PKA signaling and calcium influx though voltage-dependent calcium channels (Abou Samra *et al.*, 1987; Aguilera and Liu, 2012). In contrast to CRH, activation of calcium phospholipid-dependent pathways by VP is a weak stimulator of ACTH secretion, but it markedly potentiates the stimulatory effect of CRH on ACTH release. VP exerts this effect partly by potentiating CRH-stimulated cyclic AMP production via protein kinase C-mediated inhibition of phosphodiesterases (Abou Samra *et al.*, 1987). Pharmacological suppression of vasopressinergic activity reduces ACTH responses to acute stress, indicating that synergism between CRH and VP is important for full ACTH responsiveness during acute stress. Since VP potentiates homologous desensitization of CRHR1 through receptor phosphorylation, the peptide may also contribute to the declining phase of the ACTH response by decreasing pituitary sensitivity to CRH (Aguilera, 1998). Both CRH and VP cause transactivation of the MAP kinase pathway, which plays an important role in activating transcription factors mediating POMC transcription.

The synergism between CRH and VP on ACTH secretion and marked increases in hypothalamic parvocellular VP expression during chronic stress have suggested that VP becomes the predominant regulator of ACTH secretion during chronic stress. However, suppression of vasopressinergic activity does not prevent HPA axis hyperresponsiveness to a novel stress in chronically stressed animals (Aguilera *et al.*, 2008). V1b receptor knockout mice show impaired capacity to sustain ACTH responses to repeated stress, without affecting corticosterone responses (Lolait *et al.*, 2007). This evidence indicates that VP contributes to a full ACTH response during acute stress but that it plays only a minor role on overall HPA axis activity during chronic stress.

Vasopressin contributes to the generation of pituitary corticotropes during chronic HPA axis stimulation. The number of corticotropes in the anterior pituitary increases after long-term adrenalectomy and probably after chronic stress, and there is evidence that both CRH and VP contribute to this effect (Gulyas *et al.*, 1991; Nolan *et al.*, 2003). Administration of CRH increases the number of corticotropes *in vivo* and increases mitogenic activity in cultured pituitary cells *in vitro*. The mitogenic effect of CRH *in vitro* occurs with concentrations of the peptide submaximal for ACTH secretion, and appears to be a direct effect on corticotrope cells. Vasopressin acts as a mitogen in the pituitary, and V1 receptor antagonists prevent the increase in bromodeoxyuridine incorporation during adrenalectomy. However, the lack of co-localization of bromodeoxyuridine label in corticotropes or other hormone-producing cells in the pituitary suggests that newly produced corticotropes during chronic HPA axis stimulation originate from undifferentiated cells and not from division of existing corticotropes (Subburaju and Aguilera, 2007). This evidence indicates that VP stimulates the production of precursor cells, while CRH mediates corticotrope differentiation.

17.4.2 CRH binds to specific receptors

The hypothalamic peptide CRH initiates its actions by binding to high affinity receptors located in the plasma membrane of the pituitary corticotropes and other target tissues (Aguilera *et al.*, 2004). There are two major types of CRH receptors, both coupled to adenylate cyclase through the stimulatory guanyl nucleotide binding protein (G-protein) Gs: type 1 CRH receptors (CRHR1) are located mainly in the brain and pituitary corticotropes, and in peripheral tissues including the reproductive organs and the immune system. CRHR1 binds CRH and urocortin 1 with equal affinity but does not recognize urocortins 2 and 3. Type 2 CRH receptors (CRHR2), which have 70% homology with CRHR1, bind urocortin 1, 2 and 3, and are located mostly in the periphery. Being coupled to adenylate cyclase, both receptors activate cyclic AMP/protein kinase A-dependent pathways upon binding to their ligand. CRHR1 is the major player in the stress response and is essential for pituitary ACTH secretion and behavioral responses to stress (Bale and Vale, 2004). CRH receptors, especially CRHR1, are also present in areas of the limbic system, where they mediate behavioral and autonomic responses to stress. Interestingly, CRH neurons in the PVN contain low levels of CRHR1 mRNA and protein, and this expression markedly increases during stress (Luo *et al.*, 1994). CRHR2 located within the brain plays an important role modulating behavioral responses to stress.

An additional factor modifying the biological action of CRH is a 37 kDa non-receptor CRH binding protein (CRH-BP), first identified in blood of pregnant women and non-human primates, but also present in the brain and pituitary. This protein binds both CRH and urocortin with an affinity equal to or greater than that of the receptors, and blocks CRH-mediated

ACTH release *in vitro*. The CRH binding protein and the CRH receptors are co-localized broadly in cortical regions, in limbic structures and certain sensory relay nuclei. Mouse models of CRH-BP overexpression or deficiency have shown changes in the HPA axis and in energy balance, and behavior consistent with the hypothesis that CRH-BP plays an important *in vivo* modulatory role by regulating levels of CRH peptides available to receptors in the pituitary and brain (Westphal, 2006).

17.4.2.1 Stress regulates CRHR1 binding and mRNA levels

In addition to the prevailing levels of ligand, in several systems hormone responsiveness depends on the content of specific receptors in the target tissue. This is not the case for pituitary CRH receptors, in which changes in CRH receptor number during HPA axis manipulation correlate poorly with pituitary responsiveness (Aguilera, 1994; Aguilera *et al.*, 2004). For example, the marked downregulation (decrease in receptor number) and desensitization (decrease in CRH-stimulated adenylate cyclase activity) of CRH receptors observed following adrenalectomy are associated with increases in pituitary POMC mRNA and plasma ACTH levels (Aguilera *et al.*, 2004). Similar poor correlation between ACTH responsiveness and pituitary CRH receptor content occurs during chronic stress. This clearly shows that CRH receptor number does not determine corticotrope responsiveness. However, the severely blunted ACTH responses observed under pharmacological or genetic elimination of CRH, or CRHR1, indicate that the presence of CRH and at least a small number of CRHR1s is essential for corticotrope response to stress (Bale and Vale, 2004).

Regulation of pituitary CRH receptors during stress depends on increased exposure of the corticotropes to CRH and VP and adrenal glucocorticoids, and involves transcriptional, translational, and posttranslational mechanisms (Figure 17.8). Pituitary exposure to CRH, VP, and glucocorticoids differs according to the stress paradigm, and the pattern of secretion and interaction between these regulators is critical for CRH receptor regulation. A single injection of CRH causes transient decreases in CRHR1 mRNA (Ochedalski *et al.*, 1998; Rabadan Diehl *et al.*, 1996), while repeated daily injection of doses yielding plasma levels in the range of pituitary portal circulation increase CRHR1 mRNA and CRH binding. In contrast, osmotic minipump administration providing sustained circulating levels of CRH results in prolonged CRHR1 mRNA and binding downregulation and desensitization. The decreasing phase involves PKA and MAP kinase signaling and decreases in mRNA half-life (Moriyama *et al.*, 2005), and likely reflects receptor mRNA translation. Vasopressin potentiates CRH-induced CRH binding downregulation but a single or repeated injections increase CRHR1 mRNA levels (Aguilera, 1998; Rabadan Diehl *et al.*, 1996). Since VP is co-secreted with CRH during stress, the effect of VP probably contributes to the recovery phase of CRHR1 mRNA levels during stress.

Increases in circulating glucocorticoids cause prolonged decreases in CRH binding, *in vivo* and *in vitro* (Childs *et al.*, 1986; Hauger *et al.*, 1987; Schwartz *et al.*, 1986), but only transient reductions in CRHR1 mRNA. However, resting glucocorticoid levels are essential for rapid recovery following stress, since adrenalectomy markedly prolongs the fall of CRHR1 mRNA levels following stress of CRH injection (Luo *et al.*, 1995; Ochedalski *et al.*, 1998; Rabadan Diehl *et al.*, 1997). Consistently, injection of either CRH or glucocorticoids alone downregulates CRHR1 mRNA levels in adrenalectomized rats but their simultaneous administration accelerates recovery (Luo *et al.*, 1995; Ochedalski *et al.*, 1998; Rabadan Diehl *et al.*, 1997).

The regulatory patterns of CRHR1 mRNA in the pituitary suggest that rapid activation of translation with consequent utilization and degradation of CRHR1 mRNA is responsible for the transient decreases in CRHR1 mRNA, while transcriptional activation mediates the recovery phase of mRNA levels. In addition, the lack of correlation between CRHR1 mRNA and CRH binding suggest that the number of active receptors in the cell membrane depends largely on posttranscriptional mechanisms.

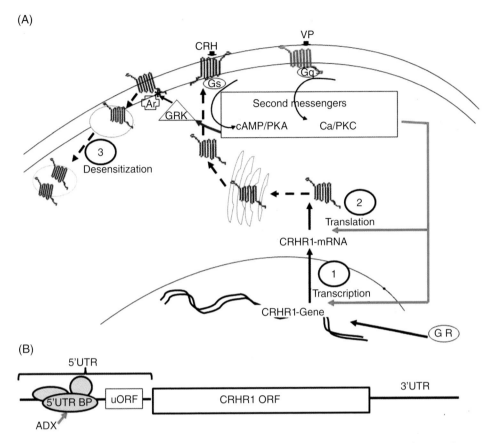

Figure 17.8 Sites of regulation of the number of CRHR1 in the pituitary. (A) The number of functional CRHR1 depends on receptor protein synthesis through transcriptional and translational activation (1 and 2), and receptor desensitization and internalization following interaction with the ligand (3). There is evidence for the roles of phosphorylation, G protein receptor kinase (GRK), and β-arrestin (Ar) in the fate of CRHR1 following CRH binding. Active V1bR gene transcription (1) is required to maintain CRHR1 mRNA levels, but posttranscriptional regulation (2) appears to be a major determinant of receptor synthesis. Based on *in vivo* studies, it is likely that CRHR1 regulation involves signaling by CRH, VP, and glucocorticoids. Ar, β-arrestin; CRHR1, corticotropin-releasing hormone type 1 receptor; GR, glucocorticoid receptor; GRK, G protein-coupled receptor kinase; VP, vasopressin. (B) Diagram of the CRHR1 mRNA showing potential mechanisms of translational regulation by the 5′ untranslated region (5′ UTR), an upstream open reading frame (uORF), and 5′ UTR binding proteins shown to increase after adrenalectomy (ADX), a condition associated with CRHR1 downregulation.

17.4.2.2 Translational regulation of CRHR1

The mechanisms controlling CRHR1 translation involve binding of proteins to the 5′ UTR of the mRNA (see Figure 17.8B). Experiments using gel shift assays have shown binding of pituitary cytosolic proteins to *in vitro* transcribed 5′ UTR of CRHR1 mRNA but not to type 1 angiotensin II receptor mRNA. In addition, protein binding to the 5′ UTR of CRHR1 mRNA increases when using cytosol from adrenalectomized rats (Wu *et al.*, 2004), suggesting that RNA binding proteins are involved in the translational regulation of mRNA for CRHR1. In addition, the 5′ UTR of CRHR1 contains a short minicistron, or upstream open reading frame (ORF), encoding a 10 amino acid putative peptide. Studies using reporter gene assays or *in vitro* translation have shown that inactivation of the upstream ORF increases translation, indicating that the upstream ORF inhibits CRHR1 translation (Xu *et al.*, 2001). Experiments using fusion constructs of the upstream ORF and green fluorescent protein show translation of the peptide encoded by the ORF. In addition, the synthetic peptide inhibits *in vitro* translation of constructs containing the CRHR1 5′UTR, suggesting that production of the peptide is part of the mechanism by which the upstream ORF inhibits translation.

17.4.2.3   Posttranslational regulation of CRHR1

Posttranslational modifications play an important role in CRHR1 stability and activity. Following translation, CRHR1s undergo o-glycosylation (Assil and Abou Samra, 2001; Flores *et al.*, 1990), a modification known to prolong the half-life of proteins. Studies using CRHR1 with mutations of the putative glycosylation sites show the requirement of three glycosylation sites for receptor binding and cylic AMP production (Assil and Abou Samra, 2001). Receptor phosphorylation plays an important role in the receptor down-regulation and desensitization observed during stress or CRH exposure. Ligand binding to CRHR1 induces changes in conformation allowing association with G protein receptor kinase (GRK) with the membrane and receptor phosphorylation (Hauger *et al.*, 2012). Receptor phosphorylation leads to recruitment of β-arrestin, receptor uncoupling from adenylate cyclase, and internalization (see Figure 17.8A). There is evidence that protein kinase C (PKC) facilitates receptor desensitization through phosphorylation. In some tissues such as the testes and placenta, CRHR1 can stimulate PKC (Hauger *et al.*, 2003). In the pituitary corticotrope, CRHR1 couples mainly to Gs and the cyclic AMP pathway. However, this effect of PKC could explain the potentiating effect of VP, a Gq coupled peptide, on CRHR1 desensitization.

17.4.3   Vasopressin binds to specific receptors

Three major vasopressin receptor subtypes encoded by different genes have been identified and cloned: V2 receptors, which are coupled to the G-protein Gs and adenylate cyclase, and mediate the antidiuretic effects of VP in the kidney; V1a receptors, present in smooth muscle, liver and brain, coupled to phospholipase C (PLC); and the pituitary V1b receptor, also linked to PLC, which is responsible for the effects of VP on ACTH secretion (Robert and Clauser, 2005). The main subtypes of VP receptors present in the brain are V1a and V1b. As with CRHR1, V1a and V1b receptors are widely distributed in the hypothalamic and limbic areas of the brain, where they mediate behavioral effects of VP (Insel, 2010; Vaccari *et al.*, 1998). The V1b receptor subtype is abundant in pituitary corticotropes where it mediates the effects of VP on HPA axis regulation (Roper *et al.*, 2011).

17.4.3.1   Regulation of pituitary V1b receptors

In contrast to CRHR1, there is a good correlation between pituitary VP binding and ACTH responsiveness to a novel stress. Acute and chronic stress cause only transient decreases in V1bR binding and mRNA, followed by consecutive increases in V1bR mRNA and VP binding. Upregulation of VP receptors during stress depends on the increased pituitary exposure to its ligand but it also requires glucocorticoids since elevated parvocellular VP in adrenalectomized rats induces sustained VP binding downregulation in spite of a recovery of V1bR mRNA levels (Aguilera *et al.*, 2008). The dissociation between VP binding and V1bR mRNA levels indicates that mRNA levels do not determine receptor content and suggests that glucocorticoids are necessary for mRNA translation. Adrenalectomy causes sustained V1bR mRNA downregulation in Brattleboro rats, which lack hypothalamic vasopressin, suggesting that vasopressin mediates V1b receptor mRNA recovery. Exogenous glucocorticoid administration downregulates pituitary vasopressin binding but increases V1b receptor mRNA and facilitates coupling of the receptor to phospholipase C, effects which contribute to the refractoriness of vasopressin actions to glucocorticoid feedback. Regulation of the pituitary V1bR involves transcriptional and translational mechanisms (Aguilera *et al.*, 2003).

17.4.3.2 Regulation of V1b receptor transcription

The rapid increases in V1bR mRNA following experimental stress strongly suggest that stress stimulates V1bR gene transcription. *In vitro* studies have provided some information about the transcriptional regulation of the V1b receptor. 5'RACE and RNAse protection analysis mapped two major putative transcription start points at -830 and -861 bp from the starting methionine. The 5' flanking region lacks a TATA box but it contains CACA repeats (CACA box), CT repeats or inverted GAGA box, several AP-1 and AP-2 sites, and a cluster of Sp1 sites upstream of the AP-2 sites. A luciferase construct containing a 2.1 kb putative promoter, and part of the 5' UTR including a first short intron, is able to drive luciferase activity in COS-7, CHO, and PC12 cell lines. The intron is essential for promoter activity (Rabadan Diehl *et al.*, 2000).

Of the cis-elements identified by computer analysis, there is evidence that the GAGA box, situated near the transcription start point, is important for activation of V1bR gene transcription (Volpi *et al.*, 2002). Deletion of 213 bp containing the GAGA box dramatically reduces promoter activity, and co-transfection of *Drosophila* GAGA binding protein with reporter constructs containing the GAGA box increases promoter activity. Similarly, transfection of *Drosophila* GAGA binding protein in cells expressing V1b receptor increases expression of the endogenous receptor. Moreover, there is evidence that endogenous pituitary nuclear proteins bind to the GAGA box of the V1bR with high association and dissociation rates. This binding activity is regulated during physiological conditions, leading to changes in V1bR expression and supporting a role for the GAGA box in the physiological regulation of V1bR transcription. Consistent with a role of the ERK mitogen-activated protein (MAP) kinase pathway regulating GAGA binding activity, VP increases GAGA binding activity through mechanisms involving MAP kinase and transactivation of the EGF receptor (Volpi *et al.*, 2006).

17.4.3.3 Regulation of V1b receptor translation

The transient pattern of VP binding downregulation with rapid recovery followed by upregulation suggests the involvement of translational mechanisms. There is evidence for the involvement of two mechanisms for regulation of V1bR translation. First, as with the CRHR1 5' UTR, the 5' UTR of the V1bR mRNA contains upstream ORFs (uORFs) with the potential of reducing the ability of ribosomes to linearly scan the V1b 5' UTR mRNA (Figure 17.9A). The proximal uORF encodes a putative peptide of 38 amino acids that can be translated *in vitro*. Co-translation of the synthetic peptide attenuates *in vitro* translation of the V1b receptor with mutant upstream ORF but not of constructs with the wild type 5'UTR, suggesting that the peptide exerts tonic inhibition of translation (Rabadan Diehl *et al.*, 2007).

The second mechanism involves cap-independent ribosome binding to an internal ribosome entry site (IRES), a feature found in a number of mRNAs with high complexity of the 5' UTR (Rabadan Diehl *et al.*, 2003). Deletions of V1bR 5' UTR fragments decreased luciferase activity, indicating that the full-length 5' UTR is required for IRES activity. Activation of PKC by phorbolesters increases IRES-mediated translation in reporter gene assays or in cells transfected with a V1bR-GFP fusion protein without affecting V1bR mRNA levels. Stimulation of IRES-dependent V1bR translation could serve as a mechanism for V1bR upregulation by VP during stress (Figure 17.9B).

17.4.4 CRH stimulates POMC transcription

Rapid POMC processing and ACTH release in response to stress require activation of POMC transcription and restoration of POMC pools in the corticotrope. The POMC gene has three exons and two introns and the peptide is encoded by part of exon 2 and exon 3. The promoter

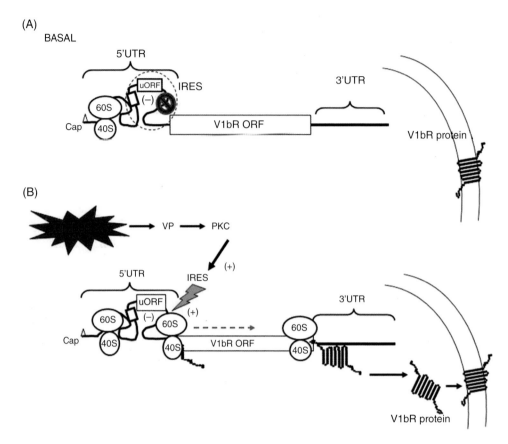

(A)

BASAL

5'UTR

uORF

IRES

(−)

60S

Cap 40S

V1bR ORF

3'UTR

V1bR protein

(B)

VP → PKC

(+)

5'UTR

IRES

uORF

(−) 60S

(+)

60S

Cap 40S

40S V1bR ORF

60S

40S

3'UTR

V1bR protein

Figure 17.9 Diagram depicting mechanisms for negative and positive regulation of translation by the 5′ UTR of the V1bR mRNA. (A) In basal conditions, the complexity of the 5′ UTR and the presence of several upstream open reading frames (uORF) could slow down cap-mediated translation and scanning of the mRNA by the translational apparatus, and account for low levels of translation of the main V1bR open reading frame (V1bR ORF). (B) Activation of neural pathways during stress can initiate translation through protein kinase C (PKC)-mediated activation of an internal ribosome entry site (IRES) present in the 5′ UTR of the V1bR mRNA. This may mediate positive regulation of V1bR translation during stress.

region has a TATA box and several response elements shown to be involved in the transcriptional regulation of the gene (Figure 17.10A). CRH induces rapid activation of POMC gene transcription in a cyclic AMP/PKA dependent manner (Boutillier *et al.*, 1998; Gagner and Drouin, 1987). The POMC promoter lacks a consensus CRE but gel shift assays have shown binding of CREB to an atypical AP1 element (Gagner and Drouin, 1987). However, the current view is that transcriptional activation of POMC is primarily mediated through activation of the orphan nuclear receptors related to NGFI-B, also called Nur77 (Maira *et al.*, 1999). Although NGFI-B mRNA is an immediate early response gene, PKA activates preexisting NGFI-B protein by promoting dephosphorylation of the NGFI-B DNA binding domain and hyperphosphorylation of its N-terminal activation function (AF1) domain. This induces NGFI-B dimer formation, followed by binding to the Nur response element in the POMC promoter (Maira *et al.*, 2003a), and recruitment of transcriptional co-activators, such as steroid receptor co-activator 2 (SRC2), also called transcriptional mediators/intermediary factor 2 (TIF2), or glucocorticoid receptor-interacting protein 1 (GRIP1). This induces recruitment of

(A)   **POMC gene**

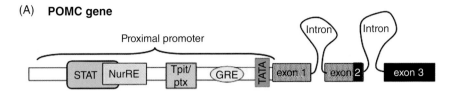

(B)   **POMC promoter activation**

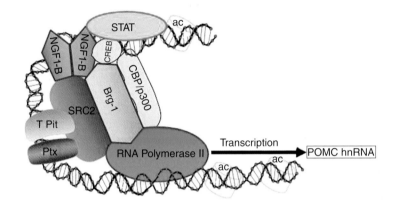

(C)   **Transcriptional repression by glucocorticoids**

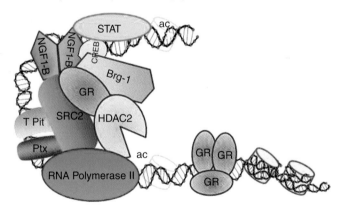

Figure 17.10 (A) Structure of the POMC gene. The coding region of POMC is shown in black in the boxes depicting exon 2 and exon 3. The 5′ flanking region corresponds to the proximal promoter and shows the main confirmed response elements involved in POMC transcriptional regulation. (B) Model of the transcriptional activation of POMC. Activation of NGFI-B by MAP kinase (see Figure 17.7) induces dimerization and binding to the Nur response element (NurRE), leading to recruitment of co-activators including steroid receptor co-activator 2 (SRC2) and Brahma-related gene 1 (Brg-1), and transcriptional proteins P300 and RNA polymerase II. The Tpit/ptx complex which is also activated by MAPK is part of the transcriptional complex. (C) Model of the transcriptional repression of POMC by glucocorticoids. Upon activation of GR by glucocorticoids, GR binds to the NGFI-B transcriptional complex in a Brg-1-dependent manner and recruits histone deacetylase 2 (HDAC2), forming a repressor complex. Recruitment of HDAC2 results in deacetylation of histone H4 throughout the POMC gene, causing transcriptional repression. There is also evidence for GR binding as a trimer to a GRE downstream of the NurRE in the POMC promoter.

the ATPase subunit of the chromatin remodeling complex Swi/Snif, Brahma-related gene 1 (Brg1), which is essential for transcriptional activity (Figure 17.10B). Phospho-CREB is also part of the Nur response element binding complex and appears to stabilize the complex, acting as a co-activator (Mynard *et al.*, 2004). Although CREB does not bind directly to the chromatin, it forms part of the NGFI-B complex and acts as a co-activator.

The pituitary corticotrope expresses other members of the Nur family, including Nur-related factor 1 (Nurr1) and neuron-derived orphan receptor 1 (NOR-1), which also undergo activation by CRH and can bind to the POMC promoter. This redundant role of Nur family members can explain the lack of a HPA axis phenotype in Nur 77 knockout mice (Jenks, 2009). A STAT binding site overlapping with the Nur response element contributes to the transcriptional activation of POMC by cytokines and it appears to be essential for optimal promoter activity *in vitro* (Mynard *et al.*, 2004). CRH also induces recruitment of SRC2 by Tpit, which is essential for POMC transcriptional activation (Maira *et al.*, 2003b) (Figure 17.10B). In contrast to the importance of VP potentiating the stimulatory effect of CRH on ACTH secretion, VP does not activate POMC transcription and inhibits stimulation by CRH (Levin *et al.*, 1989).

17.4.5 Glucocorticoids inhibit ACTH secretion and POMC transcription
The pituitary corticotrope is a well-recognized target for glucocorticoid feedback. Glucocorticoids act at two levels, first by inhibiting ACTH secretion from releasable pools, and second by repressing POMC transcription.

17.4.5.1 Glucocorticoid feedback on ACTH secretion
Glucocorticoids operate through multiple mechanisms to exert rapid and delayed inhibitory feedback effects on ACTH secretion (Jenks, 2009). Late feedback depends on genomic actions of glucocorticoids, by inhibiting POMC transcription and consequently preventing POMC synthesis. Glucocorticoids also induce the synthesis of annexin 1 in folliculostellate cells of the pituitary, which has paracrine effects in the corticoptrope. Glucocorticoids also induce annexin 1 phosphorylation and translocation to the cell surface, where it is believed to inhibit ACTH secretion by inducing actin polymerization through interaction with putative membrane receptors in the pituitary corticotrope (John *et al.*, 2008).

There is evidence that rapid feedback could depend on non-genomic effects mediated by membrane association of GR. Studies in anterior pituitary cells under constant perfusion with physiological concentrations of CRH show inhibition of CRH-stimulated ACTH secretion within 5 min exposure to basal corticosterone concentrations and recovery within 5 min of withdrawal. The same low corticosterone concentrations are able to induce association of GR with membrane fractions in heterologous cell lines expressing endogenous GR, with association and dissociation kinetics consistent with the effects on ACTH secretion. Moreover, the GR antagonist RU486, but not the transcriptional inhibitor actinomycin D, prevents the rapid inhibition of ACTH secretion induced by corticosterone (Deng *et al.*, 2014). The time frame of these effects and sensitivity to glucocorticoid concentrations within the basal range support the hypothesis that rapid glucocorticoid feedback at the pituitary corticotrope is part of the mechanism of ultradian pulse generation (Walker *et al.*, 2010).

17.4.5.2 Glucocorticoid feedback on POMC transcription
Glucocorticoid represses POMC transcription through genomic actions at the level of transcriptional initiation. Two mechanisms have been shown to mediate this effect: first, through direct binding of three GR molecules to a negative GRE in the proximal POMC promoter; second, GR inhibits POMC transcription by interfering with NGFI-B-dependent transcription, through protein-protein interaction (Martens *et al.*, 2005). Interaction of GR with

SRC2 and Brg1 within the NGFI-B transactivation complex leads to recruitment of histone deacetylase 2 (HDAC), which induces histone deacetylation of the promoter and throughout the gene and consequent transcriptional repression (Bilodeau *et al.*, 2006) (Figure 17.10C). Transcriptional repression is fast, occurring within minutes, but effects on POMC synthesis and ACTH secretion are part of the delayed phase of glucocorticoid feedback.

17.5   Cytokines and HPA axis responses to stress

Glucocorticoids play a critical role in controlling cytokine responses to immune challenges, and conversely cytokines modulate glucocorticoid production by the HPA axis. By interacting with membrane receptors linked to Janus kinase (JAK)/signal transducers and activators of transcription (STAT) signaling, cytokines such as interleukin 1 (IL-1) and tumor necrosis factor (TNF) initiate a cascade of inflammatory responses by activating nuclear factor-κB (NF-κB), a transcription factor controlling a number of proinflammatory genes. Cytokines are essential for fighting infection and building immunity but excessive increases lead to catastrophic events with cell apoptosis and ultimately death. Glucocorticoids play a key role in maintaining balanced cytokine responses. This is clear from the marked elevation in proinflammatory cytokines and increased mortality following immune challenges observed in adrenalectomized experimental animals or Addisonian patients, and reversibility by exogenous glucocorticoids. Following activation by glucocorticoids, GRs limit cytokine production and the inflammatory process by blocking the transcriptional activity of NF-κB and other transcription factors implicated in the inflammatory response, such as activating protein-1 (AP-1), and STAT family members, through protein-protein interaction. In addition, through a genomic mechanism, activated GR induces the expression of the inhibitory subunit of NF-κB (i-kB), which acts as a NF-κB antagonist (Barnes and Karin, 1997).

Cytokines in turn markedly activate HPA axis activity by acting at all three levels – the hypothalamus, pituitary, and adrenal. Cytokine signaling from the periphery can access the brain through areas outside the blood–brain barrier such as the circumventricular organs, choroidal plexus, and endothelial cells. In addition, local cytokine production by glial and ependymal cells plays a role in stimulating CRH and VP from parvocellular neurons of the PVN. At least during acute immune challenge, the major mechanism by which inflammatory signals, especially IL-1 and TNF, stimulate HPA axis activity is stimulation of hypothalamic CRH and VP. However, cytokines can activate HPA axis in the absence of CRH signaling, as occurs in mice bearing genetic ablation of CRH or CRHR1, which have impaired responses to restraint but near normal ACTH responses to an immune challenge (Turnbull *et al.*, 1999).

Interleukin 6 (IL-6) can directly stimulate the pituitary and adrenal by acting upon glycoprotein 130 (gp130) receptors in pituitary corticotropes and adrenocortical cells (Chesnokova and Melmed, 2002). Cytokine effects in the pituitary and adrenal are slower and longer lasting than HPA axis stimulation by the hypothalamic regulators CRH and VP. IL-6 synergizes the effect of CRH in activating POMC transcription through activation of the JAK/STAT pathway, leading to interaction of STAT5 with the overlapping STAT/Nur response element in the POMC promoter (Mynard *et al.*, 2004). Studies in the adrenal have shown that activation of JAK/STAT pathways also increases intracellular levels of active CREB by activating its transcription, as well as by preventing CREB proteosomal degradation through tyrosine phosphorylation (Mynard *et al.*, 2002). On the other hand, STAT signaling alone is not sufficient for stimulation of ACTH secretion of POMC transcription. Therefore, cytokines most likely interact with additional regulators in the pituitary and adrenal. Stimulation of neuropeptides such as PACAP by cytokines could provide a source of cyclic AMP essential for ACTH secretion and POMC transcription.

Cytokines also play a role in the stimulation of the HPA axis by non-immune stressors. Stressors such as restraint, foot shock, and open-field exposure increase circulating IL-6. Female mice with ablation of IL-6 show lower corticosterone responses to restraint stress (Bethin *et al.*, 2000). The sexual dimorphism in cytokine responses may explain the higher propensity of females to autoimmune disorders. Mice with IL-6 deficiency show reduced ACTH responses to restraint stress, and IL-6 injection restores responses, indicating that cytokines are important for full HPA axis responses to immune and non-immune stressors.

17.6   Perspectives

The stress response involves activation of complex networks of neural and signaling transduction pathways leading to the release of neurotransmitters and neuropeptides mediating changes in behavior, and stimulation of sympathoadrenal and adrenal glucocorticoid secretion, the end-product of HPA axis activation. Much progress has been made during the last two decades in understanding the molecular mechanisms leading to preservation of homeostasis during stress, and how the HPA axis acts as the major player in coordinating the adaptive response. However, a number of important questions have been difficult to address because of the multifactorial nature and complexity of the stress response, with interaction between multiple regulators and signaling transduction pathways at the different anatomical levels of regulation. Another difficulty in studying molecular events in CRH neurons or pituitary corticotropes arises from the heterogeneity of neurons in PVN, or anterior pituitary in which less than 10% of the cell population are corticotropes. Emerging technology, such as the creation of genetically modified animals with cell-specific targeted and conditioned gene deletion or overexpression, or cell-specific viral gene manipulation, will be helpful in resolving questions such as the physiological importance of salt-induced kinases as a switch to turn on and off CRH transcription, or the role of GR in the CRH neuron. While it is established that rapid glucocorticoid feedback involves non-genomic rapid effects, the mechanisms involved appear to differ in the hypothalamus and pituitary. Genomic and proteomic technology will be useful to determine the identity and signaling mechanisms of the receptor mediating non-genomic effects at the different target tissues. Other remaining questions are the mechanisms determining the differential regulation of CRH and VP in the same parvocellular neuron, the differential regulation of CRH in the PVN and extrahypothalamic sites, and the differential regulation of CRHR1 in the pituitary and brain. New knowledge of the molecular bases of stress adaptation and HPA axis regulation will facilitate our understanding of the pathophysiological mechanisms and identifying biomarkers and novel targets for the treatment of stress-related disorders.

References

Abou Samra, A.B., Harwood, J.P., Manganiello, V.C., *et al.* (1987) Phorbol 12–myristate 13–acetate and vasopressin potentiate the effect of corticotropin–releasing factor on cyclic AMP production in rat anterior pituitary cells. Mechanisms of action. *Journal of Biological Chemistry*, **262**(3), 1129–1136.

Aguilera, G. (1994) Regulation of pituitary ACTH secretion during chronic stress. *Frontiers in Neuroendocrinology*, **15**(4), 321–350.

Aguilera, G. (1998) Corticotropin releasing hormone, receptor regulation and the stress response. *Trends in Endocrinology and Metabolism*, **9**(8), 329–336.

Aguilera, G. and Y. Liu (2012) The molecular physiology of CRH neurons. *Frontiers in Neuroendocrinology*, **33**(1), 67–84. **[A comprehensive review of the physiological and molecular mechanism regulating the production and secretion of corticotropin-releasing hormone by parvocelular neurons of the PVN.]**

Aguilera, G., Volpi, S., Rabadan-Diehl, C., *et al.* (2003) Transcriptional and post-transcriptional mechanisms regulating the rat pituitary vasopressin V1b receptor gene. *Journal of Molecular Endocrinology*, **30**(2), 99–108. **[A review describing transcriptional and translational mechanisms regulating the vasopressin V1b receptor content in the pituitary corticotrope.]**

Aguilera, G., Nikodemova, M., Wynn, P., *et al.* (2004) Corticotropin releasing hormone receptors: two decades later. *Peptides*, **25**(3), 319–329. **[Review of the distribution in the brain and regulation of pituitary type 1 CRH receptors.]**

Aguilera, G., Subburaju, S., Young, S.F., *et al.* (2008) The parvocellular vasopressinergic system and responsiveness of the hypothalamic pituitary adrenal axis during chronic stress. *Progress in Brain Research*, **170**: 29–39.

Arnett, M., Kolber, B., Boyle. M., *et al.* (2011) Behavioral insights from mouse models of forebrain- and amygdala-specific glucocorticoid receptor genetic disruption. *Molecular and Cellular Endocrinology*, **336**(1–2), 2–5.

Assil, I.Q. and Abou Samra, A.B. (2001) N-glycosylation of CRF receptor type 1 is important for its ligand-specific interaction. *American Journal of Physiology: Endocrinology and Metabolism*, **281**(5), E1015–E1021.

Baker, M.E., Funder, J. and Kataoula, S.R. (2013) Evolution of hormone selectivity in glucocorticoid and mineralocorticoid receptors. *Journal of Steroid Biochemistry and Molecular Biology*, **137**(0), 57–70.

Bale, T.L. and Vale, W.W. (2004) CRF and CRF receptors: role in stress responsivity and other behaviors. *Annual Review of Pharmacology and Toxicology*, **44**(1), 525–557. **[A review of the characterization and physiological roles of CRH and urocorins and CRH receptors type 1 and type 2.]**

Barnes, P.J. and Karin, M. (1997) Nuclear factor-kappaB: a pivotal transcription factor in chronic inflammatory diseases. *New England Journal of Medicine*, **336**(15), 1066–1071.

Benediktsson, R., Walker, B.R. and and Edwards, C.R. (1995) Cellular selectivity of aldosterone action: role of 11 β-hydroxysteroid dehydrogenase. *Current Opinion in Nephrology and Hypertension*, **4**(1), 41–46.

Berardelli, R., Karamouzis, I., d'Angelo, V., *et al.* (2013) Role of mineralocorticoid receptors on the hypothalamus-pituitary-adrenal axis in humans. *Endocrine*, **43**(1), 51–58.

Bethin, K.E., Vogt, S.K. and Muglia L.J. (2000) Interleukin-6 is an essential, corticotropin-releasing hormone-independent stimulator of the adrenal axis during immune system activation. *Proceedings of the National Academy of Sciences of the USA*, **97**(16), 9317–9322.

Beurel, E. and Nemeroff, C. (2014) Interaction of stress, corticotropin-releasing factor, arginine vasopressin and behaviour. *Current Topics in Behavioral Neurosciences*, **18**, 67–80.

Bicknell, A.B. (2008) The tissue-specific processing of pro-opiomelanocortin. *Journal of Neuroendocrinology*, **20**(6), 692–699.

Bilodeau, S., Vallette Kasic, S. and Gautier, I. (2006) Role of Brg1 and HDAC2 in GR trans-repression of the pituitary POMC gene and misexpression in Cushing disease. *Genes & Development*, **20**(20), 2871–2886.

Boutillier, A.L., Gaiddon, C., Lorang, D., *et al.* (1998) Transcriptional activation of the proopiomelanocortin gene by cyclic AMP-responsive element binding protein. *Pituitary (Boston)*, **1**(1), 33–43.

Buttgereit, F., Wehling, M. and Burmester, G.R. (1998) A new hypothesis of modular glucocorticoid actions: steroid treatment of rheumatic diseases revisited. *Arthritis & Rheumatism*, **41**(5), 761–767.

Chapman, K., Holmes, M. and Seckl, J. (2013) 11β-hydroxysteroid dehydrogenases: intracellular gate-keepers of tissue glucocorticoid action. *Physiological Reviews*, **93**(3), 1139–1206. **[Excellent review of the properties and functions of 11-β hydroxysteroid dehydrogenase. This enzyme is a key regulator of the specificity of glucocorticoid and mineralocorticoid receptors in different tissues.]**

Chesnokova, V. and Melmed, S. (2002) Minireview: Neuro-immuno-endocrine modulation of the hypothalamic-pituitary-adrenal (HPA) axis by gp130 signaling molecules. *Endocrinology*, **143**(5), 1571–1574.

Childs, G.V., Morell, J.L., Niendorf, A., *et al.* (1986) Cytochemical studies of corticotropin-releasing factor (CRF) receptors in anterior lobe corticotropes: binding, glucocorticoid regulation, and endocytosis of [biotinyl-Ser1] CRF. *Endocrinology*, **119**(5), 2129–2142.

Conkright, M., Canettieri, G., Screaton, R., *et al.* (2003) TORCs: transducers of regulated CREB activity. *Molecular Cell*, **12**(2), 413–423. **[Review of the characterization and mechanism of action of the CREB co-activator, transducer of regulated CREB activity (TORC). This co-activator is essential for cyclic AMP/ PKA/ CREB dependent activation of CRH transcription.]**

Deng, Q., Riquelme, D., Trinh, L., *et al.* (2014) Rapid glucocorticoid feedback inhibition of ACTH secretion involves ligand-dependent membrane association of glucocorticoid receptors. *Endocrinology*, **156**, 3215–3227.

Di, S., Malcher Lopes, M., Halmos, K., *et al.* (2003) Nongenomic glucocorticoid inhibition via endocannabinoid release in the hypothalamus: a fast feedback mechanism. *Journal of Neuroscience*, **23**(12), 4850–4857.

Evans, A., Liu, Y., Mcgregor, R., *et al.* (2013) Regulation of hypothalamic corticotropin-releasing hormone transcription by elevated glucocorticoids. *Molecular Endocrinology*, **27**(11), 1796–1807.

Flores, M., Carvallo, P., and Aguilera, G. (1990) Physicochemical characterization of corticotrophin releasing factor receptor in rat pituitary and brain. *Life Sciences*, **47**(22), 2035–2040.

Funder, J. (2005) Mineralocorticoid receptors: distribution and activation. *Heart Failure Reviews*, **10**(1), 15–22.

Gagner, J.P. and Drouin, J. (1987) Tissue-specific regulation of pituitary proopiomelanocortin gene transcription by corticotropin-releasing hormone, 3′,5′-cyclic adenosine monophosphate, and glucocorticoids. *Molecular Endocrinology*, **1**(10), 677–682.

Geerling, J. and Loewy, A. (2009) Aldosterone in the brain. *American Journal of Physiology: Renal Physiology*, **297**(3), F559–F576.

Gulyas, M., Pusztai, L., Rappay, G., *et al.* (1991) Pituitary corticotrophs proliferate temporarily after adrenalectomy. *Histochemistry*, **96**(2), 185–189.

Hamm, J., Halmos, K., Muglia, L.J. and Tasker, J.G. (2010) Rapid synaptic modulation of hypothalamic neurons by glucocorticoids requires the glucocorticoid receptor. Society for Neuroscience. San Diego: Abstract 389.319.

Harrell, J., Murphy, P., Morishima, Y., *et al.* (2004) Evidence for glucocorticoid receptor transport on microtubules by dynein. *Journal of Biological Chemistry*, **279**(52), 54647–54654.

Hauger, R., Olivares Reyes, J., Braun, S., *et al.* (2003) Mediation of corticotropin releasing factor type 1 receptor phosphorylation and desensitization by protein kinase C: a possible role in stress adaptation. *Journal of Pharmacology and Experimental Therapeutics*, **306**(2), 794–803.

Hauger, R., Olivares Reyes, J., Dautzenberg, F., *et al.* (2012) Molecular and cell signaling targets for PTSD pathophysiology and pharmacotherapy. *Neuropharmacology*, **62**(2), 705–714.

Hauger, R.L., Millan, M., Catt, K.J., *et al.* (1987) Differential regulation of brain and pituitary corticotropin-releasing factor receptors by corticosterone. *Endocrinology*, **120**(4), 1527–1533.

Henley, D.E. and Lightman, S.L. (2011) New insights into corticosteroid-binding globulin and glucocorticoid delivery. *Neuroscience*, **180**(0), 1–8.

Herbert, J., Goodyer, I., Grossman, A.B., *et al.* (2006) Do corticosteroids damage the brain? *Journal of Neuroendocrinology*, **18**(6), 393–411.

Holmes, M.C., Antoni, F., Aguilera, G., *et al.* (1986) Magnocellular axons in passage through the median eminence release vasopressin. *Nature*, **319**(6051), 326–329.

Ingraham, H.A., Albert, V., Chen, R., *et al.* (1990) A family of Pou-domain and Pit-1 tissue-specific transcription factors in pituitary and neuroendocrine development. *Annual Review of Physiology*, **52**(1), 773–791.

Insel, T. (2010) The challenge of translation in social neuroscience: a review of oxytocin, vasopressin, and affiliative behavior. *Neuron*, **65**(6), 768–779.

Jenks, B. G. (2009) Regulation of proopiomelanocortin gene expression. *Annals of the New York Academy of Sciences*, **1163**(1), 17–30. **[Detailed review describing the transcription factors and molecular mechanisms regulating POMC transcription.]**

Joels, M. and de Kloet, E.R. (1994) Mineralocorticoid and glucocorticoid receptors in the brain. Implications for ion permeability and transmitter systems. *Progress in Neurobiology*, **43**(1), 1–36.

John, C., Gavins, F., Buss, N., *et al.* (2008) Annexin A1 and the formyl peptide receptor family: neuroendocrine and metabolic aspects. *Current Opinion in Pharmacology*, **8**(6), 765–776.

Johnson, L.R., Farb, C., Morrison, J.H., *et al.* (2005) Localization of glucocorticoid receptors at postsynaptic membranes in the lateral amygdala. *Neuroscience*, **136**(1), 289–299.

Khan, A., Kaminski, K., Sanchez-Watts, G., *et al.* (2011) MAP kinases couple hindbrain-derived catecholamine signals to hypothalamic adrenocortical control mechanisms during glycemia-related challenges. *Journal of Neuroscience*, **31**(50), 18479–18491.

Komatsuzaki, Y., Murakami, G., Tsurugizawa, T., *et al.* (2005) Rapid spinogenesis of pyramidal neurons induced by activation of glucocorticoid receptors in adult male rat hippocampus. *Biochemical and Biophysical Research Communications*, **335**(4), 1002–1007.

Lamas, M., Monaco, L., Zazopoulus, E., *et al.* (1996) CREM: a master-switch in the transcriptional response to cAMP. *Philosophical Transactions of the Royal Society: Biological Sciences*, **351**(1339), 561–567.

Levin, N., Blum, M. and Roberts, J.L. (1989) Modulation of basal and corticotropin-releasing factor-stimulated proopiomelanocortin gene expression by vasopressin in rat anterior pituitary. *Endocrinology*, **125**(6), 2957–2966.

Liposits, Z. and Bohn, M.C. (1993) Association of glucocorticoid receptor immunoreactivity with cell membrane and transport vesicles in hippocampal and hypothalamic neurons of the rat. *Journal of Neuroscience Research*, **35**(1), 14–19.

Liu, Y., Kamitakahara, A., Kim, A., *et al.* (2008) Cyclic adenosine 3′,5′-monophosphate responsive element binding protein phosphorylation is required but not sufficient for activation of corticotropin-releasing hormone transcription. *Endocrinology*, **149**(7), 3512–3520.

Liu, Y., Poon, V., Sanchez-Watts, G., *et al.* (2012) Salt-inducible kinase is involved in the regulation of corticotropin-releasing hormone transcription in hypothalamic neurons in rats. *Endocrinology*, **153**(1), 223–233.

Lolait, S., Stewart, L., Jessop, D., *et al.* (2007) The hypothalamic-pituitary-adrenal axis response to stress in mice lacking functional vasopressin V1b receptors. *Endocrinology*, **148**(2), 849–856.

Lowenberg, M., Verhaar, A., van den Brink, G.R., *et al.* (2007) Glucocorticoid signaling: a nongenomic mechanism for T-cell immunosuppression. *Trends in Molecular Medicine*, **13**(4), 158–163.

Luo, X., Kiss, A., Makara, G., *et al.* (1994) Stress-specific regulation of corticotropin releasing hormone receptor expression in the paraventricular and supraoptic nuclei of the hypothalamus in the rat. *Journal of Neuroendocrinology*, **6**(6), 689–696.

Luo, X., Kiss, A., Rabadan-Diehl, C., *et al.* (1995) Regulation of hypothalamic and pituitary corticotropin-releasing hormone receptor messenger ribonucleic acid by adrenalectomy and glucocorticoids. *Endocrinology*, **136**(9), 3877–3883.

Maira, M., Martens, C., *et al.* (1999) Heterodimerization between members of the Nur subfamily of orphan nuclear receptors as a novel mechanism for gene activation. *Molecular and Cellular Biology*, **19**(11), 7549–7557.

Maira, M., Martens, C., Batsche, E., *et al.* (2003a) Dimer-specific potentiation of NGFI-B (Nur77) transcriptional activity by the protein kinase A pathway and AF-1-dependent coactivator recruitment. *Molecular and Cellular Biology*, **23**(3), 763–776.

Maira, M., Couture, C., Le Martelot, G., *et al.* (2003b) The T-box factor Tpit recruits SRC/p160 co-activators and mediates hormone action. *Journal of Biological Chemistry*, **278**(47), 46523–46532.

Martens, C., Bilodeau, S., Maira, M., *et al.* (2005) Protein-protein interactions and transcriptional antagonism between the subfamily of NGFI-B/Nur77 orphan nuclear receptors and glucocorticoid receptor. *Molecular Endocrinology*, **19**(4), 885–897.

McEwen, B. (2007) Physiology and neurobiology of stress and adaptation: central role of the brain. *Physiological Reviews*, **87**(3), 873–904.

McNally, J.G., Müller, W.G., Walker, D., Wolford, R. and Hager, G.L. (2000) The glucocorticoid receptor: rapid exchange with regulatory sites in living cells. *Science*, **287**, 1262–1265.

Morgan, S., McCabe, E., Gathercole, L., *et al.* (2014) 11β-HSD1 is the major regulator of the tissue-specific effects of circulating glucocorticoid excess. *Proceedings of the National Academy of Sciences of the USA*, **111**(24), E2482–E2491.

Moriyama, T., Kageyama, K., Kasagi, Y., *et al.* (2005) Differential regulation of corticotropin-releasing factor receptor type 1 (CRF1 receptor) mRNA via protein kinase A and mitogen-activated protein kinase pathways in rat anterior pituitary cells. *Molecular and Cellular Endocrinology*, **243**(1–2), 74–79.

Myers, B., McKlveen, J. and Herman, J.P. (2014) Glucocorticoid actions on synapses, circuits, and behavior: implications for the energetics of stress. *Frontiers in Neuroendocrinology*, **35**(2), 180–196.

Mynard, V., Guignat, L., Devin Leclerc, J., *et al.* (2002) Different mechanisms for leukemia inhibitory factor-dependent activation of two proopiomelanocortin promoter regions. *Endocrinology*, **143**(10), 3916–3924.

Mynard, V., Latchoumanin, O., Guignat, L., *et al.* (2004) Synergistic signaling by corticotropin-releasing hormone and leukemia inhibitory factor bridged by phosphorylated 3',5'-cyclic adenosine monophosphate response element binding protein at the Nur response element (NurRE)-signal transducers and activators of transcription (STAT) element of the proopiomelanocortin promoter. *Molecular Endocrinology*, **18**(12), 2997–3010.

Nikodemova, M., Kasckow, J., Liu, H., *et al.* (2003) Cyclic adenosine 3',5'-monophosphate regulation of corticotropin-releasing hormone promoter activity in AtT-20 cells and in a transformed hypothalamic cell line. *Endocrinology*, **144**(4), 1292–1300.

Nolan, L.A., Thomas, C.K. and Levy, A. (2003) Enhanced anterior pituitary mitotic response to adrenalectomy after multiple glucocorticoid exposures. *European Journal of Endocrinology*, **149**(2), 153–160.

Oakley, R. and Cidlowski, J. (2013) The biology of the glucocorticoid receptor: new signaling mechanisms in health and disease. *Journal of Allergy and Clinical Immunology*, **132**(5), 1033–1044.

Ochedalski, T., Rabadan Diehl, C. and Aguilera, G. (1998) Interaction between glucocorticoids and corticotropin releasing hormone (CRH) in the regulation of the pituitary CRH receptor in vivo in the rat. *Journal of Neuroendocrinology*, **10**(5), 363–369.

Pacak, K., Palkovits, M., Kvetnansky, R., *et al.* (1995) Catecholaminergic inhibition by hypercortisolemia in the paraventricular nucleus of conscious rats. *Endocrinology*, **136**(11), 4814–4819.

Plotsky, P.M. (1991) Pathways to the secretion of adrenocorticotropin: a view from the portal*. *Journal of neuroendocrinology*, **3**(1), 1–9.

Qian, X., Droste, S., Lightman, S.L., *et al.* (2012) Circadian and ultradian rhythms of free glucocorticoid hormone are highly synchronized between the blood, the subcutaneous tissue, and the brain. *Endocrinology*, **153**(9), 4346–4353.

Rabadan Diehl, C., Kiss, A., Camacho, C., *et al.* (1996) Regulation of messenger ribonucleic acid for corticotropin releasing hormone receptor in the pituitary during stress. *Endocrinology*, **137**(9), 3808–3814.

Rabadan Diehl, C., Makara, G., Kiss, A., *et al.* (1997) Regulation of pituitary corticotropin releasing hormone (CRH) receptor mRNA and CRH binding during adrenalectomy: role of glucocorticoids and hypothalamic factors. *Journal of Neuroendocrinology*, **9**(9), 689–697.

Rabadan Diehl, C., Lolait, S. and Aguilera, G. (2000) Isolation and characterization of the promoter region of the rat vasopressin V1b receptor gene. *Journal of Neuroendocrinology*, **12**(5), 437–444.

Rabadan Diehl, C., Volpi, S., Nikodemova, M., *et al.* (2003) Translational regulation of the vasopressin v1b receptor involves an internal ribosome entry site. *Molecular Endocrinology*, **17**(10), 1959–1971.

Rabadan Diehl, C., Martinez, A., Volpi, S., *et al.* (2007) Inhibition of vasopressin V1b receptor translation by upstream open reading frames in the 5'-untranslated region. *Journal of Neuroendocrinology*, **19**(4), 309–319.

Robert, J. and Clauser, E. (2005) [Vasopressin receptors: structure/function relationships and signal transduction in target cells]. *Journal de la Societa de Biologie*, **199**(4), 351–359.

Roper, J., O'Carroll, A., Young, W.S., *et al.* (2011) The vasopressin Avpr1b receptor: molecular and pharmacological studies. *Stress*, **14**(1), 98–115.

Samarasinghe, R., Di Maio, R., Volonte, D., *et al.* (2011) Nongenomic glucocorticoid receptor action regulates gap junction intercellular communication and neural progenitor cell proliferation. *Proceedings of the National Academy of Sciences of the USA*, **108**(40), 16657–16662.

Sarabdjitsingh, R.A., Joels, M. and de Kloet, E.R. (2012) Glucocorticoid pulsatility and rapid corticosteroid actions in the central stress response. *Physiology & Behavior*, **106**(1), 73–80.

Schwartz, J., Billestrup, N., Perrin, M., *et al.* (1986) Identification of corticotropin-releasing factor (CRF) target cells and effects of dexamethasone on binding in anterior pituitary using a fluorescent analog of CRF. *Endocrinology*, **119**(5), 2376–2382.

Shepard, J., Liu, Y., Sassone Corsi, P., *et al.* (2005) Role of glucocorticoids and cAMP-mediated repression in limiting corticotropin-releasing hormone transcription during stress. *Journal of Neuroscience*, **25**(16), 4073–4081.

Spiga, F., Walker, J., Terry, J., *et al.* (2014) HPA axis-rhythms. *Comprehensive Physiology*, **4**(3), 1273–1298. [*Detailed review describing circadian and utradian rhythms of HPA axis activity, the mechanisms of pulse generation and the importance of pulsatile glucocorticoid secretion in glucocorticoid-dependent regulation.*]

Stavreva, D.A., Müller, W.G., Hager, G.L., Smith, C.L. and McNally, J.G. (2004) Rapid glucocorticoid receptor exchange at a promoter is coupled to transcription and regulated by chaperones and proteasomes. *Molecular & Cellular Biology*, **24**, 2682–2597.

Stavreva, D., Wiench, D., John, S., *et al.* (2009) Ultradian hormone stimulation induces glucocorticoid receptor-mediated pulses of gene transcription. *Nature Cell Biology*, **11**(9), 1093–1102.

Stroth, N., Liu, Y., Aguilera, G., *et al.* (2011) Pituitary adenylate cyclase-activating polypeptide controls stimulus-transcription coupling in the hypothalamic-pituitary-adrenal axis to mediate sustained hormone secretion during stress. *Journal of Neuroendocrinology*, **23**(10), 944–955.

Subburaju, S. and Aguilera, G. (2007) Vasopressin mediates mitogenic responses to adrenalectomy in the rat anterior pituitary. *Endocrinology*, **148**(7), 3102–3110.

Swanson, L.W. and Sawchenko, P.E. (1980) Paraventricular nucleus: a site for the integration of neuroendocrine and autonomic mechanisms. *Neuroendocrinology*, **31**(6), 410–417.

Swanson, L.W., Sawchenko, P., Rivier, J., *et al.* (1983) Organization of ovine corticotropin-releasing factor immunoreactive cells and fibers in the rat brain: an immunohistochemical study. *Neuroendocrinology*, **36**(3), 165–186.

Takemori, H. and Okamoto, M. (2008) Regulation of CREB-mediated gene expression by salt inducible kinase. *Journal of Steroid Biochemistry and Molecular Biology*, **108**(3–5), 287–291.

Tasker, J. and Herman, J. (2011) Mechanisms of rapid glucocorticoid feedback inhibition of the hypothalamic-pituitary-adrenal axis. *Stress*, **14**(4), 398–406.

Tauchi, M., Zhang, R., d'Alessio, D., *et al.* (2008) Role of central glucagon-like peptide-1 in hypothalamo-pituitary-adrenocortical facilitation following chronic stress. *Experimental Neurology*, **210**(2), 458–466.

Turnbull, A.V., Smith, G.W., Lee, S., *et al.* (1999) CRF type I receptor-deficient mice exhibit a pronounced pituitary-adrenal response to local inflammation. *Endocrinology*, **140**(2), 1013–1017.

Uchoa, E.T., Aguilera, G., Herman, J.P., *et al.* (2014) Novel aspects of glucocorticoid actions. *Journal of Neuroendocrinology*, **26**(9), 557–572. [*Review of the latest research on the mechanisms of glucocorticoid feedback on HPA axis acitivity and novel actions of glucocorticoids.*]

Ulrich Lai, Y. and Herman, J. (2009) Neural regulation of endocrine and autonomic stress responses. *Nature Reviews Neuroscience*, **10**(6), 397–409. **[Excellent review describing the neural pathways controlling CRH neurons in the PVN and HPA axis activity.]**

Vaccari, C., Lolait, S.J. and Ostrowski, N.L. (1998) Comparative distribution of vasopressin V1b and oxytocin receptor messenger ribonucleic acids in brain. *Endocrinology*, **139**(12), 5015–5033.

Volpi, S., Rabadan Diehl, C., Cawley, N., *et al.* (2002) Transcriptional regulation of the pituitary vasopressin V1b receptor involves a GAGA-binding protein. *Journal of Biological Chemistry*, **277**(31), 27829–27838.

Volpi, S., Liu, Y. and Aguilera, G. (2006) Vasopressin increases GAGA binding activity to the V1b receptor promoter through transactivation of the MAP kinase pathway. *Journal of Molecular Endocrinology*, **36**(3), 581–590.

Voss, T. and Hager, G. (2014) Dynamic regulation of transcriptional states by chromatin and transcription factors. *Nature Reviews Genetics*, **15**(2), 69–81. **[Review discussing cutting edge research on mechanisms of interaction between glucocorticoids and other transcription factors with the chromatin leading to chromatin remodeling and dynamic regulation of gene expression.]**

Walker, J.J., Terry, J. and Tsaneva Atanasova, K. (2010) Encoding and decoding mechanisms of pulsatile hormone secretion. *Journal of Neuroendocrinology*, **22**(12), 1226–1238.

Watts, A., Sanchez Watts, S., Liu, Y., *et al.* (2011) The distribution of messenger RNAs encoding the three isoforms of the transducer of regulated cAMP responsive element binding protein activity in the rat forebrain. *Journal of Neuroendocrinology*, **23**(8), 754–766.

Westphal, N. and Seasholtz, A.F. (2006) CRH-BP: the regulation and function of a phylogenetically conserved binding protein. *Frontiers in Bioscience*, **11**: 1878–1891.

Witchel, S. and DeFranco, D. (2006) Mechanisms of disease: regulation of glucocorticoid and receptor levels – impact on the metabolic syndrome. *Nature Clinical Practice Endocrinology & Metabolism*, **2**(11), 621–631.

Wolfl, S., Martinez, C. and Majzoub, J. (1999) Inducible binding of cyclic adenosine 3′,5′-monophosphate (cAMP)-responsive element binding protein (CREB) to a cAMP-responsive promoter in vivo. *Molecular Endocrinology*, **13**(5), 659–669.

Wu, Z., Ji, H., Hassan, A., *et al.* (2004) Regulation of pituitary corticotropin releasing factor type-1 receptor mRNA binding proteins by modulation of the hypothalamic-pituitary-adrenal axis. *Journal of Neuroendocrinology*, **16**(3), 214–220.

Xu G., Rabadan Diehl, C., Nikodemova, M., *et al.* (2001) Inhibition of corticotropin releasing hormone type-1 receptor translation by an upstream AUG triplet in the 5′ untranslated region. *Molecular Pharmacology*, **59**(3), 485–492.

CHAPTER 18

Neuroendocrine Control of Female Puberty: Genetic and Epigenetic Regulation

Alejandro Lomniczi and Sergio R. Ojeda

Division of Neuroscience, Oregon National Primate Research Center/Oregon Health & Science University, Beaverton, Oregon, USA

18.1 Introduction

Puberty is a critical developmental phase in the life of an individual as it capacitates the neuroendocrine reproductive system to perform essential functions that ensure the preservation of the species. The initiation of puberty is determined by a complex array of interactions involving genetic factors and environmental cues. These interactions begin to impact the hypothalamic-pituitary-gonadal axis long before the hormonal and somatic manifestations of puberty become apparent. Nevertheless, the final *sine qua non* event required for the pubertal process to unfold is a sustained increase in pulsatile gonadotropin-releasing hormone (GnRH) release from neurosecretory neurons of the hypothalamus. This increase is thought to be driven not by intrinsic changes in GnRH neuronal activity but rather by coordinated alterations in transsynaptic and glial input to GnRH neurons. The prevailing view summing up these interactions is that as inhibitory neurotransmission becomes weaker, both excitatory neurotransmission and facilitatory glial output gain strength, shifting the input to GnRH neurons from inhibition to stimulation.

In the last 10 years, significant progress has been made towards identifying the genetic underpinnings of this process. Single genes have been identified that are essential for puberty to occur, but it has also become apparent that the pubertal process is controlled by a diversity of genes which control various cellular functions. Emerging evidence suggests that these genes are functionally connected and, importantly, that the activity of the resulting gene networks is coordinated epigenetically. In this chapter, we will briefly describe the hormonal and transsynaptic/glia–neuron communication events underlying the advent of puberty, before discussing the genetic and epigenetic underpinnings of the pubertal process. The concept of puberty-controlling gene networks will be presented and the potential organization of these networks will be outlined. Finally, we will discuss current evidence indicating that the epigenetic regulation of these networks – and hence of puberty itself – is provided by at least two classes of modifications: changes in DNA methylation and changes in chromatin structure at genomic regions involved in the control of puberty.

Molecular Neuroendocrinology: From Genome to Physiology, First Edition. Edited by David Murphy and Harold Gainer.
© 2016 John Wiley & Sons, Ltd. Published 2016 by John Wiley & Sons, Ltd.
Companion website: www.wiley.com/go/murphy/neuroendocrinology

18.2   The hormonal changes of puberty

In the 1980s and 1990s, extensive efforts were devoted to, first, defining the hormonal changes preceding and accompanying the initiation of puberty and then identifying the neurotransmitter systems responsible for these changes (Ojeda and Terasawa, 2002). It was, however, more than 40 years ago that Boyar *et al.* (1972) introduced the now widely accepted concept that an increase in pulsatile luteinizing hormone (LH) release is the first endocrine manifestation of the initiation of puberty. This study, performed in humans, demonstrated that the amplitude of LH pulses detected in the bloodstream increases at night during early puberty. In rodents, the increase in LH pulsatility occurs in the afternoon and it is first observed by the end of the juvenile period (Ojeda and Terasawa, 2002). The imme-diate consequence of enhanced LH secretion is an increase in gonadal output of sex steroids, which is in turn responsible for the development of secondary sexual characteristics. In females, there is subsequent development of estrogen-positive feedback, which culminates with the activation of the first preovulatory surge of gonadotropins. This event essentially defines the neuroendocrine completion of the female pubertal process. Subsequent work revealed that the pubertal change in LH output is driven by an increase in pulsatile release of GnRH from the hypothalamus (Ojeda and Terasawa, 2002) (Figure 18.1). It also became evident that GnRH neurons are able to produce and release GnRH long before puberty, making it clear that they are not the ultimate responsible factor for the initiation of puberty nor do they constitute a significant obstacle for the pubertal process to be initiated earlier.

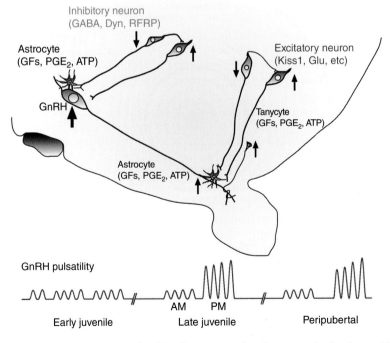

Figure 18.1 The cellular systems involved in the neuroendocrine control of puberty. The hypotha-lamic control of puberty involves excitatory and inhibitory transsynaptic inputs, in addition to facilitatory glia-to-neuron signaling. The hypothesis states that at the onset of puberty, a shift from a predominantly inhibitory (*downward arrows*) to an excitatory mode of control (*upward arrows*) results in increased GnRH release, which triggers a diurnal increase in LH secretion, the first endocrine manifestation of puberty.

18.2.1   The transsynaptic regulation of puberty

We now know that GnRH neurons are controlled by cellular systems of both neuronal and glial nature. Neuronal subsets conveying either excitatory or inhibitory inputs to GnRH neurons have been shown to act transsynaptically to advance or delay the advent of puberty (Herbison, 2006). Experimental dissection of these interactions has given rise to the concept that puberty is set in motion by a concomitant reduction in inhibitory inputs and an increase in excitatory signals impinging on GnRH neurons (Ojeda and Terasawa, 2002) (see Figure 18.1). In addition to this dual neuronal input, GnRH neurons are subjected to a glial excitatory influence, exerted in a paracrine fashion by a variety of growth factors and small molecules (Lomniczi and Ojeda, 2009) (see Figure 18.1).

18.2.1.1   Excitatory control

A sizeable fraction of the excitatory transsynaptic control of puberty is provided by glutamatergic neurons. However, a more powerful – and anatomically more circumscribed – system is supplied by two discrete subsets of hypothalamic neurons that produce a set of peptides known as kisspeptins (Pinilla *et al.*, 2012). These peptides – products of the KISS1/Kiss1 gene – are powerful stimulators of GnRH release. The critical importance of kisspeptins for puberty was demonstrated 10 years ago by studies in humans showing that loss of function of GPR54/KISS1R, the gene encoding the kisspeptin receptor, results in pubertal failure (de Roux *et al.*, 2003; Seminara *et al.*, 2003). There are two well-defined groups of kisspeptin neurons in the neuroendocrine brain. In both humans and rodents, one population is located in the arcuate nucleus (ARC) of the medial basal hypothalamus and another in the periventricular region of the anteroventral periventricular nucleus (AVPV) of rodents (Pinilla *et al.*, 2012) and the rostral periventricular area of humans (Hrabovszky *et al.*, 2010). A similar distribution has been observed in the ovine brain. While ARC kisspeptin neurons appear to be required for pulsatile GnRH release, AVPV kisspeptin neurons (at least in rodents) are needed for the preovulatory surge of gonadotropins (Pinilla *et al.*, 2012).

18.2.1.2   Inhibitory control

There are three main neuronal subsets providing inhibitory transsynaptic regulation to GnRH neurons (see Figure 18.1). They are GABAergic, opiatergic (Herbison, 2006), and RFamide-related peptide (RFRP)-containing neurons. RFRP is the mammalian ortholog of the peptide gonadotropin-inhibiting hormone (GnIH), first described in birds. RFRP neurons use the peptides RFRP1 and RFRP3 for transsynaptic communication. Both peptides are recognized by a high-affinity receptor termed GPR147 or NPFFR1, and a low-affinity receptor termed GPR74 or NPFFR2. GPR147 is expressed in GnRH neurons, suggesting that RFRP neurons can act directly on GnRH neurons to inhibit GnRH secretion (reviewed in Lomniczi *et al.*, 2013b).

GABAergic neurons have been shown to inhibit GnRH secretion by acting on neurons connected to the GnRH neuronal network via both $GABA_A$ and $GABA_B$ receptors, and by activating $GABA_B$ receptors located on GnRH neurons (Herbison, 2006; Ojeda and Terasawa, 2002). Despite these inhibitory actions, GABAergic neurons can also excite GnRH neurons directly via activation of $GABA_A$ receptors (DeFazio *et al.*, 2002). Opiatergic neurons inhibit GnRH neuronal activity by releasing different peptides that bind to a set of receptors located both on GnRH neurons and on neurons controlling GnRH secretion (Herbison, 2006; Ojeda and Terasawa, 2002). Perhaps the best recognized example of this latter type of interaction is provided by KNDy neurons of the ARC. As will be described below, these neurons produce three different peptides: kisspeptins, neurokinin B (NKB), and dynorphin. Consistent with being an opioid peptide, dynorphin inhibits GnRH secretion, and at least part of this effect is due to repression of kisspeptin release from KNDy neurons (Pinilla *et al.*, 2012).

18.3   The glial contribution

The concept of a glial involvement in the hypothalamic control of puberty was introduced in the early 1990s. Since then, a wealth of information has accumulated verifying the overall validity of this view and characterizing the cell-to-cell communication pathways involved (Lomniczi and Ojeda, 2009). It is now known that astrocytes and ependymoglial cells lining the ventrolateral surface of the third ventricle (known as tanycytes) facilitate GnRH secretion both by releasing growth factors and small molecules (such as adenosine triphosphate [ATP] and prostaglandin E_2 [PGE_2]) (see Figure 18.1) and via cell-to-cell adhesive interactions. The growth factors involved include transforming growth factor-β (TGFβ), which activates cell membrane receptors with serine-threonine kinase activity, and growth factors that bind to receptors with tyrosine kinase activity. They comprise the epidermal growth factor (EGF) family, basic fibroblast growth factor (bFGF), and insulin-like growth factor 1 (IGF-I). While bFGF and IGF-1 act directly on GnRH neurons, EGF-like peptides (such as transforming growth factor α [TGFα]) stimulate GnRH release by facilitating PGE_2 release from both astroglial cells and tanycytes (Lomniczi and Ojeda, 2009).

Glial-GnRH neuron adhesive communication, on the other hand, requires direct cell-cell contact and involves adhesion molecules with unique structural features that allow them to set in motion bidirectional intracellular signaling. The adhesion molecules thus far identified include sialylated neural cell adhesion molecule NCAM (PSA-NCAM), synaptic cell adhesion molecule 1 (SynCAM1), and receptor-like protein tyrosine phosphatase-β (RPTPβ). While PSA-NCAM and SynCAM1 signal via homophilic interactions (i.e. PSA-NCAM/PSA-NCAM and SYNCAM1/SYNCAM1), RPTPβ regulates GnRH neuronal function by interacting with contactin, a cell surface glycosylphosphatidyl inositol (GPI)-anchored protein expressed in GnRH neurons. Because the intracellular domain of all these proteins contains protein-protein interacting domains, it appears that upon recognition and binding, a bidirectional flow of information is set in motion that modifies the cellular activity of both the glial cell and GnRH neuron involved (Lomniczi and Ojeda, 2009).

18.4   Gene networks controlling puberty

The unveiling of the human genome has been followed by the fast-paced development of new tools to search and characterize the genetic underpinnings of both health and disease. The availability of these tools has allowed the identification of several genes required for puberty to occur (Sykiotis *et al.*, 2010). It has also made apparent that instead of being set in motion by a single gene, the initiation and progression of puberty requires the contribution of several genes, postulated to be organized into functionally connected networks (Lomniczi *et al.*, 2013b).

This concept notwithstanding, pubertal failure has been shown to be the inexorable outcome of inactivating mutations compromising a handful of genes involved in the control of GnRH secretion or action, including GNRHR (encoding the GnRH receptor), KISS1R (encoding the kisspeptin receptor), KiSS1 (encoding kisspeptins), TAC3 (encoding NKB), TAC3R (encoding the NKB receptor), LEP (encoding leptin, a cytokinine produced by adipocytes), and LEPR (encoding the leptin receptor), in addition to others required for GnRH neuron migration (Sykiotis *et al.*, 2010). More recently, two mutations have been described to underlie cases of familial precocious puberty. One of these mutations results in constitutive activation of KISS1R (Teles *et al.*, 2008); the other appears to implicate the lifting of an inhibitory input, because it involves inactivating mutations of MKRN3, a transcriptional repressor (Abreu *et al.*, 2013).

That additional genes contribute to the neuroendocrine control of puberty is suggested by two pieces of information. One is that the known mutations affecting puberty account for only a small percentage of individuals with pubertal disorders. The other comes from genome-wide analyses showing that sequence variations in more than 40 loci are associated with an early age at menarche (see, for instance, Sulem *et al.*, 2009, and other papers in the same issue).

If a multiplicity of genes contributes to the regulation of puberty, the developmental program governing reproductive function would be expected to contain built-in mechanisms providing both redundancy and compensatory pathways to the system. These features would ensure functional integrity of the attendant regulatory networks in the event of loss of a component. A case in point is the reestablishment of normal reproductive function following either early death of kisspeptin neurons (Mayer and Boehm, 2011) or loss of Tac3R, the NKB receptor (Yang *et al.*, 2012), both of which have been shown to be essential for puberty to occur in human subjects. The existence of such safeguard mechanisms implies that gene sets controlling puberty are organized into interactive functional networks and that these networks contain features that ensure both redundancy and compensatory plasticity to the system.

18.4.1 A blueprint for puberty-controlling gene networks

The aforementioned observations make it evident that puberty is under complex genetic control and that the timing of puberty may be determined by a host of genes regulating different aspects of the underlying biological circuitries. To begin exploring this concept, we used a combination of high throughput approaches (DNA arrays, proteomics), 'guilt by association,' 'retrospective' analysis, and systems biology approaches to identify a group of genes that, we believe, are representative members of one of the genetic networks involved in the neuroendocrine control of female puberty (Roth *et al.*, 2007). Although these genes have diverse cellular functions, they share the common feature of having been earlier identified as involved in tumor suppression/tumor formation. We refer to them as tumor-related genes (TRGs). According to our analysis, the TRG network contains a handful of 'central nodes' that provide functional integration to the network by regulating expression of subordinate genes at the transcriptional level. In turn, these subordinate genes act as conduits of network output and inlets for incoming information (Lomniczi *et al.*, 2013b) (Figure 18.2). Two of them deserve special mention: KiSS1 (previously known as suppressor of metastasis) and SynCAM1 (previously known as tumor suppressor of lung cancer, TSLC1). While KiSS1 is essential for the occurrence of puberty in mice and humans (Pinilla *et al.*, 2012), SynCAM1 plays a pivotal role in the developmental control of glia-GnRH neuron adhesive communication, and in the mechanism by which astrocytic erbB4 receptors facilitate GnRH release and control normal female reproductive function (Lomniczi *et al.*, 2013b).

The five major hubs (CDP/CUTL1/CUX1, MAF, p53, YY1, and USF2) predicted to control the TRG network at the transcriptional level are not only connected to genes encoding proteins required for intracellular signaling, and cell-to-cell communication, but also with other upper-echelon genes (OCT2, TTF1, and EAP1) postulated to be involved in the transcriptional regulation of the pubertal process (Lomniczi *et al.*, 2013b) (see Figure 18.2). These and other genes of the network are expressed in neuronal and/or glial subsets involved in the control of GnRH secretion, including GnRH neurons themselves. Recently, we showed that EAP1 is, in fact, a TRG, and that, as predicted by the draft model of the network, KiSS1 expression is under the transcriptional control of central hubs, including EAP1, YY1, CUTL1/CUX1, and TTF1 (Lomniczi *et al.*, 2013b).

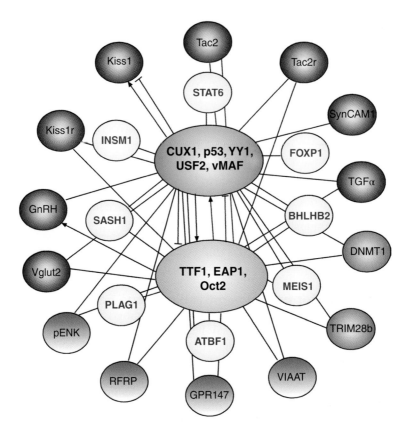

Figure 18.2 The tumor-related gene (TRG) network postulated to contribute to the hypotha-lamic control of puberty. The original draft of the network was constructed based on the presence of binding sites for transcription factors (TFs) with known TRG function within 10 kb of the 5′ flanking region of TRGs found to have increased hypothalamic expression at the time of puberty in female monkeys and rats. Based on the degree of connectivity, five TFs (MAF, CUTL1/CUX1, p53, YY1, and USF2; *green ovals*) were postulated to represent major hubs controlling the network. Additional studies showed that these central hubs are functionally connected to other transcriptional factors (TTF1, OCT2, and EAP1) postulated to be upstream regulators of puberty (*pink oval*). Subordinate genes involved in the facilitatory control of GnRH secretion (and hence of the pubertal process) are shown as red circles in the periphery of the network. Those genes involved in the inhibitory control of puberty are shown as blue circles. Notice that Kiss1 and Tac2 (playing a central role in the excitatory transsynaptic stimulation of GnRH secretion), and Syncam1 (contributing to the adhesive communication that exists between glial cells and GnRH neurons) are predicted to be subordinate genes. Subordinate transcription factors (such as STAT6, SASH1, and FOXP1) and genes encoding proteins involved in epigenetic repression (such as DNMT1 and TRIM28) are shown as an intermediate layer of nodes (*yellow color and greenish*), respectively. DNMT1 is a DNA methylating enzyme; TRIM28 (tripartite motif containing 28, an E3 SUMO protein ligase also known as KAP1) is a member of a transcriptional repressive complex involving ZNF proteins.

18.4.2   Perturbation of network function

A central theme of systems biology is that gene networks can be perturbed for experimental verification of computational predictions using either gain-of-function or loss-of-function approaches. We have used both approaches, because they can be expeditiously imple-mented in both *in vitro* and *in vivo* settings. In earlier studies, we employed transgenic mice and retroviral vectors to deliver transgenes to the hypothalamus. More recently, we have used lentiviruses to deliver either a transgene or an RNAi cargo to this region of the brain. Because lentiviruses can infect non-dividing cells (like neurons) and the viral genome is stably incorporated into the cell genome, this approach can be used to target specific brain regions for a long-lasting effect on gene expression. We have described in detail the use of this system in our hands (Dissen *et al.*, 2009). An example of a loss-of-function approach can be found in a study in which a miRNA delivered to the ARC of female rats and targeting EAP1 mRNA for silencing resulted in delayed puberty, disruption of estrous cyclicity and reduced plasma LH, follicle-stimulating hormone (FSH) and estradiol levels. When we targeted the ARC of non-human primates using the same approach, menstrual cyclicity was disrupted (for additional details see Lomniczi *et al.*, 2013b). An example of a gain-of-function approach is provided by a recent study in which Eed, a member of the polycomb group (PcG) silencing complex (see later in this chapter), was delivered (using a lentiviral vehicle) to the ARC of juvenile female rats (Lomniczi *et al.*, 2013a). The resulting phenotype of delayed puberty, disrupted estrous cyclicity, and diminished fertility gave support to the notion that the PcG complex plays a repressive role in the hypothalamic control of puberty. Altogether, these studies provide proof of principle for the notion that both gain-of-function and loss-of-function approaches applied to an *in vivo* setting are useful tools to define the function of specific genes. Importantly, when these approaches are followed by high-throughput examination of gene sets predicted to form part of a gene network, they can provide invaluable information concerning the structural organization of these networks.

18.5   Transcriptional repression: a key regulatory mechanism of prepubertal development

For years, a prevailing view to explain the initiation of puberty assumed the existence of a pubertal 'brake' (Grumbach and Styne, 1992). According to this notion, during the prepubertal period the secretory activity of GnRH neurons is under transsynaptic inhibitory control. At puberty, this inhibition would be lifted, resulting in increased GnRH release (Ojeda and Terasawa, 2002). A different, but not mutually exclusive, view is that puberty can only occur if there is activation of excitatory inputs (Ojeda, 1991). This latter concept was strongly supported by the demonstration that activation of kisspeptin neurons, which provide a significant fraction of the stimulatory inputs controlling GnRH neurons, is essential for puberty to occur (Pinilla *et al.*, 2012). Based on these and other observations, the original concept has been refined to state that a concomitant decrease in inhibitory inputs and an increase in excitatory neurotransmission are the essential underpinnings of the pubertal process (Ojeda and Terasawa, 2002).

Notwithstanding the potential importance of this transsynaptic interplay, very recent evidence suggests that the inhibitory and stimulatory control of puberty may not be exclusively provided by parallel cell-to-cell communication pathways impinging on GnRH neurons. Instead, the critical inhibitory/excitatory Yin-Yang of puberty appears to reside at a transcriptional level within neurons involved in stimulating GnRH release. The existence of a transcriptionally repressive mode controlling puberty-activating genes was initially suggested by the fact that some central nodes of the TRG network (YY1, EAP1, and the CUX isoform CUX1p120) can repress Kiss1 transcriptional activity (for details and references see Lomniczi *et al.*, 2013b). Definitive proof was provided by the demonstration that the PcG silencing complex (Schwartz and Pirrotta, 2007) prevents the premature initiation of puberty by repressing the Kiss1 gene in kisspeptin neurons of the ARC (Lomniczi *et al.*, 2013a) (see later in this chapter).

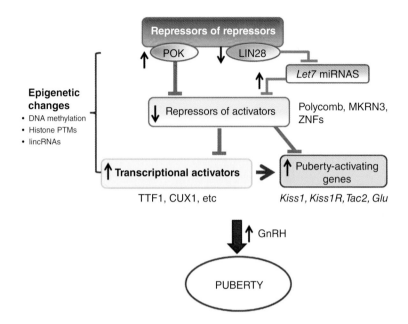

Figure 18.3 Hierarchical organization and epigenetic control of transcriptional regulatory networks postulated to control the onset of female puberty. According to this model, several layers of transcriptional repression operate in the prepubertal hypothalamus. A proximal layer is provided by the PcG complex and MKR3/ZNF proteins, acting as *first-order* repressors either in parallel or as interactive entities. A higher level of control is provided by POK proteins and LIN28. While POK proteins are predicted to act as 'repressors of repressors' by inhibiting the transcription of first-order repressors, LIN28 is postulated to act posttranscriptionally to prevent the maturation of let-7 miRNAs, which may in turn act posttranscriptionally to diminish mRNA expression of first-order repressors. Emerging evidence suggests that the inhibitory effect of POK genes increases during prepubertal development (*upward arrows*) resulting in reduced activity (*downward arrows*) of first-order repressors (PcG, MKRN3, ZNFs), and the consequent activation of puberty-activating genes (Kiss1, Tac2, TTF1, Nell2). In contrast, the inhibitory activity of LIN28 decreases at this time, resulting in increased expression of let-7 miRNAs. Because genes involved in repressing puberty-activating genes are downstream targets of let-7 miRNAs, the net outcome of increased let-7 miRNA availability would be repression of these repressors. Both the expression and actions of each component of this regulatory cascade are postulated to be regulated by epigenetic mechanisms involving changes in DNA methylation and changes in association of modified histones to gene promoters. lincRNAs, long intergenic non-coding RNAs; PTM, posttranslational modifications.

In addition to the PcG complex, recent studies in our laboratory (unpublished results) have unveiled the existence of other inhibitory systems. One of them, which appears to be located at a higher hierarchical level, is a subfamily of the BTB POZ-ZF (for Broad Complex, Tramtrack, Bric-à-brac, poxvirus and Zinc finger) group of transcriptional repressors – known as the POK family because its members have an N-terminal **POZ** domain and a C-terminal zinc finger domain of the **K**rüppel type (Kelly and Daniel, 2006). POK genes control a variety of developmental processes; their absence results in tumorigenesis and a diversity of developmental disorders (Kelly and Daniel, 2006). In our studies, we observed that hypothalamic expression of three POK genes, known as Zbtb7a/LRF, Bcl6 and Zbtb16/Plzf, increases markedly during the infantile-to-juvenile transition and then again (though less noticeably) during the juvenile-to-peripubertal transition. Using promoter assays, we also observed that these POK proteins are potent repressors of PcG gene expression, suggesting that their role in the peripubertal hypothalamus might be to repress downstream repressors (Figure 18.3), such as the PcG complex, thereby allowing disinhibition of puberty-activating genes.

As if this three-layer complexity (repression-repression-activation) were not enough, members of the zinc finger (ZNF) superfamily of transcriptional repressors (Filion *et al.*, 2006) may also be involved in the transcriptional silencing of puberty. This was earlier suggested by the finding that expression of some ZNFs declines in the primate hypothalamus at the initiation of puberty (Lomniczi *et al.*, 2013b). Conclusive evidence for this view was recently provided by the demonstration that inactivating mutations in MKRN3/ZNF127, a ZNF gene, result in precocious puberty in both girls and boys (Abreu *et al.*, 2013). MKRN3 is an intronless, maternally imprinted gene encoding a protein that in addition to several zinc finger motifs also contains a RING finger domain, suggestive of ubiquitin ligase-like activity. Hypothalamic expression of MKRN3 declines during juvenile development not only in mice (Abreu *et al.*, 2013), but also in rats and non-human primates (unpublished observations), suggesting that as puberty approaches the reproductive hypothalamus becomes free of an inhibitory influence exerted by MKRN3. That this decline in MKRN3 expression may be the consequence of an increase in a POK-dependent inhibitory input is suggested by the developmental pattern of expression of Bcl6, Zbtb7a/LRF, and Zbtb16/PLZF, which is opposite to that of MKRN3 in female rats (unpublished observations). Altogether these observations are consistent with the interpretation that MKRN3 (and perhaps other ZNFs) controls puberty, not by activating the GnRH neuronal network but instead by repressing downstream puberty-activating genes, such as Kiss1 (Lomniczi *et al.*, 2013b) (see Figure 18.3).

MKRN3 is structurally similar to EAP1 (IRF-2BPL), which as previously mentioned has been implicated in the control of puberty and reproductive function (Lomniczi *et al.*, 2013b). Because of this, MKRN3 may epitomize a class of genes that contribute to the repressive transcriptional control of puberty by directly binding to DNA (via the zinc finger domain) and interacting with other repressive proteins (via the RING finger domain).

In addition to these mechanisms of transcriptional repression, there appear to be pathways that control puberty by silencing gene expression posttranscriptionally. LIN28B exemplifies this mode of regulatory control. LIN28B encodes a protein that controls gene expression posttranscriptionally by inhibiting miRNA maturation, and hence, miRNA-dependent mRNA degradation. LIN28 targets members of the let-7 family of miRNAs by binding to the terminal loops of let-7 precursors to block processing into mature miRNAs (for references see Sangiao-Alvarellos *et al.*, 2013). An involvement of LIN28B in the control of puberty was suggested by genome-wide analysis of the human genome showing that a single nucleotide polymorphism near the LIN28B gene in human chromosome 6(q21) is associated with earlier puberty and shorter stature in girls (for references see Sangiao-Alvarellos *et al.*, 2013). Our results showed that the hypothalamic expression of Lin28 and Lin28b decreases substantially during prepubertal development in rats and non-human primates, and that let-7a and let-7b miRNA levels display the opposite pattern of expression (Sangiao-Alvarellos *et al.*, 2013). These observations raise the intriguing possibility that LIN28 acts like POK proteins, that is, by repressing downstream repressors, which in this case are members of the let-7 family of miRNAs (see Figure 18.3).It might be speculated that let-7 miRNAs may silence first-order transcriptional repressors of puberty, such as MKRN3, other ZNFs and/or PcG genes, so that as let miRNA expression increases before puberty, that of MKRN3/ZNFs/PcGs decreases (see Figure 18.3).

Altogether, the aforementioned observations suggest that the timing of mammalian puberty is controlled by gene networks endowed with overlapping layers of repressive regulation (Lomniczi *et al.*, 2013b). They are also in keeping with the recently proposed concept (Lomniczi *et al.*, 2013a) that transcriptional repression is a core component of the genomic machinery regulating puberty. We believe that this machinery is organized into functional modules containing 'activators' that set in motion key developmental milestones, and 'repressors' that prevent these events from unfolding prematurely.

18.6   Epigenetic information: an integrating mechanism of reproductive neuroendocrine development

If the existence of gene networks controlling puberty is accepted at face value, then one has to consider potential mechanisms able to coordinate the activity of these networks in a plastic and dynamic fashion during prepubertal development. These mechanisms are likely epigenetic, for many reasons. Epigenetic information is essential for a diversity of nervous system functions, including learning and memory, dendritic development, neuronal and behavioral plasticity, estrogen-induced gene expression, glial-neuronal interactions, circadian rhythms, and sexual differentiation of the brain. Epigenetic information has also been shown to display an unexpected degree of plasticity as it can rapidly and reversibly change gene expression within minutes (Metivier *et al.*, 2008).

18.6.1   Mechanisms of epigenetic control

There are two well-established- and one emerging mechanism of epigenetic control. The former are: (a) chemical modifications of the DNA via DNA methylation and hydroxymethylation, and (b) modifications of chromatin structure caused by posttranslational modifications of histone proteins that, wrapped around by two superhelical turns of DNA, make up the nucleosome, the core unit of chromatin. The most recently unveiled mechanism of epigenetic control is provided by long intergenic non-coding RNAs (lincRNAs). These transcripts do not code for any protein but are polyadenylated. They appear to serve as scaffolds that bring together chromatin remodeling complexes, targeting them to genomic regions involved in the control of gene expression. Perhaps the most well-recognized lincRNA involved in epigenetic regulation is HOTAIR, which targets PRC2 (a subcomplex of the PcG silencing complex – see later) and the LSD1-coREST repressive complex to the regulatory regions of downstream genes (Chu *et al.*, 2011). Because PRC2 catalyzes H3K27 trimethylation and LSD1-coREST demethylates H3K4me2, the net outcome of HOTAIR actions is gene silencing.

DNA methylation and hydroxymethylation are two covalent modifications of cytosine residues that mostly target the dinucleotide sequence CpG. DNA methylation consists in the addition of a methyl group to position 5 of cytosine, resulting in the formation of 5-methylcytosine (5-mC). Oxidation of 5-mC by the TET family of dio-oxygenase enzymes yields 5-hydroxymethylcytosine (5-hmC) (Tahiliani *et al.*, 2009). In general, increased DNA methylation (5-mC) is associated with gene repression, and hypomethylation (less 5-mC, more 5-hmC) is associated with transcriptional activation. The balance of 5-mC and 5-hmC at a given genomic region depends on the activity of DNA methyltransferases that generate 5-mC, and the TET enzymes that catalyze the conversion of 5-mC to 5-hmC. The methyl-transferases involved are DNA methyltransferase1 (DNMT1), which maintains basal levels of DNA methylation, and DNMT3a and DNMT3b responsible for *de novo* methylation of both unmethylated and hemimethylated DNA. Although 5-mC and 5-hmC co-exist throughout the genome, their relative abundance varies in different regions. Thus, 5-hmC has been found to be associated with euchromatin (i.e. chromatin in the open state) and enriched in promoter regions of active genes, whereas 5-mC displays an opposite pattern (Ficz *et al.*, 2011).

Posttranslational histone modifications, on the other hand, alter the N-terminus tails of the four histones (H2A, H2B, H3, and H4) that make up the protein core of nucleosomes (Kouzarides, 2007). These modifications include acetylation, methylation, phosphorylation, ubiquitination, and sumoylation (Kouzarides, 2007). Acetylation by acetyltransferases (HATs), deacetylation by histone deacetylases (HDACs), and methylation by methyltransferases (HMTs) are perhaps the best characterized. Again, some generalizations can be made: acetylation is associated with activation of transcription; deacetylation with gene silencing (Kouzarides, 2007). The functional consequences of the histone methylation status are more complex. For instance, H3 methylation of lysine 9 and 27 (H3K9me and H3K27me) is usually seen in silenced genes whereas trimethylation of histone 3 (H3) at lysine 4 (H3K4me3) is a feature of active transcription (Kouzarides, 2007).

Importantly, changes in DNA methylation and histone modifications are not dissociated events as they usually work in sync to regulate gene expression (Cedar and Bergman, 2009). For instance, H3K4me2/3 prevents DNA methylation, and H3K9me facilitates DNMT3a and DNMTb recruitment to target genes. DNMTs do not work in isolation; upon recruitment, they associate with HDACs, bringing about gene silencing. In addition to enzymes instituting these histone modifications, there is a similarly complex set of enzymes that reverse them (Kouzarides, 2007). Together, 'writers' and 'erasers' (Borrelli *et al.*, 2008) allow epigenetic 'readers' (such as PRC1, another subcomplex of the PcG complex) to regulate gene activity with a degree of flexibility and developmental plasticity not provided by DNA sequence. Because complex biological processes require integrative mechanisms of gene regulation displaying these attributes, it would appear logical to assume that the role of epigenetics in the regulation of puberty is substantial.

18.6.2   Epigenetic regulation of puberty

Recent studies have begun to address the contribution of epigenetics to the control of neuroendocrine reproductive function. Although not directly dealing with puberty itself, a recent study showed that changes in DNA methylation of the GnRH promoter accompany the increase in GnRH expression that occurs during embryonic development of non-human primates (Kurian *et al.*, 2010). Another report demonstrated that estrogen-positive feedback results in increased acetylation of histone 3 in the kiss1 promoter of AVPV kisspeptin neurons, but not ARC kiss1 neurons (Tomikawa *et al.*, 2012).

The first study directly addressing the contribution of epigenetics to puberty revealed that the PcG silencing complex plays a major role in this process (Lomniczi *et al.*, 2013a). The PcG system is composed of two main (PRC1, PRC2) and one accessory (PhoRC) repressive complexes (termed **P**olycomb **R**epressive **C**omplexes, PRCs) (Schwartz and Pirrotta, 2007). Prominent within the PRC1 complex are the members of the CBX group. They receive their name because they contain a conserved chromodomain (CBX) at their amino terminus. Although there are several mammalian CBX proteins (CBX2, 4, 6, 7, and 8), different Cbx genes are expressed in different cells (Schwartz and Pirrotta, 2007). In mammals, the PRC2 complex includes four core subunits: enhancer of zeste 1 or 2 (EZH1, EZH2), suppressor of zeste (Suz12), and the WD40 domain proteins EED and P55. The PhoRC complex has been well characterized in *Drosophila* and shown to contain two proteins, Pho and its homolog Phol, which bind directly to DNA (Schwartz and Pirrotta, 2007). In mammals, Yy1 performs some of the functions of these proteins. However, YY1 has both repressive and activating functions (Schwartz and Pirrotta, 2007).

PcG-mediated gene silencing requires H3K27me3, a histone modification catalyzed by PRC2. This mark is then used as a landing pad for the CBX components of PRC1, which bind to H3K27me3 via their chromodomain to form a repressive complex (Schwartz and Pirrotta, 2007). YY1 has been shown to recruit both PRC2 and PRC1 proteins, in addition to H3K27me3 to gene promoters to silence transcription. Although CBX proteins are core components of the repressive PRC1 complex, very recent studies revealed the existence of six different non-canonical PRC1 complexes that do not contain CBX proteins (Gao *et al.*, 2012). One of them, containing the protein RYBP (RING1/YY1 binding protein), was shown to be critical for PRC1-dependent monoubiquitination of histone 2A at lysine 119 (Tavares *et al.*, 2012). This modification (H2AK119ub1) appears to inhibit gene expression by both suppressing RNA polymerase activity in bivalent promoters and preventing H3K4 methylation (Di and Helin, 2013), a feature of activated genes.

In our studies (Lomniczi *et al.*, 2013a), we interrogated the hypothalamus during the juvenile and peripubertal stages of female rat reproductive maturation. We selected three developmental phases for inquiry: early juvenile (day 21), late juvenile (day 28), and the day of the first preovulatory surge of gonadotropins (days 30–36). Day 21 corresponds to the initiation of the juvenile period (Ojeda and Terasawa, 2002), day 28 to the beginning of puberty, and the first preovulatory surge of gonadotropins to the completion of puberty. We consider day 28 as the initiation of puberty in female rats, because it is at this time that a diurnal change in pulsatile LH release is first detected (Ojeda and Terasawa, 2002).

We reasoned that comparing day 28 to day 21 would be enlightening because of the changes in pulsatile LH release that occur at this time, and the notion that pulsatile GnRH secretion is driven by KNDy neurons of the ARC (for a review and references see Pinilla *et al.*, 2012). It is believed that NKB released by KNDy neurons acts in a paracrine/autocrine fashion on other KNDy neurons via TAC3R receptors to stimulate kisspeptin release. In turn, dynorphin inhibits NKB release (and hence kisspeptin release) via a phase-delayed, recurrent inhibitory feedback. These features make the Kiss1 and Tac2 genes unique members of a class of 'puberty-activating' genes, because both are active within the same neurons, and their protein products are functionally connected. Using the Kiss1 gene as a prototype, we showed that two essential components of the PcG complex (Eed and Cbx7) are expressed in kisspeptin neurons of the ARC, and, more importantly, that their encoded proteins are associated with the 5' flanking region of the Kiss1 gene (Lomniczi *et al.*, 2013a) (Figure 18.4). The results showed that at the end of juvenile development, methylation of the Eed and Cbx7 promoters increases in the ARC concurrent with decreased expression of both genes. These changes were accompanied by eviction of EED and CBX proteins from the Kiss1 promoter, and a notable change in the chromatin status of the Kiss1 promoter, with displayed increased levels of acetylated H3 (H3K9,14ac) and trimethylated H3 (H3K4me3), two modifications associated with gene activation (see Figure 18.4). As predicted by these changes, Kiss1 expression increased in the ARC at this time.

Unexpectedly, the content of H3K27me3, a histone modification catalyzed by PRC2 (Wang *et al.*, 2008), decreased significantly later during peripubertal maturation, on the day of the first preovulatory surge of gonadotropins. This developmental pattern suggests that association of H3K27me3 and H3K4me3 to the Kiss1 promoter behaves as predicted by the hypothesis of bivalent domains, i.e. both marks co-exist in the regulatory region of genes 'poised' for activation in response to incoming inputs (Bernstein *et al.*, 2006).

The study of Lomniczi *et al.* (2013a) also demonstrated that systemic administration of 5-azacytidine, an inhibitor of *de novo* DNA methylation, prevented both the peripubertal decline in Cbx7 and Eed mRNA expression, and the eviction of CBX7 and EED from the Kiss1 promoter. It also prevented the association of activating histone marks to the promoter, and the increase in Kiss1 expression that occurs at the end of juvenile development. These observations are consistent with the interpretation that if an increase in DNA methylation of PcG promoters is prevented, eviction of EED/CBX7 occupancy of the Kiss1 promoter fails to occur, and this diminishes the accessibility of activating histone marks to the promoter. The importance of PcG-mediated silencing for the timing of puberty was demonstrated by experiments in which EED was overexpressed in the ARC of early juvenile rats (via lentiviral-mediated delivery). In these animals, exogenous EED was recruited to the Kiss1 promoter, and Kiss1 expression was reduced as evidenced by the lower number of immunoreactive kisspeptin neurons in the ARC and a reduction in Kiss1 mRNA levels. Pulsatile GnRH release was blunted, puberty was delayed, and estrous cyclicity was disrupted. Importantly, the number of pups delivered by animals receiving EED in the ARC was markedly reduced, suggesting that if the repressive influence of the PcG complex on kisspeptin neurons is maintained beyond juvenile days, fertility is compromised (Lomniczi *et al.*, 2013a).

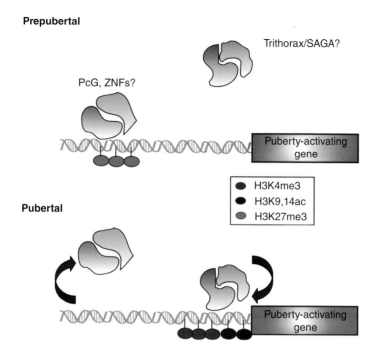

Figure 18.4 The transcriptional repression of puberty-activating genes. This model predicts the existence of a dual mechanism of transcriptional regulation underlying the developmental changes in expression of genes that facilitate pubertal development. According to this concept, the transcriptional activity of these genes (Kiss1, Tac2, Nell2, TTF1, and others) is repressed during prepubertal development by the Polycomb silencing complex and perhaps by ZNF proteins, such as MKRN3. As puberty approaches, both PcG and ZNF proteins are evicted from, and the content of H3K27me3 is reduced in, the regulatory regions controlling puberty-activating genes. As this change unfolds, transcriptional activators are recruited to these regions, resulting in enhanced gene expression. A strong candidate for this activational role is the Trithorax activating complex, which both catalyzes the methylation of histone 3 at lysine 4 (H3K4me2/3, an activating histone mark) and binds to promoter DNA containing this mark. Another potential candidate is the SAGA complex, which binds to DNA associated with H3K4me2/3 to activate gene expression.

The prepubertal increase in H3K4me3 abundance at the Kiss1 promoter is extremely interesting because it implies the recruitment of an activating complex concomitant to the loss of PcG inhibition (see Figure 18.4). A likely candidate for this role is the Trithorax complex because these proteins are known to antagonize PcG silencing by catalyzing H3K4 trimethylation and H3 acetylation (Shilatifard, 2012). By doing so, the Trithorax complex may provide the necessary transactivational component to puberty-activating genes at the time when the inhibitory influence of repressive epigenetic information is waning (see Figure 18.4). The potential importance of the Trithorax complex in the control of puberty is suggested by the finding that inactivating mutations of CHD7, a chromatin remodeling protein that antagonizes PcG action by binding to H3K4me2 and H3K4me3 via its chromodomain, are associated with hypothalamic hypogonadism in humans (Kim *et al.*, 2008). Notwithstanding the potential importance of Trithorax genes as transcriptional activators of the pubertal process, the alternative contribution of other transactivating protein complexes known to 'read' the H3K4me3 mark needs careful consideration. The SAGA complex, which binds to the DNA of H3K4me2/3 containing promoters via the subunit Sgf29 (Vermeulen *et al.*, 2010), is an attractive candidate for this role (see Figure 18.4).

An additional aspect made evident by the study of Lomniczi *et al.* is that the decrease in PcG expression that antedates the onset of puberty is not an estrogen (E2)-dependent phenomenon. Indeed, there are no canonical estrogen-responsive elements (EREs) in PcG promoters. Moreover, E2 action is associated with gene activation instead of gene repression, as evidenced by the ability of E2 to induce demethylation of DNA and loss of H3K9me2/3 from E2-target promoters (Metivier *et al.*, 2008; Perillo *et al.*, 2008). While E2 may not be responsible for the dissociation of PcG proteins from the Kiss1 promoter in kisspeptin neurons of the ARC, it may elicit epigenetic modifications affecting either other puberty-related genes or the Kiss1 gene itself expressed in AVPV neurons. The former case is suggested by studies showing that E2 triggers fluctuations in DNA methylation (Metivier *et al.*, 2008), induces the formation of co-activating complexes containing histone acetyltransferases and histone methyltransferases, and enhances gene transcription by inducing demethylation of H3K9me2/3 (Perillo *et al.*, 2008). Expression of ERα itself is regulated by DNA methylation (Issa *et al.*, 1994).

A role for E2 in the epigenetic control of AVPV kisspeptin neurons was recently demonstrated by Tomikawa and colleagues who showed that E2 increases acetylated H3 content of the Kiss1 promoter in the AVPV, but reduces acetylated H3 in the ARC (Tomikawa *et al.*, 2012). Furthermore, E2 increased ERα binding to the Kiss1 promoter in the AVPV, but not the ARC. In agreement with Lomniczi *et al.* (2013a), the study of Tomikawa and colleagues did not detect changes in Kiss1 promoter DNA methylation in either the AVPV or ARC, suggesting that DNA methylation is not an epigenetic change regulating Kiss1 promoter activity. Interestingly, there is an estrogen-responsive enhancer in the 3'-region of the Kiss1 gene, which operates in the AVPV but not the ARC (Tomikawa *et al.*, 2012), suggesting an epigenetic contribution to E2-positive feedback. Whether the inhibitory effect of estrogen on ARC Kiss1 expression involves an epigenetic component remains to be elucidated.

Changes in epigenetic information are likely to affect simultaneously several components of the gene networks controlling puberty. For instance, POK proteins recruit HDACs to a repressor complex to silence gene expression (Dhordain *et al.*, 1997), and some POZ-ZF proteins recognize methylated CpG dinucleotides for binding and repression of gene transcription (Filion *et al.*, 2006).

18.7 Perspectives

The above discussed observations are consistent with the view that gene sets organized into functionally connected networks reside at the heart of the neuroendocrine control of mammalian puberty (see Figure 18.3). They also suggest that the pubertal process is held in check by a multilayered repressive mechanism of epigenetic regulation that operates within the boundaries of the participating networks. The Polycomb group of transcriptional silencers provides a proximal repressive layer that controls downstream 'puberty-activating' genes epitomized by Kiss1 (see Figure 18.3 and 18.4). In turn, expression of PcG genes is controlled by changes in DNA methylation. We speculate that members of the ZNF family of transcriptional silencers, typified by MKRN3, function at a hierarchical level similar to that of the PcG complex to also repress puberty-activating genes (see Figure 18.3). An upstream layer of repressive control is provided by the POK family of transcriptional silencers, whose members recognize and bind preferentially to methylated regions controlling gene expression. Although POK genes may mostly act as 'repressors of repressors,' it is also possible that they may repress puberty-activating genes as well. Finally, LIN28 may contribute to the overall process by acting as a repressor of repressors but at a posttranscriptional level, preventing the maturation of let-7 miRNAs (see Figure 18.3). Because expression of let-7 miRNAs increases in the hypothalamus before puberty, it is possible that they silence transcriptional repressors that target puberty-activating genes.

These observations give the initial impetus to further explore the contribution of epigenetics to the hypothalamic control of puberty. As is always the case, they provide some answers but also raise a variety of questions. Among these, the issue of the cell-specific make-up of the epigenetic machinery stands out as the most important issue to be addressed in years to come. We do not know if different neuronal populations, such as kisspeptin, glutamatergic or GABAergic neurons, have a different composition of transcriptional repressors. Or perhaps they express the same repressors – and it is only the ability of each repressor to recruit additional silencing partners that determines the final output. Contemporary techniques for the isolation of specific cell populations coupled with genome-wide characterization of histone and DNA methylation landscapes may provide an effective tool to undertake this endeavor. The nature of the transcriptional activators that counterbalance the repressive effect of transcriptional silencers is also unknown, although in the case of the PcG system, the balancing counterpart might be provided by the Trithorax and/or SAGA complexes. Additional questions concern the potential existence of cell-specific developmental changes in transcriptional repressor/activator expression, and the nature of the factors responsible for those changes. If different puberty-controlling cell populations in the hypothalamus have their own epigenetic landscape, it would be important to learn about what determines this cellular specificity and how developmental changes in potentially diverse epigenetic land-scapes are established. A critical question is the identity of the pathways by which very different environmental inputs (nutrition, toxins, light, social interactions) are conveyed to and detected by the hypothalamic epigenome. Finally, little, if anything, is known about the role of microRNAs and lincRNAs in the hypothalamic control of puberty. Future studies are needed to address this issue.

Undoubtedly, much research is needed to provide answers to these and other questions. Presumably, these investigative efforts will result in the identification of epigenetic defects as an underlying cause of idiopathic precocious and delayed puberty in humans.

Acknowledgments

This research was supported by grant IOS IOS1121691 from the National Science Foundation (USA).

References

Abreu, A.P., Dauber, A., Macedo, D., *et al.* (2013) Central precocious puberty caused by mutations in the imprinted gene MKRN3. *New England Journal of Medicine*, **368**, 2467–2475. [*This report demonstrated for the first time the contribution of a transcriptional repressor to the neuroendocrine control of human puberty.*]

Bernstein, B.E., Mikkelsen, T.S., Xie, X., *et al.* (2006) A bivalent chromatin structure marks key developmental genes in embryonic stem cells. *Cell*, **125**, 315–326.

Borrelli, E., Nestler, E.J., Allis, C.D. and Sassone-Corsi, P. (2008) Decoding the epigenetic language of neuronal plasticity. *Neuron*, **60**, 961–974.

Boyar, R., Finkelstein, J., Roffwarg, H., *et al.* (1972) Synchronization of augmented luteinizing hormone secretion with sleep during puberty. *New England Journal of Medicine*, **287**, 582–586. [*This report provided the first evidence that a diurnal, sleep-related increase in pulsatile LH release is the first neuroendocrine manifestation of the initiation of puberty.*]

Cedar, H. and Bergman, Y. (2009) Linking DNA methylation and histone modification, patterns and paradigms. *Nature Reviews Genetics*, **10**, 295–304.

Chu, C., Qu, K., Zhong, F.L., Artandi, S.E. and Chang, H.Y. (2011) Genomic maps of long noncoding RNA occupancy reveal principles of RNA-chromatin interactions. *Molecular Cell*, **44**, 667–678.

de Roux, N., Genin, E., Carel, J.C., *et al.* (2003) Hypogonadotropic hypogonadism due to loss of function of the KiSS1-derived peptide receptor GPR54. *Proceedings of the National Academy of Sciences USA*, **100**, 10972–10976. [*This is one of the two original reports demonstrating that mutations in KISS1R/GPR54, the gene encoding the kisspeptin receptor, results in pubertal failure and hypothalamic hypogonadism in humans.*]

DeFazio, R.A., Heger, S., Ojeda, S.R. and Moenter, S.M. (2002) Activation of A-type g-aminobutyric acid receptors excites gonadotropin-releasing hormone neurons. *Molecular Endocrinology*, **16**, 2872–2891.

Dhordain, P., Albagli, O., Lin, R.J., *et al.* (1997) Corepressor SMRT binds the BTB/POZ repressing domain of the LAZ3/BCL6 oncoprotein. *Proceedings of the National Academy of Sciences USA*, **94**, 10762–10767.

Di, C.L. and Helin, K. (2013) Transcriptional regulation by Polycomb group proteins. *Nature Structural & Molecular Biology*, **20**, 1147–1155.

Dissen, G.A., Lomniczi, A., Neff, T.L., *et al.* (2009) In vivo manipulation of gene expression in non–human primates using lentiviral vectors as delivery vehicles. *Methods*, **49**, 70–77. [*This publication describes in detail the use of lentiviruses as delivery vehicles for either gain-of-function or loss-of-function approaches.*]

Ficz, G., Branco, M.R., Seisenberger, S., *et al.* (2011) Dynamic regulation of 5-hydroxymethylcytosine in mouse ES cells and during differentiation. *Nature*, **473**, 398–402.

Filion, G.J., Zhenilo, S., Salozhin, S., *et al.* (2006) A family of human zinc finger proteins that bind methylated DNA and repress transcription. *Molecular & Cellular Biology*, **26**, 169–181.

Gao, Z., Zhang, J., Bonasio, R., *et al.* (2012) PCGF homologs, CBX proteins, and RYBP define functionally distinct PRC1 family complexes. *Molecular Cell*, **45**, 344–356.

Grumbach, M.M. and Styne, D.M. (1992) Puberty, ontogeny, neuroendocrinology, physiology, and disorders, in Williams Textbook of Endocrinology, 8th edn (eds J.D. Wilson, D.W. Foster), W.B. Saunders, Philadelphia, pp. 1139–1221.

Herbison, A.E. (2006) Physiology of the gonadotropin-releasing hormone neuronal network, in Physiology of Reproduction, 3rd edn (ed. J.D. Neill), Academic Press/Elsevier, San Diego, pp. 1415–1482.

Hrabovszky, E., Ciofi, P., Vida, B., *et al.* (2010) The kisspeptin system of the human hypothalamus, sexual dimorphism and relationship with gonadotropin-releasing hormone and neurokinin B neurons. *European Journal of Neuroscience*, **31**, 1984–1998.

Issa, J.P., Ottaviano, Y.L., Celano, P., *et al.* (1994) Methylation of the oestrogen receptor CpG island links ageing and neoplasia in human colon. *Nature Genetics*, **7**, 536–540.

Kelly, K.F. and Daniel, J.M. (2006) POZ for effect – POZ-ZF transcription factors in cancer and development. *Trends in Cell Biology*, **16**, 578–587.

Kim, H.G., Kurth, I., Lan, F., *et al.* (2008) Mutations in CHD7, encoding a chromatin-remodeling protein, cause idiopathic hypogonadotropic hypogonadism and Kallmann syndrome. *American Journal of Human Genetics*, **83**, 511–519.

Kouzarides, T. (2007) Chromatin modifications and their function. *Cell*, **128**, 693–705.

Kurian, J.R., Keen, K.L. and Terasawa, E. (2010) Epigenetic changes coincide with in vitro primate GnRH neuronal maturation. *Endocrinology*, **151**, 5359–5368.

Lomniczi, A. and Ojeda, S.R. (2009) A role for glial cells of the neuroendocrine brain in the central control of female sexual development, in Astrocytes in (Patho)Physiology of the Nervous System (eds V. Parpura, P. Haydon), Springer, New York, pp. 487–511.

Lomniczi, A., Loche, A., Castellano, J.M., *et al.* (2013a) Epigenetic control of female puberty. *Nature Neuroscience*, **16**, 281–289. [*This is the first report demonstrating a role for epigenetics in the hypothalamic control of puberty.*]

Lomniczi, A., Wright, H., Castellano, J.M., Sonmez, K. and Ojeda, S.R. (2013b) A system biology approach to identify regulatory pathways underlying the neuroendocrine control of female puberty in rats and nonhuman primates. *Hormones & Behavior*, **64**, 175–186.

Mayer, C. and Boehm, U. (2011) Female reproductive maturation in the absence of kisspeptin/GPR54 signaling. *Nature Neuroscience*, **14**, 704–710.

Metivier, R., Gallais, R., Tiffoche, C., *et al.* (2008) Cyclical DNA methylation of a transcriptionally active promoter. *Nature*, **452**, 45–50.

Ojeda, S.R. (1991) The mystery of mammalian puberty, how much more do we know? *Perspectives in Biology and Medicine*, **34**, 365–383.

Ojeda, S.R. and Terasawa, E. (2002) Neuroendocrine regulation of puberty, in Hormones, Brain and behavior (vol **4**) (eds D. Pfaff, A. Arnold, A. Etgen, *et al.*), Elsevier, New York, pp. 589–659.

Perillo, B., Ombra, M.N., Bertoni, A., *et al.* (2008) DNA oxidation as triggered by H3K9me2 demethylation drives estrogen-induced gene expression. *Science*, **319**, 202–206.

Pinilla, L., Aguilar, E., Dieguez, C., Millar, R.P. and Tena-Sempere, M. (2012) Kisspeptins and reproduction, physiological roles and regulatory mechanisms. *Physiological Review*, **92**, 1235–1316.

Roth, C.L., Mastronardi, C., Lomniczi, A., *et al.* (2007) Expression of a tumor-related gene network increases in the mammalian hypothalamus at the time of female puberty. *Endocrinology*, **148**, 5147–5161.

Sangiao-Alvarellos, S., Manfredi-Lozano, M., Ruiz-Pino, F., *et al.* (2013) Changes in hypothalamic expression of the Lin28/let-7 system and related microRNAs during postnatal maturation and after experimental manipulations of puberty. *Endocrinology*, **154**, 942–955.

Schwartz, Y.B. and Pirrotta, V. (2007) Polycomb silencing mechanisms and the management of genomic programmes. *Nature Reviews Genetics*, **8**, 9–22.

Seminara, S.B., Messager, S., Chatzidaki, E.E., *et al.* (2003) The GPR54 gene as a regulator of puberty. *New England Journal of Medicine*, **349**, 1614–1627. [***This is one of the two original reports demonstrating that mutations in KISS1R/GPR54, the gene encoding the kisspeptin receptor, results in pubertal failure and hypothalamic hypogonadism in humans.***]

Shilatifard, A. (2012) The COMPASS family of histone H3K4 methylases, mechanisms of regulation in development and disease pathogenesis. *Annual Review of Biochemistry*, **81**, 65–95.

Sulem, P., Gudbjartsson, D., Rafnar, T., *et al.* (2009) Genome-wide association study identifies sequence variants on 6q21 associated with age at menarche. *Nature Genetics*, **41**, 734–738. [***This is one of a series of independent reports published in the same issue of* Nature Genetics *showing an association between common DNA sequence variation and the age of puberty in human females.***]

Sykiotis, G.P., Pitteloud, N., Seminara, S.B., Kaiser, U.B. and Crowley, W.F. Jr (2010) Deciphering genetic disease in the genomic era, the model of GnRH deficiency. *Science Translational Medicine*, **2**, 32rv2.

Tahiliani, M., Koh, K.P., Shen, Y., *et al.* (2009) Conversion of 5-methylcytosine to 5-hydroxymethylcytosine in mammalian DNA by MLL partner TET1. *Science*, **324**, 930–935.

Tavares, L., Dimitrova, E., Oxley, D., *et al.* (2012) RYBP-PRC1 complexes mediate H2A ubiquitylation at polycomb target sites independently of PRC2 and H3K27me3. *Cell*, **148**, 664–678.

Teles, M.G., Bianco, S.D., Brito, V.N., *et al.* (2008) A GPR54-activating mutation in a patient with central precocious puberty. *New England Journal of Medicine*, **358**, 709–715.

Tomikawa, J., Uenoyama, Y., Ozawa, M., *et al.* (2012) Epigenetic regulation of Kiss1 gene expression mediating estrogen-positive feedback action in the mouse brain. *Proceedings of the National Academy of Sciences USA*, **109**, E1294–E1301.

Vermeulen, M., Eberl, H.C., Matarese, F, *et al.* (2010) Quantitative interaction proteomics and genome–wide profiling of epigenetic histone marks and their readers. *Cell*, **142**, 967–980. [***This report comprehensively characterizes the histone landscape of the human genome.***]

Wang, Z., Zang, C., Rosenfeld, J.A., *et al.* (2008) Combinatorial patterns of histone acetylations and methylations in the human genome. *Nature Genetics*, **40**, 897–903. [***This report offers a detailed, genome-wide characterization of the combinatorial features of posttranslationa histone modifications in the human genome.***]

Yang, J.J., Caligioni, C.S., Chan, Y.M. and Seminara, S.B. (2012) Uncovering novel reproductive defects in neurokinin B receptor null mice, closing the gap between mice and men. *Endocrinology*, **153**, 1498–1508.

CHAPTER 19

Oxytocin, Vasopressin, and Diversity in Social Behavior

Lanikea B. King and Larry J. Young

Center for Translational Social Neuroscience, Department of Psychiatry and Behavioral Sciences, Yerkes National Primate Research Center, Emory University, Atlanta, Georgia, USA

19.1 Introduction

Neuroendocrine factors play a critical role in regulating many aspects of social behaviors, from individual recognition of conspecifics to the formation of social bonds. There is a remarkable diversity in social behavioral phenotype across vertebrate species, even among closely related species. Two neuropeptides, oxytocin (OXT) and arginine vasopressin (AVP), or their non-mammalian homologs (Figure 19.1), have been studied most intensively for their role in modulating social cognition and sociosexual behaviors. In this chapter we focus on OXT and AVP, collectively known as neurohypophysial peptides, and their receptors (NHPRs), and examine how they modulate a rich diversity of social behavior in an often species-specific manner. Following a brief overview of the roles of OXT and AVP in regulating social behavior, we explore the evolution of the OXT and AVP family of neuropeptides and highlight the evolutionarily conserved role of these peptides in regulating social behavior. While some aspects of the OXT and AVP systems are evolutionarily very ancient and well conserved, there is a remarkable evolutionary plasticity in the receptors for these peptides in terms of their neural expression. We consider this plasticity in NHPR expression to be critical to diversity in social behaviors across and within species. Detailed knowledge of NHPR diversity comes from elegant research in monogamous prairie voles. A section on voles illustrates the neural mechanisms by which OXT and AVP regulate behaviors associated with monogamy. The chapter concludes with a highlight of recent research in humans demonstrating remarkable parallels between the roles of these peptides in animals and in our own species.

19.2 Oxytocin, vasopressin, and social behavior

Oxytocin is best known for its role in initiating labor and delivery through regulating uterine contractions, and stimulating milk letdown during nursing. But mammalian mothers do more than deliver and provide sustenance to their offspring – they provide nurturing care, and in some cases develop mother–infant bonds. This caring in parturient mothers contrasts with virgin females in many species, which ignore or attack infants. OXT guides this shift in behavior. Central OXT injections facilitate onset of maternal behavior while antagonists for the OXT receptor (OXTR) delay onset in rats. Furthermore,

Molecular Neuroendocrinology: From Genome to Physiology, First Edition. Edited by David Murphy and Harold Gainer.
© 2016 John Wiley & Sons, Ltd. Published 2016 by John Wiley & Sons, Ltd.
Companion website: www.wiley.com/go/murphy/neuroendocrinology

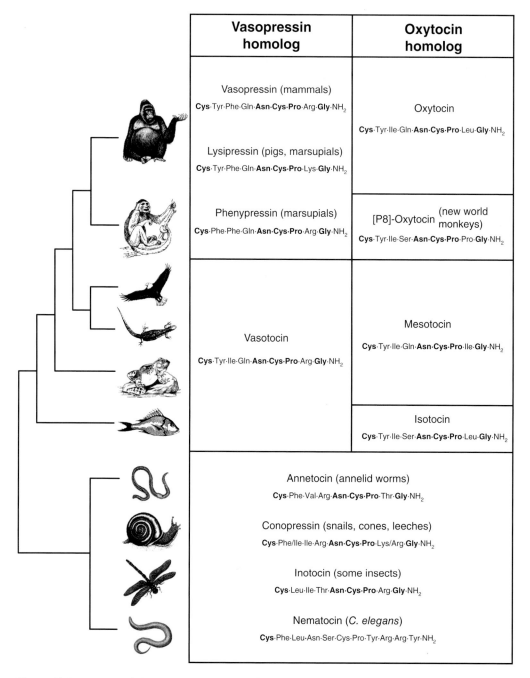

	Vasopressin homolog	Oxytocin homolog
	Vasopressin (mammals) **Cys**·Tyr·Phe·Gln·**Asn**·**Cys**·**Pro**·Arg·**Gly**·NH₂ Lysipressin (pigs, marsupials) **Cys**·Tyr·Phe·Gln·**Asn**·**Cys**·**Pro**·Lys·**Gly**·NH₂	Oxytocin **Cys**·Tyr·Ile·Gln·**Asn**·**Cys**·**Pro**·Leu·**Gly**·NH₂
	Phenypressin (marsupials) **Cys**·Phe·Phe·Gln·**Asn**·**Cys**·**Pro**·Arg·**Gly**·NH₂	[P8]-Oxytocin (new world monkeys) **Cys**·Tyr·Ile·Ser·**Asn**·**Cys**·**Pro**·Pro·**Gly**·NH₂
	Vasotocin **Cys**·Tyr·Ile·Gln·**Asn**·**Cys**·**Pro**·Arg·**Gly**·NH₂	Mesotocin **Cys**·Tyr·Ile·Gln·**Asn**·**Cys**·**Pro**·Ile·**Gly**·NH₂
		Isotocin **Cys**·Tyr·Ile·Ser·**Asn**·**Cys**·**Pro**·Leu·**Gly**·NH₂
	Annetocin (annelid worms) **Cys**·Phe·Val·Arg·**Asn**·**Cys**·**Pro**·Thr·**Gly**·NH₂ Conopressin (snails, cones, leeches) **Cys**·Phe/Ile·Ile·Arg·**Asn**·**Cys**·**Pro**·Lys/Arg·**Gly**·NH₂ Inotocin (some insects) **Cys**·Leu·Ile·Thr·**Asn**·**Cys**·**Pro**·Arg·**Gly**·NH₂ Nematocin (*C. elegans*) **Cys**·Phe·Leu·Asn·Ser·Cys·Pro·Tyr·Arg·Arg·Tyr·NH₂	

Figure 19.1 Neuropeptide homologs of oxytocin and vasopressin. Modified with permission from Donaldson and Young (2008).

Oxtr knockout (KO) mice are impaired in maternal retrieval (Ross and Young, 2009). Dams also aggressively defend offspring against intruders. OXT, as well as AVP, regulates maternal aggression in rodents (Bosch and Neumann, 2012).

Rats and mice are promiscuously maternal. In ungulates, however, mothers live in herds and deliver offspring that are immediately mobile. Thus mothers not only have to become motivated to care for infants, but they must selectively care for their own young. To achieve this, ungulates, like sheep, develop a strong mother–infant bond. OXT is released during labor in the mother's brain, and infusions of OXT into the brain of steroid-primed ewes lead to the development of a bond between the female and a novel lamb, through modulation of olfactory processing. These studies and many others in animal models demonstrate that the same molecule that regulates the peripheral physiology of reproduction also coordinates the onset of maternal responsiveness and bonding (Ross and Young, 2009).

Although best known for a role in female reproduction, OXT plays a more general role in social cognition, as revealed by more recent studies in knockout mice. Mice distinguish each other by smell, a process referred to as social recognition. Male *Oxt* KO mice fail to recognize mice that they have previously encountered. However, a central infusion of OXT restores social recognition abilities. Thus, OXT plays a critical role in the neural processing of social information, which is key to many aspects of social relationships, including mother–infant bonding and, as we will see, pair bonding (Donaldson and Young, 2008; Ross and Young, 2009). Indeed, one of the fundamental processes by which OXT is thought to modulate social behavior is by enhancing the salience of social cues, which in rodents is primarily olfactory. OXT also appears to mediate the rewarding aspects of social interactions through its interaction with the serotonin system in mice (Dolen *et al.*, 2013). It is likely that these two fundamental processes, social information processing and social reward, contribute significantly to many of the more complex behavioral roles of OXT.

Vasopressin was named for its constricting effects of the vascular system. AVP is also referred to as antidiuretic hormone because of its role in regulating water retention in the kidney. Like OXT, AVP also modulates many aspects of social behavior, particularly male-typical social behaviors. For instance, two of the first behavioral effects of central AVP to be described were scent marking behavior and territorial aggression in hamsters. Why did AVP evolve a role in regulating territorial behaviors? One intriguing possibility is that its role in regulating water balance, and hence urine production, became linked behaviorally and neurobiologically to scent marking of territory, which many species achieve through urination (Freeman and Young, 2013).

Like OXT, AVP is more generally involved in social information processing as well. AVP receptor 1a (*Avpr1a*) KO mice also display social amnesia, and selectively expressing AVRP1A in the lateral septum of KO mice restores social recognition abilities. As we will see later, its role in regulating territorial behavior and social information processes led to new social behavioral roles for AVP in some monogamous species, including pair bonding in prairie voles and perhaps humans. We will now explore some evolutionary aspects of the neuroendocrine regulation of social behavior.

19.2.1 Neurohypophysial peptide gene structure, evolution, and conserved function

The genes encoding OXT and AVP share a common ancestral sequence, which underwent a duplication yielding the two new genes just after the emergence of the vertebrate lineage. This original peptide is estimated to be at least 700 million years old and neuropeptide homologs of OXT and AVP have been discovered in distantly related invertebrates, including cephalopods, insects, and nematodes (Gruber, 2014). All peptides in the OXT/AVP family

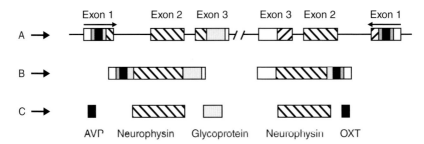

Figure 19.2 Oxytocin and vasopressin genes: structure and processing. Gene structure in the DNA (A), preprohormone mRNA (B), final processed protein products including oxytocin and vasopressin. On line A, boxed regions indicate the locations of the exons. Shaded or hatched regions indicate coding regions. Vasopressin (AVP) and oxytocin (OXT) neurohypophysial peptides are shown in black. Reprinted with permission from *Kaplan and Sadock's Comprehensive Textbook of Psychiatry*, 9th edn. Lippincott, Williams & Wilkins (www.lww.com).

share a cyclical structure and most are nine-amino acid peptides, with some examples of length variants, like nematocin (see Figure 19.1). In vertebrates, the *Oxt* and *Avp* genes are adjacent to one another and transcribed towards each other. The gene encodes a prohormone that is cleaved by a protease to yield the OXT or AVP peptide, neurophysin, and other peptide products (Figure 19.2).

The general physiological functions of the OXT/AVP family are conserved from invertebrates to vertebrates. Invertebrate homologs stimulate egg-laying behavior in leeches and earthworms, analogous to the regulation of parturition by OXT in mammals, and activate water balance reflexes in sea squirts, flour beetles, locusts, and leeches, analogous to the antidiuretic properties of AVP homologs in vertebrates. The involvement in regulating sociosexual behaviors emerged early in invertebrate evolution as well. Neurohypophysial peptides activate stereotypical mating behaviors in leeches and snails. In *C. elegans*, nematocin plays a role in sensory processing involved in mating, suggesting that very early in evolution this peptide family became involved in regulating social interactions and social information processing (Gruber, 2014).

19.2.2 Conservation in neural expression

In mammals, OXT and AVP are primarily synthesized by neurons in the paraventricular nucleus (PVN) and supraoptic nucleus (SON) of the hypothalamus, although other populations can be found. OXT and AVP neurons project primarily, but not exclusively as we shall see, to the posterior pituitary, where they secrete peptide into the blood for distribution to peripheral organs. Across taxa of the animal kingdom, there is an extraordinary conservation in the distribution or nature of neurohypophysial peptide-producing neurons (Knobloch and Grinevich, 2014).

The phylogenetic stability of the neuropeptide system appears to stem from a conservation of cell type-specific regulatory factors in the neurosecretory cells producing the peptide and of the *cis*-regulatory elements (*cis*-REs) surrounding the gene. In a detailed comparison between zebrafish (*Danio rerio*) and an annelid nereidid (*Platynereis dumerilii*), Tessmar-Raible and colleagues show that neurosecretory cells expressing AVP homologs in both species share common transcription factors, micro-RNA, and embryonic migratory patterns (Tessmar-Raible *et al.*, 2007). Even in the annelid, these neuroendocrine cells share anatomical features with the vertebrate hypothalamus such as ease of access to the vasculature and as a site of information integration.

In vertebrates, the transcriptional regulation of neurohypophyseal peptide genes is remarkably conserved across species. In a series of experiments, Venkatesh and colleagues isolated the gene encoding the pufferfish (*Fugu rubripes*) homolog of OXT, isotocin, and generated transgenic rodents that expressed both the endogenous rodent and transgenic Fugu peptides (Venkatesh *et al.*, 1997). In both rats and mice, the Fugu isotocin gene was expressed specifically within OXT-containing neurons in the hypothalamus. In addition, the isotocin gene responded to increased salt concentrations at the same rate as OXT. This work remarkably demonstrates that a 5 kb region of the Fugu isotocin gene contains sufficient information for the gene to not only express in the correct secretory cells, but to respond to an environmental signal in an appropriate manner in a rodent as well. Similar results were seen for AVP and its fish homolog, vasotocin.

19.3 Oxytocin and vasopressin in the vertebrate brain

The best characterized secretory OXT neurons are the hypothalamic magnocellular neurons of the (PVN), supraoptic nucleus (SON) and accessory nuclei (AN) that lies between PVN and SON (Knobloch and Grinevich, 2014; Ross and Young, 2009). These neurons project axons to the posterior pituitary, where they release OXT into circulation. The magnocellular neurons are also the sites of co-localization between rodent OXT and the fish isotocin discussed earlier and thus represent the most highly conserved population of OXT-expressing cells. Parvocellular neurons in the PVN project primarily to the hindbrain and brainstem. Magnocellular neurons also provide OXT innervation to the forebrain. It is likely that these forebrain projections of magnocellular neurons are the major source of OXT regulating social behavior. These neurons release OXT both peripherally and centrally in response to vaginocervical stimulation during parturition and mating and to nipple stimulation during nursing. Oxytocin expression is not sexually dimorphic under normal physiological conditions, in contrast to AVP expression.

Like OXT, AVP is also synthesized in magnocellular neurons in the PVN and SON as well as parvocellular neurons in the PVN, although largely in separate neurons from OXT. The magnocellular neurons project to and release AVP from the posterior pituitary into the bloodstream in response to osmotic or blood pressure variation. In addition, AVP is expressed in the suprachiasmatic nucleus of the hypothalamus, where it may play a role in circadian rhythms. However, it is small parvocellular neurons in the medial amygdala and bed nucleus of the stria terminalis that project to forebrain regions and are likely to be the source of behaviorally relevant AVP in the vertebrate brain. The expression of AVP in these neurons is androgen dependent, and consequently sexually dimorphic. Males have more AVP-producing neurons in these areas and more dense AVP fibers in projection sites, including the lateral septum and ventral pallidum (de Vries, 2008; Kelly and Goodson, 2014).

19.4 OXT and AVP receptors

In mammals, there are four neurohypophysial peptide receptors: a single OXT receptor (OXTR) and three subtypes of AVP receptor (AVPR1A, AVPR1B, and AVPR2). OXTR is expressed in the brain and in the periphery. AVPR2 is expressed in the kidney and regulates the antidiuretic properties of AVP. AVPR1B is expressed in the anterior pituitary and restricted brain areas. However, AVPR1A is widely expressed in the forebrain and is the receptor most often linked to the regulation of social behavior (Donaldson and Young, 2008). It should be noted that OXT can bind to AVPRs and AVP can bind to OXTR, as would be expected from their structural similarity.

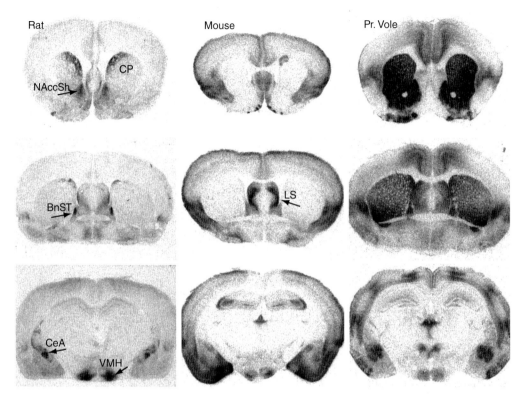

Figure 19.3 Receptor autoradiograms illustrating the distribution of OXT receptor binding sites in the forebrain of rat, mouse, and prairie vole. Note that while the distribution of receptors is conserved in some brain regions, such as the lateral septum (LS), bed nucleus of the stria terminalis (BnST), central nucleus of the amygdala (CeA), and ventromedial nucleus of hypothalamus (VMH), several striking species differences are apparent. For example, in the rat striatum, OXT receptors are restricted to the dorsal caudate putamen (CP) and the shell of the nucleus accumbens (NAccSh), while OXT receptors are abundant throughout the prairie vole striatum and absent in the mouse striatum. Reprinted with permission from *Knobil and Neill's Physiology of Reproduction*, 3rd edn, Elsevier.

In stark contrast to the conservation in peptide expression across vertebrates, the brain expression patterns of OXTR and AVPR1A vary remarkably across species. Figure 19.3 illustrates such a pattern for OXTR between rats, mice, and prairie voles. As we will see below, even closely related species within a genus exhibit this neural diversity and demonstrate that it may be linked to species-specific behaviors. Furthermore, individual variation in brain OXTR and AVPR1A distribution within a species has been associated with variation in social behavior. This diversity in neural expression patterns is a hallmark feature of OXTR and AVPR1A that clearly sets them apart from other receptor systems, including sex steroid receptors and neurotransmitter systems such as dopamine. In contrast to the receptor differences, there are few species differences in brain peptide projections. It appears that OXT or AVP can be released locally in brain regions and then diffuse to the nearby receptor-expressing neurons. This paracrine type of transmission allows for the evolution of receptor distribution without necessitating co-evolution of neuropeptide-producing neurons.

Figure 19.4 The monogamous prairie vole. Prairie voles form lifelong pair bonds after mating and develop an affiliative attachment with each other and selective aggression against unfamiliar voles. Prairie vole families can be philopatric, with multiple generations living in a single burrow, where males and older pups assist in caring for pups. Photo by Todd H. Ahern.

19.4.1   OXT and AVP receptors and pair bonding

Prairie voles (*Microtus ochrogaster*) are one of the 3–5% of mammal species that share a monogamous reproductive strategy (Figure 19.4). In the wild, prairie voles form a robust pair bond, establishing an affiliative attachment that generally lasts over the lifetime of the animals. Prairie vole males and females sometimes engage in extra-pair copulations, making them socially rather than genetically monogamous. In addition to this bonding behavior, prairie voles live in philopatric family units where offspring delay dispersal from the nest. Male and female prairie voles share parental care responsibilities and older pups alloparentally care for younger siblings as well. Both sexes develop selective aggression towards unfamiliar conspecifics after bonding. These species-specific behaviors are generated by a brain hardwired for affiliation and attachment towards the family unit and aggression towards conspecifics outside of it.

Prairie voles offer a distinct advantage in understanding neuroendocrine and other networks underlying their unique social behaviors as closely related vole species are not monogamous, allowing comparative opportunities. Prairie voles belong to the rodent genus *Microtus*, which is highly differentiated taxonomically and has higher rates of genome evolution than other mammal genera. Despite these signs of diversity, vole species are often nearly indistinguishable by appearance alone. Social behavior, on the other hand, is diverse amongst vole species. Comparative genetic and neurobiological studies using the prairie, meadow (*Microtus pennsylvancius*), and montane (*Microtus montanus*) voles have provided insights into the neurobiology and evolution of social organization.

19.4.2   OXT and AVP influence pair bonding

In laboratory settings, the occurrence of a prairie vole pair bond is measured using the partner preference test and a test for selective aggression toward novel conspecifics. Before testing, adult subjects are co-habited with members of the opposite sex. Co-habitation time can be varied depending on the needs of the experiment, with shorter co-habitations used when testing a hypothesis about enhancement of pair bond formation and longer co-habitations used when testing a hypothesis focused on blocking pair bond formation. Following the co-habitation period, subjects can be tested for pair bond formation.

The partner preference test is illustrated in Video Clip 1 (see Supplemental section). The partner preference test uses a three-chamber apparatus, with two outer chambers containing a partner and a stranger stimulus vole, each tethered in opposite chambers to restrict their movement to a minimal area. The partner is the animal with which the subject was previously co-habitated while the stranger is a novel vole sharing the gender and reproductive experience of the partner. The subject is allowed to freely roam the apparatus for 3 hours. The primary measure is immobile huddling next to either stimulus animal, and automated video analysis systems now allow for automatic scoring of the partner preference test. Prairie voles of both sexes spend the majority of the test in social contact, while meadow voles spend most of the time in the empty chamber. A partner preference is considered to have formed when voles spend twice as much time huddling next to the partner compared to the stranger. Mating facilitates pair bonding, but partner preferences can develop in the absence of mating with extended co-habitation periods. A 24-hour co-habitation with mating is generally sufficient to stimulate a significant partner preference in males. Females are able to form a partner preference during a shorter co-habitation. However, there is considerable variation in these parameters across labs.

Pair bond formation also causes an increase in aggressive behaviors targeted at novel animals, particularly in males, which is likely reflective of mate-guarding behavior. Thus pair bonding in males has an element reminiscent of territoriality. Both of these tests capture a unique aspect of the changes in prairie vole social behavior as a consequence of pair bonding. Many experiments using one or other of these tests have revealed important biological factors for pair bonding in laboratory experiments.

Both OXT and AVP play critical roles in prairie vole pair bonding. Central injection of either peptide into the brain results in facilitated pair bonding during a truncated co-habitation in the absence of mating. Conversely, blocking endogenous OXT and AVP signaling using central infusions of antagonists for OXTR and AVRP1A prevents prairie vole pair bond formation. Both peptides influence male and female bonding, although most studies suggest that OXT plays a more important role in females, while AVP plays a more important role in males. However, recent unpublished evidence from our laboratory demonstrates that OXTR neurotransmission is essential for male pair bonding as well. There is less evidence for a role for endogenous AVP in the regulation of female pair bonding.

19.4.3 Species differences in social behavior are patterned by receptor diversity

As mentioned, prairie voles are particularly useful as a model for studying social behavior due to the availability of non-monogamous *Microtus* species for comparative purposes. Indeed, when viewing anatomical data for OXTR and AVRP1A, the brains of different *Microtus* species appear very different.

Relative to montane and meadow voles, prairie voles exhibit high OXTR density differences in several brain regions, particularly the nucleus accumbens (NAcc) (Figure 19.5A). OXTR signaling in both the NAcc and prefrontal cortex (PFC) is required for prairie vole pair bonding. Injections of an OXTR antagonist into either region in females prevent mating-induced partner preference formation (Figure 19.5C). Both of these regions are part of the mesolimbic reward pathway (MLR) and receive intensive dopaminergic projections from the ventral tegmental area (VTA). OXT signaling is also required in the prairie vole NAcc for alloparental behavior in adult virgin females, a behavior relevant to the unique prairie vole social structure (Ross and Young, 2009).

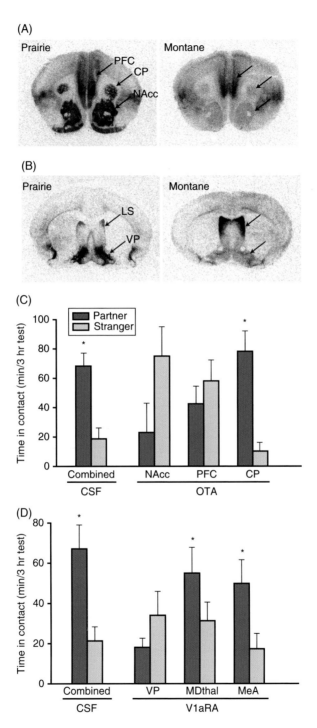

Figure 19.5 Receptor diversity reveals brain regions underpinning species-specific social behavior. Autoradiograms of coronal brain sections reveal binding densities of oxytocin receptor (A) and vasopressin receptor 1a (B). Relative to montane voles, monogamous prairie voles express oxytocin receptor at higher density in the nucleus accumbens (NAcc) and caudate putamen (CP) (A) and vasopressin receptor 1a at higher density in the ventral pallidum (VP) (B). Infusions of oxytocin antagonist into the NAcc or prefrontal cortex (PFC) but not CP prevented partner preference in female prairie voles (C). Infusions of vasopressin 1a antagonist into the VP but not mediodorsal thalamus (MDThal) or medial amygdala (MeA) prevented blocked partner preference in male prairie voles (D). CSF, cerebrospinal fluid; LS, lateral septum; OTA, oxytocin receptor antagonist; V1aRA, vasopressin receptor 1a antagonist. Modified with permission from Young and Wang (2004).

AVRP1A distributions also differ between vole species. Compared to montane and meadow voles, prairie voles express AVRP1A differentially in the ventral pallidum (VP) (Figure 19.5B). The VP is also a key region in the MLR pathway, serving as an output nucleus for the NAcc. Again, the species-specific expression of AVRP1A revealed a key region for AVP-mediated pair bonding. Injection of an AVRP1A antagonist into the VP, but not medial amygdala or thalamus, blocks partner preference in male prairie voles (Figure 19.5C). Remarkably, elevating AVRP1A density in the VP of meadow voles to levels resembling prairie voles using viral vector gene transfer allows the normally promiscuous species to form a partner preference (Lim *et al.*, 2004).

Primate species have a higher rate of social monogamy than other mammals. A general pattern first determined in voles is repeated in the primates: monogamous species express OXTR or AVRP1A in MLR regions. A monogamous titi monkey (*Callicebus cupreus*) expresses AVRP1A but not OXTR in the NAcc (Freeman *et al.*, 2014). Another monogamous primate, the common marmoset (*Callithrix jaccus*), expresses OXTR and AVRP1A in the NAcc and VP, respectively. The rhesus macaque (*Macaca mulatta*), which is not monogamous, does not express either receptor in the NAcc or VP. It is of interest to note that all primates analyzed so far express OXTR in brain regions involved in visual processing and attention such as the superior colliculus and nucleus basalis of Meynert, as well as auditory processing regions. Freeman *et al.* hypothesize that this pattern may be adaptive for primates, allowing OXT to directly modulate brain regions involved in visual and auditory processing, sensory modalities clearly more important to primates than rodents (Freeman *et al.*, 2014).

Bird species also exhibit quantitative diversity in neuropeptide expression that has been linked to social behavior. After the suggestion of Kelly and Goodson (2014) to maintain simplicity in nomenclature, homologs of OXTR and AVRP1A, such as the closest homolog of OXTR in birds, the vasotocin 3 receptor, will be referred to using the mammalian nomenclature. Birds are more commonly monogamous than mammals but they also exhibit a diverse range of preferences for sociality, which leads to flocking in some species. A site of particular interest for bird sociality is the lateral septum (LS). In a survey of territorial and gregarious species, a consistent pattern emerges. Gregarious species, such as the zebra finch (*Taeniopygia guttata*), express OXTR in the dorsal portions of the LS. More isolated territorial species, on the other hand, lack dorsal LS OXTR but may express some receptor in the ventral LS. This socially segregated distribution is indeed functional. Injection of an OXTR antagonist into the LS of female zebra finches lowers sociality preference (Goodson, 2013). Although considerable diversity in density and regional distribution is found across species, the LS as a whole has a conserved presence of both OXTR and AVRP1A expression among vertebrates and researchers are now generating theories on general roles the region may play in various social behaviors (Kelly and Goodson, 2014).

Social behavior is also influenced by other neuromodulator and transmitter systems (Figure 19.6). For example, pair bonding requires dopamine, endogenous opioids, and other neuropeptides. In the case of dopamine, for example, concurrent OXT and D2 receptor activation is required in the NAcc (Burkett and Young, 2012). This result enables a more elegant understanding of pair bonding. During copulation, dopamine and OXT are both released in females. D2 receptors mediate reinforcing properties of the experience while OXTR increases the salience of partner-associated cues. AVP and dopamine interact in a similar fashion in males. This likely results in strengthening of connections between neural circuits encoding the olfactory cues of the partner with circuits encoding reward. Thus, pair bonding may occur in prairie voles as a more general conditioned partner preference through reliance on a species-specific OXTR or AVPR1A mechanism.

For a brief overview of the role of OXT and AVP in regulating pair bonding, see: www. youtube.com/watch?v=EoweLVvR7-8. For a more extensive presentation of the vole work and parallels with humans studies, see: www.youtube.com/watch?v=Iko7UXC-M94.

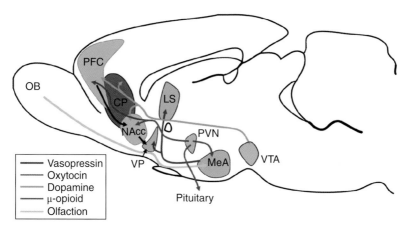

Figure 19.6 A model of a pair bonding network in the prairie vole brain. During mating, oxytocin and vasopressin are released from cells in the paraventricular nucleus (PVN) and dopamine is released from the ventral tegmental area (VTA). Neuropeptidergic innervation is widespread in the forebrain and receptors for the ligands are localized in key regions involved in pair bonding. In females, concurrent oxytocin receptor and dopamine receptor D2 activation in the NAcc is required for partner preference. Additional oxytocin receptor activation in the prefrontal cortex (PFC) is also needed. Male prairie voles require vasopressin receptor 1a activation in the ventral pallidum (VP) and lateral septum (LS). All of these regions are part of the mesolimbic reward system which regulates learning and reinforcement in the brain. Finally, μ-opioid signaling in the caudate putamen (CP) signals reward. During the activation of these reinforcement pathways, social information about the sexual partner enters through the olfactory system into the medial amygdala (MeA), which relays this information to other areas. Oxytocin and vasopressin enhance the processing of this social information, allowing efficient dopamine-gated association with the rewarding properties of sex. Thus the prairie vole pair bond is a highly selective conditioned partner preference. OB, olfactory bulb. Modified with permission from Young *et al.* (2005).

19.5   Neuropeptide receptor expression contributes to individual differences in behavior

Neuropeptide receptor distributions are highly variable not only between species but also between individuals within a species. In particular, brain regions that tend to show the highest interspecies variability are the same regions that show within-species variability. There is evidence that this individual variation in receptor expression contributes to individual differences in social behavior. Prairie voles exhibit high levels of intraspecific variation in behavior and receptor expression across environmental context as well as between individuals in a population (Cushing *et al.*, 2001; Phelps and Young, 2003; Ross and Young, 2009). Experiments in the lab and field have yielded important information on the gene expression/behavior relationship between individuals.

In the laboratory, genetic tools and discrete behavioral analysis have allowed for an intimate dissection of the relationship between OXTR and AVPR1A densities in specific brain regions and prairie vole social behaviors. Viral vectors offer a powerful tool to manipulate the expression of a gene. Vectors containing a copy of a gene cause ectopic expression in cells within the vicinity of an injection site, leading to an upregulation of density in the case of OXTR and AVRP1A. Other vectors containing a silencing RNA (siRNA) allow knockdown, or reduction in endogenous expression, by exploiting RNA regulatory mechanisms. These tools, in conjunction with autoradiography to map densities, allow for behavioral variation to be associated with higher or lower levels of NHPRs.

Initial work with affiliative and parental behaviors sparked interest in a relationship with NHPR density. Natural variation in OXTR NAcc density is positively correlated with alloparental care in female prairie voles (Ross and Young, 2009). Viral vector OXTR upregulation in the NAcc during the juvenile period, but not during adulthood, facilitates female alloparental behaviors. The differences in timing above suggest that some NHPR density effects occur by modulating development of social behavioral circuits. AVRP1A variation between individuals is also associated with differences in affiliation and paternal care. Upregulating AVRP1A density in the VP of male prairie voles increases VP neural responses to a social encounter and enhances affiliation. Densities of both OXTR and AVRP1A in the LS differ between high and low investigating males. Density variation in multiple regions can influence individual differences in affiliation.

NHPR density has been well investigated for its role in pair bonding. As stated previously, AVRP1A upregulation in the meadow vole VP allows the normally promiscuous species to exhibit partner preferences. AVRP1A density variation across the brain is associated with the propensity to form a partner preference in male prairie voles (Hammock and Young, 2005). More specifically, upregulation of AVRP1A in the prairie vole VP enhances individual propensity to form a partner preference. Conversely, knockdown of AVRP1A expression in the VP prevents partner preference in males, demonstrating a direct function for endogenous AVRP1A density in pair bonding (Barrett *et al.*, 2013). Viral vector injections ectopically increasing OXTR density in the NAcc facilitate partner preference formation in female prairie voles (Ross *et al.*, 2009). These studies on pair bonding have focused on NHPR density in the MLR pathway and reinforce the notion that NHPR involvement in prairie vole partner preference revolves around facilitating the association between individual social cues and reward signals.

Outside the laboratory, in a semi-natural field setting, prairie voles can be tracked with radio telemetry collars to observe sociosexual behaviors in a complex spatial and social environment. As mentioned, prairie voles in the wild are socially monogamous and engage in extra-pair copulations. Telemetry allows individual territory formation to be analyzed in order to identify individual mating tactics. Individual males and females with overlapping territory can be considered pair bonded or, in this case, residents, as opposed to wanderers that overlap with numerous other animals. Embryos can be collected from females following a field experiment, allowing parentage to be assessed with genetic techniques to determine mating success. Thus attachment and monogamy can be studied in complex environments, allowing insights into how circuits discovered in the laboratory operate in close to natural conditions (Phelps, 2010).

Field results do not follow perfectly from laboratory models. Male voles that adopt a resident mating tactic, and presumably have pair bonded with a female, have higher OXTR density in the NAcc but have no differences in AVRP1A in the VP as might be expected. Females that have paired in the field have higher AVRP1A density in the VP than single females, but do not differ in OXTR density in the NAcc. Interestingly, these results show that generally, NHPRs in the MLR pathway are still key, but in the field, males and females have a different sensitivity to the two peptides relative to the laboratory tests.

Field research allows analyses not possible in the laboratory. For example, there are complex relationships between NHPR density and reproductive success – whether a vole was able to mate and produce embryos. Mesolimbic reward pathway regions seem not to play a role in this measure of reproductive success. Instead, other regions that exhibit NHPR density variation are associated. In males, AVRP1A density in the posterior cingulate and lateral dorsal thalamus is lower in unpaired wandering males that successfully mate, relative to wandering males that did not mate. Mated wanderers also had less OXTR density in the hippocampus and septohippocampal nucleus. These NHPR density-behavior relationships arise in a network of brain regions that are important for spatial cognition. Thus, OXTR and

AVRP1A density change the sensitivity with which spatial reasoning networks receive information about social context. This allows voles with optimal individual NHPR levels to better navigate a particular sociospatial environment. Indeed, a common observation for successfully reproducing wanderer males is that their territory overlaps with more individual voles than unmated wanderers. Ophir *et al.* (2012) suggest that specific combinations of NHPR densities allow wandering animals to maximize the success of their mating strategy by increasing intrusions into other males' territory.

This section has aimed to illustrate that variation in NHPR density weights the importance of social information to different neural regions in core survival networks, generating a diversity of social strategies that allow greater flexibility over evolutionary time.

19.6   How diversity in receptor expression is achieved

So far, we have focused on NHPR densities and their influence on behavior. Receptor density is most often measured with autoradiography, which reveals receptor molecules in the cell membrane. Generally, a higher density of binding indicates a greater number of receptors; more receptors occur from increased expression of the gene, or less mRNA instability. In vole species, this appears to be the case: OXTR and AVRP1A density are significantly correlated with mRNA levels. Both AVRP1A and OXTR density distributions have been related to transcriptional mechanisms.

Mechanisms that control gene transcription can be thought of in terms of two broad classes: transcription factors and non-coding *cis*-REs. Transcription factors are specialized proteins that bind to *cis*-REs in DNA in order to modulate gene expression. These *cis*-REs occur commonly in promoters near genes, and at other sites, such as enhancers that can be further from protein-coding regions. Mutations in *cis*-REs are the most common cause of phenotypic differences between species (Wittkopp and Kalay, 2012), and thus are likely to contribute to the evolutionary diversity exhibited by NHPRs.

The prairie vole NHPR distributions help differentiate them from other *Microtus* species. The promoter regions of both *Oxtr* and *Avpr1a* contain *cis*- that generate expression in brain regions where prairie vole receptor distributions occur. Mice transgenic for 5 kb of prairie vole *Oxtr* promoter expressed an ectopic label in many brain areas that contain *Oxtr* mRNA in prairie voles. This powerful genetic technique revealed that a minimal portion of promoter does indeed contain *cis*-REs that drive gene expression in certain brain regions. The *Avpr1a* promoter also contains important *cis*-REs. Mice similarly transgenic for minimal components of the prairie vole *Avpr1a* gene, including 2.2 kb of promoter sequence, exhibited a unique AVPR1A distribution. The transgenic mice expressed *Avpr1a* in regions where wild-type mice normally lack the gene. The *cis*-REs contained in the transgenic construct were so potent that the mutant mice exhibited increased affiliative behavior in response to an AVP injection, which wild-type mice did not.

Some *cis-RE* sequences are conserved across species and these motifs can be analyzed in genomic areas suspected of harboring *cis*-REs. Sequence analysis of the *Oxtr* promoter between prairie and montane voles identified few differences in putative *cis*-REs. The *Avpr1a* promoter on the other hand contains a stark difference between meadow and prairie voles. There is an approximately 600 bp complex microsatellite in the prairie vole *Avpr1a* promoter. Microsatellites are highly unstable and polymorphic genetic features. Montane voles lack 400 bp of the microsatellite. The prairie vole microsatellite demonstrates direct functional properties in *in vitro* luciferase assays either when deleted or when compared to the meadow vole microsatellite. The *Avpr1a* promoter microsatellite contains *cis*-REs that contribute to differences in gene regulation between species.

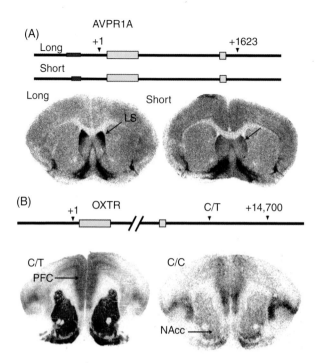

Figure 19.7 Individual diversity in OXTR and AVPR1A. Autoradiograms of coronal brain sections from individual prairie voles. The *Avpr1a* promoter microsatellite (*in red*) predicts individual differences in AVPR1A in many regions including the lateral septum (LS) (A). OXTR also exhibits high individual variation in expression, particularly in the nucleus accumbens (NAcc), while other regions such as the prefrontal cortex (PFC) are more stable between individuals (B). The *Oxtr* gene differs in structure from Avpr1a: it lacks a microsatellite but has larger non-coding regions, leading to a longer total gene sequence.

Interestingly, the prairie vole *Avpr1a* microsatellite is also polymorphic in length between individuals (Hammock and Young, 2005). These length variants are also functional in a luciferase assay. Genotype-based breeding to generate prairie voles that were homozygous for long or short alleles revealed that the length of the microsatellite was associated with complex AVRP1a density differences throughout the brain (Figure 19.7). The microsatellite alleles were also associated with differences in behavior, including variation in partner preference formation. Males with long alleles formed a partner preference after a truncated co-habitation while males with short alleles did not. The prairie vole *Avpr1a* promoter microsatellite therefore exerts a robust influence on gene expression, such that it can also predict differences in social behavior.

A detailed analysis of the specific contributions of *cis*-REs in the vole microsatellite was performed using transgenic mice. Knock-in mice were created using homologous recombination, replacing 3.4 kb of mouse DNA with prairie vole *Avpr1a* promoter sequence (Donaldson and Young, 2013). In addition, three separate lines also had unique microsatellite content: short prairie vole, long prairie vole or meadow vole microsatellites. The long prairie vole microsatellite increases AVRP1A density in three brain regions relative to both wild-type background and the meadow vole microsatellite. Incredibly, the long prairie vole microsatellite increases dentate gyrus AVPR1A density relative to the short. This experiment conclusively demonstrates that the *Microtus* microsatellite contains *cis*-REs and that instability in the microsatellite is likely a mechanism that creates diversity in receptor expression patterns and likely social behavior.

These increasingly detailed genetic studies have dissected one particular *cis-RE* that contributes to prairie vole social diversity through regulation of *Avpr1a* in the brain. Vole species provide natural diversity while proving tractable to a combination of advanced behavioral and genetic techniques. Vole genetic resources are now catching up with more traditional models like the mouse, including sequence information and transgenic techniques (McGraw and Young, 2010). The experiments above suggest a strategy to identify sources of diversity in social behavior and then to narrow down influences to the level of brain regions and/or DNA sequences. For example, mechanisms regulating *Oxtr* expression in the brain have proven difficult to explain, whether as a result of *cis*-REs or otherwise. Prairie voles exhibit high individual variation in OXTR in the NAcc, which may be subject to as yet unknown *cis*-REs. However, recent preliminary studies demonstrate that a single nucleotide polymorphism (SNP) in the 3' untranslated region of the prairie vole *Oxtr* robustly predicts OXTR binding density in the NAcc and caudate putamen, suggesting that this individual variation in expression is the result of polymorphic *cis*-REs yet to be discovered.

19.7 Translational implications for OXTR and AVRP1A

The biomedical community has substantial interest in uncovering biological mechanisms that influence human social cognition. This is in part due to a growing concern with psychiatric disorders with deficits in social cognition, namely autism spectrum disorders (ASD) (Modi and Young, 2012; Young *et al.*, 2002). Work with animal models has provided a strong conceptual platform on which to base hypotheses about human social behavior. Of course, human research necessarily allows far less intrusive methods, which is a problem for OXT and AVP research, since large molecules such as OXT cannot pass from the periphery across the blood–brain barrier. Intranasal OXT (IN-OXT) delivery was proposed as a solution, as the nasal epithelium has a weak barrier.

Human IN-OXT research has progressed rapidly over the past decade since the first effects were seen. Two major research themes have emerged: social salience and prosociality (Bartz *et al.*, 2011), although other categories have been proposed. Prosocial effects include increases in trust and generosity, to name just two. Enhancements in social saliency include increased gaze to the eyes and lead to improved abilities to detect emotion in faces and a selective increase in social recognition. The effects parallel nicely the work reviewed above in rodents and suggest that the role of OXT in regulating the processing of social information and the salience of social information is evolutionarily conserved from rodent to man, even across sensory modalities. Brain networks underlying these behavioral effects are now being elucidated. Increased eye gaze, as stimulated by IN-OXT, is associated with activation of the superior colliculus and its connectivity with the amygdala, suggesting a modulation of a circuit for directing visual saccades to salient features. This finding is particularly interesting when considering that the superior colliculus is a site of conserved *OXTR* expression in primates, potentially including humans (Freeman *et al.*, 2014).

Basic processes like these may lead to significant effects on human social behavior. In an experiment on human attachment, IN-OXT caused male subjects, if they were in a monogamous relationship, to maintain a longer distance from a novel female. The treatment also selectively increases the attractiveness of a partner's photograph and increased NAcc and VTA activation, suggesting that IN-OXT stimulates MLR pathway signaling in humans (Scheele *et al.*, 2013).

IN-OXT has a robust effect on human social cognition and behavior. Researchers are now developing strategies to combine IN-OXT or drugs that stimulate endogenous central OXT release with social behavioral training to improve social symptoms in ASD (Modi and

Young, 2012). Early results suggest that IN-OXT can indeed improve social cognitive processing in ASD. The ability of IN-OXT to increase eye gaze, emotion detection, and social recognition holds in ASD patients. IN-OXT also enhances social reciprocity in high-functioning autistic subjects (Andari *et al.*, 2010).

Central *OXTR* gene expression has not been characterized in humans. However, social behavior has been associated with SNPs in the *OXTR* gene. Studies focused on ASD etiology found numerous SNPs in *OXTR* that are associated with ASD diagnosis in patients or endo-phenotypes relating to social interaction or communication in both typically developing subjects or patients. We have postulated that a general function of *OXTR* is to increase salience of social information, and deficits in that process could help explain the genetic associations with ASD mentioned above. Along those lines, studies with infants that have or would later be diagnosed with ASD found deficient gaze to socially relevant features such as eyes or biological motion. It has been suggested that this early loss of attention towards social information denies the brain opportunities to learn about the social environment, potentially leading to further deficits. Interestingly, then, *OXTR* variation may contribute to early gaze deficits, as SNPs in the gene are associated with individual variation in social cognition at 18 months (Wade *et al.*, 2014).

Variation in *OXTR* is related to human social cognition in adulthood. SNPs in the gene are associated with complex human behaviors such as prosocial dispositions, trait empathy, and even pair bonding. These effects may stem in part from a role for *OXTR* in basic processing of visual and auditory information. Skuse *et al.* revealed a profound effect of *OXTR* after testing ASD patients, their parents, and siblings in a facial recognition memory task (Skuse *et al.*, 2014). The recognition ability of all subjects was modulated by the genotype of *OXTR* SNP rs237887. Within each group, i.e. proband or parents, subjects with the A/A genotype of the SNP were significantly more competent in the recognition task than those with the G/G genotype while the A/G genotype responded intermediately. This study revealed a clear linear effect of an *OXTR* SNP on social cognition, with the genotype explaining 10% of individual behavioral variation.

All of the *OXTR* SNPs discussed in the above section are in non-coding regions such as introns, and thus may occur inside, or associate with, polymorphic *cis*-REs where they can modulate expression of the gene. Indeed, a *cis*-RE has been identified in a human *OXTR* intron. *OXTR* expression is reduced in the cortex of ASD patients, who also have increased *OXTR* methylation, an epigenetic mark associated with transcriptional suppression. Another *OXTR* SNP exhibits allelic expression imbalance in human brain tissue. Such imbalance is a specific effect of heterozygosity in *cis*-REs actively involved in expression of a gene. Thus, this result provides direct evidence of association between *OXTR* SNPs and expression of the receptor. It is tempting to speculate that such allelic effects on expression could lead to linear associations between SNPs and *OXTR* density in brain regions responsible for visual processing and thus to robust modulation of social cognition like that reported by Skuse and colleagues.

The glut of findings regarding the influence of *OXTR* SNPs on social cognition would benefit from animal models providing more detailed information on mechanisms by which *OXTR* density is regulated in the brain. This information has so far proven elusive. In the case of *AVRP1A*, however, work with prairie voles has provided detailed insights into mechanisms by which a promoter microsatellite containing *cis*-REs generates diversity of the receptor in the brain (Donaldson and Young, 2013; Hammock and Young, 2005). Intranasal AVP enhances recognition of emotional faces and sexual cues, suggesting that *AVRP1A* activation modulates social salience in the human brain similarly to *OXTR*. The primate *AVRP1A* promoter also contains unstable microsatellites that vary in length between individuals. Using these variants as markers, *AVRP1A* has been associated with altruism, pair bonding, and autism in humans (Donaldson and Young, 2008). Long alleles of the RS3 microsatellite are associated with elevated *AVRP1A* expression in the brain.

As noted, humans share some *AVRP1A* microsatellites in common with other primates. The RS3 microsatellite has received much of the focus in human studies and its length polymorphisms have a profound impact on how *AVRP1A* influences human social cognition. The chimpanzee (*Pan troglodytes*) is polymorphic for a deletion of RS3. Most chimpanzees are therefore either homozygous for the RS3 deletion, or heterozygous, meaning they have only a single genomic copy of the RS3 microsatellite. Research in voles suggests that rich genetic diversity will lead to interesting individual differences in behavior. Indeed, the RS3 deletion polymorphism associates with a number of chimpanzee social behavioral measures, including personality traits such as dominance and conscientiousness and receptive joint attention (Hopkins *et al.*, 2012, 2014).

19.8 Perspectives

The OXT/AVP family of neuropeptides has a long evolutionary history of regulating peripheral physiology as well as sociosexual behaviors that are in some way linked to those physiological processes. The conservation spans from sensory processing of cues from the mate in *C. elegans*, to olfactory bonds between prairie vole partners, to visual attraction to lovers in humans. The synthesis of the peptide systems in the brain is remarkably conserved over evolutionary time. By stark contrast, extraordinary diversity in receptor expression patterns contributes to diversity in social behavioral phenotype across species and between individuals of a species. Genetic studies in humans suggest that similar diversity in transcriptional regulation, not protein structure, underlie individual differences in some social behavioral traits. Prairie voles have provided an excellent model organism to elucidate the neural mechanisms by which OXT and AVP regulate complex behaviors such as pair bonding. They also provide an excellent opportunity for future studies to explore the molecular mechanisms by which polymorphisms in neuropeptide genes generate diversity in social cognition and behavior.

References

Andari, E., Duhamel, J.R., Zalla, T., Herbrecht, E., Leboyer, M. and Sirigu, A. (2010) Promoting social behavior with oxytocin in high-functioning autism spectrum disorders. *Proceedings of the National Academy of Sciences USA*, **107**(9), 4389–4394.

Barrett, C.E., Keebaugh, A.C., Ahern, T.H., Bass, C.E., Terwilliger, E.F. and Young, L.J. (2013) Variation in vasopressin receptor (Avpr1a) expression creates diversity in behaviors related to monogamy in prairie voles. *Hormones & Behavior*, **63**(3), 518–526. **[Direct manipulation of endogenous neurohypophysial receptor expression.]**

Bartz, J.A., Zaki, J., Bolger, N. and Ochsner, K.N. (2011) Social effects of oxytocin in humans: context and person matter. *Trends in Cognitive Science*, **15**(7), 301–309.

Bosch, O.J. and Neumann, I.D. (2012) Both oxytocin and vasopressin are mediators of maternal care and aggression in rodents: from central release to sites of action. *Hormones & Behavior*, **61**(3), 293–303.

Burkett, J.P. and Young, L.J. (2012) The behavioral, anatomical and pharmacological parallels between social attachment, love and addiction. *Psychopharmacology (Berlin)*, **224**(1), 1–26.

Cushing, B.S., Martin, J.O., Young, L.J. and Carter, C.S. (2001) The effects of peptides on partner preference formation are predicted by habitat in prairie voles. *Hormones & Behavior*, **39**(1), 48–58.

de Vries, G.J. (2008) Sex differences in vasopressin and oxytocin innervation of the brain. *Progress in Brain Research*, **170**, 17–27.

Dolen, G., Darvishzadeh, A., Huang, K.W. and Malenka, R.C. (2013) Social reward requires coordinated activity of nucleus accumbens oxytocin and serotonin. *Nature*, **501**(7466), 179–184.

Donaldson, Z.R. and Young, L.J. (2008) Oxytocin, vasopressin, and the neurogenetics of sociality. *Science*, **322**(5903), 900–904.

Donaldson, Z.R. and Young, L.J. (2013) The relative contribution of proximal 5 flanking sequence and microsatellite variation on brain vasopressin 1a receptor (Avpr1a) gene expression and behavior. *PLoS Genetics*, **9**(8), e1003729. [**Demonstrates function of a cis-regulatory element on neural gene expression in vivo.**]

Freeman, S.M. and Young, L.J. (2013) Oxytocin, vasopressin, and the evolution of mating systems in mammals, in *Oxytocin, Vasopressin and Related Peptides in the Regulation of Behavior* (eds E. Choleris, D.W. Pfaff and M. Kavaliers, M), Cambridge University Press, New York, pp. 128–147.

Freeman, S.M., Walum, H., Inoue, K., *et al.* (2014) Neuroanatomical distribution of oxytocin and vasopressin 1a receptors in the socially monogamous coppery titi monkey (Callicebus cupreus). *Neuroscience*, **273C**, 12–23.

Goodson, J.L. (2013) Deconstructing sociality, social evolution and relevant nonapeptide functions. *Psychoneuroendocrinology*, **38**(4), 465–478.

Gruber, C.W. (2014) Physiology of invertebrate oxytocin and vasopressin neuropeptides. *Experimental Physiology*, **99**(1), 55–61.

Hammock, E.A. and Young, L.J. (2005) Microsatellite instability generates diversity in brain and sociobehavioral traits. *Science*, **308**(5728), 1630–1634. [**First study relating a genetic regulatory feature with diversity in brain and behavior.**]

Hopkins, W.D., Donaldson, Z.R. and Young, L.J. (2012) A polymorphic indel containing the RS3 microsatellite in the 5 flanking region of the vasopressin V1a receptor gene is associated with chimpanzee (Pan troglodytes) personality. *Genes, Brain and Behavior*, **11**(5), 552–558.

Hopkins, W.D., Keebaugh, A.C., Reamer, L.A., Schaeffer, J., Schapiro, S.J. and Young, L.J. (2014) Genetic influences on receptive joint attention in chimpanzees (Pan troglodytes). *Scientific Reports*, **4**, 3774.

Kelly, A.M. and Goodson, J.L. (2014) Social functions of individual vasopressin-oxytocin cell groups in vertebrates: What do we really know? *Frontiers in Neuroendocrinology*, **35**, 512–529. [**Review on distinct oxytocin and vasopressin neuronal populations.**]

Knobloch, H.S. and Grinevich, V. (2014) Evolution of oxytocin pathways in the brain of vertebrates. *Frontiers in Behavioral Neuroscience*, **8**, 31.

Lim, M.M., Wang, Z., Olazabal, D.E., Ren, X., Terwilliger, E.F. and Young, L.J. (2004) Enhanced partner preference in a promiscuous species by manipulating the expression of a single gene. *Nature*, **429**(6993), 754–757. [**Receptor diversity between species contributes to behavioral differences.**]

McGraw, L.A. and Young, L.J. (2010) The prairie vole: an emerging model organism for understanding the social brain. *Trends in Neuroscience*, **33**(2), 103–109.

Modi, M.E. and Young, L.J. (2012) The oxytocin system in drug discovery for autism: animal models and novel therapeutic strategies. *Hormones & Behavior*, **61**(3), 340–350.

Ophir, A.G., Gessel, A., Zheng, D.J. and Phelps, S.M. (2012) Oxytocin receptor density is associated with male mating tactics and social monogamy. *Hormones & Behavior*, **61**(3), 445–453.

Phelps, S.M. (2010) From endophenotypes to evolution: social attachment, sexual fidelity and the avpr1a locus. *Current Opinion in Neurobiology*, **20**(6), 795–802.

Phelps, S.M. and Young, L.J. (2003) Extraordinary diversity in vasopressin (V1a) receptor distributions among wild prairie voles (Microtus ochrogaster): patterns of variation and covariation. *Journal of Comparative Neurology*, **466**(4), 564–576.

Ross, H.E. and Young, L.J. (2009) Oxytocin and the neural mechanisms regulating social cognition and affiliative behavior, *Frontiers in Neuroendocrinology*, **30**(4), 534–547. [**Review on the role of oxytocin in maternal, affiliative, and attachment behaviors.**]

Ross, H.E., Freeman, S.M., Spiegel, L.L., Ren, X., Terwilliger, E.F. and Young, L.J. (2009) Variation in oxytocin receptor density in the nucleus accumbens has differential effects on affiliative behaviors in monogamous and polygamous voles. *Journal of Neuroscience*, **29**(5), 1312–1318.

Scheele, D., Wille, A., Kendrick, K.M., *et al.* (2013) Oxytocin enhances brain reward system responses in men viewing the face of their female partner. *Proceedings of the National Academy of Sciences USA*, **110**(50), 20308–20313.

Skuse, D.H., Lori, A., Cubells, J.F., *et al.* (2014) Common polymorphism in the oxytocin receptor gene (OXTR) is associated with human social recognition skills. *Proceedings of the National Academy of Sciences USA*, **111**(5), 1987–1992. [**Robust correlation between an oxytocin receptor SNP and human social recognition.**]

Tessmar-Raible, K., Raible, F., Christodoulou, F., *et al.* (2007) Conserved sensory-neurosecretory cell types in annelid and fish forebrain: insights into hypothalamus evolution, *Cell*, **129**(7), 1389–1400.

Venkatesh, B., Si-Hoe, S.L., Murphy, D. and Brenner, S. (1997) Transgenic rats reveal functional conservation of regulatory controls between the Fugu isotocin and rat oxytocin genes. *Proceedings of the National Academy of Sciences USA*, **94**(23), 12462–12466.

Wade, M., Hoffmann, T.J., Wigg, K. and Jenkins, J.M. (2014) Association between the oxytocin receptor (OXTR) gene and children's social cognition at 18 months. *Genes, Brain and Behavior*, **13**, 603–610.

Wittkopp, P.J. and Kalay, G. (2012) Cis-regulatory elements: molecular mechanisms and evolutionary processes underlying divergence. *Nature Reviews Genetics*, **13**(1), 59–69.

Young, L. and Alexander, B. (2012) *The Chemistry Between Us: Love, Sex and the Science of Attraction*. Penguin, New York.

Young, L.J. and Wang, Z. (2004) The neurobiology of pair bonding. *Nature Neuroscience*, **7**(10), 1048–1054.

Young, L.J., Pitkow, L.J. and Ferguson, J.N. (2002) Neuropeptides and social behavior: animal models relevant to autism. *Molecular Psychiatry*, **7**(Suppl 2), S38–39.

Young, L.J., Murphy Young, A.Z. and Hammock, E.A. (2005) Anatomy and neurochemistry of the pair bond. *Journal of Comparative Neurology*, **493**(1), 51–57.

Glossary

3′ untranslated region (3′ UTR): a portion of a messenger RNA (mRNA) preceded by the translation termination codon and followed by the poly-A tail. Does not encode a protein but plays a regulatory role governing mRNA half-life, translation efficiency, etc. The 3′ UTR is a main target of miRNAs, where the majority of microRNA binding sites are located.

5′ untranslated region (5′ UTR): the region of the mRNA directly upstream from the initiation codon of the main open reading frame encoding the protein, also known as RNA leader sequence. This region plays a role in the regulation of protein translation, and has a modified 5′ end with a methylated guanosine, called 5′ cap, necessary for ribosome entry

5-methylcytosine (5-mC): an epigenetic modification of DNA that usually occurs at CpG dinucleotides; the 5-mC modification usually correlates with repressed gene expression.

Ablation (of neurons): eliminating (killing) neurons by mechanical, chemical or genetic means.

Absorption: process where the virus attaches to the cell surface through interactions of its proteins with receptors on the cell surface.

Adipocyte: individual, fully differentiated fat cell.

Adipose peptides: peptides first identified in adipose tissues.

Adipose tissue: fat tissue. Found in a number of different depots around the body such as visceral adipose tissue or subcutaneous tissue. Usually used in reference to white adipose tissue, as opposed to brown adipose tissue.

Adrenocorticotropin (ACTH): signals to melanocortin 2 receptors on the adrenal gland to stimulate steroidogenesis.

Agouti mice: a strain of yellow and obese mice with a mutation in the promoter of the agouti gene, leading to dysregulated and ectopic expression of agouti. Agouti is normally only produced in the coat of mice and is expressed only during a small window of development. When it isn't expressed, α-MSH, which is constitutively produced, signals to the MC1R (melanocortin 1-receptor) on melanocytes making eumelanin, or black pigmentation. When agouti is expressed, it competes off α-MSH and antagonizes the MC1R, resulting in phaeomelanin production, which produces a bleaching effect and results in a yellow color. So in the coat of a wild-type mouse, the individual strands of hair are normally dark brown with a small subapical yellow band. When agouti is overexpressed, there is a complete lack of black pigmentation, thereby leading to the yellow phenotype. In the mutant mice, agouti is also ectopically expressed in the brain, where it competes off α-MSH and antagonizes the MC4R, resulting in obesity.

Agouti-related peptide (AgRP): an 132-amino acid peptide, co-synthetized with neuropeptide Y in neurons of the hypothalamic arcuate nucleus (as well as in many other brain areas and peripheral tissues). AgRP is considered as a powerful anorexic peptide,

Molecular Neuroendocrinology: From Genome to Physiology, First Edition. Edited by David Murphy and Harold Gainer.
© 2016 John Wiley & Sons, Ltd. Published 2016 by John Wiley & Sons, Ltd.
Companion website: www.wiley.com/go/murphy/neuroendocrinology

inducing hyperphagic obesity. To achieve this effect AgRP acts as a competitor of melanocortins in binding to some types of their receptors.

Alternative splicing: the process whereby a single gene gives rise to multiple mRNA isoforms by differential intron-exon shuffling.

Anorexigenic: reducing appetite (as opposed to **orexigenic**, which stimulates appetite); typically used to describe a hormone or neuropeptide.

Antagomir: synthetic target of an miRNA used to silence the effect of an endogenous miRNA.

Anterograde tracing: following projections from a neuron cell body to the ending of their axons or to the synapses.

Arcuate nucleus: hypothalamic nucleus, located in the mediobasal hypothalamus, composed of two types of neurons, expressing AgRP/NPY or POMC/CART and considered as a feeding center.

Assembly: process by which the parts of a virus (protein and genome) are put together to form a full new virus particle.

ATP-dependent chromatin remodeling: process mediated by large multi-subunit molecular machines that uses ATP energy to reorganize nucleosome structures, often by sliding the nucleosome to a new position on the DNA.

Bilaterian: animals demonstrating bilateral symmetry.

Biosynthesis: a multi-step enzymatic process that drives the production of novel biological compounds, e.g. neuropeptides.

Bisulfite conversion: the deamination of non-methylated cytosine bases to uracil by treatment with sodium bisulfite ($NaHSO3$); 5-mC bases are resistant to bisulfite conversion.

BLAST: the **B**asic **L**ocal **A**lignment **S**earch **T**ool is an algorithm for comparing primary biological sequence information at the level of DNA (deoxyribonucleotide sequence), RNA (ribonucleotide sequence) or protein (amino acid sequence). A BLAST search enables a researcher to compare a query sequence with a library or database of sequences, and identify library sequences that resemble the query sequence above a certain threshold.

Budding (exocytosis)/lysis: enveloped viruses take part of the membrane with them as they are released from the cell in a process called budding. Non-enveloped viruses must break open or lyze the cells to release their new virus particle.

Capsid: the protein shell which surrounds and protects viruses. The capsid is made up of repeating protein subunits called capsomers.

cas9: a nuclease, one of many CRISPR-related genes that encode proteins involved in acquiring, processing or interfering with foreign DNA or RNA in bacteria with this form of adaptive immunity. cas9 is most frequently used and has been modified for optimal use as a genome engineering nuclease. A modified version which is limited to cutting a single strand of the double-stranded DNA target, making a nick.

Cerebellins: a group of structurally related peptides with neuropeptide and growth factor-like properties of which the prototype was originally found in the cerebellum.:

Channelrhodopsin: a family of light-activated cation channels found in aquatic algae. When expressed in neurons or other excitable cells, these channels allow optical activation of electrical activity.

Chemokine: a member of a large class of secreted peptides that induce chemotaxis ('cell movement') of nearby cells.:

ChIP-chip: a technology (also known as ChIP-on-chip) that combines chromatin immunoprecipitation ('ChIP') with DNA microarray ('chip'). Like regular ChIP, ChIP-on-chip is used to investigate interactions between proteins and DNA *in vivo*. Specifically, within the context of a particular cell, tissue or organ at a specific point in time, it enables the identification of all the binding sites of a particular DNA-binding protein within the genome.

ChIP-Seq: similar to ChIP-chip, but interacting DNA motifs are read out by high-throughput parallel sequencing.

Chromatin: the complex of DNA, histones, RNA, and other proteins that comprise the structural basis of chromosomes.

Chromatin immunprecipitation (ChIP): a type of immunoprecipitation technique used to investigate the interaction between proteins and DNA in a cell, tissue or organ. ChIP can be used to determine whether specific proteins or protein isoforms are associated with specific genomic regions. Briefly, the method is as follows: protein and associated chromatin are temporarily bonded (cross-linked), the DNA-protein complexes are then sheared into ~500 bp DNA fragments by sonication, cross-linked DNA fragments associated with the protein(s) of interest are selectively immunoprecipitated using an appropriate protein-specific antibody, and the associated DNA fragments are purified and identified by PCR, hybridization (as in Chip-chip) or sequencing (ChIP-Seq).

Cis-regulatory element: a non-coding region of DNA that regulates gene expression by binding transcription factors.

Classic neuropeptide: a neuropeptide stored in secretory granules and released when the neuron is stimulated and the granule fuses with the plasma membrane. Most scientists in the field use the term 'neuropeptide' to describe this type of molecule, and the term 'classic' is a new concept to distinguish this group from other types of peptides that can signal between cells.

Clathrin-mediated endocytosis: is mediated by small (approx. 100 nm in diameter) vesicles that have a morphologically characteristic crystalline coat made up of a complex of proteins that are mainly associated with the cytosolic protein clathrin. Clathrin-coated vesicles are found in virtually all cells and form domains of the plasma membrane termed clathrin-coated pits.

Clustered regularly interspaced short palindromic repeats (CRISPR): the adaptive immunity gene loci of many bacteria and archeae have a structure with repeated elements that alternate with exogenous sequence that has been captured and will provide a template for guide RNAs.

Co-activators: proteins lacking a DNA binding domain that allows or potentiates gene transcription by interacting with transcription factors.

Complementary endogenous RNA (ceRNA) hypothesis: a theory that coding and non-coding RNAs compete for a microRNA based on their sequence complementarity to an miRNA, which leads to establishment of a complex network of co-regulation.

Co-repressors: proteins that interact with transcription factors, reducing their ability to recruit RNA polymerase, thus blocking transcription.

Corticotropin-releasing hormone (CRH): 41-amino acid peptide located in the paraventricular hypothalamic nucleus. CRH is a releasing hormone, driving the activity of the pituitary-adrenocortical axis, thereby executing hormonal changes during response to stress. Besides the location in the hypothalamus, CRH is expressed in neurons of many other brain regions, such as the central amygdala, bed nucleus of stria terminalis, and cortex.

CRE recombinase: an enzyme that recombines the DNA at specific recognition sequences called loxP sites.

Cyclotides: plant-derived peptides, with an unusual structural topology of a cystine-knotted disulfide network and a head-to-tail cyclized backbone.

Cytopathic effect: virus-induced damage to cells that alters their microscopic appearance. This can be changes in shape, size, or through the development of intracellular changes.

Deuterostomian: coelemate animals whose embryogenesis involves the first opening (the blastopore) developing into the anus.

Dicer: a key endoribonuclease governing one of the last steps of microRNA maturation, involved in cleavage of pre-miRNA precursor into mature miRNA.

Direct neuropeptide: a new term to describe the traditional role for a neuropeptide; a molecule released from a cell which then stimulates a neighboring cell by binding to a cell surface receptor.

DNA footprinting: technique to detect protein-DNA interactions using an enzyme to cut DNA, followed by analysis of the resulting cleavage pattern to identify the footprint that the protein protects.

DNA methylation: chemical modification of the DNA consisting of the reversible addition of a methyl group to position 5 of cytosines in the DNA sequence.

DNA methyltransferase (DNMT): an enzyme that catalyzes the addition of a methyl group to the 5-carbon position of cytosine bases.

DNA microarrays: collection of DNA probes spotted onto a solid surface; used in transcriptomic studies to simultaneously measure the expression levels of mRNAs in a cell or tissue.

DNAse I footprinting: a technique that can map specific interaction between a protein and a particular DNA sequence by virtue of the fact that a protein bound to DNA may protect that DNA from enzymatic cleavage (that is, the protein leaves a 'footprint'). This makes it possible to locate a protein-binding site on a particular DNA molecule. The method uses an enzyme, deoxyribonuclease I, to cut a radioactively end-labeled DNA. Gel electrophoresis is then used to detect the resulting cleavage pattern.

Double strand break (DSB): a cut in both the strands of duplex DNA.

DRE recombinase: an enzyme that recombines DNA at specific recognition sequences called rox sites.

Electrophoretic mobility shift assay (EMSA): otherwise known as band shifting or gel retardation, this is an electrophoresis technique used to study protein-DNA interactions. Labeled double-stranded probes are incubated with the protein or protein mixture of interest and then fractionated through a neutral polyacrylamide gel that maintains protein DNA associations. Relative to the unbound probe control, a protein-DNA complex will move more slowly through the gel; that is, its position in the gel will be shifted, or retarded, due to the associated protein.

Electrospray ionization (ESI): a method for vaporization and ionization of molecules from liquid samples prior to introduction into the mass spectrometer. ESI often forms multiply charged ions.

Encyclopedia of DNA Elements (ENCODE): a multi-centered, worldwide community resource project started in 2003 by the National Human Genome Research Institute (USA), aiming to identify all functional elements of the human genome by using primarily NGS methodologies and allowing free, public access to the data managed by the ENCODE Data Coordination Center.

Endocytosis: an energy-using process by which cells absorb molecules (such as proteins) by engulfing them. It is used by all cells of the body because most substances important to them are large polar molecules that cannot pass through the hydrophobic plasma or cell membrane. This is the process that some viruses use to pass the cell membrane.

Endosome: early endosomes are the first station of the endocytic pathway. Early endosomes are often located in the periphery of the cell, and receive most types of vesicles coming from the cell surface. They have a characteristic tubulovesicular structure and a mildly acid pH. They are principally sorting organelles where many ligands dissociate from their receptors in the acid pH of the lumen, and from which many of the receptors recycle to the cell surface.

Energy homeostasis: energy balance, or a matching of energy (measured in joules or calories) intake with energy expenditure.

Enhancer element: a region of the gene that binds activating transcription factors.

Enveloped: viruses, which are made by budding off of the infected cell. During this budding, they take part of the cell lipid membrane with them. This lipid membrane is the viral envelope. The lipid membrane can easily dry out, therefore enveloped viruses are not very stable.

Epigenetic mark: a modifying moiety that carries an epigenetic signal; examples include methylation of DNA and modification of histones.

Epigenetics: the study of any potentially stable and, ideally, heritable change in gene expression or cellular phenotype that occurs without changes in DNA sequence.

Extracellular fluid (ECF): the ECF (or extracellular fluid volume) is that proportion of the total body fluid located outside cells.

Flip recombinase: an enzyme that recombines DNA at specific recognition sequences called frt sites.

Function-related plasticity: activity-dependent but reversible changes in the morphology, electrical properties, and biosynthetic and secretory activities of a cell.

G protein-coupled receptor (GPCR): a large family of 7-transmembrane protein receptors. GPCRs undergo conformational changes following the binding of their cognate ligand(s) and consequently activate their associated cytoplasmic G-protein(s) that mediate an intracellular signaling cascade. The two main signaling pathways activated by GPCRs are the cAMP and phosphatidylinositol pathways. Based on similarity in sequence and function, GPCR family members are classified into six groups (class A–F).

Gain of function (GOF) and loss of function (LOF): analysis of various mutants of the PCs allowed the identification of common LOF variants and rather rare GOF ones. This is so far especially evident for PC1/3 and PCSK9. The future will tell if the other PCs also exhibit specific single point mutations allowing the analysis of their consequences *in vivo*.

Gene duplication: an important evolutionary process in which there is a duplication of DNA fragments, genes, chromosomes or even entire genomes. A duplicated gene may undergo mutations and have a redundant or complementary function to its original gene, may gain a different function or lose function.

Gene family: a group of genes related to each other by structural and functional features.

Gene networks: a set of genes organized hierarchically and containing both central 'nodes,' which provide both coordination and control, and subordinate nodes responsible for the integrated output of the network.

Gene ontology: a universal classification system of gene functions and other attributes that uses a controlled vocabulary.

Gene promoter: the region of a gene, usually immediately 5' to the transcriptional start site, which recruits multiple transcription factors.

Genome mining: a term that has been used to describe the exploitation of genomic information for the discovery of new processes, targets, and products.

Germline transgenesis: the alteration of the genome in those developmental lineages of the organism that give rise to the germ cells (eggs or sperm) such that the genetic change will be transmitted to subsequent generations, wherein it will be present in all somatic and germline cells.

Ghrelin: ghrelin is secreted by X/A-like endocrine cells from the stomach and proximal small intestine, and is the only orexigenic gut-derived hormone acting centrally to modulate feeding. There are two major forms of mature ghrelin, acyl and desacyl ghrelin, the former of which is believed to be the predominant active form, but is present at much lower concentrations within the circulation. Acyl ghrelin possesses a unique posttranslational modification of O-noctanoylation at serine 3 that is essential for its ability to signal. Circulating ghrelin levels have a characteristic preprandial increase and postprandial decrease, giving support for a role in meal initiation and/or meal preparation. The ghrelin receptor is a typical G protein-coupled receptor that belongs to the rhodopsin-like 7-transmembrane domain receptor family.

Gonadotropin-releasing hormone (GnRH): a 10-amino acid peptide, expressed in neurons located in the medial septum, diagonal band of Broca and medial preoptic area. GnRH is a releasing hormone that triggers the activity of the hypothalamic-pituitary-gonadal axis, controlling sexual competence.

Granin: a member of a family of proteins with neuropeptide-like features without clear-cut biological activities. They are thought to function largely to chaperone neuropeptides in the secretory pathway.

guide RNA (gRNA): a 20 bp RNA sequence that guides the cas9 nuclease to a specific target sequence. The gRNA, crRNA, and cas9 nuclease form a complex that scans the genome, recognizes the sequence that is complementary to the gRNA and then cuts the target at that site.

Halorhodopsin: a family of light-activated chloride pumps found in bacteria. When expressed in neurons or other excitable cells, these pumps allow optical inhibition of electrical activity.

Hedonic: in addition to fulfilling one's metabolic need, the mesolimbic pathways that participate in the reward aspects of food intake also play an important role in regulating feeding behavior. The major mesolimbic reward pathway includes neurons located in the ventral tegmental area of the midbrain and projecting predominantly to the nucleus accumbens, prefrontal cortex, and amygdala. It is generally accepted that the mesolimbic pathway is implicated in behaviors associated with reward anticipation, motivation or incentive salience, associative learning, and reinforcement, so-called hedonic behavior.

Heteronuclear RNA (hnRNA): primary transcript or messenger RNA precursor, which is the exact copy of the DNA template before splicing of the intronic regions.

Histone: a small, highly conserved basic protein, found in the chromatin of all eukaryotic cells.

Histone modification: posttranslational addition or removal of epigenetic marks from histones; includes methylation, acetylation, phosphorylation, ubiquitination, sumoylation, and the removal of these marks.

Holometabolous insects: a term used for insect groups, for example beetles, ants and flies, that undergo complete metamorphism to describe their specific type of development which includes four life stages, each with its own morphology (embryo or egg, larva, pupa and imago or adult).

Homeostasis: the ability of an organism or a biological system to preserve internal stable conditions within a certain range. Neuroendocrine organs play a central role in maintaining homeostasis in a living organism.

Homologous recombination (HR): process in which a donor DNA strand with sequence that matches the target site invades that target site and by a process of insertion replaces the endogenous sequence.

Hypermethylation: increase in the level of DNA methylation in a population of cells relative to a reference or normal sample; may be used to describe a specific nucleotide or a group of nucleotides.

Hyperphagia: An abnormally increased appetite that can lead to obesity, usually associated with defects in the hypothalamus.

Hypomethylation: decrease in the level of DNA methylation in a population of cells relative to a reference or normal sample; may be used to describe a specific nucleotide or a group of nucleotides.

Hypophagia: pathogenic undereating.

Hypothalamic-pituitary-adrenal axis (HPA axis): major endocrine response to stress involving feedforward and feedback influences between hypothalamic corticotropin-releasing hormone, pituitary adrenocorticotropic hormone, and adrenal glucocorticoids. The HPA axis plays a key role in restoring homeostasis by regulating body processes, such as energy storage and expenditure, immune system, mood and emotions, energy storage and expenditure, feeding and reproductive behavior.

Hypothalamo-neurohypophysial system (HNS): neuroendocrine system consisting of neurons, whose cell bodies are located in the supraoptic and paraventricular nuclei, whilst their axons run towards the neurohypophysis, where they terminate, releasing arginine vasopressin and oxytocin into the general circulation.

Hypothalamus: region of the forebrain, composed of nuclei (in mammals ~32 pairs) and fiber tracts. While the anterior, mediobasal, and lateral hypothalamus contains mainly peptidergic and neuroendocrine neurons, the posterior hypothalamus contains classic non-peptidergic neurons. Despite the small size (in humans about 1 cm³), the hypothalamus executes an enormous number of endocrine, autonomic, metabolic, and behavioral functions. Among them are neurons controlling food intake, sleep-wakefulness cycle, food intake, reproductive competence and sexual drive, etc.

Immunohistochemistry (ICH): localization of an antigen in cells of tissue sections (usually 3–40 µm thick) with an antibody raised specifically to that antigen. This primary antibody can be directly labeled with a fluorescent dye or traced with a fluorescent or chromagenically (for example, biotin)-labeled secondary antibody (an antibody that binds to the primary antibody) to determine the pattern of antigen immunoreactivity under a fluorescence or light microscope.

Indel: a disruption in the DNA at the target site of a programmable nuclease. This can be an insertion or a deletion, hence an IN/DEL. This change in the DNA results from the repair of a double-strand break by endogenous DNA repair mechanisms.

Indirect neuropeptide: a new term to describe peptides that have biological activity when applied to cells, but this activity is due to inhibition of enzymes that degrade direct neuropeptides. Several indirect neuropeptides have been reported, but it is not known if they are functional *in vivo*.

***In situ* hybridization histochemistry (ISSH):** method by which radiolabeled (usually ³⁵S) or chromogen-labeled DNA or RNA probes are bound to thin (usually 10–20 µm) tissue sections to anatomically localize mRNA expression after exposure to X-ray film and/or photographic emulsion.

Internal ribosome entry site (IRES): structural feature found in genes with a complex 5′ UTR able to initiate translation in a cap-independent manner when activated by signaling transduction pathways.

Intracellular fluid (ICF): the liquid found within cells.

***In vitro, ex vivo* and *in vivo* models:** the distinction between the models used to define the specificity and characteristics of the processing reaction are either tested *in vitro* in test tubes, in cell lines (*ex vivo*) or in animals and/or humans (*in vivo*). The use of KO mice or knockdown procedures in *Xenopus* and zebrafish allowed the assessment of some of the *in vivo* functions of the PCs and their comparison across phylae. This also allows the estimation of the degree of functional redundancy of the PCs *in vivo*.

Isotocin: an oxytocin-like peptide found in most fishes.

Label-free quantitation: approaches used to determine peptide/protein abundances from the intensity of their signal, the frequency of their detection, or from the number of molecular fragmentation events from each ion inside the tandem mass spectrometer in unit of time.

Large dense-core vesicle (LDCV): LDCVs are formed in the endoplasmic reticulum and Golgi apparatus and appear at electron microscopic level as structures having an external membrane, a light halo, and a dense core. The dense core contains tightly packed neurohormones oxytocin and vasopressin, together with transport peptides such as neurophysins. The size of LDCVs is 250–300 nm.

Latent: viruses which are in a dormant state and not causing active infection but are ready to be activated by an outside signal to become infectious.

Ligand: a molecule that binds to specific targets or receptors and activates signaling processes/cascades.

Liquid chromatography (LC): a two-phase process to fractionate a sample into purified bands based on the physicochemical properties of the analytes making up the sample and their interaction with stationary and mobile phases. For peptidomics measurements, a reversed phase liquid chromatography system is used where the stationary phase is relatively non-polar and the mobile phase is a part aqueous, moderately polar mobile phase. In this case, the more polar peptides elute earlier.

Long non-coding RNA (lncRNA) a type of non-protein coding RNA of length over 200 nt.

lox site: the 20 base pair recognition sequence for the CRE recombinase.

Mass spectrometry: an analytical method for structural characterization of molecules based on measurement of their mass-to-charge ratio *(m/z)*. An MS measurement requires a process to vaporize and ionize the sample, and then a mass analyzer to characterize the *m/z* of the components of the sample. As the instrument performance specifications depend on the specific ionization approaches and mass analyzer used, different combinations have different applications.

Mass spectrometry imaging (MSI): a method of tissue analysis that allows chemical mapping with spatial resolution. In most approaches, laser desorption/ionization is performed at a specific location on the tissue, and then the location is rastered across the tissue. Signals from individual ions at each raster spot are compiled to make ion distribution images; individual ion images can be superimposed with each other and combined with optical images.

Mass-to-charge ratio *(m/z)*: a dimensionless physical quantity that reflects the ratio between the molecular or atomic mass number *m* and the charge number *z* of the ion.

Matrix-assisted laser desorption/ionization (MALDI): a vaporization and ionization method suitable for analysis of peptides and other biomolecules by mass spectrometry. Briefly, the tissue or samples are incorporated into a UV-absorbing matrix mixed with the analyte. Upon irradiation from a laser, some of the analyte molecules are vaporized, ionized, and then introduced into the appropriate mass spectrometer.

Median eminence: an elongated neurohemal system, which contains blood capillaries and neurosecretory cell endings from the hypothalamus. It connects to the hypothalamo-hypophysial portal system and facilitates the transfer of secreted neuropeptides between the hypothalamus and the adenohypophysis.

Melanocortin 4 receptor (MC4R): one of the five melanocortin receptors, the MC4R is expressed almost exclusively within the central nervous system. Its endogenous ligands are α- and β-MSH. When it is activated within the hypothalamus, particularly in the paraventricular nucleus, it results in a reduction of food intake. However, it is also expressed and activated in other areas of the brain, where it controls autonomic output. Mutation in MC4R in both man and mouse leads to dominantly inherited obesity, hyperphagia, severe hyperinsulinemia (in childhood in humans), and increased linear growth.

Melanocortins: the pro-opiomelanocortin-derived peptides ACTH, α-, β-, and γ-MSH.

Melanocyte-stimulating hormone (MSH): there are free isoforms (α-, β-, and γ) of MSH known, which are composed of 13, 18, and 11 amino acids, respectively. In addition to the regulation of skin color in non-mammalian species, MSH is a powerful anorexic peptide, acting on specific receptors, antagonizing with AgRP.

Mesolimbic reward pathway: the network of brain regions involved with the processing of reward and reinforcement, responding to stimuli such as sex and drugs of abuse.

Mesotocin: an oxytocin-like peptide found in most non-mammalian vertebrates except for fish. **Metabolic syndrome:** a complex disorder of energy storage and utilization, sometimes synonymous with prediabetes. Diagnosis is based on the presence of three out of the five following conditions: abdominal obesity, hypertension, hyperglycemia, hypertriglyceridemia, low HDL.

Microarray: the use of high-throughput hybridization technology for transcriptomic profiling.

MicroRNA profiling: methodology used to detect and quantify simultaneously all known miRNA species in a particular species, organ, tissue or cell.

MicroRNA-Seq: type of NGS technology able to sequence the entire population of microRNAs in a given sample, permitting quantitative and qualitative analysis of microRNAs, including discovery of new species of microRNA.

Monogenic syndrome: otherwise known as a Mendelian genetic syndrome, where a mutation in a single gene causes a phenotype which is passed on in Mendelian fashion.

Nascent-Seq: a variant of RNA-Seq that provides a direct measure of transcription as opposed to simply transcript levels.

Neuroendocrine system: the system that integrates communication between the nervous system and the periphery through hormones.

Neuropeptides: endogenous compounds composed of 3–50 amino acids that are derived from larger precursor proteins by enzymatic cleavages; they act as neuromodulators and hormones in nervous and neuroendocrine systems.

Neurosecretion: the cellular principle of storage and regulated secretion of peptides by neurons.

Neurosecretory preoptic area (NPO): a region in the brain of non-mammalians. It is a part of the hypothalamus and contains neurosecretory cell bodies.

Neurotropin: growth factor-like protein or peptide with competence to support neuronal survival, proliferation or differentiation

Next-generation sequencing (NGS): recently developed methodology for fast, cheap, parallel sequencing of whole genomes, epigenomes, transcriptomes, etc., allowing an unprecedented insight into cell-wide responses.

Non-classic neuropeptide: a new term to describe peptides that are produced in the cytosol of the cell and secreted by non-conventional mechanisms which have not been determined. Although several non-classic neuropeptides have been found, it has not been proven that they are functional *in vivo*.

Non-enveloped/naked: viruses which do not have a lipid envelope. Their outer covering is a protein shell which makes these viruses very stable. All gastrointestinal viruses are non-enveloped and they can survive the harsh acidic environment of the stomach and intestine.

Non-homologous end joining (NHEJ): the resolution of a break in the DNA that results in an indel. It is mediated by the cell's endogenous repair mechanism. It results in insertion or deletion of nucleotides to the site of the break.

Nucleocapsid: the protein shell that wraps around the viral nucleic acid.

Nucleosome: the central unit of chromatin, composed of an octet of histone proteins containing two copies each of H2A, H2B, H3, and H4 and wrapped around by two superhelical turns of DNA of about 147 bp in length.

Obligate intracellular parasite: an organism that cannot multiply without the help of a specific host (i.e. a virus).

Oncoviruses: viruses which cause the formation of tumors.

Opsins (optognetics): a family of microbial proteins, which includes light-gated ion channels (channelrhodopsin-1 and -2) and light-activated ion pumps (halorhodopsin). Illumination with blue light (action spectrum ~480 nm) induces the opening of ion channels, allowing exchange of extracellular and intracellular ions, resulting in depolarization of membrane and initiation of exocytosis of transmitters and peptides from synaptic and extrasynaptic sites. In contrast, illumination of ion pumps with red light (action spectrum ~580 nm) leads to changes of conformation of the pump, inducing influx of Cl⁻, resulting in hyperpolarization of cell membrane and preventing exocytosis of transmitters and peptides.

Optogenetic actuators: light-sensitive ion channels and pumps whose expression can be genetically targeted to specific cell types and used to control the electrical activity of the targeted cells. Prominent examples include channelrhodopsin-2 from the aquatic algae *Chlamydomonas* and halorhodopsin from *Natronomonas* bacteria.

Optogenetic sensors: fluorescent proteins that change their fluorescence properties in response to various cellular signals. Expression of these sensors can be genetically targeted to specific cell types and used to monitor the electrical activity of the targeted cells. Prominent examples include the GCaMPs, which are fluorescent reporters of intracellular calcium ion concentration, and ArcLight, a fluorescent reporter of cellular membrane potential.

Optrode: a device that resembles an electrode but utilizes an optical measurement technique. For example, an optical fiber optrode can be used to detect and measure fluorescent material within the brain. The optrode consists of two parts: an excitation optical fiber, linked to a laser, and a recording optical fiber feeding into a photomultiplier tube.

Orexigenic: stimulating appetite (as opposed to **anorexigenic** – which reduces appetite).

Orexin (also known as hypocretin): 28- and 33-amino acid peptide isoforms of orexin exist. Orexin is expressed preferentially in the lateral hypothalamus and controls sleep-wakefulness cycle, appetite, and mood.

Ortholog: locus in two species that is derived from a common ancestral locus by a speciation event.

Oxytocin (OXT): a 9-amino acid peptide (nonapeptide), which is a member of the nonapeptide family together with vasopressin and homologs of non-mammalian species. OXT neurons project to the posterior pituitary lobe and are released into the systemic brain circulation, initiating and maintaining uterine contraction (during labor) and milk let-down (during lactation) in females and erection in males. Additionally, OXT has become the most popular neuropeptide, which supports prosocial behavior and ameliorates symptoms of socially relevant pathologies (such as autism spectrum disorder). In this regard, the pathways of the mechanism of action of OT require further research.

Pair bond: a selective attachment to a conspecific, formed after mating or prolonged contact. A partner preference is a laboratory test for the presence of a pair bond.

Paraventricular nucleus (PVN): the most studied nucleus of the hypothalamus, containing magnocellular neurons expressing oxytocin and vasopressin, parvocellular neurons, expressing releasing hormones and preautonomic peptidergic neurons projecting to brainstem and spinal cord.

Peptidergic neurons: neurons which, together with classic neurotransmitters such as GABA or glutamate, also express peptides, which play a role of neurohormones or neuromodulators rather than as fast transsynaptic signaling.

Peptidergic signaling: a functional complex consisting of a cell that synthesizes and releases a peptide mediator, a cell that responds to that peptide by a certain physiological change, and the process of transferring the peptide from the site of synthesis to the site of action. In particular, we use the term peptidergic signaling for pathways that are mediated by peptides, their endogenous receptors and associated signaling molecules, which commonly belong to the family of G protein-coupled receptors.

Peptides: short chains of amino acid monomers linked by peptide (amide) bonds.

Peptidomics: the global high-throughput measurement, sequencing, and identification of the endogenous peptides present in biological samples.

Persistent infections: a virus that maintains its infection within a cell without lysis or death of the cell. This can last from a few weeks to the rest of the host's life.

Photoinhibition: the use of light-activated ion channels or pumps to enable optical inhibition of cellular electrical activity. One popular means of photoinhibiting neurons is via the light-activated chloride ion pump halorhodopsin.

Photostimulation: the use of light-activated ion channels to enable optical stimulation of cellular electrical activity. One popular means of photostimulating neurons is via the light-activated cation channel channelrhodopsin-2.

Pioneer factors: proteins that can penetrate condensed chromatin to pioneer recruitment of secondary co-factors that remodel the chromatin to allow other TFs access.

Pituitary (hypophysis): an endocrine gland regulated by the hypothalamus. The anterior pituitary (adenohypophysis) receives an endocrine input from the hypothalamus and in turn releases hormones into the blood circulation. The posterior pituitary (neurohypophysis) contains oxytocinergic and vasopressinergic nerve endings in which these hormones are stored and released.

Pituitary adenoma: a benign tumor of the pituitary. Pituitary adenomas are the most commonly occurring intracranial neoplasms. They are often hormonally active and can be defined based on a hormone they release. For example, corticotropic adenomas secrete adrenocorticotropic hormone, gonadotropic adenomas secrete luteinizing hormone and follicle-stimulating hormone, lactotropic adenomas secrete prolactin, and somatotropic adenomas secrete growth hormone.

Polycomb complex: epigenetic repressor of gene transcription that plays a critical role in eliciting and maintaining gene silencing at key developmental events.

Posttranslational histone modifications: alterations in the compositing of histone proteins that package the genome; these modifications are brought about by a variety of enzymes that either add or remove chemical moieties from the amino-terminal tail of histones.

Posttranslational modification: the additional of a small molecule or peptide onto a protein after its translation is complete; modifications are usually reversible and regulatory.

Precursor protein: a biologically inactive protein that needs to be cleaved by enzymes to release biologically active peptides.

Processing: the enzymatic process of enzymatic cleavage or modification of substrate proteins to produce biologically active peptides.

Promoter bashing: a term used to describe the mapping of specific regulatory domains within a promoter/enhancer. Specific point mutations or deletions are made in the promoter/enhancer and the effect on transcription is then measured using a reporter assay.

Pro-opiomelanocortin (POMC): a complex propeptide that is posttranslationally processed by prohormones, to produce a number of biologically active peptides including the melanocortin peptides ACTH, α-, β-, and γ-MSH.

Proprotein convertases: this is a family of nine subtilisin-like serine proteases that cleave multiple substrates either at basic residues (PC1/3, PC2, furin, PC4, PACE4, PC5/6, and PC7) or at non-basic residues (SKI-1/S1P). PCSK9, the last member of the family, only cleaves itself once in the ER and remains associated with its inhibitory prosegment and remains an inactive enzyme that binds and enhances the degradation of its target molecules by escorting them to lysosomes.

Protein-DNA array: a technique that enables the simultaneous assay of multiple protein-DNA interactions in a single experiment. Nuclear extracts are incubated with labeled double-stranded oligonucleotides corresponding to known specific protein-binding sites. Unbound and bound oligonucleotides are then separated, and the latter are then dissociated from protein and hybridized to the same complementary sequences on the array. The degree of hybridization corresponds to the amount of a DNA binding protein in the original nuclear extract.

Protospacer adjacent motif (PAM): short sequence, triplet or quadruplet that is required for recognition of the target nucleic acid. In different species this sequence can vary. For *S. pyogenes* it is NGG, though NAG will also allow some binding. This sequence is present in the target DNA but not in the CRISPR locus, so those sequences are protected from 'autoimmunity.'

Protostomean: coelemate animals whose embryogenesis involves the first opening (the blastopore) developing into the mouth.

Provirus: the retrovirus genome which is inserted into the host cell chromosome.

Pseudotyping: process of adding a glycoprotein from a different virus to a viral vector. For example, making lentivirus particles which use the VSV-G glycoprotein and not the retrovirus env.

Puberty-activating genes: genes involved in the stimulatory control of puberty.

Reads per kilobase of transcript per million mapped reads (RPKM): a numerical transformation used to describe RNA-Seq data.

Receptor: protein, lipid, or sugars which are recognized by the virus and allow its attachment to the cell.

Receptor autoradiography (ARG): the incubation of radiolabeled (usually iodinated or tritiated) receptor ligands on tissue sections (usually 15–30 μm thick), followed by washing and exposure to X-ray film, to determine the distribution of receptor binding sites.

Receptor diversity: differences in gene expression at a given brain region that occurs between species or individuals within a species.

Regulated secretion: a form of release of vesicularly stored peptides in cells that is triggered by specific stimuli.

Reporter assay: a technique that enables the facile quantification of the capacity of a promoter/enhancer sequence of a gene to drive transcription. This is achieved by cloning the putative promoter/enhancer upstream of the coding sequences of a protein that can be readily assayed, for example luciferase. This artificial reporter construct can then be introduced into cells, usually in culture, and the activity of the promoter/enhancer can be indirectly assessed by measuring the reporter.

Retrograde tracing: following projections from axonal terminal or synapse to neuron cell body.

Retrotransposon: a frequent element of eukaryotic genomes with self-replicating capabilities. Retrotransposons comprise nearly half of the human genome and are transcribed into an RNA intermediate, which is reverse transcribed into DNA and inserted back into a genome, typically in a different location from the original. The function of retrotransposons is largely unclear but recently, retrotransposons have been implicated in microRNA production.

RNA-Seq: use of high-throughput sequencing techniques for transcriptomic profiling.

Secretory pathway: the cellular transition of subcellular compartments that transport peptides and proteins to the plasma membrane and release the content in the extracellular space.

Selectivity dilemma: many neuropeptides mediate their distinct function by signaling through more than one G protein-coupled receptor (for instance, there are four receptors for OXT and AVP). The receptors often share high sequence homology, in particular of the extracellular binding domains, and the peptide ligands are structurally very similar. From a pharmacological viewpoint, this results in significant cross-reactivity when administering a non-selective drug, which may lead to severe side effects. The discovery and design of selective ligands for peptide receptors can overcome this problem.

SELEX: a combinatorial technique for producing DNAs that bind specifically and with high affinity to a DNA-binding protein of interest.

Signaling transduction: chain of biochemical events triggered by activation of a receptor following binding of its specific ligand.

Signal peptide: a 20–25 amino acid peptide domain located at the N-terminus of newly synthesized proteins that allows the protein to be pulled into the lumen of the ER, the first compartment of the secretory pathway. Thus all secreted proteins and precursors of neuropeptides are characterized by a signal peptide.

Silencer element: a region of the gene that binds repressing transcription factors.

Single cell sequencing: a sequencing method for whole genome or whole transcriptome sequencing of a single cell; involves an amplification step.

Social salience: increased signal-to-noise for social stimuli specifically, for example manipulations of oxytocin and vasopressin can be shown to selectively influence responses to social information while not impacting responses to non-social information.

Somatic transgenesis: gene transfer into somatic cells (usually utilizing viral vectors) such that only the individual organism is affected. The germline is not affected unless specifically targeted.

Stable isotopic labeling: a mass spectrometry-based quantitation method that employs distinct non-radioactive isotopic tags for peptides that allow them to be distinguished based on their mass differences. By comparing the relative abundance in samples of two (or more) distinct tags, relative quantitation becomes possible.

STAT3: a transcription factor that, amongst others, is a downstream effector of leptin receptor signaling.

Stereotaxic technique: with respect to viral vectors, this technique allows researchers to precisely insert a needle into the brain of an animal to infuse a viral solution into defined brain areas, which is estimated by 3D coordinates (axes X, Y, and Z).

Stress: state of threatened homeostasis.

Stress response: coordinated neuroendocrine and peripheral responses to a threat or perceived threat to homeostasis leading to restoration of the internal environment equilibrium.

Subtractive hybridization: a procedure that increases the effective concentration of induced sequences expressed in an experimental RNA population (target) but not in a control RNA population (driver).

Suppression PCR: when complementary sequences are present on each end of a single-stranded cDNA, PCR amplification is suppressed. This is because, during each primer annealing step, the hybridization kinetics strongly favor the formation of a pan-like secondary structure that prevents primer annealing.

Suppression subtractive hybridization-polymerase chain reaction (SSH-PCR): a technique that couples subtractive hybridization with suppression PCR to selectively amplify cDNAs corresponding to genes that differ in expression between two cell populations.

Supraoptic nucleus (SON): the paired nucleus located at lateral sides of the optic chiasm in the basal hypothalamus. The SON is composed of magnocellular oxytocin and vasopressin neurons.

Sympathoadrenal axis: autonomic response to stress involving descending neural pathways resulting in the release of catecholamines from the adrenal medulla to the circulation. The sympathoadrenal system is essential for the physical demands of the fight or flight responses.

TALENuclease: Transcription Activator Like is a synthetic programmable nuclease that uses the nuclease domain of the Fok I endonuclease fused to a DNA recognition structure. Each domain of XX amino acids uses a dipeptide to bind a single nucleotide in the recognition target.

tBLASTn: a variant of BLAST that compares a protein query against all the six reading frames of a nucleotide sequence database.

Transcript: the RNA product of gene transcription – may be either coding for a protein (mRNA) or non-coding (ncRNA – a functional RNA molecule that is not translated into a protein).

Transcription: first step of gene expression in a cell in which activation of the gene promoter initiates copying of the DNA into RNA by the enzyme RNA polymerase.

Transcriptional activation: an increase in the activity of a gene elicited by transcription factors that recognize and bind to specific DNA sequences in regulatory regions of the genome.

Transcriptional repression: a reduction in the activity of a gene imposed by proteins known as transcription factors that recognize and bind to specific DNA sequences in regulatory regions of the genome.

Transcription factors: proteins that following activation by signaling transduction molecules (or ligand for intracellular receptors) bind to specific DNA sequences (response elements) in the promoter and form transcription regulation complexes by recruiting co-activators, co-repressors, histone acetylases or deacetylases, kinases, other DNA-modifying enzymes, and RNA polymerase.

Transcriptome: the full complement of transcripts produced in the cell or tissue under investigation.

Translation: process in which ribosomes scan the mRNA until they find the initiation codon and start recruiting amino acids for protein synthesis until they encounter the stop codon.

Transsynaptic communication: the means by which neurons communicate to each other using a variety of substances known as neurotransmitters and neuromodulators. These substances are released into the cleft of specialized structures known as synapses.

Tropism/host range: the specific cells/tissues that viruses are able to infect. If a virus can infect a large number of cell/tissue types then they are considered to have a broad tropism/ host range. If they can only infect one or two types of cells/tissues, they are said to have a restrictive tropism/host range.

Uncoating: the outer coating of the virus (envelope/capsid) is removed and the viral genome is released into the cytoplasm.

Upstream open reading frames or minicistrons: small open reading frames containing initiation and termination codons located in the 5′ UTR, with the potential of being translated into peptides able to regulate translation.

Urbilaterian: the last common ancestor of the bilaterians.

Vasopressin: 9-amino acide peptide (nonapeptide), synthesized by magnocellular hypothalamic neurons in the PVN and SON and released from the posterior pituitary lobe into the systemic blood circulation. The main peripheral target of VP is the kidney. Acting on collecting ducts of kidney vasopressin induces reabsorption of water from primary urine. Besides magnocellular neurons, VP is synthesized in parvocellular PVN neurons expressing CRH, potentiating action of the CRH on ACTH cells of the pituitary, especially during chronic stress. In addition, VP is synthesized in various neurons of the brain, including the medial preoptic area, medial amygdala, and bed nucleus of stria terminalis, where it modulates parental and social behavior.

Viral genome: the nucleic acid that holds the virus genetic information. This can be either RNA or DNA (but not both) and can exist as either double stranded or single stranded.

Viral vector: a biological tool that exploits a virus to deliver a gene of interest into a target population of cells. The viral carrier uses endogenous mechanisms to enter the cell and nucleus and then hijack transcriptional mechanisms to express its experimental gene.

Zinc finger nuclease: a synthetic programmable nuclease that contains DNA recognition domains fused to the nuclease moiety of the FokI endonuclease. Zn finger domains are among the most common DNA binding protein structures in eukaryotes and are named for the Zn cation that is bound by the protein.

Index

Note: Page numbers in *italics* refer to Figures; those in **bold** to Tables.